Texts and
Monographs
in Physics

Roberto Fernández
Jürg Fröhlich
Alan D. Sokal

Random Walks, Critical Phenomena, and Triviality in Quantum Field Theory

With 26 Figures

Springer-Verlag Berlin Heidelberg GmbH

Dr. Roberto Fernández
Professor Jürg Fröhlich
Institut für Theoretische Physik, ETH Hönggerberg
CH-8093 Zürich, Switzerland

Professor Alan D. Sokal
Department of Physics, New York University
4, Washington Place, New York, NY 10003, USA

Editors

ISBN 978-3-662-02868-1

Library of Congress Cataloging-in-Publication Data. Fernández, R. (Roberto), 1951-. Random Walks, critical phenomena, and triviality in quantum field theory / R. Fernández, J. Fröhlich, A. D. Sokal. p. cm. -- (Texts and monographs in physics). Includes bibliographical references and index. 1. Quantum
ISBN 978-3-662-02868-1 ISBN 978-3-662-02866-7 (eBook)
DOI 10.1007/978-3-662-02866-7
field theory. 2. Random walks (Mathematics). 3. Critical phenomena (Physics). I. Fröhlich, J. (Jürg.), 1946-. II. Sokal, A. D. (Alan D.), 1955-. III. Title. IV. Series. QC174.45.F46 1991 530.1'43--dc20 91-40850

Originally published by Springer-Verlag Berlin Heidelberg New York in 1992
Softcover reprint of the hardcover 1st edition 1992

Typesetting: Camera ready by authors
Production Editor: P. Treiber

55/3140-5 4 3 2 1 0 - Printed on acid-free paper

Foreword

Simple random walks — or equivalently, sums of independent random variables — have long been a standard topic of probability theory and mathematical physics. In the 1950s, non-Markovian random-walk models, such as the self-avoiding walk,were introduced into theoretical polymer physics, and gradually came to serve as a paradigm for the general theory of critical phenomena. In the past decade, random-walk expansions have evolved into an important tool for the rigorous analysis of critical phenomena in classical spin systems and of the continuum limit in quantum field theory. Among the results obtained by random-walk methods are the proof of triviality of the φ^4 quantum field theoryin space-time dimension $d \underset{(-)}{>} 4$, and the proof of mean-field critical behavior for φ^4 and Ising models in space dimension $d \underset{(-)}{>} 4$. The principal goal of the present monograph is to present a detailed review of these developments. It is supplemented by a brief excursion to the theory of random surfaces and various applications thereof.

This book has grown out of research carried out by the authors mainly from 1982 until the middle of 1985. Our original intention was to write a research paper. However, the writing of such a paper turned out to be a very slow process, partly because of our geographical separation, partly because each of us was involved in other projects that may have appeared more urgent. Meanwhile, other people and we found refinements and extensions of our original results, so that the original plan for our paper had to be revised. Moreover, a preliminary draft of our paper grew longer and longer. It became clear that our project — if ever completed — would not take the shape of a paper publishable in a theoretical or mathematical physics journal. We therefore decided to aim for a format that is more expository and longer than a research paper, but too specialized to represent something like a textbook on equilibrium statistical mechanics or quantum field theory.

This volume reviews a circle of results in the area of critical phenomena in spin systems, lattice field theories and random-walk models which are, in their majority, due to Michael Aizenman, David Brydges, Tom Spencer and ourselves. Other people have also been involved in these collaborations, among whom one should mention Jan Ambjørn, Carlos Aragão de Carvalho, David Barsky, Anton Bovier, Sergio Caracciolo, Bergfinnur Durhuus, Paul Federbush, Giovanni Felder, Ross Graham, Gerhard Hartsleben, Thordur Jónsson and Antti Kupiainen. This work, carried out between roughly 1979 and the present, was much inspired by Symanzik's deep article "Euclidean Quantum Field Theory", which appeared in the proceedings of the 1968 Varenna school on "Local Quantum Theory" [494]. Papers by Schrader [451, 454, 452] and Sokal [481] devoted to an analysis of φ_d^4 theory in $d \geq 4$ space-time dimensions, and work by Dvoretsky, Erdös, Kakutani, Taylor and Lawler [327, 158, 164, 350, 351, 353, 352] on intersection properties of Brownian paths and simple random walks and on self-avoiding walks, provided much additional stimulation for the work described in this volume.

Symanzik's article was fundamental, because it showed that Euclidean φ^4 field theory can be represented as a gas of weakly self-avoiding random paths and loops. This is the main approach explored in this book. The articles by Dvoretsky-Erdös-Kakutani, Erdös-Taylor, and Lawler were important, because they showed that Brownian paths do not intersect in dimension $d \geq 4$ and that (weak) self-avoidance is an irrelevant constraint in dimension $d \underset{(\sim)}{>} 4$.[1] Jim Glimm and Tom Spencer contributed much to making these ideas and results popular among mathematical physicists.

It appeared that, with the ideas of Symanzik and the results of Erdös, Taylor and Lawler at hand, one might be able to prove a no-interaction ("triviality") theorem for continuum φ_d^4 theory in space-time of dimension $d \underset{(\sim)}{>} 4$. It was pointed out in Sokal's thesis [481] that, for $d > 4$, it would be enough to prove a tree-diagram bound on the connected four-point function to conclude triviality. The right version of such a bound was first proven by Aizenman [4, 5], using a "random-current representation" to prove new correlation inequalities. Shortly afterwards, Fröhlich [213] proved a slightly different version of this bound, using Symanzik's random-path representation which also yields suitable correlation inequalities. Their proofs were based, in an essential way, on the intuition that random walks in dimension $d > 4$ do not intersect.

These results triggered much activity in developing random-current and random-walk representations into a systematic tool for proving new correlation inequalities which yield rigorous qualitative and quantitative information on critical phenomena in lattice field theories and lattice spin systems. The best results emerged for Ising models (sharpness of the phase transition, behavior of the magnetization, bounds on critical exponents) and for the superrenormalizable φ_d^4 field theoriesin $d = 2$ and 3 dimensions (simple new proofs of existence and nontriviality in the single-phase region, positivity of the mass gap, and asymptoticity of perturbation theory). These ideas and methods were later extended to percolation theory and to the theory of polymer chains, branched polymers, lattice animals and random surfaces. In comparison to renormalization-group constructions, the random-walk and random-current methods are somewhat soft and less quantitative. They have, however, the advantage of closely following simple and appealing intuitions, being mathematically precise and at the same time technically rather simple.

[1] See also the important recent work of Brydges and Spencer [93], Slade [471, 473, 472] and Hara and Slade [297, 293, 294] on self-avoiding walks in dimension $d > 4$.

This book has three parts of rather different flavor. Part I is an overview of critical phenomena in lattice field theories, spin systems, random-walk models and random-surface models. We show how field-theoretic techniques can be used to investigate the critical properties of random-walk models, and how random-walk representations can be used to analyze the critical properties of lattice field theories and spin systems. We review the basic concepts involved in the construction of scaling limits and in the renormalization group. The emphasis is on ideas and concepts, not on complete mathematical arguments (which can, however, be found in literature quoted). Chapter 7, co-authored by Gerhard Hartsleben, contains a fairly detailed exposition of random-surface theory and its various applications in statistical physics, quantum field theory and two-dimensional quantum gravity.

In Part II we systematize three different random-walk representations: the Brydges-Fröhlich-Spencer (BFS) representation of lattice spin systems, the Aizenman (ARW) representation of the Ising model, and polymer-chain models generalizing the self-avoiding walk. We introduce a common framework and show that all the relevant results follow from just two properties of the weights: repulsiveness "on the average" between walks, and attractiveness (or noninteraction) between nonoverlapping walks. This provides a unified explanation for most of the correlation inequalities obtained in Chapter 12. Unfortunately, Part II is somewhat technical; it will be of interest mainly to readers with specific research interests in the random-walk representation.

In Part III we present a fairly systematic survey of what can be proven with random-walk methods about critical exponents and the scaling (continuum) limit for φ^4 lattice field theories and Ising spin systems. In particular, we give a detailed explanation of the triviality theorems for φ^4_d models in space-time dimension $d > 4$ and $d = 4$. We have tried hard to make Part III accessible to readers who may not have read Part II.

A detailed overview of Parts II and III can be found in Chapter 8. We remark that Parts II and III contain some previously unpublished results.

Our book is certainly not complete or exhaustive. Many things are only sketched, others are not treated at all. For example, the new construction of superrenormalizable φ^4_d ($d < 4$), based on random-walk representations and correlation inequalities, is reviewed only briefly in this book (Section 6.3). Results on percolation theory, the superrenormalizable Edwards model of polymer chains, branched polymers, ... are not included at all.

If one thinks of all the progress that the renormalization group has made in recent years, or if one thinks of the revolution that has recently occurred in our understanding of two-dimensional conformal field theory and hence of critical phenomena in two-dimensional statistical systems, one might argue that the time for a book like ours has passed. However, we feel that some of the ideas and methods explained in this book are interesting enough, or at least pretty enough, to justify our attempt to collect and preserve them in the present form. All the basic ideas underlying the main results reviewed in this book express simple and appealing (physical or mathematical) intuitions. This feature distinguishes this book from some of the recent scientific production in mathematical and theoretical physics which is growing more and more abstract and mathematically sophisticated. We have the modest hope that our book is at least agreeable scientific entertainment.

This book would never have seen the light of the day without the work of Michael Aizenman, David Brydges and Tom Spencer, with whom we have had the pleasure of several collaborations and countless interesting and important discussions. Chapter 7 is the result of a collaboration with Gerhard Hartsleben, whose deep insight helped to clarify our understanding of random-matrix models and non-perturbative quantum gravity. He contributed with enormous enthusiasm and devoted his best effort to the preparation and writing of this chapter. Finally, Aernout van Enter, Takashi Hara and Gordon Slade read and made helpful comments on parts of the manuscript. We wish to express our sincere thanks to all these people, and to the many colleagues and friends whose work and ideas have helped to determine the shape of this book.

Princeton, New Brunswick, Austin,
Roma, Zürich and New York,
December ~~1987~~ ~~1988~~ ~~1989~~ ~~1990~~ 1991

Roberto Fernández
Jürg Fröhlich
Alan D. Sokal

Contents

List of Tables

List of Figures

Part I

Critical phenomena, quantum field theory, random walks and random surfaces: Some perspectives

1. General introduction

1.1 Phenomenology of phase transitions and critical phenomena

We start by describing the different kinds of phase transitions encountered in the study of condensed matter in thermal equilibrium.

Consider a system like H_2O in thermal equilibrium at some pressure p which is neither very low nor very high. For this system, there is a temperature $T_0 = T_0(p)$ at which two phases of H_2O, liquid and vapor, coexist. If one plots e.g. the equilibrium density of H_2O at pressure p as a function of temperature (T), it is discontinuous at $T = T_0(p)$. For $T < T_0(p)$, the density is large ("liquid"), while for $T > T_0(p)$ the density is very small ("vapor"). Moreover, the evaporation of water at $T = T_0(p)$ requires energy: heat must be pumped into the system. Let Δv be the difference in molar volume (inverse density) between the liquid and vapor phases at $T = T_0(p)$, and let ΔQ be the heat necessary to convert one mole of liquid into vapor at $T = T_0(p)$. Then the dependence of T_0 on p is related to Δv and ΔQ by the *Clausius-Clapeyron equation*

$$\frac{dT_0}{dp} = T_0 \frac{\Delta v}{\Delta Q} . \tag{1.1}$$

What we are describing here is a typical example of a *first-order (discontinuous) phase transition.*

Lines of first-order transitions in the (T, p)-plane often end in a *critical point* (T_c, p_c). This is so for the H_2O liquid-vapor system [Figure 1.1(a)]. The density difference between the liquid and vapor phases, $\Delta\varrho \equiv \varrho_L - \varrho_V$, which is defined along the phase-transition line, vanishes as the critical point is approached. Thus, in a tiny vicinity of (T_c, p_c), there is no clear distinction between liquid and vapor; moreover, the system exhibits fluctuations on essentially all length scales, and correlations between local measurements in distant regions of the system decay quite slowly in the separation distance (power-law decay, as compared to the exponential decay encountered in the highly disordered vapor phase). If we plot the density difference $\varrho_L - \varrho_V$ as a function of T, then it is continuous but non-differentiable at T_c: in fact, $\varrho_L - \varrho_V$ has a power-law singularity of the form [Figure 1.1(b)]

$$\varrho_L - \varrho_V \sim |T_c - T|^{\widehat{\beta}}$$

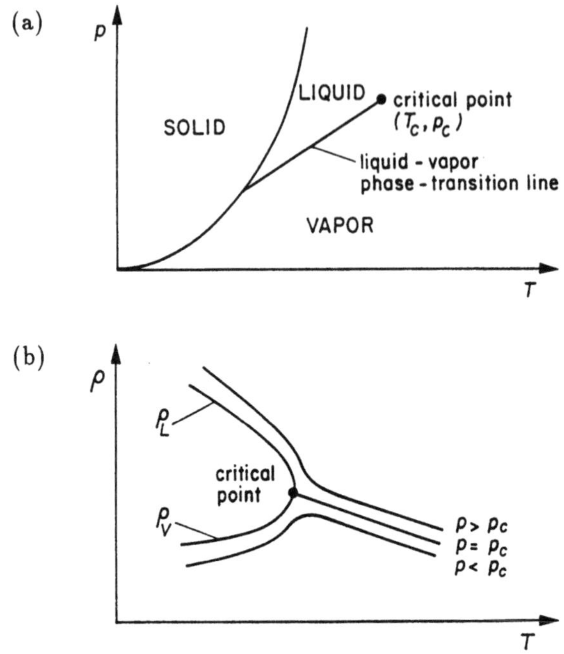

Fig. 1.1. Qualitative phase diagrams for the H_2O system. **(a)** Phase-transition lines in the (T, p)-plane. Only the liquid-vapor phase transition will concern us in this book. **(b)** Qualitative behavior of the density ϱ as a function of T at $p < p_c$, $p = p_c$ and $p > p_c$ near the liquid-vapor critical point

where experimentally the *critical exponent* $\widehat{\beta}$ is found to be approximately 0.32. The transition encountered at pressure p_c and temperature T_c is called a *continuous transition* (in our example, a second-order transition, because second derivatives of the free energy are divergent).

Our aim in this book is to analyze statistical systems in thermal equilibrium near critical points. We would like to understand, for example, how the density ϱ behaves as a function of T and p in a neighborhood of the critical point (T_c, p_c). Of course, we are not capable of determining the exact behavior of $\varrho(T,p)$ for H_2O, which is after all a very complicated quantum-mechanical system. However, a very striking simplification occurs in the critical region (and only there): it is found experimentally that many different fluids, with radically different chemical properties and radically different values of T_c and p_c, nevertheless have *identical critical behavior*, i.e. the function $\varrho(T,p)$ is *universal* (modulo trivial changes of scale) in a small neighborhood of (T_c, p_c). Even more strikingly, it is found that many other physical systems, such as binary alloys and anisotropic ferromagnets, have the same critical behavior as do the fluids. This phenomenon is called *universality*; physical systems are divided, according to their critical behavior, into a relatively small number of classes, called *universality classes*. Roughly speaking, the universality class of a system depends only on rather general attributes of the system, e.g. its spatial dimensionality, the symmetry group of its order parameter, the short- or long-ranged nature of the interactions, and so on. Thus, not knowing the exact dynamics of a statistical system need not prevent us from predicting its exact critical behavior.

The phenomenon of universality is explained, at least in general terms, by the *renormalization group*. In this approach, the critical behavior of statistical systems is governed by a fixed point of some renormalization transformation and the behavior of that transformation in a small neighborhood of the fixed point. Consider, for example, a situation in which the unstable manifold of the renormalization-group fixed point is n-dimensional. Then all families of statistical systems depending on n thermodynamic parameters (e.g. temperature, pressure, concentrations, etc.) which, under the renormalization transformation, are attracted towards the unstable manifold of that fixed point tend to exhibit the *same critical behavior* (i.e. have the same critical exponents). These notions will be made more precise in Chapter 4; see also [520, 379, 24, 136].

Thus, it may happen that a complicated liquid-vapor system occurring in the "real world", which we are not able to theoretically investigate directly, exhibits the same critical behavior as some simple model system. In this book we study some of the simplest statistical-mechanical models, such as lattice gases, using precise mathematical methods. The physical relevance of our results on the critical behavior of such idealized systems stems from the fact that real systems of experimental interest may belong to the same universality class and thus exhibit critical behavior identical to that in the models we are able to analyze. (If there were nothing like universality, then the study of very special, idealized systems would be an exercise without physical interest.)

The simplest lattice gas is the *Ising model.* Consider a simple (hyper-) cubic lattice $\mathbb{Z}^d$. Each site $j \in \mathbb{Z}^d$ may either be empty or else occupied by one "atom" (multiple occupancy of a site is forbidden). This is described by associating to each $j \in \mathbb{Z}^d$ a random variable n_j taking the values 0 or 1. Atoms on neighboring sites are assumed to attract each other, but there is no interaction between atoms on sites which are not nearest neighbors. The chemical potential of a single atom is denoted by μ. Thus, the Hamiltonian of the system confined to a bounded region $\Lambda \subset \mathbb{Z}^d$ is given by

$$H_\Lambda = -4J \sum_{\langle i,j \rangle \subset \Lambda} n_i n_j + \mu \sum_{j \in \Lambda} n_j, \tag{1.2}$$

where $\langle i, j \rangle$ denotes a pair of nearest-neighbor sites in $\mathbb{Z}^d$ (each pair taken *once*), and $J > 0$ is a coupling constant describing the strength of the attraction.

Setting

$$\sigma_j = 2n_j - 1 \tag{1.3}$$

the Hamiltonian becomes

$$\begin{aligned} H_\Lambda &= -J \sum_{\langle i,j \rangle \subset \Lambda} \sigma_i \sigma_j + \left(\frac{\mu}{2} - 2dJ\right) \sum_{j \in \Lambda} \sigma_j + \text{const}(\Lambda) \\ &\equiv -J \sum_{\langle i,j \rangle \subset \Lambda} \sigma_i \sigma_j - h \sum_{j \in \Lambda} \sigma_j + \text{const}(\Lambda)\,. \end{aligned} \tag{1.4}$$

But now we may reinterpret $\sigma_j = \pm 1$ as a spin variable (spin up: $\sigma_j = 1$; spin down: $\sigma_j = -1$), J as an exchange coupling and h as an external magnetic field acting on the spins of the system.

Thus, certain liquid-vapor systems may exhibit the same critical behavior as certain anisotropic ($\sigma_j = \pm 1$), ferromagnetic ($J > 0$) spin systems. Moreover, that critical behavior is believed to be independent of the precise choice of the interaction between different spins, or of the lattice structure. If, for example, the term

$$-J \sum_{\langle i,j \rangle \subset \Lambda} \sigma_i \sigma_j$$

is replaced by

$$- \sum_{\{i,j\} \subset \Lambda} J(i-j) \sigma_i \sigma_j,$$

where $J(i-j) > 0$ is rapidly decreasing as $|i-j| \to \infty$, then, after taking the infinite-volume limit $\Lambda \uparrow \mathbb{Z}^d$, the critical behavior of the modified model is expected to be the same as the critical behavior of the model with Hamiltonian (1.4).

For these reasons, it is of considerable physical interest to study the model with Hamiltonian (1.4) — the *nearest-neighbor ferromagnetic Ising model* — in detail and with mathematical precision. From now on, we shall use magnetic language — speaking of spins, magnetic fields, magnetic susceptibility and so forth — but the reader should keep in mind that our results can be applied also to other systems, such as liquid-vapor systems and binary alloys.

The equilibrium distribution for the Hamiltonian (1.4) at inverse temperature β is, by definition,

$$d\mu_\Lambda(\sigma) = Z_\Lambda^{-1} e^{-\beta H_\Lambda(\sigma)} , \tag{1.5}$$

where the normalization function (partition function) Z_Λ is

$$Z_\Lambda = \sum_{\{\sigma\}} e^{-\beta H_\Lambda(\sigma)} . \tag{1.6}$$

We now wish to take the infinite-volume limit $\Lambda \uparrow \mathbb{Z}^d$. It can be shown [309, 274] that the probability measures μ_Λ have a weak limit as $\Lambda \uparrow \mathbb{Z}^d$; such a limit is called an infinite-volume Gibbs measure for the system (1.4). It can furthermore be shown [309] that the finite-volume Gibbs free energy densities

$$\Phi_\Lambda(\beta, h) = -\frac{1}{\beta|\Lambda|} \log Z_\Lambda(\beta, h) \tag{1.7}$$

have a limit as $\Lambda \uparrow \mathbb{Z}^d$; this limit is called the (infinite-volume) Gibbs free energy density $\Phi(\beta, h)$. It can be proven that $\Phi(\beta, h)$ is independent of boundary conditions. It is not hard to see that $\Phi(\beta, h)$ is a *concave* function of β and h.

Thermodynamic quantities are the derivatives of $\Phi(\beta, h)$ with respect to β and/or h:

- the magnetization $M = -\dfrac{\partial \Phi}{\partial h}$
- the magnetic susceptibility $\chi = \dfrac{\partial M}{\partial h} = -\dfrac{\partial^2 \Phi}{\partial h^2}$
- the entropy $S = \beta^2 \dfrac{\partial \Phi}{\partial \beta}$
- the internal energy $U = \Phi + \beta^{-1} S + hM$
- the specific heat $C_h = -\beta^3 \dfrac{\partial^2 \Phi}{\partial \beta^2}$

By a formal calculation using (1.5)–(1.6), these thermodynamic quantities can be related to correlation functions in the infinite-volume Gibbs measure:

$$M = \langle \sigma_0 \rangle \tag{1.8}$$

$$\chi = \sum_x \langle \sigma_0 ; \sigma_x \rangle \tag{1.9}$$

$$U = -J \sum_{|i|=1} \langle \sigma_0 \sigma_i \rangle \tag{1.10}$$

$$C_h = \beta^2 J^2 \sum_{\substack{x,i,j \\ |i| = |j| = 1}} \langle \sigma_0 \sigma_i ; \sigma_x \sigma_{x+j} \rangle \tag{1.11}$$

where we have used the convenient notation $\langle A;B\rangle \equiv \langle AB\rangle - \langle A\rangle\langle B\rangle$. Relations like (1.8)–(1.11) are called *sum rules* (or *fluctuation-dissipation theorems*). The rigorous proof of such relations is not entirely trivial, as it involves an interchange of differentiation with the infinite-volume limit; for details, see [479].

The Hamiltonian (1.4) has a symmetry,

$$\sigma_j \to -\sigma_j, \qquad h \to -h$$

that is also encountered in real ferromagnets (though not in real fluids). Thinking about the consequences of this symmetry, it is not surprising that phase transitions occur only at zero magnetic field. This is actually a rigorous theorem due to Lee and Yang [363, 360].[1] There is a line of first-order transitions $\{h=0,\ 0\le T<T_c\}$ in the (T,h)-plane [Figure 1.2(a)]. A convenient parameter to study these transitions is the magnetization, $M = M(T,h)$, which is the analogue of the density in liquid-vapor systems. The behavior of $M(T,h)$ is shown in Figure 1.2(b).[2] In particular, let us define the *spontaneous magnetization*

$$M_0(T) \equiv \lim_{h\downarrow 0} M(T,h)\,. \tag{1.12}$$

The spontaneous magnetization vanishes for $T>T_c$ and is strictly positive for $T<T_c$; it is expected (and in some examples actually proven [12, 523, 388]) to be continuous at $T=T_c$. Its qualitative behavior is therefore as indicated in Figure 1.2(c). Moreover, the point $(h=0, T=T_c)$ is a critical point, in the sense that the correlation length diverges there [31, 445, 259, 464, 372, 96, 6, 10]. (All these facts will be discussed more fully in Chapter 14.)

Our main goal in this book is to study the behavior of ferromagnets, in particular those in the universality class of the nearest-neighbor ferromagnetic Ising model, in the vicinity of the critical point $(h=0, T=T_c)$. One would like to show, for example, that the thermodynamic quantities behave as a power law in the vicinity of the critical point, e.g.

$$M(T_c,h) \sim |h|^{1/\delta} \tag{1.13}$$
$$M_0(t) \sim (-t)^{\widehat{\beta}} \tag{1.14}$$
$$\chi(t) \sim |t|^{-\gamma} \tag{1.15}$$

for suitable exponents $\delta, \widehat{\beta}, \gamma$; here $t \equiv T - T_c$. In (1.13)–(1.15), the statement

$$f(x) \sim x^\lambda \tag{1.16}$$

means that there are finite, strictly positive constants f_- and f_+, such that

$$f_- x^\lambda \le f(x) \le f_+ x^\lambda \tag{1.17}$$

as $x \downarrow 0$.

[1] See [371] for a list of references through 1981.

[2] We remark that in an actual experiment, carried out in a finite time span, the magnetization $M(T,h)$ will exhibit hysteresis when h is varied at fixed temperature $T<T_c$. Hysteresis is a dynamic phenomenon which is beyond the scope of equilibrium statistical mechanics; it will not be pursued in this book.

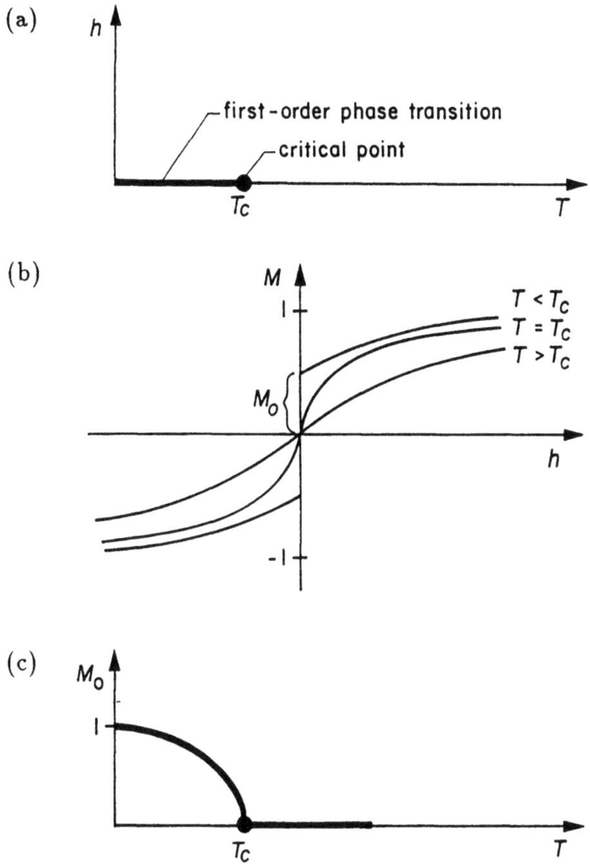

Fig. 1.2. Phase transition and critical behavior in a magnetic system. **(a)** Phase diagram. **(b)** Magnetization M as a function of magnetic field h. **(c)** Spontaneous magnetization M_0 as a function of temperature T

The exponents δ, $\widehat{\beta}$, γ are so-called *critical exponents*.[3] The expected behavior of the Ising model (and other related models) depends on the spatial dimensionality d:

(a) For $d > 4$, (1.13)–(1.15) are expected to hold, with the critical exponents taking their *mean-field values*

$$\delta = 3, \quad \widehat{\beta} = \frac{1}{2}, \quad \gamma = 1. \tag{1.18}$$

At the end of this section we give a brief discussion of one version of mean-field theory.

(b) For $d = 4$, the mean-field critical behavior is expected to be modified by multiplicative logarithmic corrections.

(c) For $d < 4$, (1.13)–(1.15) are expected to hold, with the critical exponents satisfying the strict inequalities

$$\delta > 3, \quad \widehat{\beta} < \frac{1}{2}, \quad \gamma > 1. \tag{1.19}$$

Some of the reasoning underlying these expectations will be presented in the following chapters. Moreover, for a variety of models related to the Ising model, some of these expectations can be rigorously proven [388, 5, 213, 10, 290, 298], as we discuss in detail in Chapter 14.

Of course, in a laboratory, all that is available to us are finite samples of three-dimensional or approximately two-dimensional ferromagnets. But in theory one can study d-dimensional infinite ferromagnets, where d is an arbitrary natural number[4], and that study is useful to understand the properties of real three-dimensional ferromagnets. Moreover, the four-dimensional case is relevant to quantum field theory, where space-time now plays the role of "space" (see Section 1.5).

At first sight, it is quite surprising that the critical exponents δ, $\widehat{\beta}$ and γ are independent of dimension above four dimensions. But the reader who makes it to the end of Part III will not be surprised any more. For $d < 4$, the exponents δ, $\widehat{\beta}$ and γ do depend on d, but not on the details of the model: within rather wide classes of models (and real physical systems), the critical exponents are *universal*.

While there are calculations of critical exponents for $d < 4$ in terms of power series in various expansion parameters (e.g. $\varepsilon = 4 - d$) which yield very precise

[3] In this book we denote by $\widehat{\beta}$ the critical exponent that is traditionally called β, in order to avoid confusion with the inverse temperature $\beta = 1/(kT)$.

[4] Physicists like to go further, and imagine arbitrary real (or even complex!) dimensions. There have been two different approaches: one based on analytic continuation ([485, 518, 66, 31, 278] and references therein), which preserves Euclidean symmetry but has problems with reflection positivity; and another based on the use of fractal lattices [250, 431, 165, 166], which preserves reflection positivity but has problems with Euclidean symmetry and universality.

values, the mathematical status of these series is still rather mysterious. Our methods are too soft to shed much light on this question, but they are good enough to produce various rigorous bounds on critical exponents which are not entirely without interest. The situation in two dimensions is special. Thanks to important recent advances in two-dimensional conformal field theory [58, 104, 103, 209, 526], one might not be too far from a reasonably complete classification of all possible critical behavior in two-dimensional statistical systems. But these astonishing and profound developments are beyond the scope of our book.

Let us conclude this discussion by giving a brief sketch of one approach to mean-field theory, the so-called Landau-Ginzburg theory. We consider the above example in which the Gibbs free energy Φ is a function of only two parameters, β and h. The Landau-Ginzburg theory is based on the assumption that in the vicinity of the critical point, the Helmholtz free energy $F(\beta, M)$ — which is the Legendre transform of the Gibbs free energy $\Phi(\beta, h)$ — is a smooth function of $t \equiv T - T_c = (k\beta)^{-1} - (k\beta_c)^{-1}$ and M. [This assumption is *false* in dimension $d \leq 4$, which is why mean-field theory is incorrect in that case.] By general arguments, it can be concluded that, following a smooth change of variables, F takes the form

$$F(t, M) = \frac{1}{4}M^4 + \frac{1}{2}tM^2 \tag{1.20}$$

— or, more precisely, $F(t, M)$ is the convex hull of the graph of this function [so as not to violate the convexity requirements for F]. The thermodynamic prescription $h = \partial F/\partial M$ implies that for $h \neq 0$ the magnetization must be the solution of the equation

$$M_{LG}^3 + tM_{LG} = h \tag{1.21}$$

with the additional constraint that $M^2 \geq -t$. In particular, by setting $t = 0$ we conclude that (1.13) holds with $\delta = 3$. For $h = 0$, the magnetization M_{LG} vanishes for $t > 0$; for $t < 0$, the form of the convex hull of (1.20) implies that the magnetization can be any number between $-\sqrt{-t}$ and $\sqrt{-t}$. The line $(t < 0, h = 0)$ is a line of first-order phase transitions (the first derivative of the Gibbs free energy with respect to h is discontinuous); the physical interpretation is that at each of its points there are coexisting phases. The extreme values $M_{LG} = \pm\sqrt{-t}$ correspond to pure phases. Hence (1.14) holds with $\widehat{\beta} = 1/2$. In an analogous manner one obtains the value $\gamma = 1$ for the remaining exponent in (1.18). For further discussion of mean-field theory, see Sections 2.3 and 14.2.1.

1.2 Multicritical points

In the preceding section we have described the simplest phase-transition situation, in which there are two thermodynamic parameters (β and h). This corresponds to a "single-component" or "pure fluid" system, characterized by a phase diagram of the type of Figure 1.2(a): a line where two phases *coexist* — the line of first-order phase transitions — terminating in a point — the critical (end)point — where the two phases *become identical*.

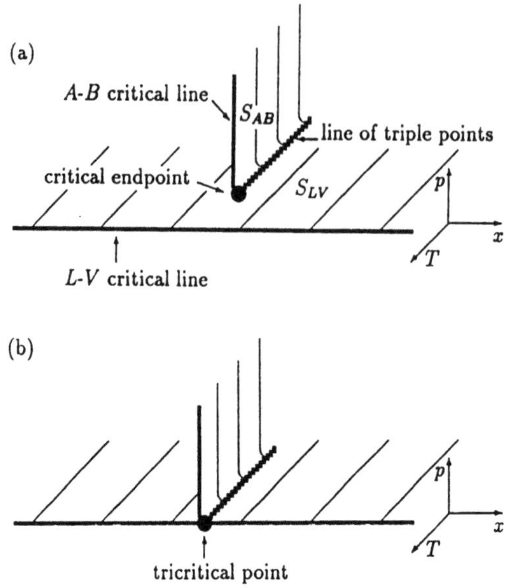

Fig. 1.3. Phase diagram of a binary mixture: **(a)** The generic situation; **(b)** "lucky accident"

More complicated behavior can arise in models involving $n \geq 3$ thermodynamic parameters [273, 280, 365, 277, 276, 511, 354, 341]. Consider, for example, a binary mixture, like water and oil. Such a system can exist in three distinct phases: a high-density phase rich in substance A ("liquid A"), a high-density phase rich in substance B ("liquid B"), and a low-density phase ("vapor"). To describe the state of such a system we need three thermodynamic parameters (preferably "fields" in Griffiths' nomenclature [280, 277]), for instance pressure, temperature and the difference between the chemical potentials of the two species A and B. (Another choice could be temperature and the two chemical potentials.) With this third parameter, the liquid-vapor coexistence line of Figure 1.2(a) expands into a coexistence surface, and the critical point becomes a critical curve limiting that surface. These manifolds correspond respectively to the surface S_{LV} and the L–V critical line in Figure 1.3. Diagrams like the one in Figure 1.2(a) are obtained for the sections $x = \text{const}$.

Moreover, besides the liquid-vapor coexistence surface, there appears a liquid-liquid coexistence surface — S_{AB} in Figure 1.3 — bounded by its own (liquid-liquid) critical curve. The curve where the two coexistence surfaces intersect (note that the surface S_{AB} does not cross to the other side of S_{LV}) indicates *three-phase coexistence* (liquid-liquid-vapor), that is, it is a curve of *triple points*. The endpoint of this curve is a critical endpoint at which, in the generic situation [Figure 1.3(a)], two phases become identical in presence of a different (vapor) phase — the "spectator phase" [204]. If the system is enclosed in a test tube [277], then for parameters with values along the line of triple points, the system will exhibit two meniscusses: the bottom one separating A

from B and the top one separating the liquid from the vapor. If the system is brought along this line towards the critical endpoint, the arrival to the latter will be signaled by the disappearance of the A–B meniscus.

A more dramatic possibility [Figure 1.3(b)] is that this critical endpoint of the liquid-liquid critical curve also belong to the liquid-vapor critical curve, that is, that the two critical curves intersect. At such an intersection point, the *three phases* would *become identical* (both meniscusses would disappear at the same time); such a point is termed a *tricritical point.* In that extreme case, the phase diagram of the mixture would consist of three two-dimensional manifolds, corresponding to the A–V, A–B and B–V coexistence surfaces intersecting at a common line of triple points which is bounded by a tricritical point.

The occurrence of a tricritical point in a binary mixture is, however, extremely rare (except in the presence of additional symmetry): the intuitive argument is that two lines do not intersect in three dimensions except by a "lucky accident" [277]. In order to see generic tricritical behavior, we need to consider instead a *ternary* mixture, that is, a system characterized by *four* thermodynamic parameters. Such a mixture will, in general, exhibit two-phase coexistence on a variety formed by several three-dimensional manifolds (= hypersurfaces) in the four-dimensional parameter space. The boundaries of these hypersurfaces define two-dimensional manifolds of critical points where two phases become identical; and the intersections of these hypersurfaces form two-dimensional manifolds of triple points. The boundaries of the latter form one or more curves of tricritical points.

In general, one expects that for a system with n thermodynamic parameters [$(n-1)$-ary mixture] there is a variety of dimension $n-1$ on which two phases can coexist, formed by several manifolds whose intersections determine a variety of dimension $n-2$ on which three phases can coexist, and so on; this is the so-called *Gibbs phase rule*. Each $(n-k)$-dimensional manifold of $(k+1)$-phase coexistence is bounded by an $(n-k-1)$-dimensional manifold of $(k+1)$-critical points.

A simple phenomenological approach to these phenomena is based on a Landau-Ginzburg ansatz [277]. For instance, the simplest such situation can be deduced from a Helmholtz free energy

$$F(t,x,M) = \frac{1}{6}M^6 + \frac{x}{4}M^4 + \frac{t}{2}M^2 \tag{1.22}$$

(more precisely, F is the convex hull of this formula). The coefficient $1/6$ (and the symmetry $M \to -M$) are chosen so that a tricritical point does appear in the resulting phase diagram. The pure phases correspond to the absolute minima of $F - hM$ as a function of M for fixed x, t, h. The phase diagram for this system is shown in Figure 1.4. There are three surfaces in (x,t,h)-space along which two phases coexist. These three surfaces intersect in a common line ($x < 0, t = \frac{3}{16}x^2, h = 0$), along which three phases coexist (line of triple points). The boundaries of the two-phase-coexistence surfaces are curves of critical points (two phases become identical), which arise when the bump between

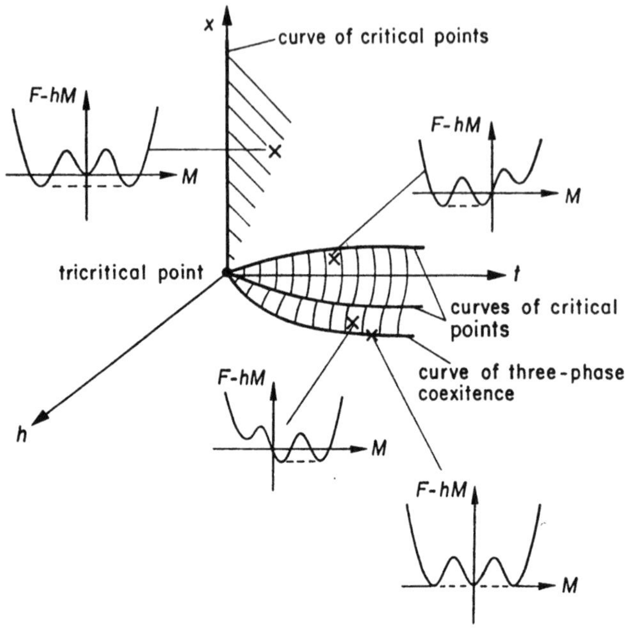

Fig. 1.4. Phase diagram corresponding to the Landau-Ginzburg ansatz (1.22). The curve of three-phase coexistence ($x < 0, t = \frac{3}{16}x^2, h = 0$) resembles the keel of a canoe. On this curve three surfaces (here shaded) of two-phase coexistence meet: two of these surfaces correspond to the sides of the canoe, and one is the part of the vertical plane terminating in the keel. Each of these surfaces terminates in a curve of critical points (here bold). These three curves of critical points meet at the tricritical point

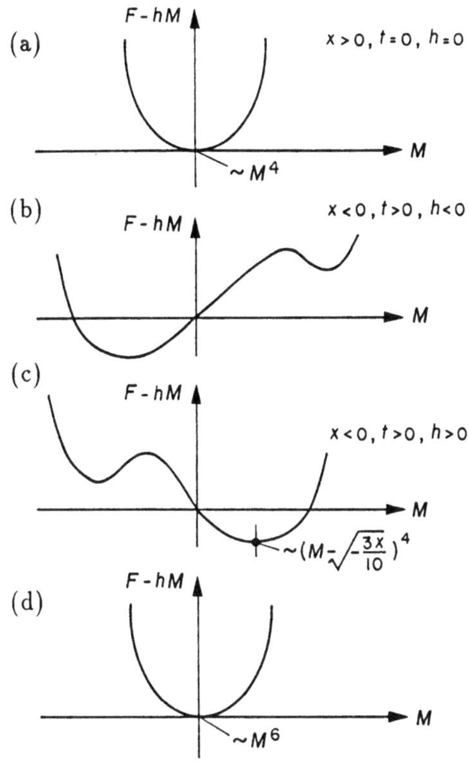

Fig. 1.5. Shape of the function $F(M) - hM$ on the critical curves and at the tricritical point [see Figure 1.4]. **(a)** Critical curve $x < 0, t = 0, h = 0$. **(b)** Critical curve $x < 0, t = \frac{9}{20}x^2, h = -\frac{6}{25}\sqrt{\frac{3}{10}}(-x)^{5/2}$. **(c)** Critical curve $x < 0, t = \frac{9}{20}x^2, h = +\frac{6}{25}\sqrt{\frac{3}{10}}(-x)^{5/2}$. **(d)** Tricritical point $x = t = h = 0$

the two minima degenerates into a single fourth-order minimum [Figure 1.5(a)–(c)]; these curves are

- $x > 0,\ t = 0,\ h = 0$ [corresponding to a minimum at $M = 0$]
- $x < 0,\ t = \frac{9}{20}x^2,\ h = \pm\frac{6}{25}\sqrt{\frac{3}{10}}(-x)^{5/2}$
  [corresponding to a minimum at $M = \pm(-\frac{3}{10}x)^{1/2}$]

The three curves of critical points intersect at the *tricritical point* $(x, t, h) = (0, 0, 0)$, which corresponds to a single sixth-order minimum [Figure 1.5(d)].

The mean-fieldcan be easily predictions for tricritical exponents read off from (1.22). Indeed, the hardest problem is inventing names for the plethora of exponents arising from different paths of approach to the tricritical point [275]. These mean-field predictions are expected to be valid in dimension $d > 3$, with multiplicative logarithmic corrections for $d = 3$. More generally, at an n-critical point (corresponding to a Landau-Ginzburg polynomial of order $2n$), the upper critical dimension is expected to be $d_c = 2n/(n-1)$.

1.3 Spin systems, quantum field theory and random walks: An overview

One of the main purposes of Part I is to explain the mathematical relations and connections between the following three areas in physics:

(A) Classical lattice spin systems, with emphasis on their critical behavior.

(B) Relativistic quantum field theory — in particular, the Euclidean(imaginary-time) formulation.

(C) Random walks, intersection properties of random walks, and gases of interacting random walks.

The realization that (A), (B) and (C) are intimately related is an important and deep idea which is at the basis of enormous progress in understanding critical phenomena in statistical mechanics and in constructing models of relativistic quantum fields.

The connections between (A) and (B) were explored in the pioneering work of Symanzik [493, 494] and Nelson [405, 406, 408, 407] on Euclidean quantum field theory, in the work of Wilson and others on the renormalization group [323, 516, 517, 520, 324, 325, 317, 106, 202, 77, 83, 136, 505, 500, 24], and most recently in the study of conformal field theory [436, 396, 58, 104, 103, 209, 105]. We discuss these connections in more detail in Section 1.5 and Chapters 3 and 4. The upshot is that massless (resp. massive) quantum field theories can be obtained as the scaling limits of statistical-mechanical models which are at (resp. approaching) a critical point. Conversely, the large-scale behavior of a statistical-mechanical system at (resp. near) criticality is described by the conformal quantum field theory corresponding to the renormalization-group fixed point to which the system is attracted (resp. by the non-conformal-invariant quantum field theories corresponding to the unstable manifold of that fixed point).

Conformal field theories are heavily constrained, and one may therefore expect that there are not too many of them. If made precise, this idea provides a concrete version of universality. Indeed, the program of understanding critical properties of statistical systems by studying conformal field theories has been very successful in two dimensions. In more general situations, the renormalization group has been an extremely versatile tool.

The connection between (A)/(B) and (C) was first elucidated by Symanzik [493, 494], and in recent years it has been developed into a precise tool for studying lattice spin systems and their continuum (scaling) limits. In this approach [92, 213, 5], lattice spin systems are represented as gases of interacting random walks. The basic intuition, which is made precise in these papers, is that (in certain cases) the critical behavior of the spin system can be analyzed in terms of the intersection properties of the associated random walks. For example, the results of Erdös and Taylor [164] and Lawler [351] on simple random walks, together with Symanzik's loop-gas representation of φ^4 field theory [494], suggested a no-interaction ("triviality") theorem for φ^4 field theoryin dimension $d \geq 4$. Such a theorem was later proven for $d > 4$ (together with partial results for $d = 4$) by methods which follow closely the random-walk intuition [5, 213]. This circle of ideas is the central theme of this book. We give an overview in Chapters 5 and 6.

In the next three sections, we introduce the main models of lattice spin systems, relativistic quantum fields and random walks, and outline the principal problems concerning them which will occupy us in the remainder of this book.

1.4 Classical lattice spin systems

In this section we consider lattice systems of N-component classical "spins" $\boldsymbol{\varphi}_x = (\varphi_x^{(1)}, \ldots, \varphi_x^{(N)}) \in \mathbb{R}^N$ for $x \in \mathbb{Z}^d$. The *a priori* distribution of the spin $\boldsymbol{\varphi}_x$ is given by a spherically symmetric finite measure dP_x on $\mathbb{R}^N$:

$$dP_x(\boldsymbol{\varphi}_x) = g(\boldsymbol{\varphi}_x^2)\, d\boldsymbol{\varphi}_x\,, \tag{1.23}$$

where

$$\boldsymbol{\varphi}_x^2 \equiv \sum_{\alpha=1}^{N} \varphi_x^{(\alpha)^2} \tag{1.24}$$

and $d\boldsymbol{\varphi}_x$ is Lebesgue measure on $\mathbb{R}^N$. The key example is the (isotropic) $\boldsymbol{\varphi}^4$ *model*

$$g(\boldsymbol{\varphi}^2) = \exp\left[-\frac{\lambda}{4}\boldsymbol{\varphi}^4 - \frac{B}{2}\boldsymbol{\varphi}^2\right] \qquad (\lambda \geq 0) \tag{1.25}$$

which includes as limiting cases the *Gaussian model* ($\lambda = 0$)

$$g(\boldsymbol{\varphi}^2) = \exp\left[-\frac{B}{2}\,\boldsymbol{\varphi}^2\right], \tag{1.26}$$

which is exactly solvable, and the *N-vector model*

$$g(\boldsymbol{\varphi}^2) = \delta(\boldsymbol{\varphi}^2 - 1) \tag{1.27}$$

(obtained by letting $\lambda \to \infty$ with $B = -\lambda$). Special cases of the N-vector model are the *Ising model* ($N = 1$), the *plane-rotator* (XY) *model* ($N = 2$), and the *classical Heisenberg model* ($N = 3$). The Hamiltonian is given formally by

$$H(\boldsymbol{\varphi}) = -\tfrac{1}{2}\sum_{x,y} J_{xy}\, \boldsymbol{\varphi}_x \cdot \boldsymbol{\varphi}_y - \sum_x \mathbf{h} \cdot \boldsymbol{\varphi}_x\,. \tag{1.28}$$

For simplicity we shall take the magnetic field $\mathbf{h}$ to point in the 1-direction, i.e. $\mathbf{h} = (h, 0, \ldots, 0)$. We shall usually also assume that the pair interactions J are *ferromagnetic, i.e.* $J_{xy} \geq 0$ for all x, y.

The formal expression (1.28) is, however, ill-defined due to the infinite-volume sums (which are almost surely divergent). We therefore define first the system in a finite region $\Lambda \subset \mathbb{Z}^d$: the Hamiltonian is

$$H_\Lambda(\varphi) = -\tfrac{1}{2} \sum_{\substack{\{x,y\}: \\ \{x,y\}\cap\Lambda \neq \varnothing}} J_{xy}\, \varphi_x \cdot \varphi_y - \sum_{x\in\Lambda} \mathbf{h}\cdot\varphi_x \tag{1.29}$$

and the equilibrium (Boltzmann-Gibbs) distribution at inverse temperature β is

$$d\mu_\Lambda(\varphi) = Z_\Lambda^{-1} e^{-\beta H_\Lambda(\varphi)} \prod_{x\in\Lambda} dP_x(\varphi_x)\, dB_{\Lambda^c}(\varphi)\ . \tag{1.30}$$

Here dB_{Λ^c} is a probability measure on the configurations $\{\varphi_x\}_{x\in\Lambda^c}$ of spins outside Λ, which fixes some boundary condition, and Z_Λ is a normalization factor (the *partition function*) chosen so that $\int d\mu_\Lambda(\varphi) = 1$. It is then necessary to construct the infinite-volume (thermodynamic) limit

$$d\mu(\varphi) = \lim_{\Lambda\uparrow\mathbb{Z}^d} d\mu_\Lambda(\varphi) \tag{1.31}$$

(with a suitable notion of convergence). This is a conventional problem treated in many excellent books; see e.g. [447, 274, 309].

The equilibrium properties of such spin systems in the infinite-volume limit can be derived from a study of the *correlation functions*

$$\langle \varphi_{x_1}^{(\alpha_1)} \cdots \varphi_{x_n}^{(\alpha_n)} \rangle \equiv \int \left(\prod_{j=1}^{n} \varphi_{x_j}^{(\alpha_j)} \right) d\mu(\varphi)\ , \quad n = 1,2,3,\ldots\ . \tag{1.32}$$

In a pure phase, the correlation functions factorize as the points $x_1, \ldots, x_n$ move apart, so it is useful to introduce the *truncated* (or *connected*) *correlation functions*

$$\langle \varphi_{x_1}^{(\alpha_1)} \rangle^T \equiv \langle \varphi_{x_1}^{(\alpha_1)} \rangle \tag{1.33}$$

$$\langle \varphi_{x_1}^{(\alpha_1)} \varphi_{x_2}^{(\alpha_2)} \rangle^T \equiv \langle \varphi_{x_1}^{(\alpha_1)} \varphi_{x_2}^{(\alpha_2)} \rangle - \langle \varphi_{x_1}^{(\alpha_1)} \rangle \langle \varphi_{x_2}^{(\alpha_2)} \rangle \tag{1.34}$$

$$\begin{aligned}
\langle \varphi_{x_1}^{(\alpha_1)} \varphi_{x_2}^{(\alpha_2)} \varphi_{x_3}^{(\alpha_3)} \rangle^T &\equiv \langle \varphi_{x_1}^{(\alpha_1)} \varphi_{x_2}^{(\alpha_2)} \varphi_{x_3}^{(\alpha_3)} \rangle - \langle \varphi_{x_1}^{(\alpha_1)} \rangle \langle \varphi_{x_2}^{(\alpha_2)} \varphi_{x_3}^{(\alpha_3)} \rangle \\
&\quad - \langle \varphi_{x_2}^{(\alpha_2)} \rangle \langle \varphi_{x_1}^{(\alpha_1)} \varphi_{x_3}^{(\alpha_3)} \rangle - \langle \varphi_{x_3}^{(\alpha_3)} \rangle \langle \varphi_{x_1}^{(\alpha_1)} \varphi_{x_2}^{(\alpha_2)} \rangle \\
&\quad + 2 \langle \varphi_{x_1}^{(\alpha_1)} \rangle \langle \varphi_{x_2}^{(\alpha_2)} \rangle \langle \varphi_{x_3}^{(\alpha_3)} \rangle \\
&\equiv \langle \varphi_{x_1}^{(\alpha_1)} \varphi_{x_2}^{(\alpha_2)} \varphi_{x_3}^{(\alpha_3)} \rangle - \langle \varphi_{x_1}^{(\alpha_1)} \rangle^T \langle \varphi_{x_2}^{(\alpha_2)} \varphi_{x_3}^{(\alpha_3)} \rangle^T \\
&\quad - \langle \varphi_{x_2}^{(\alpha_2)} \rangle^T \langle \varphi_{x_1}^{(\alpha_1)} \varphi_{x_3}^{(\alpha_3)} \rangle^T - \langle \varphi_{x_3}^{(\alpha_3)} \rangle^T \langle \varphi_{x_1}^{(\alpha_1)} \varphi_{x_2}^{(\alpha_2)} \rangle^T \\
&\quad - \langle \varphi_{x_1}^{(\alpha_1)} \rangle^T \langle \varphi_{x_2}^{(\alpha_2)} \rangle^T \langle \varphi_{x_3}^{(\alpha_3)} \rangle^T
\end{aligned} \tag{1.35}$$

and in general

$$\left\langle \prod_{i=1}^{n} \varphi_{x_i}^{(\alpha_i)} \right\rangle^T \equiv \left\langle \prod_{i=1}^{n} \varphi_{x_i}^{(\alpha_i)} \right\rangle - \sum_{\substack{\mathcal{P}\,\in\,\text{partitions} \\ \text{of } \{1,\ldots,n\}}} \prod_{I\in\mathcal{P}} \left\langle \prod_{i\in I} \varphi_{x_i}^{(\alpha_i)} \right\rangle^T . \tag{1.36}$$

The truncated correlation functions are also called *Ursell functions* and are denoted $u_n^{(\alpha_1,\ldots,\alpha_n)}(x_1,\ldots,x_n)$. We also use a convenient semicolon notation

$$\begin{aligned}\langle A;B\rangle &\equiv \langle A\,B\rangle - \langle A\rangle\,\langle B\rangle \\ \langle A;B;C\rangle &\equiv \langle A\,B\,C\rangle - \langle A\rangle\,\langle B\,C\rangle - \langle B\rangle\,\langle A\,C\rangle - \langle C\rangle\,\langle A\,B\rangle \\ &\qquad +2\langle A\rangle\,\langle B\rangle\langle C\rangle \\ &\text{etc.}\end{aligned}$$

for truncated correlation functions of arbitrary observables A, B, C, ...

For simplicity let us first restrict our attention to one-component models ($N = 1$). Then, away from a critical point, the truncated correlation functions are expected to decay exponentially in the spatial separation $|x_i - x_j|$. In particular, the two-point function is expected to decay as

$$\langle \varphi_x;\varphi_y\rangle \underset{|x-y|\to\infty}{\sim} e^{-|x-y|/\xi}\ . \tag{1.37}$$

The quantity ξ is called the *correlation length*, and is defined more precisely as

$$\xi \equiv \limsup_{|x|\to\infty} \frac{-|x|}{\log\langle\varphi_0;\varphi_x\rangle}\ . \tag{1.38}$$

It is also common to use the inverse correlation length

$$m = \xi^{-1}\ , \tag{1.39}$$

which in quantum field theory (Section 1.5) has the interpretation of a *mass*.

Other important quantities are the *magnetization*

$$M \equiv \langle\varphi_0\rangle \tag{1.40}$$

and the *susceptibility*

$$\chi \equiv \sum_x \langle\varphi_0;\varphi_x\rangle\ . \tag{1.41}$$

A formal computation using the Hamiltonian (1.28) shows that

$$\chi = \frac{\partial M}{\partial h}\ . \tag{1.42}$$

The validity of this relation in the infinite-volume limit is a nontrivial fact, called the "fluctuation-dissipation relation" (or "sum rule") [274].

When $\langle\varphi_x\rangle = 0$, an important role will be played by the truncated four-point correlation function, which in this case simplifies to

$$\begin{aligned}u_4(x_1,x_2,x_3,x_4) \;=\; &\langle\varphi_{x_1}\varphi_{x_2}\varphi_{x_3}\varphi_{x_4}\rangle \;-\; \langle\varphi_{x_1}\varphi_{x_2}\rangle\langle\varphi_{x_3}\varphi_{x_4}\rangle \\ &-\; \langle\varphi_{x_1}\varphi_{x_3}\rangle\langle\varphi_{x_2}\varphi_{x_4}\rangle \;-\; \langle\varphi_{x_1}\varphi_{x_4}\rangle\langle\varphi_{x_2}\varphi_{x_3}\rangle\ ,\end{aligned} \tag{1.43}$$

and by its summed ("zero-momentum") version

$$\overline{u}_4 \equiv \sum_{x_2,x_3,x_4} u_4(x_1,x_2,x_3,x_4) \,. \tag{1.44}$$

In particular, the *dimensionless renormalized four-point coupling constant*

$$g \equiv \frac{-\overline{u}_4}{\chi^2 \xi^d} \tag{1.45}$$

will be crucial in analyzing the triviality or non-triviality of quantum field theories (Section 1.5 and Chapter 15).

Among the important questions one might wish to ask about such spin systems are:

(a) Does the system exhibit phase transitions? Are they accompanied by symmetry-breaking? Consider, for example, the spontaneous magnetization $M_0(\beta) \equiv \lim_{h\downarrow 0} M(\beta,h)$. It is easy to show that when β is small enough (i.e. the temperature is high enough), $M_0(\beta) = 0$. Now increase the value of β. Is there some β_0 such that for $\beta > \beta_0$, $M_0(\beta) > 0$?

(b) Is there a critical point $\beta_c \le \beta_0$ such that

$$\begin{aligned} &\xi \uparrow \infty \text{ as } \beta \uparrow \beta_c \text{ with } h = 0 \,; \\ &\xi \uparrow \infty \text{ as } h \downarrow 0 \text{ with } \beta = \beta_c \,? \end{aligned} \tag{1.46}$$

If so, is $\beta_c = \beta_0$?

If a critical point exists one can ask:

(c) What does the approach to the critical point look like? Do the thermodynamic quantities behave as power laws near the critical point? If so, one may introduce *critical exponents*:

$$\langle \varphi_x \varphi_y \rangle_{\beta=\beta_c, h=0} \underset{|x-y|\to\infty}{\sim} |x-y|^{-(d-2+\eta)} \tag{1.47}$$

$$\xi \sim (\beta_c - \beta)^{-\nu} \tag{1.48}$$

$$\chi \sim (\beta_c - \beta)^{-\gamma} \tag{1.49}$$

etc.

One may then ask:

(d) What are the values of these and other exponents? Why are $\gamma \ge 1$, $\nu \ge \frac{1}{2}$, $\eta \ge 0$ for short-range J? Why are $\gamma = 1$, $\nu = \frac{1}{2}$, $\eta = 0$ for short-range J in dimension $d > 4$? Why is $\gamma = (2-\eta)\nu$?

(e) How does the dimensionless renormalized coupling g behave, for $h = 0$ and $\beta \uparrow \beta_c$?

Answers to (a) and (b) are sketched briefly in Chapter 2. A fairly detailed treatment of (c), (d) and (e) appears in Chapter 14. This will be one of the main topics to be discussed in this book.

In the multi-componentcase ($N \geq 2$), these definitions must be adapted to distinguish between components parallel to the magnetic field ("longitudinal") and perpendicular to it ("transverse"). Thus, we define

$$\xi_{\parallel}^{-1} = m_{\parallel} = \lim_{x\to\infty} -\frac{1}{|x|} \log\langle\varphi_0^{(1)};\varphi_x^{(1)}\rangle \tag{1.50}$$

$$\xi_{\perp}^{-1} = m_{\perp} = \lim_{x\to\infty} -\frac{1}{|x|} \log\langle\varphi_0^{(\alpha)};\varphi_x^{(\alpha)}\rangle \qquad \text{any } \alpha = 2,\ldots,N \tag{1.51}$$

$$\chi_{\parallel} \equiv \sum_x \langle\varphi_0^{(1)};\varphi_x^{(1)}\rangle \tag{1.52}$$

$$\chi_{\perp} \equiv \sum_x \langle\varphi_0^{(2)};\varphi_x^{(2)}\rangle \qquad \text{any } \alpha = 2,\ldots,N\ . \tag{1.53}$$

We remark that multi-component models below the critical temperature (at zero magnetic field) have additional subtleties due to the Goldstone modes [264, 502, 24, 361]. In this book we shall usually focus on one-component spins ($N = 1$), but shall sometimes indicate results for $N = 2$ (plane rotator) and "$N = 0$" (which we shall later see is equivalent to the self-avoiding walk).

1.5 Relativistic quantum field theory and Euclidean field theory

Relativistic quantum field theory is an attempt towards combining quantum mechanics and the special theory of relativity into one mathematically consistent (and hopefully physically correct) theory. It is a particular type of quantum theory in which the observables are *fields* and the invariance group contains that of *special relativity* (the Poincaré group).

Let us first recall the general structure of quantum mechanics:

- The (pure) *states* of the system are the unit rays of a complex Hilbert space $\mathcal{H}$.
- The *observables* are self-adjoint operators on $\mathcal{H}$.
- The *expectation value* of an observable A in a state ψ is $(\psi, A\psi)$.
- *Symmetries* of the system are represented by unitary (or sometimes antiunitary) operators on $\mathcal{H}$.

In particular, a *relativistic* quantum theory should carry a (projective) representation $U(a, \Lambda)$ of the proper orthochronous Poincaré group[5]

$$\mathcal{P}_+^{\uparrow} = ISO(d-1,1)^{\uparrow}\ ,$$

which is the invariance group of d-dimensional Minkowski (special-relativistic) space-time. A quantum *field* theory should include observables $\varphi(\mathbf{x}, t)$ which

[5] See [488, 321] for a detailed description of the Poincaré group.

represent the strength of some field (e.g. the electric field) measured at spatial location $\mathbf{x}$ and time t. (We use the Heisenberg picture.)

There is really only one subtlety: It turns out that a relativistic quantum field is too singular at short distances to be a pointwise-defined operator-valued *function* $\varphi(\mathbf{x},t)$; indeed, the physical impossibility of measuring a quantum field at a space-time point was noticed already in the 1930's by Bohr and Rosenfeld [69].[6] Rather, a quantum field is an operator-valued *distribution* in the sense of Schwartz. Physically this means that the only sensible observables are *smeared fields* $\varphi(f)$, where f is a smooth test function; heuristically one makes the interpretation

$$\varphi(f) \quad \text{“=”} \quad \int \varphi(\mathbf{x},t) f(\mathbf{x},t)\, d\mathbf{x}\, dt\ . \tag{1.54}$$

We are now ready to write down a set of properties which any physically sensible relativistic quantum field theory ought to possess. A mathematically precise statement of these properties was first given by Gårding and Wightman [514]. We shall first state the *Gårding-Wightman axioms*, and then comment on their physical meaning. For simplicity we restrict attention to the case of a single Hermitian scalar field. Henceforth we use the notation $x = (x^0, x^1, \ldots, x^{d-1}) \equiv (t,\mathbf{x}) \in \mathbb{R}^d$.

(GW0) *States.* The states of the system are the unit rays of a separable complex Hilbert space $\mathcal{H}$. There is a distinguished state Ω, called the *vacuum.*

(GW1) *Fields and temperedness.* There exists a dense subspace $D \subset \mathcal{H}$, and for each test function f in the Schwartz space $\mathcal{S}(\mathbb{R}^d)$ there exists an (unbounded) operator $\varphi(f)$ with domain D, such that:

a) For all $\psi_1, \psi_2 \in D$, the map $f \mapsto (\psi_1, \varphi(f)\psi_2)$ is a tempered distribution.

b) For real-valued f, the operator $\varphi(f)$ is Hermitian, i.e. $(\psi_1, \varphi(f)\psi_2) = (\varphi(f)\psi_1, \psi_2)$ for all $\psi_1, \psi_2 \in D$.

c) The vacuum Ω belongs to D, and $\varphi(f)$ leaves D invariant, i.e. $\psi \in D$ implies that $\varphi(f)\psi \in D$.

d) The set D_0 of finite linear combinations of vectors of the form $\varphi(f_1)\ldots\varphi(f_n)\Omega$ with $n \geq 0$ and $f_1,\ldots,f_n \in \mathcal{S}(\mathbb{R}^d)$ is dense in $\mathcal{H}$.

(GW2) *Relativistic covariance.* There is a strongly continuous unitary representation $U(a,\Lambda)$ of the proper orthochronous Poincaré group $\mathcal{P}_+^\uparrow$, such that

a) $U(a,\Lambda)$ leaves D invariant, i.e. $\psi \in D$ implies $U(a,\Lambda)\psi \in D$.

b) $U(a,\Lambda)\Omega = \Omega$ for all $(a,\Lambda) \in \mathcal{P}_+^\uparrow$.

c) $U(a,\Lambda)\varphi(f)U(a,\Lambda)^{-1} = \varphi(f_{(a,\Lambda)})$, where

$$f_{(a,\Lambda)}(x) \equiv f(\Lambda^{-1}(x-a))\ . \tag{1.55}$$

[6] See [513, 522, 109] for proofs that a relativistic quantum field (in space-time dimension $d \geq 2$) cannot be a pointwise-defined operator.

(GW3) *Spectral condition.* The joint spectrum of the infinitesimal generators of the translation subgroup $U(a, \mathbf{1})$ is contained in the forward light cone $\overline{V}_+ = \{p \equiv (p^0, \mathbf{p}) \in \mathbb{R}^d \colon p^0 \geq |\mathbf{p}|\}$.

(GW4) *Locality.* If f and g have spacelike-separated supports, then $\varphi(f)$ and $\varphi(g)$ commute, i.e. $[\varphi(f)\varphi(g) - \varphi(g)\varphi(f)]\psi = 0$ for all $\psi \in D$.

We now make some brief remarks on the mathematical and physical meaning of the axioms:

(GW0) The separability of $\mathcal{H}$ is actually a consequence of (GW1d) (see [321, p. 64]).

(GW1) The smeared fields $\varphi(f)$ are in general *unbounded* operators[7]; for this reason, they cannot sensibly be defined on the whole Hilbert space $\mathcal{H}$, but only on a dense subspace D. We leave D unspecified, but remark that D clearly contains D_0, and that all of the physical information of interest to us (i.e. vacuum expectation values) is contained in the restriction of the operators $\varphi(f)$ to D_0. So there would be little loss of generality in assuming that D *equals* D_0.

The interpretation of the smeared fields as physical observables requires that the operators $\varphi(f)$ be *self-adjoint* (on a suitable domain) and not merely Hermitian.[8] This assumption is *not* contained in the Gårding-Wightman axioms; it is noteworthy how much information can be gained without it. Self-adjointness of the field operators can, however, usually be proven in specific models (see e.g. [146, 260]).

(GW2c) says that φ transforms as a *scalar* field under Poincaré transformations; it can be written heuristically as

$$U(a, \Lambda)\varphi(x)U(a, \Lambda)^{-1} = \varphi(\Lambda x + a) \ . \tag{1.56}$$

If φ were a non-gauge tensor (or spinor) field, the appropriate generalization of (1.56) would be

$$U(a, \Lambda)\varphi(x)U(a, \Lambda)^{-1} = R(\Lambda^{-1})\,\varphi(\Lambda x + a) \tag{1.57}$$

where $R(\Lambda)$ is a finite-dimensional representation of the proper orthochronous Lorentz group $\mathcal{L}_+^\uparrow = SO(d-1, 1)^\uparrow$ (or its universal covering group). However, gauge theories require considerable changes in the Gårding-Wightman framework [489].

(GW3) It is physically natural to assume that the energy H (which is the infinitesimal generator of time translations) is bounded below. This, together

[7] See [514, pp. 135–136] for a discussion of the reasons why the smeared fields $\varphi(f)$ cannot be bounded operators, at least in the *bosonic* case. In particular, already for the scalar free field the operators $\varphi(f)$ are unbounded, in any spacetime dimension $d \geq 1$.

[8] See [441, 442] for definitions of "self-adjoint", "essentially self-adjoint" and "Hermitian", and for the main theorems concerning self-adjoint operators.

with the existence of the Poincaré group representation $U(a, \Lambda)$, *implies* the spectral condition (GW3).

(GW4) It is a fundamental principle of special relativity that signals cannot be transmitted faster than the speed of light. Thus, measurements of fields in spacelike-separated regions of space-time cannot interfere with one another; in quantum mechanics this means that the associated operators should commute. (GW4) is a weak statement of this principle; for a stronger version, which can be verified in some models, see [146, 260].

Remark. Usually axioms (GW0)–(GW4) are supplemented by the following additional axiom:

(GW5) *Uniqueness of the vacuum.* The only vectors in $\mathcal{H}$ left invariant by all the translation operators $U(a, \mathbf{1})$ are the scalar multiples of Ω.

(GW5) essentially says that we are working in a "pure phase". Any theory satisfying (GW0)–(GW4) can be decomposed as a "direct integral" of theories in which (GW5) holds [211]. For simplicity we omit (GW5) from our version of the Gårding-Wightman axioms.

The next step is to define the vacuum expectation values of products of field operators

$$W_n(f_1, \ldots f_n) = (\Omega, \varphi(f_1) \ldots \varphi(f_n)\Omega) \ , \tag{1.58}$$

known as the *Wightman distributions.* Heuristically (1.58) can be written as

$$W_n(x_1, \ldots, x_n) = (\Omega, \varphi(x_1) \ldots \varphi(x_n)\Omega) \ ; \tag{1.59}$$

henceforth we shall write equations in this way and leave it to the reader to supply the necessary smearing. The Wightman distributions of a relativistic quantum field theory have the following properties [488, 321]:

(W0) *Temperedness.* The W_n are tempered distributions, i.e. $W_n \in \mathcal{S}'(\mathbb{R}^{nd})$, satisfying the hermiticity condition

$$W_n(x_1, \ldots, x_n) = W_n(x_n, \ldots, x_1)^* \tag{1.60}$$

where * denotes complex conjugation.

(W1) *Poincaré invariance.*

$$W_n(\Lambda x_1 + a, \ldots, \Lambda x_n + a) = W_n(x_1, \ldots, x_n) \tag{1.61}$$

for all $(a, \Lambda) \in \mathcal{P}_+^\uparrow$.

(W2) *Positive definiteness (Wightman positivity).* For any finite sequence $\{f_n\}_{n=0}^k$ of complex-valued test functions $f_n \in \mathcal{S}(\mathbb{R}^{nd})$, we have

$$\sum_{n,m=0}^{k} W_{n+m}(\overline{f}_n \otimes f_m) \geq 0 \,. \tag{1.62}$$

Here we have defined

$$\overline{f}_n(x_1, \ldots, x_n) = f_n(x_n, \ldots, x_1)^* \tag{1.63}$$

and

$$(g_n \otimes g_m)(x_1, \ldots, x_{n+m}) = g_n(x_1, \ldots, x_n)\, g_m(x_{n+1}, \ldots, x_{n+m}) \,. \tag{1.64}$$

(W3) *Spectral condition.*The Fourier transform

$$\widetilde{W}_n(p_1, \ldots, p_n) = \int e^{i\sum_{j=1}^{n} p_j \cdot x_j} W_n(x_1, \ldots, x_n)\, dx_1 \ldots dx_n \tag{1.65}$$

is a tempered distribution supported in the set

$$\Big\{(p_1, \ldots, p_n)\colon \;\; \sum_{j=1}^{n} p_j = 0 \;\;\text{ and }\; \sum_{j=1}^{k} p_j \in \overline{V}_+ \;\;\text{ for } k = 1, \ldots, n-1\Big\} \,. \tag{1.66}$$

[Here $p \cdot x \equiv p^0 x^0 - \mathbf{p} \cdot \mathbf{x}$.]

(W4) *Locality.* If x_j and x_{j+1} are spacelike-separated, then

$$W_n(x_1, \ldots, x_j, x_{j+1}, \ldots, x_n) = W_n(x_1, \ldots, x_{j+1}, x_j, \ldots, x_n) \,. \tag{1.67}$$

Most of these properties are straightforward consequences of the corresponding Gårding-Wightman axioms. The positive definiteness property (W2) expresses the positive definiteness of the norm in the Hilbert space $\mathcal{H}$; more precisely, it says that $\|P(\varphi)\Omega\|^2 \geq 0$, where $P(\varphi)$ is a polynomial in the smeared fields.

The first fundamental result of axiomatic quantum field theory is the *Wightman reconstruction theorem* [512, 488, 321]: it says that the properties (W0)–(W4) are sufficient for the reconstruction of a relativistic quantum field theory satisfying (GW0)–(GW4). In other words, all the essential information about a relativistic quantum field theory is contained in its Wightman distributions. This is an important simplification, as the numerical-valued quantities W_n are easier to work with than the operator-valued fields φ.

One key consequenceof (W0)–(W4) is that the Wightman distributions W_n are boundary values of analytic functions (the so-called Wightman functions) defined in a rather large complex domain. Heuristically, the reasoning [488, 321] is as follows: the W_n can be written as

$$W_n(x_1, \ldots, x_n) = (\Omega, \varphi(0) e^{iP\cdot(x_2 - x_1)} \varphi(0) e^{iP\cdot(x_n - x_{n-1})} \varphi(0) \Omega) \tag{1.68}$$

where $U(a, \mathbf{1}) = e^{ia \cdot P}$. By the spectral condition, W_n can be analytically continued to the domain

$$\mathcal{T}_n = \{(x_1, \ldots, x_n) \colon \operatorname{Im}(x_j - x_{j-1}) \in V_+ \text{ for } j = 2, \ldots, n\} \tag{1.69}$$

where V_+ is the open forward light cone; this domain is called the *forward tube.* Then a nontrivial argument using Lorentz invariance and locality [488, 321] implies that the W_n can be analytically continued to a yet larger domain, the *permuted extended forward tube* $\mathcal{T}_n^{\mathrm{pe}}$. The importance of $\mathcal{T}_n^{\mathrm{pe}}$ for our purposes is that it contains the *non-coincident Euclidean region*

$$\begin{aligned} \mathcal{E}_n \;\; = \;\; \{(x_1, \ldots, x_n) \colon & \; t_1, \ldots, t_n \text{ pure imaginary}, \; \mathbf{x}_1, \ldots, \mathbf{x}_n \text{ real}, \\ & \text{and } x_i \neq x_j \text{ for } i \neq j\} \end{aligned} \tag{1.70}$$

where $x_i \equiv (t_i, \mathbf{x}_i)$. This permits us to introduce the *Euclidean Green's functions*, or *Schwinger functions*, defined by

$$S_n(x_1, \ldots, x_n) = W_n((it_1, \mathbf{x}_1), \ldots, (it_n, \mathbf{x}_n)) \tag{1.71}$$

for non-coinciding arguments $x_1, \ldots, x_n \in \mathbb{R}^d$. In other words, *the vacuum expectation values of a Gårding-Wightman quantum field theory can be analytically continued to imaginary time.*

The Schwinger functions S_n are, of course, analytic functions of their arguments, but we have the right to forget temporarily about their analyticity and consider them simply as *distributions* belonging to $\mathcal{S}'_{\neq}(\mathbb{R}^{nd})$, the space of tempered distributions defined at non-coinciding arguments [419, 420]. The key properties of the Schwinger distributions were elucidated by Osterwalder and Schrader [419, 420]:

First let $\mathcal{S}_+(\mathbb{R}^{nd})$ be the set of all those functions in $\mathcal{S}(\mathbb{R}^{nd})$ with support in the region $\{(x_1, \ldots, x_n) \colon 0 < x_1^0 < x_2^0 < \ldots < x_n^0\}$. Next let θ be the reflection in the hyperplane $x^0 = 0$, i.e.

$$\theta(x^0, \mathbf{x}) = (-x^0, \mathbf{x}) \,. \tag{1.72}$$

Let reflection act on functions $f_n \in \mathcal{S}(\mathbb{R}^{nd})$ by

$$(\Theta f_n)(x_1, \ldots, x_n) = f_n(\theta x_1, \ldots, \theta x_n) \,. \tag{1.73}$$

Finally, recall the definitions (1.63) and (1.64) of the operations $^-$ and $\otimes$. Then the Schwinger distributions of a relativistic quantum field theory have the following properties:

(OS0) *Temperedness.* The S_n are tempered distributions at noncoinciding arguments, i.e. $S_n \in \mathcal{S}'_{\neq}(\mathbb{R}^{nd})$, satisfying the reality condition

$$S_n(\theta x_1, \ldots, \theta x_n) = S_n(x_1, \ldots, x_n)^* \,. \tag{1.74}$$

(OS1) *Euclidean invariance.*

$$S_n(Rx_1 + a, \ldots, Rx_n + a) = S_n(x_1, \ldots, x_n) \qquad (1.75)$$

for all $(a, R) \in ISO(d)$.

(OS2) *Osterwalder-Schrader (reflection) positivity.* For any finite sequence $\{f_n\}_{n=0}^k$ of complex-valued test functions $f_n \in \mathcal{S}_+(\mathbb{R}^{nd})$, we have

$$\sum_{n,m=0}^{k} S_{n+m}(\Theta \overline{f}_n \otimes f_m) \geq 0 \,. \qquad (1.76)$$

(OS3) *Permutation symmetry.* The S_n are symmetric in their arguments.

Most of these properties are straightforward to prove. The Euclidean invariance of the S_n is the obvious translation of the Poincaré invariance of the W_n. Osterwalder-Schrader positivity for the S_n is the counterpart of Wightman positivity for the W_n. The symmetry of the S_n is a consequence of the locality property (GW4): heuristically, *all* noncoincident Euclidean points are spacelike-separated one from another.

The deep theorem proven by Osterwalder and Schrader is the converse: the properties (OS0)–(OS3) are *sufficient* (modulo a weak technical condition) for the construction of a Gårding-Wightman relativistic quantum field theory having the given $\{S_n\}$ as its Schwinger functions. The technical condition, which we denote (OS0′), is a bound on the growth of the S_n as $n \to \infty$. The properties (OS0′)–(OS3) are called the *Osterwalder–Schrader axioms*, and the deep theorem is the *Osterwalder-Schrader reconstruction theorem*A detailed statement of the OS axioms together with a proof of the reconstruction theorem can be found in [419, 420, 418].

Thus, the Euclidean (Osterwalder-Schrader) and Minkowski (Gårding-Wightman) formulations of quantum field theory are mathematically equivalent; the Osterwalder-Schrader reconstruction theorem guarantees free passage from either formulation to the other. But the Euclidean approach has several advantages which make it more convenient in applications:

• *Euclidean vs. Poincaré invariance.*The Euclidean inner product $x \cdot y = x^0y^0 + \mathbf{x} \cdot \mathbf{y}$ is positive-definite, while the Minkowski inner product $x \cdot y = x^0y^0 - \mathbf{x} \cdot \mathbf{y}$ is indefinite. This has two major consequences:

(a) The rotation group $SO(d)$ is compact, while the Lorentz group $SO(d-1,1)$ is noncompact.

(b) The Helmholtz operator $-\Delta + m^2$ is elliptic, while the Klein-Gordon operator $\Box + m^2$ is hyperbolic. The greater simplicity of elliptic as compared to hyperbolic problems is well known. Thus, the Helmholtz equation $(-\Delta + m^2)\varphi = j$ has a unique tempered solution, while the Klein-Gordon equation $(\Box + m^2)\varphi = j$ has multiple solutions which need to be distinguished by boundary conditions at temporal infinity (retarded, advanced, etc.). In perturbation theory this is familiar: the Euclidean propagator $(p^2 + m^2)^{-1}$ has an unambiguous meaning, while the Minkowski propagator is ambiguous and requires an $i\varepsilon$ prescription.

• *Symmetry of the Schwinger functions.* The Schwinger functions are symmetric under permutation of their arguments, while the Wightman distributions are not. This raises a tantalizing possibility: suppose that the Schwinger functions $S_n(x_1, \ldots, x_n)$ could be written as the expectation values of some "Euclidean field" $\varphi_E(x)$ in some "Euclidean vacuum vector" Ω_E belonging to a "Euclidean Hilbert space" $\mathcal{H}_E$:

$$S_n(x_1, \ldots, x_n) \stackrel{?}{=} (\Omega_E, \varphi_E(x_1) \ldots \varphi_E(x_n)\Omega_E) \ . \tag{1.77}$$

Then the symmetry of the Schwinger functions would mean that the Euclidean fields *commute* (unlike the Minkowski fields, which commute only at spacelike separations). Hence the operators $\varphi_E(x)$ could be simultaneously diagonalized, i.e. simultaneously represented as multiplication operators on some probability space, and the Schwinger functions would be moments of a probability measure:

$$S_n(x_1, \ldots, x_n) \stackrel{?}{=} \int \varphi_E(x_1) \ldots \varphi_E(x_n)\, d\mu(\varphi_E) \ . \tag{1.78}$$

But then Euclidean field theory would be a branch of *classical probability theory*, or in physical terms, *classical statistical mechanics*!

This dream turns out to be only slightly over-optimistic. The situation is essentially the following: Under mild but nontrivial regularity conditions, the Schwinger functions of a Gårding-Wightman quantum field theory can be written as moments of a *complex* measure [70]. However, simple examples [461] show that the S_n cannot in general be written as moments of a *positive* measure (as is of course necessary for the probabilistic interpretation of φ_E). Rather, the representation (1.78) with a *positive* measure μ is a special property (called *Nelson-Symanzik positivity*) which holds for certain quantum field theories and fails for others. Nevertheless, the most important examples of quantum field theories do possess (at least formally) Nelson-Symanzik positivity, as we now proceed to show.

Example 1.5.1. *Free scalar field.* In Minkowski space, the free scalar quantum field of mass m is an operator-valued distribution $\varphi(x)$ satisfying the Klein-Gordon equation

$$(\Box + m^2)\varphi(x) = 0 \tag{1.79}$$

and having c-number commutators

$$[\varphi(x), \varphi(y)] = -i\Delta(x-y) \ , \tag{1.80}$$

where

$$\Delta(x-y) = \frac{i}{(2\pi)^{d-1}} \int dp\; e^{-ip\cdot(x-y)}\; \mathrm{sgn}(p^0)\, \delta(p^2 - m^2) \tag{1.81}$$

with $p \cdot x \equiv p^0x^0 - \mathbf{p} \cdot \mathbf{x}$ and $p^2 \equiv (p^0)^2 - \mathbf{p}^2$. The Wightman distributionsare given recursively by

$$W_n(x_1,\ldots,x_n) = \begin{cases} \sum_{j=2}^{n} W_2(x_1,x_j)\, W_{n-2}(x_2,\ldots,\not{x}_j,\ldots,x_n) & n \text{ even} \\ 0 & n \text{ odd} \end{cases} \tag{1.82}$$

where

$$W_2(x,y) = -i\Delta^+(x-y) \equiv \frac{1}{(2\pi)^{d-1}} \int dp\, e^{-ip\cdot(x-y)}\, \theta(p^0)\, \delta(p^2-m^2) \tag{1.83}$$

and $\not{x}_j$ means to omit x_j. The Schwinger functions are [462, Chapter III]

$$S_n(x_1,\ldots,x_n) = \begin{cases} \sum_{j=2}^{n} S_2(x_1,x_j) S_{n-2}(x_2,\ldots,\not{x}_j,\ldots,x_n) & n \text{ even} \\ 0 & n \text{ odd} \end{cases} \tag{1.84}$$

where

$$S_2(x,y) = \frac{1}{(2\pi)^d} \int dp\, \frac{e^{ip\cdot(x-y)}}{p^2+m^2} \tag{1.85}$$

and $p\cdot x$ now denotes the *Euclidean* inner product $p^0x^0 + \mathbf{p}\cdot\mathbf{x}$, and similarly for p^2. But (1.84) are precisely the moments of a Gaussian measure with mean zero and covariance $S_2 = (-\Delta+m^2)^{-1}$:

$$S_n(x_1,\ldots,x_n) = \int \varphi_E(x_1)\ldots\varphi_E(x_n)\, d\mu_0(\varphi_E)\ . \tag{1.86}$$

This shows that the free field is Nelson-Symanzik positive.

Formally the Gaussian measure μ_0 is given by

$$d\mu_0(\varphi_E) = \mathcal{N}_0\ \exp\left[-\int\left[\frac{1}{2}(\nabla\varphi_E)^2 + \frac{m^2}{2}\varphi_E^2\right]dx\right]\mathcal{D}\varphi_E \tag{1.87}$$

where $\mathcal{N}_0$ is a normalization and $\mathcal{D}\varphi_E \equiv \prod_x d\varphi_E(x)$ is an uncountably infinite product of Lebesgue measures. *Rigorously*, the individual pieces in the formal expression (1.87) make no mathematical sense, but the measure μ_0 is nevertheless a well-defined probability measure on $\mathcal{S}'(\mathbb{R}^d)$, the space of tempered distributions [462, Chapters I and III]. The support properties of μ_0 are not, however, what one would expect naively from (1.87): if $d \geq 2$, then with μ_0-probability 1, the Euclidean field φ_E is *not* a pointwise-defined function (much less a once-differentiable function with square-integrable gradient!); rather, it is a more singular distribution [109].

Remark. A *generalized free field* is, by definition, a quantum field theory whose Wightman functions are given by (1.82) for some choice of W_2. Equivalently, it is a theory whose Schwinger functions are given by (1.84) for some choice of S_2. If the theory is Nelson-Symanzik positive (as it is for S_2 not too singular), this means that the representing measure μ is Gaussian. The *free field of mass m* is the special case in which W_2 and S_2 are given by (1.83) and (1.85), respectively. See [462] for details.

Example 1.5.2. *Interacting scalar fields.*One natural way to attempt to obtain interacting (i.e. non-free) quantum fields is to perturb the free field. For example, one might attempt to construct a non-Gaussian measure μ by perturbing the Gaussian measure μ_0, and then define the Schwinger functions as moments of μ. Of course, it is necessary to satisfy the Osterwalder-Schrader axioms. The most delicate property is Osterwalder-Schrader positivity: perturbations which involve nonlocal interactions in the "time" direction are likely to destroy OS positivity. On the other hand, by Euclidean invariance, locality in the "time" direction implies locality in the spatial directions as well. So we are led to consider *local* perturbations of the free field,

$$d\mu(\varphi_E) = \mathcal{N} \exp\left[-\int \left[\frac{1}{2}(\nabla\varphi_E)^2 + \frac{m^2}{2}\varphi_E^2 + V(\varphi_E)\right] dx\right] \mathcal{D}\varphi_E , \tag{1.88}$$

where V is, for example, a semibounded polynomial of degree > 2. Readers familiar with conventional Lagrangian quantum field theory [310] will recognize (1.88) as the Euclidean functional integral obtained by canonical quantization of the theory with Minkowski-space Lagrangian density

$$\mathcal{L} = \frac{1}{2}(\partial_\mu\varphi)^2 - \frac{m^2}{2}\varphi^2 - V(\varphi) . \tag{1.89}$$

This is the theory whose Minkowski-space field formally satisfies the nonlinear operator-valued wave equation (Heisenberg equation of motion)

$$(\Box + m^2)\varphi(x) = -V'(\varphi(x)) . \tag{1.90}$$

The simplest example is the φ^4 *theory*, corresponding to $V(\varphi) = \lambda\varphi^4/4$.

Unfortunately, the formal expressions (1.88) and (1.90) are plagued with troubles at short distances whenever $d \geq 2$. For example, we know that with μ_0-probability 1, the Euclidean field φ_E is not a pointwise-defined function; this means that the perturbation term $V(\varphi_E)$ is ill-defined. Likewise, the Minkowski field φ is not a pointwise-defined operator-valued function, so the nonlinear perturbation $V'(\varphi)$ is ill-defined. These troubles make themselves felt already in perturbation theory, which is plagued with *ultraviolet divergences.*

The fundamental task of mathematical quantum field theory is, therefore, to make sense of formal expressions like (1.88). The procedure for doing this is called *renormalization theory*, and it involves three steps:

1) *Regularization.* Impose an ultraviolet cutoff Λ (or equivalently, a short-distance cutoff $a = \Lambda^{-1}$) so that the formal expression (1.88) is a well-defined measure. This can be done in various ways. Of course, all of these regularizations break either Euclidean invariance or OS positivity or both.

2) *Renormalization.* Allow the parameters in (1.88) [the so-called *bare parameters*] to be functions of Λ. For example, in the case of the φ^4 theory we consider

$$d\mu^{\Lambda}(\varphi_E) = \mathcal{N}(\Lambda)\ \exp\left[-\int\Big[\frac{A(\Lambda)}{2}(\nabla\varphi_E)^2 + \frac{B(\Lambda)}{2}\varphi_E^2 + \frac{\lambda(\Lambda)}{4}\varphi_E^4\Big]\, dx\right]\ \mathcal{D}\varphi_E \tag{1.91}$$

with $A(\Lambda)$, $B(\Lambda)$ and $\lambda(\Lambda)$ being arbitrary functions of Λ [with $\lambda(\Lambda) \geq 0$ in order that the measure (1.91) be well-defined].

3) *Continuum limit.* Take $\Lambda \to \infty$. For each choice of $A(\Lambda)$, $B(\Lambda)$, $\lambda(\Lambda)$ we ask:

a) Do the Schwinger functions $S_n^{\Lambda}(x_1, \ldots, x_n)$ converge?

b) If so, do the limiting Schwinger functions $S_n(x_1, \ldots, x_n)$ satisfy the Osterwalder-Schrader axioms?

c) If so, what are the properties of the resulting theory? In particular, is it a free (or generalized free) field, i.e. is the corresponding measure $d\mu(\varphi_E)$ Gaussian?

One convenient regularization is to replace continuous space $\mathbb{R}^d$ by a lattice $a\mathbb{Z}^d$ of lattice spacing a. The gradient $\nabla\varphi_E$ and the integral $\int dx$ are then replaced by their discrete analogues:

$$\begin{aligned} (\nabla_\mu\varphi_E)(x) &\longrightarrow a^{-1}\,[\varphi_E(x+ae_\mu) - \varphi_E(x)] \\ \int_{\mathbb{R}^d} dx &\longrightarrow a^d \sum_{x\in a\mathbb{Z}^d} \end{aligned}$$

The ultraviolet-cutoff φ^4 measure (1.91) is then recognized as nothing other than the φ^4 lattice spin model with nearest-neighbor interaction, (1.25)/(1.28), with the identifications

$$\lambda_{\text{lat}} = a^d\lambda_{\text{FT}} \tag{1.92}$$

$$B_{\text{lat}} = a^d B_{FT} + 2da^{d-2}A_{\text{FT}} \tag{1.93}$$

$$J_{xy} = \begin{cases} a^{d-2}A_{\text{FT}} & \text{if } x, y \text{ are nearest neighbors} \\ 0 & \text{otherwise} \end{cases} \tag{1.94}$$

$$h = 0 \tag{1.95}$$

between lattice and continuum quantities. We then wish to take the continuum limit $a \to 0$ so that physical correlation lengths (i.e. inverse masses) stay bounded away from zero when distances are measured *in centimeters.* But this means that the correlation length is going to infinity when measured *in lattice spacings.* Hence, the bare parameters must be chosen so as to approach the critical surface of the φ^4 lattice model. In other words, *the problem of renormalization in quantum field theory is equivalent to the problem of critical phenomena in classical statistical mechanics!*

One convenient measure of the non-Gaussianness of a continuum quantum field theory is the truncated four-point Schwinger function

$$\begin{aligned} S_4^T(x_1,x_2,x_3,x_4) \;\equiv\; & S_4(x_1,x_2,x_3,x_4) - S_2(x_1,x_2)\,S_2(x_3,x_4) \\ & -S_2(x_1,x_3)\,S_2(x_2,x_4) - S_2(x_1,x_4)\,S_2(x_2,x_3)\;, \end{aligned}$$

and its zero-momentum value

$$\overline{S}_4^T \equiv \int dx_2 dx_3 dx_4\, S_4^T(x_1,x_2,x_3,x_4)\;. \tag{1.96}$$

Clearly, if a quantum theory is Gaussian, then $\overline{S}_4^T = 0$ [cf. (1.84)]. Conversely, for a large class of quantum field theories including the φ^4 model (1.91), it can be shown that $\overline{S}_4^T = 0$ *implies* that the theory is Gaussian [411, 52]. Now, it is not too hard to see that the dimensionless renormalized four-point coupling constant of the lattice theory,

$$g \equiv \frac{-\overline{u}_4}{\chi^2 \xi^d}\;, \tag{1.97}$$

converges in the continuum limit (at least formally) to the corresponding ratio of continuum quantities. Therefore, at least for a massive continuum limit (one in which the physical correlation lengths stay bounded in centimeters), one can test for the *triviality* (Gaussianness) of the continuum limit by asking in the lattice theory whether $g \longrightarrow 0$ as the critical point is approached. We discuss these issues in detail in Chapter 15.

1.6 Some random-walk models

In this section we introduce some random-walk models which, at first sight, bear no obvious relation with spin systems or quantum field theory — in fact, their definition is purely geometric (and probabilistic). However, as will be seen, these models exhibit critical phenomena closely analogous to those found in spin systems; for this reason, they can serve as a simplified arena in which to test methods (both heuristic and rigorous) for dealing with critical phenomena. The random-walk models also have direct application as models of polymer molecules [121]. Most importantly, suitable *generalizations* of the models defined in this section arise as random-walk representations of spin systems, in which spin-system correlation functions are represented as sums over random walks (see Chapter 6). Moreover, connected correlation functions in the spin system can in some cases be bounded in terms of the *intersection properties* of the associated random walks. In this fashion, questions about the critical behavior of the spin system can sometimes be reduced to geometric questions about intersections of random walks, which in favorable cases can be answered nonperturbatively.

We consider some lattice $\mathbb{L}$, typically $\mathbb{Z}^d$ or a finite subset of $\mathbb{Z}^d$. Its elements are called sites. Unoriented pairs $\{x, y\}$ of sites in $\mathbb{L}$ with $x \neq y$ are called bonds;

oriented pairs (x, y) are called steps with initial site x and final site y. A *walk* ω on $\mathbb{L}$ is an ordered sequence $(\omega(0), \omega(1), \ldots, \omega(k))$ of sites in $\mathbb{L}$ $(k \geq 0)$. We call $|\omega| \equiv k$ the length of ω, $b(\omega) \equiv \omega(0)$ its initial site, and $e(\omega) \equiv \omega(k)$ its final site. If $b(\omega) = x$ and $e(\omega) = y$ we write $\omega\colon x \to y$. The *support* of a walk ω is defined by

$$\mathrm{supp}(\omega) \;=\; \{x \in \mathbb{L}\colon \omega(j) = x \text{ for some } j\}. \tag{1.98}$$

A random-walk model is defined by assigning to every n-tuple of walks $\omega_1, \ldots, \omega_n$ $(n \geq 0)$ a statistical weight $\varrho^{[n]}(\omega_1, \ldots, \omega_n) \geq 0$, with $\varrho^{[0]} = 1$. Typically, these weights will have the form

$$\varrho^{[n]}(\omega_1, \ldots, \omega_n) = \left(\prod_{i=1}^{n} J^{\omega_i}\right) \exp[-U_n(\omega_1, \ldots, \omega_n)] \;, \tag{1.99}$$

where

$$J^{\omega} = \prod_{s=0}^{|\omega|-1} J_{\omega(s),\omega(s+1)} \;. \tag{1.100}$$

Here $J_{xy} \geq 0$ is the "activity" of the step (x, y), and U_n is the interaction energy of the walks(which may be $+\infty$ for forbidden configurations). U_n is assumed to be symmetric in its arguments. The interaction is called *repulsive* if

$$U_{n+m}(\omega_1, \ldots, \omega_{n+m}) \geq U_n(\omega_1, \ldots, \omega_n) + U_m(\omega_{n+1}, \ldots, \omega_{n+m}) \;, \tag{1.101}$$

that is,

$$\varrho^{[n+m]}(\omega_1, \ldots, \omega_{n+m}) \leq \varrho^{[n]}(\omega_1, \ldots, \omega_n)\; \varrho^{[m]}(\omega_{n+1}, \ldots, \omega_{n+m}) \;. \tag{1.102}$$

Many properties of the random walks can be studied by considering *generating functions* (which we call "kernels"):

$$K(x_1 x_2 | \ldots | x_{2n-1} x_{2n}) \;\equiv \sum_{\substack{\omega_1\colon x_1 \to x_2 \\ \vdots \\ \omega_n\colon x_{2n-1} \to x_{2n}}} \varrho^{[n]}(\omega_1, \ldots, \omega_n) \;, \tag{1.103}$$

and in particular

$$K(x, y) = \sum_{\omega\colon x \to y} \varrho^{[1]}(\omega) \;. \tag{1.104}$$

The *Green functions* for the model are defined by summing the kernels over all possible pairings:

$$S_{2n}(x_1, \ldots, x_{2n}) = \sum_{\text{pairings } \pi} K(x_{\pi(1)} x_{\pi(2)} | \ldots | x_{\pi(2n-1)} x_{\pi(2n)}) \;. \tag{1.105}$$

We now suppose that $\mathbb{L} = \mathbb{Z}^d$ and assume that the weights $\varrho^{[n]}$ are translation-invariant, i.e.

$$\varrho^{[n]}(\omega_1, \ldots, \omega_n) = \varrho^{[n]}(\omega_1 + a, \ldots, \omega_n + a) \tag{1.106}$$

for all n and all $a \in \mathbb{Z}^d$. Here $\omega + a$ is the walk visiting the sites $\omega(0) + a, \ldots, \omega(N) + a$ $(N = |\omega|)$. For simplicity, we shall also assume that $\varrho^{[n]}$ has the form (1.99)–(1.100), with $J_{x+a,y+a} = J_{xy}$ for all $a \in \mathbb{Z}^d$. The model is well-defined for activities $J = \{J_{xy}\} \geq 0$ in the set

$$\mathcal{J} = \{J\colon\ K(x_1x_2| \ldots |x_{2n-1}x_{2n}) < \infty \text{ for all } n,\ x_1, \ldots, x_{2n}\} \tag{1.107}$$

Clearly $\mathcal{J}$ is a decreasing set, i.e. if $J \in \mathcal{J}$ and $0 \leq J'_{xy} \leq J_{xy}$ for each x, y, then $J' \in \mathcal{J}$. The upper boundary $\mathcal{J}_{\text{crit}}$ of $\mathcal{J}$ (suitably defined) is typically a *critical surface* in the sense that

$$\chi \equiv \sum_x K(0, x) \tag{1.108}$$

diverges as J approaches this boundary. Moreover, the "correlation length"

$$\xi \equiv \limsup_{|x| \to \infty} \frac{-|x|}{\log K(0, x)} \tag{1.109}$$

is typically finite for J in the interior of $\mathcal{J}$, and diverges for $J \longrightarrow \mathcal{J}_{\text{crit}}$.

As an example, consider *simple* (or *ordinary*) *random walk*. In this model

$$J_{xy} = \begin{cases} J & \text{if } x \text{ and } y \text{ are nearest neighbors in } \mathbb{Z}^d \\ 0 & \text{otherwise} \end{cases} \tag{1.110}$$

and

$$U_n(\omega_1, \ldots, \omega_n) = 0 \qquad \text{for all } n. \tag{1.111}$$

Then the critical surface is given by

$$J^* = \frac{1}{2d}\,. \tag{1.112}$$

Moreover, if we define

$$m^2 \equiv J^{-1} - 2d\,, \tag{1.113}$$

then the kernels are given by

$$K(x_1x_2| \ldots |x_{2n-1}x_{2n}) = J^{-n} \prod_{k=1}^{n} \left[(-\Delta + m^2)^{-1}\right]_{x_{2k-1}x_{2k}}\,, \tag{1.114}$$

where Δ is the finite-difference Laplacian, and $[(-\Delta + m^2)^{-1}]_{xy}$ is the Green function of $-\Delta + m^2$. The critical surface corresponds to $m = 0$. A central remark for this book is that the correlation functions of the Gaussian model (1.26) [or what is essentially equivalent, the free Euclidean field (1.84)] are given precisely by a sum over pairings of the kernel (1.114).

$$\langle \varphi_{x_1}, \ldots, \varphi_{x_{2n}} \rangle = \sum_{\text{pairings } \pi} K(x_{\pi(1)}x_{\pi(2)}| \ldots |x_{\pi(2n-1)}x_{\pi(2n)})\,. \tag{1.115}$$

That is, from (1.105) and (1.115)

$$\langle \varphi_{x_1}, \ldots, \varphi_{x_{2n}} \rangle = S_{2n}(x_1, \ldots, x_{2n}) \,. \tag{1.116}$$

This *fundamental identity* shows that the simple random walk can serve as a *random-walk representation* of the Gaussian model. In Chapters 6 (Part I) and 9–10 (Part II) we will show how more complicated random-walk models can serve as random-walk representations of more general spin systems. Formulas like (1.116) justify somehow to refer to the Green functions S_{2n} as the $2n$*-point correlations* of the model.

A much more difficult example is the *self-avoiding walk* (SAW). In this model, $U_n(\omega_1, \ldots, \omega_n) = +\infty$ whenever there is a site $j \in \mathbb{Z}^d$ that is visited by more than one walk or more than once by some walk ω_i; otherwise $U_n(\omega_1, \ldots, \omega_n) = 0$. In the simplest (nearest-neighbor) case, J_{xy} is as in (1.110). In this model, J^* is strictly larger than $1/(2d)$; in fact, it is not hard to show that $1/(2d-1) \leq J^* \leq 1/d$.

Typical problems for the random-walk models introduced here are the following:

(a) Given a subset $S \subset \mathbb{Z}^d$ and a site $x \notin S$, what is the probability that a walk ω, starting at x with weight $\varrho^{[1]}(\omega)$ given by (1.99)–(1.100), visits S?

In particular, we can pose this question in the "scaling limit":

(a′) Replacing x by $\overline{\xi}(J)x$ and S by $\overline{\xi}(J)S$, where $\overline{\xi}$ is the least integer $\geq \xi$, how does the probability defined in (a) behave, as $J \to \mathcal{J}_{\text{crit}}$?

(b) What is the typical *end-to-end distance*of a walk ω, as J approaches $\mathcal{J}_{\text{crit}}$? To answer this question, one might analyze the quantity

$$\langle \omega(N)^2 \rangle = \frac{\sum\limits_{\substack{\omega:\, |\omega|=N \\ b(\omega)=0}} \varrho^{[1]}(\omega)\, \omega(N)^2}{\sum\limits_{\substack{\omega:\, |\omega|=N \\ b(\omega)=0}} \varrho^{[1]}(\omega)} \tag{1.117}$$

or

$$\langle e(\omega)^2 \rangle = \frac{\sum\limits_{\omega:\, b(\omega)=0} \varrho^{[1]}(\omega)\, e(\omega)^2}{\sum\limits_{\omega:\, b(\omega)=0} \varrho^{[1]}(\omega)} \tag{1.118}$$

and try to prove a behavior of the form

$$\langle \omega(N)^2 \rangle \sim N^{2\nu} \tag{1.119}$$

$$\langle e(\omega)^2 \rangle \sim \operatorname{dist}(J, \mathcal{J}_{\text{crit}})^{-2\nu} \tag{1.120}$$

for some critical exponent ν.The quantity $d_H = 1/\nu$ can be interpreted as a *Hausdorff dimension* for walks ω with weight $\varrho^{[1]}(\omega)$.

(c) Consider n walks, $\omega_1, \ldots, \omega_n$, with weight $\varrho^{[n]}(\omega_1, \ldots, \omega_n)$, starting at sites $\overline{\xi}x_1, \ldots, \overline{\xi}x_n$, $x_i \neq x_j$ for $i \neq j$. What is the behavior of the probability that $\omega_1, \ldots, \omega_n$ intersect in a common site $z \in \mathbb{Z}^d$, as J approaches $\mathcal{J}_{\text{crit}}$? How does it depend on n and on d?

(d) Consider two walks ω_1 and ω_2, starting at 0, with weight $\varrho^{[2]}(\omega_1, \omega_2)$. How does the probability that 0 is the *only* common site of ω_1 and ω_2 behave as J approaches $\mathcal{J}_{\text{crit}}$?

Answers to (a)–(c), and a partial answer to (d), are known for simple random walks, where they are non-trivial (see Chapter 5). In essentially all other cases, there are at best heuristic, though conceivably exact, answers. Such answers will turn out to be useful in analyzing the critical behavior of spin systems.

2. Phase transitions and critical points in classical spin systems: A brief survey

In this chapter[1], we describe some fairly recent results on phase transitions and critical points in classical lattice spin systems. We emphasize the analysis of explicit models and quantitative information on such models. A complementary point of view is developed in [309]. See also [447, 469].

2.1 Existence of phase transitions

For a large class of models (including all those considered in this book),it is relatively easy to prove that there is no phase transition at high temperature (β small) or large magnetic field ($|h|$ large); this can be done either by expansion methods [459] or by the Dobrushin uniqueness theorem [465]. A deep fact is that for φ^4 and N-vector models (at least for $N \leq 3$), phase transitions (if any) can only occur at *zero magnetic field*: this is a consequence of the Lee-Yang theorem (see the references in [371]).So we are reduced to the problem of determining whether a phase transition occurs as β is varied at $h = 0$.

At present there are basically three or four general methods to establish rigorously the existence of phase transitions in lattice systems of statistical mechanics:

(a) Exact solutions. This technique applies only to a limited class of models, such as one-dimensional systems with finite-range interactions[2], the two-dimensional nearest-neighbor Ising model (in zero magnetic field), the eight-vertex models, etc. In recent years, the interest in exact solutions has been revived through the work of Baxter [369, 57], Jimbo, Miwa, Sato and coworkers [448, 314, 113, 315, 313], Faddeev and collaborators [470, 343, 169, 167, 168], the Landau-Institute group [443], and others [499, 25, 26, 27, 426]. Exact solutions tend to provide a fairly detailed description of the phase transition, including quantitative information, but often somewhat obscure the physical mechanisms leading to the transition. We shall not discuss any exact solutions in the following.

[1] Parts of this chapter are adapted from [228].

[2] These models have *no* phase transition.

(b) Energy-entropy (Peierls-type) arguments. In its most general form this method can be viewed as a way of reinterpreting spin systems as gases of "topologically stable" defects in an ordered medium [392, 395], and then analyzing phase transitions in defect gases by estimating defect energies and entropies. Examples of topological defects are Bloch walls (Peierls contours) in the Ising model, vortices in the XY model, and magnetic monopoles in the $U(1)$ lattice gauge theory. This method can be applied to study thermodynamic phases in which the defect gas is dilute.

The original Peierls argument [428] was invented in 1936 to analyze the Ising model, and it was made rigorous by Griffiths [268] and Dobrushin [132] in the 1960's. We give here a brief account of the argument. Consider the Ising model (Section 1.4) defined by the Hamiltonian

$$H_\Lambda(\varphi) = \sum_{\substack{x,y\in\Lambda \\ |x-y|=1}} (1-\varphi_x\varphi_y) \tag{2.1}$$

for finite $\Lambda \subset \mathbb{Z}^d$. For simplicity, let us consider the case $d = 2$. If the spins outside Λ are fixed in the position $\varphi_x = +1$ for all $x \in \Lambda^c$ ("+ boundary conditions"), we expect that at low temperatures most of the spins in Λ will be aligned with those on the boundary. Indeed, if an "island" of "−" spins were to form, the energy of the configuration would rise by an amount

$$2 \times \text{number of bonds joining opposite spins} \approx 2 \times \text{perimeter of the island.}$$

Therefore, the probability of occurrence of any particular island of perimeter r is damped by a factor $e^{-2\beta r}$. At $\beta = \infty$ ($T = 0$), this argument shows that such islands are strictly forbidden; however, at nonzero temperatures, there is another factor to be taken into account: the entropy. Even when a particular island of perimeter r has small probability, the probability of occurrence of *some* island of perimeter r may not be negligible due to the large number n_r of possible such islands:

$$P_r \equiv \text{Prob(at least one island of perimeter } r) \sim n_r\, e^{-2\beta r} .$$

If we write $n_r = e^{S_r}$ we see that the probability P_r depends on the balance of the entropy gain S_r and the energy cost $2\beta r$:

$$P_r \sim e^{S_r - 2\beta r} .$$

This is a typical energy-entropy argument. At high temperature ($\beta \ll 1$) entropy wins, and arbitrarily large islands of "overturned spins" are possible: the system is disordered. On the other hand, at low temperature ($\beta \gg 1$) the energy cost is expected to be the dominant factor, and only small and sparse islands are possible: the system is ordered. These two facts imply the existence of some temperature at which the transition from order to disorder takes place, that is, the occurrence of a phase transition.

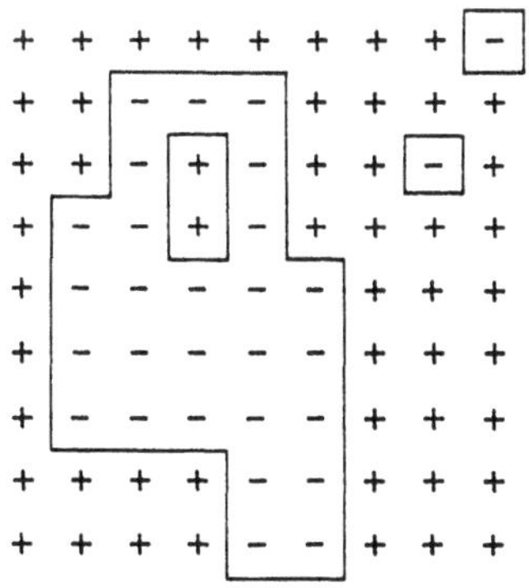

Fig. 2.1. Peierls contours

This argument can be formalized as follows. (For simplicity we consider dimension $d = 2$; the extension to $d > 2$ is straightforward.) Let $\Omega_\Lambda = \{-1, 1\}^\Lambda$ be the space of configurations of Ising spins on the finite subset $\Lambda \subset \mathbb{Z}^2$. We use "+ boundary conditions", that is, we fix $\varphi_x = +1$ for all $x \in \Lambda^c$. For each configuration $\varphi \in \Omega_\Lambda$, we draw curves separating the regions of "+" and "−" spins: that is, for each nearest-neighbor pair x, y with $\varphi_x \neq \varphi_y$ we draw a unit segment perpendicular to the bond $\langle x, y\rangle$ (see Figure 2.1). These curves are closed (because of the + boundary conditions) and can be decomposed into connected components called *contours*. By fixing a rule to "chop off" the corners where four faces meet, the map from configurations to families of contours can be made well-defined. Let us call *compatible* the families of contours coming from configurations; they satisfy certain constraints, in particular they must be nonoverlapping. The map from configurations to compatible families of contours is one-to-one, since the entire configuration can be determined starting from the + boundary spins and working inwards. Finally, the energy of each configuration can be calculated by adding to the energy of the ground state $\varphi \equiv 1$ an energy $2|\Gamma|$ for each contour Γ, where $|\Gamma|$ denotes the length of the contour. We can summarize this construction by saying that the configurations are mapped into a *gas of contours*, characterized by an exclusion interaction and an activity $e^{-2\beta|\Gamma|}$.

The heart of Peierls argument is the observation that the statistics of the Ising model can be related to the simpler statistics of this gas of contours. Consider, for example, the probability that at some chosen site in Λ, say the origin, the spin value is -1. This is the probability that the origin belongs to an island of "overturned spins". For this to happen, there must be at least one contour Γ surrounding the origin:

$$\begin{aligned}\mathrm{Prob}_\Lambda^+(\varphi_0 = -1) &\leq \mathrm{Prob}_\Lambda^+(\text{there exists a contour } \Gamma \text{ surrounding } 0)\\ &\leq \sum_{\substack{\text{Contours } \Gamma\\ \text{around } 0}} \mathrm{Prob}_\Lambda^+[\Gamma] \qquad (2.2)\end{aligned}$$

The "+" indicates the boundary condition being used. The probability $\mathrm{Prob}_\Lambda^+[\Gamma]$ can be written purely in terms of the ensemble of contours:

$$
\begin{aligned}
\mathrm{Prob}_\Lambda^+[\Gamma] &= \frac{\displaystyle\sum_{\substack{\varphi\in\Omega_\Lambda:\ \varphi \text{ has} \\ \Gamma \text{ as a contour}}} e^{-\beta H_\Lambda^+(\varphi)}}{\displaystyle\sum_{\varphi\in\Omega_\Lambda} e^{-\beta H_\Lambda^+(\varphi)}} \\
&= \frac{\displaystyle\sum_{\substack{\{\Gamma,\Gamma_1,\Gamma_2,\dots\} \\ \text{compatible}}} e^{-2\beta|\Gamma|}\, e^{-2\beta\sum|\Gamma_i|}}{\displaystyle\sum_{\substack{\{\Gamma_1,\Gamma_2,\dots\} \\ \text{compatible}}} e^{-2\beta\sum|\Gamma_i|}}
\end{aligned} \tag{2.3}
$$

The crucial remark is that if $\{\Gamma, \Gamma_1, \dots, \Gamma_n\}$ is a compatible family of contours, then $\{\Gamma_1, \dots, \Gamma_n\}$ is also a compatible family. Indeed, to obtain the corresponding spin configuration it suffices to take the configuration with contours $\{\Gamma, \Gamma_1, \dots, \Gamma_n\}$ and flip all spins inside Γ. This operation preserves all the contours nested inside Γ (and obviously those outside Γ), but removes Γ because the spins on both sides of it now have the same sign. Therefore, in (2.3) there is a one-to-one correspondence between terms in the numerator and a subset of terms in the denominator. It follows that

$$\mathrm{Prob}_\Lambda^+[\Gamma] \;\le\; e^{-2\beta|\Gamma|} \tag{2.4}$$

and hence, by (2.2),

$$\mathrm{Prob}_\Lambda^+(\varphi_0 = -1) \;\le\; \sum_{l\ge 4} n_l\, e^{-2\beta l} \tag{2.5}$$

where $n_l = \#\{\text{contours } \Gamma \text{ surrounding the origin with } |\Gamma| = l\}$. It is a simple combinatorial exercise to show that $n_l \le 4l^2 3^{l-1}$, hence

$$\mathrm{Prob}_\Lambda^+(\varphi_0 = -1) \le \frac{4}{3}\sum_{l\ge 4} l^2 (3e^{-2\beta})^l \le \text{const} \times e^{-8\beta} < \frac{1}{2} \tag{2.6}$$

for β sufficiently large, *uniformly in Λ*. This proves that islands of "−" spins are very unlikely at low temperature, or alternatively that the magnetization

$$\langle\varphi_0\rangle^+ \;=\; 1 - 2\mathrm{Prob}_\Lambda^+(\varphi_0 = -1)$$

becomes nonzero at low enough temperatures. The symmetry $\varphi \to -\varphi$, present in the Hamiltonian (2.1), is broken, signaling the occurrence of a phase transition.

The Peierls argument was generalized in the early 1970's by Minlos, Pirogov and Sinai [469] to systems in which the ground states are not related by a symmetry of the Hamiltonian (and hence one does not have an operation analogous to "flipping the spins", to remove contours). Glimm, Jaffe and Spencer first applied it to quantum field models, introducing a new technique to analyze contour probabilities [261]. Furthermore they combined a Peierls argument with expansion methods, permitting the estimation of small fluctuations around defect configurations [262, 263]. Their ideas were systematized and extended in [219, 307, 308, 490, 210] and many other papers.

More recently, it has been shown that the basic elements of the Peierls argument, namely energy-entropy considerations, can be applied to rigorously analyze a wide class of model systems equivalent to defect gases, including models with long-range interactions and massless phases [224, 227, 226]. In particular, one can set up Peierls-type arguments in systems with continuous (but *abelian*) symmetry groups. Such techniques combine entropy (i.e. combinatorial) estimates for suitably constructed blocks of defects with some kind of "block-spin integration", borrowed from the renormalization group, which serves to exhibit self-energies of defects.

We now briefly describe some general elements of the simplest kind of Peierls argument in this general setting. Consider a physical system whose configurations can be described by a classical spin-field φ. We suppose for the moment that φ is defined on continuous space $\mathbb{R}^d$ (rather than on a spatial lattice $\mathbb{Z}^d$), takes values in a compact manifold M (e.g. S^{N-1}, the unit sphere in $\mathbb{R}^N$) and is continuous except on surfaces of codimension ≥ 1.

Consider, as an example, a configuration φ which is continuous except on a manifold H_k of dimension $k \leq d-1$. The space of all configurations φ: $\mathbb{R}^d \backslash H_k \to M$ can be decomposed into homotopy classes labelled by the elements of the homotopy groups $\pi_{d-k-1}(M)$. A configuration φ labelled by a non-trivial element of $\pi_{d-k-1}(M)$ is called a topological defect of dimension k.

The idea is now to interpret the equilibrium configurations of the spin field φ as equilibrium configurations of a *gas of interacting topological defects*. The locus of a defect δ_k in this gas, corresponding to a non-trivial element $g_k \in \pi_{d-k-1}(M)$, is a closed bounded surface Σ_k of dimension k. In the following we assume that all homotopy groups of M are discrete.

It turns out that the main features of the statistical mechanics of defect gases can often be described by an energy-entropy argument that parallels the one given above for the Ising case. The energy of a defect δ_k corresponding to a non-trivial element $g_k \in \pi_{d-k-1}(M)$ can be estimated by

$$E(\delta_k) \gtrsim \varepsilon(g_k)|\Sigma_k| \;, \tag{2.7}$$

where $\varepsilon(g_k)$ is the self-energy density of the defect and $|\Sigma_k|$ is the k-dimensional area of Σ_k. After introducing some coarse-graining (e.g. replacing continuum models by lattice models) one can argue that the entropy $S(g_k, l)$ of the class of all defects labelled by g_k whose loci contain a given point, e.g. the origin, and have area $|\Sigma_k| = l$ is estimated by

$$S(g_k, l) \leq c(g_k) \cdot l \;, \tag{2.8}$$

where $c(g_k)$ is a geometrical constant. The density $\varrho(g_k, l)$ of such defects is then proportional to

$$\varrho(g_k, l) \sim e^{-\beta E(\delta_k)+S(g_k,l)} \lesssim e^{[-\beta\varepsilon(g_k)+c(g_k)]l} , \qquad (2.9)$$

provided that the interactions between distinct defects are, in some sense, weak. Formula (2.9) suggests that when the inverse temperature β decreases below the point

$$\beta(g_k) \approx c(g_k)/\varepsilon(g_k) , \qquad (2.10)$$

defects labelled by g_k condense, and there are, with high probability, infinitely extended defects of type g_k. One expects, therefore, that there is a phase transition, as β is varied through $\beta(g_k)$, from an ordered (few defects) to a disordered (extended defects) phase.

The energy-entropy argument sketched in (2.7)–(2.10) has been applied to a large class of lattice spin systems with abelian symmetry groups[3], to actually *prove* that a transition occurs. This may sound confusing, because the notion of a "topological defect" does not make sense when one considers spin configurations on a lattice. It turns out, however, that in models with abelian symmetry groups one can use a duality transformation (Fourier transformation on the group) to exhibit what in the continuum limit corresponds to topological defects. Since this will presumably sound rather vague, we now briefly describe two examples.

(1) *The Ising model.*The same argument described above works in any dimension. In this example $M = \{-1, 1\}$ and the Hamiltonian is given by (2.1). The defects are the Peierls contours, i.e. $(d-1)$-dimensional closed connected surfaces in the dual lattice separating a domain where φ takes the value $+1$ from a domain where it takes the value -1. By (2.1), the energy of a contour is equal to its $(d-1)$-dimensional area. The interactions between defects are given by an exclusion principle. An analogue of bound (2.6) holds in any dimension with $c = c(d)$ being a geometric constant:

$$\mathrm{Prob}_\Lambda^+(\varphi_0 = -1) \le \mathrm{const} \times \sum_{l=2d}^{\infty} l^{2d} e^{-\beta l} c^l \le \mathrm{const} \times e^{-2d\beta} < \frac{1}{2} \qquad (2.11)$$

uniformly in Λ if β is large enough, and thus $\langle\varphi_0\rangle^+ > 0$ for large β. This shows that in zero magnetic field ($h = 0$) and for large β there is a spontaneous magnetizationin the direction imposed by the boundary conditions. It is not hard to show that for small β there is no spontaneous magnetization (the equilibrium state in the thermodynamic limit is unique for small β).Thus there is a phase transition.

(2) *The two-component rotor (classical XY) model.*In this model $M = S^1$, $dP_x(\varphi_x)$ is Lebesgue measure on S^1, and the Hamiltonian is given by

$$H_\Lambda(\varphi) = \sum_{\substack{x,y\in\Lambda \\ |x-y|=1}} (1 - \varphi_x \cdot \varphi_y) = \sum_{\substack{x,y\in\Lambda \\ |x-y|=1}} [1 - \cos(\theta_x - \theta_y)] , \qquad (2.12)$$

where θ_x is the angle parametrizing the unit vector φ_x.

[3]Or non-abelian but *discrete* symmetry groups.

Since $\pi_1(S^1) = \mathbb{Z}$, $\pi_i(S^1) = 0$ for $i > 1$, the defects of this model are labelled by an integer and their loci have codimension 2. They are called *vortices*. In order to study the transitions in this model, the idea is to invent a rigorous version of the energy-entropy argument (2.7)–(2.10) for the gas of vortices equivalent to the rotator model. The equivalence between the rotator model and a vortex gas can be seen by Fourier series expansion of the equilibrium state $d\mu_\beta(\boldsymbol{\varphi})$ in the angular variables $\{\theta_x\}$ and subsequent application of the Poisson summation formula; see e.g. [229, 226]. The problem that one meets when one tries to analyze the vortex gas is that there are Coulomb interactions, which have extremely long range, between individual vortices. In *three or more dimensions*, these interactions turn out to be quite irrelevant, and the arguments (2.7)–(2.10) can be made rigorous. One concludes from (2.9) that, for large β, the density of vortices is small, i.e. the number of defects per unit volume in each equilibrium configuration $\boldsymbol{\varphi}$ is very small. Therefore one expects that, in the average, $\boldsymbol{\varphi}$ has a fixed direction, i.e.

$$\langle \boldsymbol{\varphi}(x) \rangle = \mathbf{M}(\beta) \neq 0 \, , \tag{2.13}$$

for large β; $\mathbf{M}(\beta)$ is determined by the boundary conditions. These arguments are made rigorous in [226, 229] (a slightly non-trivial task). On the other hand, it is well known that for small β,

$$\langle \boldsymbol{\varphi}(x) \rangle = 0 \, . \tag{2.14}$$

Therefore, for $d \geq 3$ there is a ferromagnetic phase transition.

In *two dimensions*, the vortices are point-like objects. The interaction between two vortices of strength q_1 and q_2, respectively, separated by a distance l is approximately given by

$$-\frac{q_1 q_2}{2\pi} \ln l \tag{2.15}$$

which is the Coulomb potential between two point charges, q_1 and q_2, in two dimensions. Suppose now that $q_1 = 1$, $q_2 = -1$. The entropy S of the class of configurations of a $+$ vortex and a $-$ vortex separated by a distance l, within some distance $\sim l$ from the origin, is given by

$$e^S \approx \text{ const} \times l^3 \, . \tag{2.16}$$

Thus, for $\beta > 8\pi$,

$$e^{-\beta E} e^S \approx \text{ const } \times (l+1)^{3-(\beta/2\pi)} \tag{2.17}$$

is summable in l. This means that configurations of one vortex of strength $+1$ and one vortex of strength -1, separated by a finite distance, are thermodynamically *stable*. In fact, it can be shown by a rather difficult inductive construction [224], extending over an infinite sequence of length scales, that for sufficiently large values of β all vortices can be arranged in finite, neutral clusters of finite diameter and finite density. The conditions characterizing those clusters are scale-invariant. This construction thus involves idea of scale-invariance and self-similarity. Furthermore, it requires successive integrations over "fluctuations" on ever larger length scales — a device reminiscent of renormalization-group methods.

For small β, vortices unbind and form a plasma. Such Coulomb plasmas are studied rigorously in [89, 91, 306, 177]. Thus, one expects a phase transition as β is varied. It is non-trivial to show that the transition just described in the two-dimensional vortex gas corresponds, in the two-dimensional classical XY model, to one from a small-β phase in which $\langle \varphi_0 \cdot \varphi_x \rangle$ has exponential fall-off in $|x|$ to a large-β phase in which $\langle \varphi_0 \cdot \varphi_x \rangle$ falls off like an inverse power (≤ 1) of $|x|$ as $|x| \to \infty$. This is proven rigorously in [224].

We now proceed to discussing the third general method in the theory of phase transitions.

(c) Infrared bounds (rigorous spin-wave theory) [222, 218]. This method, which originated in [222], is rather general and is the only known method which gives satisfactory results in models where the spin takes values in a non-linear manifold and the symmetry group is non-abelian. We describe it in terms of an example, the ferromagnetic nearest-neighbor N-vector model: Let φ be a lattice spin with N components, and let the *a priori* single-spin measure be the uniform measure on the unit sphere,

$$dP_x(\varphi_x) = \delta(|\varphi_x|^2 - 1)\, d^N \varphi_x \,, \tag{2.18}$$

The Hamiltonian is

$$H_\Lambda(\varphi) = \beta \sum_{\substack{x,x' \in \Lambda \\ |x-x'|=1}} (1 - \varphi_x \cdot \varphi_{x'}) \; - \; h \sum_{x \in \Lambda} \varphi_x^{(1)} \tag{2.19}$$

($\beta > 0$), and let $d\mu(\varphi)$ be an equilibrium state as in (1.30)/(1.31). For $h \neq 0$, $d\mu$ is believed to be unique (within some class of boundary conditions); this has been proven for $N \leq 3$. Let $\triangle$ be a large finite (hyper-) cube, and define the random variable

$$\varphi(\triangle) = \frac{1}{|\triangle|} \sum_{x \in \triangle} \varphi_x \tag{2.20}$$

(the average magnetization in $\triangle$), where $|\triangle|$ is the volume of $\triangle$. The basic idea of spin-wave theory is that, for large β,

$$\varphi(\triangle) \approx M \mathbf{e}_1 + \delta\varphi(\triangle) \tag{2.21}$$

where $\mathbf{e}_1$ is the unit vector in the 1-direction (the direction of the magnetic field), M is the magnetization, and $\delta\varphi(\triangle)$ is the fluctuation of $\varphi(\triangle)$ around $M\mathbf{e}_1$, which one expects to be $\sim \beta^{-1/2}$ for equilibrium configurations at low temperatures (large β).

These ideas can be formalized as follows: in a pure phase (e.g. $h \neq 0$), consider the truncated two-point function

$$G(x) \equiv \langle \varphi_0 \cdot \varphi_x \rangle - \langle \varphi_0 \rangle^2 \,, \tag{2.22}$$

and let $\widetilde{G}(p)$ be its Fourier transform, which is a function on the d-dimensional torus $B = [-\pi, \pi]^d$ (the first Brillouin zone). By using the so-called transfer-matrix method, Fröhlich, Simon and Spencer [222, 218] have shown that

$$0 \leq \widetilde{G}(p) \leq N\beta^{-1} \left[2d - 2\sum_{i=1}^{d} \cos p_i\right]^{-1} . \tag{2.23}$$

The upper bound in (2.23) is called an *infrared* (or *spin-wave*) *bound*; its proof was inspired by the Källen-Lehmann spectral representation of the two-point function in relativistic quantum field theory. [Note that the upper bound in (2.23) would be *equality* in the massless Gaussian model.] By Fourier transformation, we obtain

$$0 \leq G(0) \leq N\beta^{-1} I_d , \tag{2.24}$$

where

$$I_d \equiv (2\pi)^{-d} \int_B d^d p \left[2d - 2\sum_{i=1}^{d} \cos p_i\right]^{-1} . \tag{2.25}$$

We note that I_d is divergent for $d \leq 2$, but finite for $d > 2$ (with $I_d \sim d^{-1}$ for large d).

By (2.18) it is obvious that

$$\langle \varphi_0 \cdot \varphi_0 \rangle = 1 . \tag{2.26}$$

Thus for $\beta > NI_d$,

$$M(\beta, h)^2 = \langle \varphi_0 \rangle^2 = 1 - G(0) \geq 1 - N\beta^{-1} I_d > 0 \tag{2.27}$$

uniformly in $h \neq 0$, i.e.

$$M_0(\beta) \equiv \lim_{h \downarrow 0} M(\beta, h) > 0 . \tag{2.28}$$

On the other hand, it is easy to prove that $M_0(\beta) = 0$ for sufficiently small β. It follows that, if $d > 2$, there is a phase transition as β is varied.

The infrared bound (2.23) implies that

$$\delta\varphi(\triangle) \lesssim \sqrt{N\beta^{-1} |\triangle|^{(2-d)/d}} , \tag{2.29}$$

in accordance with heuristic ideas based on spin-wave theory. Note that for $d \leq 2$, this upper bound does *not* become small as the volume $|\triangle|$ tends to ∞. This suggests that there should be no spontaneous magnetization when $d = 1$ or 2. Indeed, for models with continuous symmetry ($N \geq 2$), this conjecture is correct: there is no spontaneous magnetization and no continuous symmetry breaking in $d \leq 2$. This is the well-known Mermin-Wagner theorem [393, 391, 386]; see also [220] for a proof which formalizes the above fluctuation argument.

The results reported here extend to a large class of spin systems, but the hypotheses required for the known proofs of the infrared bound (2.23) impose serious limitations on the class of Hamiltonians for which (2.23) is known to be valid [222, 218].

(d) Renormalization group. A fourth method would consist of performing renormalization-group transformations (Chapter 4) on the state of a statistical system and grinding out information on its equilibrium phase diagram by proving convergence of the nth iterate of a renormalization-group transformation applied to an equilibrium state to some high- or low-temperature fixed points, as $n \to \infty$. This strategy has been followed in [11, 196] and especially [238, 86, 87]. But its rigorous implementations are still rather limited.

Often, a combination of methods (b)–(d), rather than a "pure method", leads to success. For all further details we refer the reader to the literature.

We conclude this section by mentioning some recent results on the structure of the space of translation-invariant equilibrium states in the Ising model ($N = 1$) and the two-component rotor model ($N = 2$):

For $h \neq 0$, or for $h = 0$ but β so small that $M(\beta) = 0$, the translation-invariant equilibrium state is *unique* [360, 220]. Next, suppose that $h = 0$, $M_0(\beta) \neq 0$ (i.e. there is a non-zero spontaneous magnetization) and that β is a point of continuity of the internal energy density, $-\partial(\beta\Phi)/\partial\beta$. (Since $\beta\Phi(\beta)$ is concave in β, this is true for all except perhaps countably many values of β.) Then:

(i) In the Ising model, there exist precisely two extremal translation-invariant equilibrium states $\langle\cdot\rangle_\pm$ with

$$0 < \langle\varphi_0\rangle_+ = -\langle\varphi_0\rangle_- \,. \tag{2.30}$$

See [358, 359]. A deeper result, due to Aizenman [3] and Higuchi [303], is that in the *two-dimensional* Ising model this is true for *all* $\beta > \beta_c$, *without* assuming translation invariance.

(ii) In the $N = 2$ rotor model, the extremal translation-invariant equilibrium states are parametrized by an angle $\theta \in [0, 2\pi)$ and are carried one into another by global spin rotations; in particular,

$$\langle\boldsymbol{\varphi}_0\rangle_\theta = \left|\langle\boldsymbol{\varphi}_0\rangle_0\right| \begin{pmatrix} \cos\theta \\ \sin\theta \end{pmatrix} \,. \tag{2.31}$$

For the proof, see [220].

The proofs of the results mentioned here are rather unintuitive, although they involve some clever ideas.

2.2 Existence of critical points

The next step, after proving the existence of a phase transition, is to obtain information about its nature: is it discontinuous (first-order), or is it continuous (passing through a critical point)? If the latter, one would like to compute (or at least to prove bounds on) the critical exponents.

Almost all available *rigorous* results concerning the existence of critical points and critical exponents concern the ferromagnetic φ^4 and N-vector models with one or two components (or in some cases up to four components). We therefore restrict our review to these models.

The first rigorous results on the existence of critical points and estimates on critical exponents were proven by Glimm and Jaffe [256]. Let $d \geq 2$, and let β_c be defined by

$$\beta_c = \sup\{\beta\colon\ \xi(\beta) \equiv \xi(\beta, h = 0) < \infty\}\ . \tag{2.32}$$

Then Rosen [445] and Glimm and Jaffe [259] have shown that $\beta_c < \infty$, and that $\xi(\beta) \longrightarrow \infty$ as $\beta \uparrow \beta_c$. It has also been shown [259] that the magnetic susceptibility $\chi(\beta)$ diverges as $\beta \uparrow \beta_c$. Simplified proofs were later given in [464, 372, 96].

It has recently been proven [6, 10] that, in a class of Ising-type models, the transition at $(h = 0, \beta = \beta_c)$ is sharp, in the sense that

- For $\beta < \beta_c$, $\xi(\beta) < \infty$ and $M_0(\beta) = 0$.
- For $\beta = \beta_c$, $\xi(\beta) = \infty$.
- For $\beta > \beta_c$, $M_0(\beta) > 0$.

Moreover, for the Ising model, it can be shown that

$$M_0(\beta) \downarrow 0 \ \text{ as } \beta \downarrow \beta_c\ , \tag{2.33}$$

at least in dimensions $d = 2$ (exact solution [523]) and $d \geq 4$ [12]. Unfortunately, for some "silly" technical reasons, (2.33) is not yet rigorously established in three dimensions. See [12] and Chapter 14 of Part III.

2.3 Critical exponents in some exactly solvable models and standard approximations

Undoubtedly the most important exactly solvable model in statistical mechanics is the Gaussian model (1.26)/(1.28): it serves as a reference model which has "classical" critical exponents, and as a starting point for perturbation expansions. All these properties arise from the fact that Gaussian integrals are trivial to perform. The truncated correlation functions are (see e.g. [424])

$$\langle \varphi_x^{(\alpha)} \rangle^T = \widetilde{G}(0) h^{(\alpha)} \tag{2.34}$$

$$\langle \varphi_x^{(\alpha)} \varphi_y^{(\beta)} \rangle^T = \delta_{\alpha\beta} (2\pi)^{-d} \int\limits_{[-\pi,\pi]^d} dp \, e^{ip\cdot(x-y)} \widetilde{G}(p) \tag{2.35}$$

$$\langle \varphi_{x_1}^{(\alpha_1)} \ldots \varphi_{x_n}^{(\alpha_n)} \rangle^T = 0 \qquad \text{for } n \geq 3 \tag{2.36}$$

where

$$\widetilde{G}(p) = \widetilde{E}(p)^{-1} \equiv \left[B - \widetilde{J}(p)\right]^{-1} \tag{2.37}$$

$$\widetilde{J}(p) = \sum_x e^{ip\cdot x} J_{0x} \tag{2.38}$$

and we are assuming translation invariance $J_{xy} = J_{x+a,y+a}$. Note that $\widetilde{E}(p)$ is precisely the energy per unit volume of a spin wave of wavenumber p. In particular, for short-range (e.g. nearest-neighbor) ferromagnetic interactions, $\widetilde{J}(p)$ reaches its maximum at $p = 0$ and behaves near $p = 0$ like

$$\widetilde{J}(p) = \widetilde{J}(0) - cp^2 + o(p^2) \; .$$

Criticality occurs when $B = \widetilde{J}(0)$, at which point the susceptibility $\chi = \widetilde{G}(0)$ and the correlation length ξ diverge. The critical exponents are easily found to be

$$\gamma = 1 \, , \quad \nu = \frac{1}{2} \, , \quad \eta = 0 \, , \quad \alpha = \max\left(0, 2 - \frac{d}{2}\right) \, , \quad \alpha_{\text{sing}} = 2 - \frac{d}{2} \, .$$

Here α and α_{sing} are, respectively, the exponents for the full specific heat and for its singular part (see Chapter 13). In this model the specific heat is given essentially by the "bubble integral"

$$C_H \sim \int\limits_{[-\pi,\pi]^d} dp \, \widetilde{G}(p)^2 \, ,$$

which will play a key role in our analysis of critical behavior for Ising and φ^4 models (Chapters 14 and 15). Note, finally, that the Gaussian model is *undefined* when $B < \widetilde{J}(0)$ [and when $h \neq 0$ with $B = \widetilde{J}(0)$], since the relevant Gaussian integrals are divergent (even in finite volume). Thus, unlike other statistical-mechanical models, *the Gaussian model has no low-temperature phase* (or critical isotherm).

Another solvable case is the limit $N \to \infty$ of an isotropic N-component spin model (e.g. the N-vector model) with the field strengths scaled as $\boldsymbol{\varphi}^2 \sim N$. This limit also serves as a starting point for a systematic expansion in powers of $1/N$. The basic idea is to Fourier-expand the *a priori* single-spin distribution (1.23)/(1.24) by introducing an auxiliary variable a_x conjugate to φ_x^2:

$$g_N(\varphi_x^2) \equiv e^{-NV(\varphi_x^2/N)} = \int_{-\infty}^{\infty} da_x \, e^{ia_x\varphi_x^2} \, \tilde{g}_N(a_x)$$

where, for example,

$$\tilde{g}_N(a_x) = \begin{cases} \exp\left[\frac{-N}{\lambda}a^2 + \frac{iBN}{\lambda}a\right] & \varphi^4 \text{ model} \\ \exp(-iNa) & N\text{-vector model} \end{cases}$$

The φ integral is then Gaussian, and hence can be performed, yielding a determinant; the a integral is then carried out by the saddle-point method, yielding an asymptotic expansion in powers of $1/N$. For more details, see [77, 378, 348]. The critical exponents at $N = \infty$ are found to be

$$\gamma = \left\{ \begin{array}{ll} \frac{2}{d-2} & \text{for } 2 < d < 4 \\ 1 & \text{for } d \geq 4 \end{array} \right\}, \quad \nu = \frac{\gamma}{2}, \quad \eta = 0,$$
$$\hat{\beta} = \frac{1}{2}, \quad \delta = \left\{ \begin{array}{ll} \frac{d+2}{d-2} & \text{for } 2 < d < 4 \\ 3 & \text{for } d \geq 4 \end{array} \right\}, \quad \alpha_{\text{sing}} = \frac{4-d}{d-2}. \tag{2.39}$$

(modulo logarithms in $d = 4$). Moreover, critical exponents can be expanded (non-rigorously) in powers of $1/N$; see [77, 378].

Finally, we remark that the $N \to \infty$ limit of the N-vector model is equivalent (in at least some aspects [486]) to the *spherical model* defined in the early 1950's by Berlin and Kac [62]. In our opinion, the original spherical model is of mainly historical interest; in this book, we sometimes use the term "spherical model" simply as a shorthand for "$N \to \infty$ limit of the N-vector model".

The most famous nontrivial exactly solved model in statisticalmechanics is the two-dimensional nearest-neighbor Ising model (in zero magnetic field). The free energy was first computed by Onsager [417] in a 1944 paper which marked the beginning of the modern theory of critical phenomena. The spontaneous magnetization was computed in 1952 by Yang [524]. Previously, Kaufman and Onsager computed a few two-point correlations [330]; while other correlation functions have been computed in recent years as well [387, 1, 316, 421, 422]. The critical exponents are

$$\gamma = \frac{7}{4}, \quad \nu = 1, \quad \eta = \frac{1}{4}, \quad \beta = \frac{1}{8}, \quad \delta = 15, \quad \alpha = 0\,(\log).$$

For reviews, see [388, 57, 456, 487, 438].

The φ^4 model can be expanded formally in a perturbation series around the Gaussian model, i.e. a Taylor expansion in powers of λ. For details, see e.g. [424]. For $d > 4$, one obtains an asymptotic expansion for the critical temperature which is finite order-by-order; the critical exponents coincide with those of the Gaussian model ($\gamma = 1$, $\nu = \frac{1}{2}$, $\eta = 0$, $\alpha = 0$) but now include the low-temperature and critical-isotherm exponents

$$\widehat{\beta} = \frac{1}{2} \quad , \quad \delta = 3 \; .$$

Thus, for $d > 4$, the $\lambda\varphi^4$ term in the Hamiltonian is *irrelevant* in the critical region; this will be made precise by the renormalization group (Chapter 4). On the other hand, for $d \leq 4$ perturbation theory predicts its own downfall: the coefficients in the perturbation series are *infrared-divergent* (i.e. divergent due to long distances or small momenta) when the critical point is approached; and the divergences get stronger at higher orders in the series. This is a signal of the fact that, when $d < 4$, the critical temperature *cannot* be expanded in powers of λ [85] and that the critical exponents are *not* those of the Gaussian model.

Rather, the critical exponents take on nontrivial values which are difficult to compute; approximate values can be obtained by series extrapolation [237, 287], renormalization-group-improved perturbation theory [77, 24], or Monte Carlo simulation [65]. In this book we take on the more modest task of proving rigorous nonperturbative *upper and lower bounds* on critical exponents.

Another important approximation method is *mean-field theory*, which can be derived in several different ways:

1) *Variational approach* (see e.g. [424]). The Boltzmann-Gibbs distribution is the probability measure which maximizes the entropy minus $\beta\times$ the mean energy. By carrying out this maximization over a restricted set of probability measures, we obtain an approximate solution of the model. Mean-field theory is obtained by using *product measures* (i.e. measures in which the spins are statistically independent). The result is a self-consistent equation for the magnetization, e.g. for the Ising model

$$M_x = \tanh\left[h_x + \sum_y J_{xy} M_y\right] \; .$$

(This equation can alternatively be derived by focussing on one spin and replacing the effect of all the other spins by their mean values, hence the name "mean field".) From this equation one deduces the critical exponents

$$\gamma = 1 \; , \quad \widehat{\beta} = \frac{1}{2} \; , \quad \delta = 3 \; , \quad \alpha = 0 \tag{2.40}$$

— the so-called *mean-field* (or *classical*) exponents. Of course, the neglect of correlations between spins is a rather terrible approximation in the critical region, so these predicted exponents should not be taken too seriously. The remarkable thing is that these exponents are *correct* for $d > 4$: in this case the effects of fluctuations are not so severe.

2) *Landau-Ginzburg theory* (see [487]). Here one simply *postulates* that the Legendre transform of the free energy, written as a function of the magnetization M and temperature T, is analytic (or at least sufficiently-many-times differentiable) in a neighborhood of $M = 0$, $T = T_c$. One then imposes the known qualitative behavior and deduces the critical exponents (2.40). Mathematically,

this is catastrophe theory [29]; physically, it is wishful thinking. Surprisingly, the predictions of Landau-Ginzburg theory are essentially *correct* for $d > 4$, as we discuss in Sections 14.2–14.5.

The disadvantage of these two approaches to mean-field theory is that they cannot be systematically improved. The third method avoids this failing:

3) *Systematic expansion.* Mean-field theory can be derived as the zeroth order in a systematic perturbation expansion around the Gaussian model [77]. This perturbation expansion is finite order-by-order in $d > 4$, but infrared-divergent order-by-order in $d \leq 4$. This suggests that the mean-field exponents may be correct for $d > 4$ and incorrect for $d < 4$; and this can be rigorously proven (see Chapter 14).

2.4 Bounds on critical exponents

Though it is in general difficult to compute the exact critical exponents of a nontrivial statistical-mechanical model, some information can be obtained through rigorous inequalities involving one or more critical exponents. Indeed, the proof of critical-exponent inequalities has become a minor industry.

Inequalities on critical exponents are best classified, and their physical meaning understood, by examining the conditions under which they become equality. In this way we obtain six classes of critical-exponent inequalities (see Section 13.2 for definitions of exponents):

(a) Those that become equality in mean-field theory (or what is almost the same, for the Gaussian model). *Examples*: $\gamma \geq 1$, $\nu \geq \frac{1}{2}$, $\eta \geq 0$, $\widehat{\beta} \leq \frac{1}{2}$, $\widehat{\beta}(\delta - 1) \geq 1$, $\alpha \leq (2 - \frac{d}{2})\gamma$. These inequalities are related to (but weaker than) the fact that the deviation from Gaussianness is one-sided: for example, $\gamma \geq 1$ is a consequence of the Lebowitz inequality $u_4 \leq 0$, while $\eta \geq 0$ is a consequence of the infrared bound (2.23).

(b) Those that become equality in the $N \to \infty$ limit of an N-component model. *Example*: $\gamma \leq 2/(d-2)$. This inequality expresses the fact that the "bubble sum" becomes exact as $N \to \infty$: all other classes of diagrams are suppressed by powers of $1/N$ [518, 110, 283].

(c) Those that become equality under the thermodynamic scaling hypothesis (see [487, 289]). *Examples*: $\alpha' + 2\beta + \gamma' \geq 2$, $\alpha' + \beta(\delta+1) \geq 2$. These inequalities express the general thermodynamic properties (e.g. convexity) of the free energy.

(d) Those that become equality under the correlation scaling hypothesis (see [487, 289]). *Example*: $\gamma \leq (2-\eta)\nu$. These inequalities are related to (but weaker than) the existence of the scaling limit (see Chapter 14).

(e) Those that become equality under the hyperscaling hypothesis (see [509, 323, 198, 201, 510, 200, 288]). *Examples*: $d\nu - 2\Delta_4 + \gamma \geq 0$, $d\nu' \geq \gamma' + 2\beta$, $d\nu \geq 2 - \alpha$, $d(\delta-1)/(\delta+1) \geq 2 - \eta$. These inequalities express deep facts about the nature of critical fluctuations; they will play a prominent role in Chapter 15. For example, the dimensionless renormalized coupling constant g behaves as

$g \sim (\beta_c - \beta)^{d\nu - 2\Delta_4 + \gamma}$; therefore, the inequality $d\nu - 2\Delta_4 + \gamma \geq 0$ states that g stays bounded as $\beta \uparrow \beta_c$, while the hyperscaling relation $d\nu - 2\Delta_4 + \gamma = 0$ states that g stays nonzero in this limit (and hence that the corresponding continuum quantum field theory is non-Gaussian).

(f) Those that become equality under no reasonable condition. *Examples:* $\eta \leq 1$, $\Delta_4 \geq \gamma$.

Unfortunately, many of these inequalities can only be proven for rather restricted classes of models (e.g. Ising, plane-rotator, φ^4), even though their validity is presumably more general. Much of Chapter 14 is devoted to inequalities for critical exponents.

3. Scale transformations and scaling (continuum) limits in lattice spin systems

In this chapter[1] we discuss a method for constructing continuum Euclidean field theories as scaling (continuum) limits of lattice spin systems approaching a critical point. We show what the existence of a scaling limit gives us information about the critical behavior of the lattice spin system.

In order to simplify our discussion, we consider a one-component spin field φ on the lattice $\mathbb{Z}^d$. Let $\{d\mu_\beta(\varphi)\}$ be a family of equilibrium states parametrized by β. (For simplicity, we imagine that β is the only thermodynamic parameter that is varied, but there could also be dependence on a magnetic field h or on other parameters.) We assume that each $d\mu_\beta(\varphi)$ is translation-invariant. As in Section 1.4, we define the correlation functions as the moments of $d\mu_\beta$, i.e.

$$\langle \varphi_{x_1} \cdots \varphi_{x_n} \rangle_\beta = \int \prod_{k=1}^{n} \varphi_{x_k} \, d\mu_\beta(\varphi) \; . \tag{3.1}$$

By a trivial re-definition of φ it is always possible to assume that

$$\langle \varphi_x \rangle_\beta = 0 \; . \tag{3.2}$$

In the following we are interested in analyzing the long-distance limit of the correlation functions defined in (3.1), and in relating existence and properties of this limit to the behavior of the equilibrium state and the correlations as β approaches a critical point β_c [defined as in (1.46)]. We assume that, for $\beta < \beta_c$, the state $\langle \, \cdot \, \rangle_\beta$ is extremal invariant (i.e. $d\mu_\beta$ is ergodic under the action of lattice translations) and that $m(\beta)$ is positive, i.e. $\langle \varphi_x \varphi_y \rangle_\beta$ tends to 0 exponentially fast as $|x-y| \to \infty$, with decay rate denoted $m(\beta)$; see (1.38)/(1.39). Furthermore, we assume that $m(\beta)$ tends to 0 continuously as $\beta \uparrow \beta_c$. As mentioned in Section 2.2, these assumptions are known to hold in the Ising and plane-rotator models, and more generally in φ^4 models with 1 or 2 components.

We now define the *scaled correlations*

$$G_\theta(x_1, \ldots, x_n) \equiv \alpha(\theta)^n \langle \varphi_{\theta x_1} \cdots \varphi_{\theta x_n} \rangle_{\beta(\theta)} \; , \tag{3.3}$$

where $1 \le \theta < \infty$, each x_j belongs to the set

[1]This chapter is adapted from [228].

$$\mathbb{Z}^d_{\theta^{-1}} \equiv \{y\colon\ \theta y \in \mathbb{Z}^d\}\ , \tag{3.4}$$

and $\beta(\theta) < \beta_c$ and $\alpha(\theta)$ are functions of the scale parameter θ which one tries to choose in such a way that a non-trivial limit, as $\theta \to \infty$, exists. In the models mentioned above it suffices to impose the following *renormalization condition*: For $0 < |x-y| < \infty$,

$$0 < \lim_{\theta\to\infty} G_\theta(x,y) \equiv G^*(x-y) < \infty\ . \tag{3.5}$$

It turns out that, in our class of models, (3.5) suffices to show that the limit

$$G^*(x_1,\ldots,x_n) = \lim_{\theta_i\to\infty} G_{\theta_i}(x_1,\ldots,x_n) \tag{3.6}$$

exists for some sequence $\{\theta_i\}$, and that $G^*(x_1,\ldots,x_n)$ is a translation-invariant distribution, for all $n = 3, 4, \ldots$.

Let us henceforth assume that a scaling limit satisfying (3.5) does exist; we shall show that from this hypothesis, along with some natural regularity conditions, we can derive information on the critical exponents of the underlying lattice model. Note first that the non-vanishing of G^* at $x - y \neq 0$ implies that the lattice theories $\langle\ \cdot\ \rangle_{\beta(\theta)}$ must be approaching a critical point,

$$\beta(\theta) \uparrow \beta_c \quad \text{as } \theta \to \infty\ . \tag{3.7}$$

If the limiting correlation $G^*(x-y)$ is required to have exponential fall-off,

$$G^*(x-y) \sim e^{-m^*|x-y|} \quad \text{as } |x-y| \to \infty \tag{3.8}$$

(with $m^* > 0$), it is natural to impose that the underlying lattice theories have similar behavior, i.e.

$$\theta m(\beta(\theta)) \to m^* \quad \text{as } \theta \to \infty\ . \tag{3.9}$$

If $m(\beta)$ is known to satisfy a scaling law

$$m(\beta) \sim (\beta_c - \beta)^\nu \quad \text{as } \beta \uparrow \beta_c \tag{3.10}$$

[cf. (1.48)], then (3.9) and this scaling law imply that we must choose $\beta(\theta)$ such that

$$\beta_c - \beta(\theta) \sim \theta^{1/\nu} \quad \text{as } \theta \to \infty\ . \tag{3.11}$$

Assuming now that $G^*(x-y)$ is integrable at zero, it follows that

$$\chi^* \equiv \int d^dx\, G^*(x) \tag{3.12}$$

is finite; and it is natural to expect that the rescaled lattice susceptibilities

$$\chi_\theta \equiv \sum_{x\in\mathbb{Z}^d_{\theta^{-1}}} \theta^{-d} G_\theta(0,x) \tag{3.13}$$

converge to χ^* as $\theta \to \infty$. By (3.3),

$$\chi_\theta = \alpha(\theta)^2 \theta^{-d} \chi(\beta(\theta)) . \tag{3.14}$$

If $\chi(\beta)$ satisfies a scaling law

$$\chi(\beta) \sim (\beta_c - \beta)^{-\gamma} \quad \text{as } \beta \uparrow \beta_c \tag{3.15}$$

[cf. (1.49)], then we conclude that $\alpha(\theta)$ must be chosen as

$$\alpha(\theta)^2 \sim \theta^{d-(\gamma/\nu)} \quad \text{as } \theta \to \infty . \tag{3.16}$$

By assuming a little more regularity in the limit (3.5), we can relate (3.16) to the critical exponent η defined in (1.47). Indeed, let us assume that the *same* choice of $\alpha(\theta)$ works for any value of $m^* \geq 0$. [That is, the "field-strength renormalization" is the same in the massive and massless scaling limits.] Then, imposing (3.5) with $\beta(\theta) \equiv \beta_c$ (i.e. $m^* = 0$), we conclude immediately that $\alpha(\theta)$ must be chosen as

$$\alpha(\theta)^2 \sim \theta^{d-2+\eta} \quad \text{as } \theta \to \infty . \tag{3.17}$$

Comparing (3.16) and (3.17), we deduce the *scaling relation*

$$\gamma = (2 - \eta)\nu . \tag{3.18}$$

It is worth remarking that in our class of models, the infrared bound (2.23) holds, from which one can deduce that, for $d > 2$,

$$0 \leq \langle \varphi_0 \varphi_x \rangle_\beta \leq c_d \beta^{-1} |x - y|^{2-d} \tag{3.19}$$

(at least for one- or two-component fields; see [481]); here c_d is a universal constant. On the one hand, it is immediate from (3.19), evaluated at $\beta = \beta_c$, that

$$\eta \geq 0 . \tag{3.20}$$

On the other hand, it follows from (3.5) and (3.19) that we must take

$$\alpha(\theta)^2 \geq \text{const} \times \theta^{d-2} \quad \text{as } \theta \to \infty . \tag{3.21}$$

Note that (3.20) and (3.21) are equivalent if (3.17) holds.

We can apply a similar argument to correlation functions of the field φ^2. Define the rescaled φ^2–φ^2 correlation function

$$G_\theta^{(\varphi^2)}(x,y) \equiv \alpha_2(\theta)^2 \langle \varphi_{\theta x}^2 ; \varphi_{\theta y}^2 \rangle_{\beta(\theta)} , \tag{3.22}$$

where we have introduced a new field-strength renormalization factor $\alpha_2(\theta)$. Assuming that with the *same* choice of $\beta(\theta)$, and for a suitable choice of $\alpha_2(\theta)$, we have

$$0 < \lim_{\theta \to \infty} G_\theta^{(\varphi^2)}(x,y) \equiv G^{(\varphi^2)*}(x-y) < \infty , \tag{3.23}$$

for $0 < |x-y| < \infty$, then an argument analogous to the preceding one shows that we must choose

$$\alpha_2(\theta)^2 \sim \theta^{d-(\alpha/\nu)} , \tag{3.24}$$

where α is the critical exponent defined by

$$\sum_x \langle \varphi_0^2 ; \varphi_x^2 \rangle_\beta \sim (\beta_c - \beta)^{-\alpha} \text{ as } \beta \uparrow \beta_c . \tag{3.25}$$

And assuming that this choice of $\alpha_2(\theta)$ works also for $m^* = 0$, we conclude that

$$\alpha = (4 - d - \eta_{\varphi^2})\nu , \tag{3.26}$$

where η_{φ^2} is the critical exponent defined by

$$\langle \varphi_0^2 ; \varphi_x^2 \rangle_{\beta_c} \sim |x|^{-(2d-4+\eta_{\varphi^2})} \text{ as } |x| \to \infty . \tag{3.27}$$

Let us mention, however, that this reasoning assumes that $G^{(\varphi^2)*}(x-y)$ is integrable at zero, which we expect to fail in dimension $d \geq 4$.

In summary, we have shown (modulo some regularity assumptions) that the critical-exponent ratios γ/ν and α/ν, which describe the *approach* to the critical point $\beta \uparrow \beta_c$, can be computed from the decay as $|x| \to \infty$ of correlation functions *at* the critical point β_c.

Remark. Plausible (but by no means airtight) physical arguments [508, 326] imply that the *hyperscaling relation*

$$d\nu = 2 - \alpha \tag{3.28}$$

ought to hold, at least in dimension $d < 4$. If hyperscaling holds, then (3.26) becomes

$$\nu = \frac{2}{4 - \eta_{\varphi^2}} \tag{3.29}$$

$$\alpha = 2 - \frac{2d}{4 - \eta_{\varphi^2}} \tag{3.30}$$

so that the critical exponents ν and α (and not only their ratio) can be determined from the critical correlations. Note that although the hyperscaling relation $d\nu = 2 - \alpha$ has not yet been proven, it has been proven [319, 480] that

$$d\nu \geq 2 - \alpha . \tag{3.31}$$

We now must focus our attention on the question of why we are interested in the large-scale behavior of a lattice spin system, i.e. in studying the limit where $\theta \to \infty$. Here are some answers.

1) Suppose we are able to *construct* the limiting rescaled correlation functions, $G^*(x_1, \ldots, x_n) \equiv \lim_{\theta\to\infty} G_\theta(x_1, \ldots, x_n)$, such that the renormalization

conditions (3.5) and (3.9) hold. Then we must have, in particular, a way of determining the functions $\beta(\theta)$ and $\alpha(\theta)$. But, as we have just shown, the choice of $\beta(\theta)$ determines the critical exponent ν, and the choice of $\alpha(\theta)$ determines η. Thus an explicit *construction* of the $\theta \to \infty$ limit determines, in principle, the critical exponents ν, γ and η.

2) As our derivation of relation (3.18) shows, proving merely the *existence* of a $\theta \to \infty$ limit yields non-trivial relations between critical exponents.

3) But perhaps the main interest in constructing the limits, $G^*(x_1, \ldots, x_n)$, of the rescaled correlation functions comes from the fact that *these limits may be the Euclidean Green's functions of a relativistic quantum field theory*, i.e.

$$G^*(x_1, \ldots, x_n) \equiv S_n(x_1, \ldots, x_n) \ , \tag{3.32}$$

for some quantum field theory satisfying the Gårding-Wightman axioms (GW0)–(GW4). Indeed, in the models considered above, this is is the case provided that the distributions $G^*(x_1, \ldots, x_n)$ are invariant under simultaneous rotations of their arguments. [Even if rotation-invariance were to fail, the G^*'s would still be the Euclidean Green's functions of a non-Lorentz-invariant quantum field theory.]

This observation is at the basis of understanding *universality* in the theory of critical phenomena: Critical phenomena in lattice systems can be studied quantitatively precisely by analyzing the scaling limits of their correlation functions which are the Euclidean Green functions of some Euclidean Field Theory. But the manifold of Euclidean Field Theories has a simple structure that is quite well understood, at least conjecturally. In particular, in $d > 2$ dimensions it is believed to be *finite-dimensional.* It contains special points corresponding to scale-covariant (and hence, in general, conformally covariant [195, 383, 455, 376, 435]) Euclidean Field Theories. In our class of examples, these points, i.e. the Conformal Field Theories, are *isolated.* In $d \geq 4$ dimensions, there is a single conformal field theory describing critical behavior in the class of lattice systems considered in this section: the massless, free (Gaussian) field theory. In three dimensions, there appear to exist just two conformal field theories describing critical phenomena in our class of lattice systems: the massless free field theory, appropriate for the description of tri-critical behavior, and critical $\lambda\varphi_3^4$-theory, appropriate for the description of critical behavior. In two dimensions, there is an *infinite, discrete* series of conformal field theories, the *A-series of minimal models* [58, 209, 104, 103], which describe n-critical behavior, for $n \geq 2$, in the class of lattice systems considered in this section.

All critical exponents can be calculated, in principle, from Green functions of conformal field theories, as explained above. This and the above remarks on the structure of conformal field theories explain what universality in the theory of critical phenomena means and where it comes from.

For some scaling (= continuum) limits of the models introduced in Section 1.4 in two and three dimensions and of the two-dimensional Ising model it has been shown (see *e.g* [260, 229, 387, 316, 1, 421, 422]), that the distributions G^* are the Euclidean Green's functions of relativistic quantum field theories satisfying all Wightman axioms (W0)–(W4).

4. Construction of scaling limits: the renormalization group

In this chapter[1] we sketch a specific method for constructing scaling ($\equiv$ continuum) limits, $G^*(x_1, \ldots, x_n)$, of rescaled correlations, $G_\theta(x_1, \ldots, x_n)$, as $\theta \to \infty$, namely the Kadanoff block spin transformations. They shall serve as a typical example of "renormalization group transformations."

Of course, there are many other incarnations of the renormalization group strategy, including ones in the context of dynamics, in particular the Feigenbaum theory [180, 181, 349, 112]. One of them, the Gallavotti-Nicolò tree expansion [232, 434] — which is very close to Wilson's original version [516, 517, 519, 520] — is briefly sketched in this chapter.

We also indicate how mathematical control of (Wilson) renormalization group transformations enables one to calculate critical exponents.

4.1 Block spin transformations

We define a function $\mathcal{K} \equiv \mathcal{K}^\varepsilon$ on $\mathbb{R}^d$ as follows:

$$\mathcal{K}(y) = \begin{cases} \varepsilon^{-d} & -\frac{\varepsilon}{2} \leq y^\mu \leq \frac{\varepsilon}{2} \ , \ \mu = 1, \ldots, d \\ 0 & \text{otherwise} \end{cases} \ , \tag{4.1}$$

where $y = (y^1, \ldots, y^d) \in \mathbb{R}^d$ and ε is an arbitrary positive number. Let

$$\mathcal{K}_x(y) = \mathcal{K}(y - \varepsilon x) \ , \quad x \in \mathbb{Z}^d \ . \tag{4.2}$$

Let $G_\theta(x_1, \ldots, x_n)$ be a rescaled correlation function defined in (3.3). Then

$$\begin{aligned} G_\theta(\mathcal{K}_{x_1}, \ldots, \mathcal{K}_{x_n}) &= \sum_{y_1, \ldots, y_n \in \mathbb{Z}^d_{\theta^{-1}}} G_\theta(y_1, \ldots, y_n) \prod_{k=1}^n \theta^{-d} \mathcal{K}_{x_k}(y_k) \\ &= (\alpha(\theta)\theta^{-d})^n \sum_{z_1, \ldots, z_n \in \mathbb{Z}^d} \langle \varphi_{z_1} \cdots \varphi_{z_n} \rangle_{\beta(\theta)} \cdot \prod_{k=1}^n \mathcal{K}_{x_k}(\theta^{-1} z_k) \ . \end{aligned} \tag{4.3}$$

[1]Sections 4.1 and 4.2 are adapted from [228].

We now set

$$\theta = \theta_m = \varepsilon^{-1} L^m , \tag{4.4}$$

where L is some positive integer and $m = 1, 2, 3, \ldots$, and define

$$r_m(\varphi_x) = \alpha(\varepsilon^{-1} L^m) \cdot L^{-dm} \sum_{\substack{z \in \mathbb{Z}^d \\ -\frac{1}{2} \le L^{-m} z^\mu - x^\mu \le \frac{1}{2}}} \varphi_z , \tag{4.5}$$

$x \in \mathbb{Z}^d$, $\mu = 1, \ldots, d$. Then

$$G_{\theta_m}(\mathcal{K}_{x_1}, \ldots, \mathcal{K}_{x_n}) = \langle r_m(\varphi_{x_1}) \ldots r_m(\varphi_{x_n}) \rangle_{\beta(\theta_m)} . \tag{4.6}$$

Let $d\mu(\varphi)$ be an arbitrary, translation-invariant, finite, positive measure on the space $\mathbb{R}^{\mathbb{Z}^d}$ of all configurations $\{\varphi_x : x \in \mathbb{Z}^d\}$. We define a transformation R_m of μ by the equation

$$\int \prod_{k=1}^{n} r_m(\varphi_{x_k}) d\mu(\varphi) = \int \prod_{k=1}^{n} \varphi_{x_k} d(R_m \mu)(\varphi) , \tag{4.7}$$

for all $x_1, \ldots, x_n$ in $\mathbb{Z}^d$, $n = 1, 2, 3, \ldots$.

Note that r_m (resp. R_m) consists of a transformation increasing the scale size (taking the average over all spins in a block) followed by a (in the present example linear) coordinate transformation in spin space. Furthermore, we note that if μ is ergodic then so is $R_m \mu$.

In order to arrive at an interesting concept we now suppose that $\alpha(\theta)$ is proportional to some power of θ, i.e.

$$\alpha(\theta)^2 \sim \theta^{d-2+\eta} , \tag{4.8}$$

for some η. We then define

$$r(\varphi_x) = L^{(\eta-d-2)/2} \sum_{\substack{z \in \mathbb{Z}^d \\ -\frac{1}{2} \le L^{-1} z^\mu - x^\mu \le \frac{1}{2}}} \varphi_z . \tag{4.9}$$

Then

$$r_m(\varphi_x) = \alpha(\varepsilon^{-1}) \underbrace{r \circ \ldots \circ r}_{m \text{ times}}(\theta(x)) . \tag{4.10}$$

Let $R\mu$ be the unique measure such that

$$\int \prod_{k=1}^{n} r(\varphi_{x_k}) d\mu(\varphi) = \int \prod_{k=1}^{n} \varphi_{x_k} d(R\mu)(\varphi) , \tag{4.11}$$

for all $x_1, \ldots, x_n$, $n = 1, 2, 3 \ldots$. [Note, R maps ergodic measures to ergodic ones.] Then

$$d(R_m \mu)(\varphi) = d(\underbrace{R \circ \ldots \circ R}_{m \text{ times}} \mu)(\alpha(\varepsilon)\varphi) \equiv d(R^m \mu)(\alpha(\varepsilon)\varphi) . \tag{4.12}$$

If we now choose $d\mu = d\mu_\beta$, where $\{\mu_\beta\}$ is a family of Gibbs states of our spin system indexed by β we obtain, setting $\beta = \beta(\theta_m)$,

$$\begin{aligned} G^*(\mathcal{K}_{x_1}, \ldots, \mathcal{K}_{x_n}) &= \lim_{m\to\infty} G_{\theta_m}(\mathcal{K}_{x_1}, \ldots, \mathcal{K}_{x_n}) \\ &= \lim_{m\to\infty} \int \prod_{k=1}^{n} \varphi_{x_k} d(R^m \mu_{\beta(\theta_m)})(\alpha(\varepsilon)\varphi) , \end{aligned} \tag{4.13}$$

provided the limit exists.

In order to prove existence of the limit in (4.13), one must analyze the transformation R on (the boundary of) a suitably chosen cone of finite measures. In particular, one has to construct fixed points of R, study the spectrum of the linearization of R at the fixed points (the linearization of R acts on a linear space of measurable (or continuous, or analytic) functions of spin configurations, φ), and construct the stable and unstable manifolds of R near a fixed point. We shall discuss some examples below.

Remarks.

1) By (4.8)–(4.11), the transformation $R \equiv R_\eta$ depends on the exponent η. The condition that the limit in (4.13) exist and be non-trivial fixes η.

2) We shall see that the critical exponents ν and γ are determined by positive eigenvalues > 1 of the linearization of R_η at the appropriate fixed point of R_η.

3) It is usually expected that if a measure μ is a Gibbs measure for some Hamilton function H, then $R_\eta\mu$ is again a Gibbs measure. This, however, is not true in general. But if it is true on a suitably chosen space of Gibbs states then R_η uniquely determines a transformation $\mathcal{R}_\eta$ acting on a space of (equivalence classes of) Hamilton functions, or interactions. The simplifying feature of this set-up is that the derivative of $\mathcal{R}_\eta$ acts on the linear hull of the same space.

4) Below, we shall briefly indicate how these ideas are applied to dynamics.

4.2 Fixed points of block spin transformations, stable and unstable manifolds, critical exponents

Let M be some cone of finite measures, μ, on some space of spin configurations $\varphi = \{\varphi_j\}_{j\in\mathbb{Z}^d}$. Let R_η be a renormalization (block spin) transformation acting on M, as discussed in Section 4.1. (One ought to assume probably that M can be given a differentiable structure such that the action of R_η on M is smooth.) Of particular interest are the fixed points, μ^*, of R_η. It is usually not so hard to convince oneself that there exists at least one fixed point. Supposing, for example, that the spins are real-valued and that R_η is given by (4.11), it is easy to see that R_η has at least a one-dimensional manifold of fixed points, μ_t^*, $t \in \mathbb{R}$, which are *Gaussian* measures. Gaussian measures are uniquely characterized by

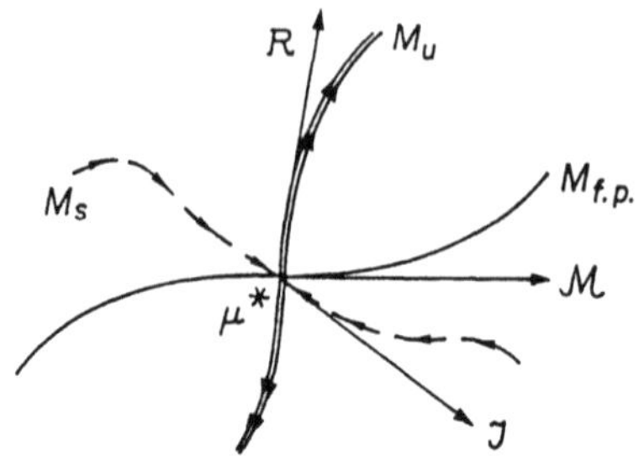

Fig. 4.1. Manifold M of measures in the vicinity of a fixed point μ^*. Unstable manifold = M_u; stable manifold = M_s; relevant perturbations = $\mathcal{R}$; irrelevant perturbations = $\mathcal{I}$; marginal perturbations = $\mathcal{M}$

their mean and their covariance. Here the mean of μ_t^* if set to 0, the covariance is of the form $e^t C^*$, where

$$\int d\mu_{t=0}^*(\varphi)\varphi_x\varphi_y \equiv C^*(x,y) = c^*(x-y) \sim |x-y|^{2-d-\eta} \,. \tag{4.14}$$

See [468, 467] and the references given there. (Non-Gaussian fixed points have been constructed, too, but no non-Gaussian fixed points interesting for statistical physics appear to be known, in the sense of rigorous mathematics, except in the two-dimensional models [58, 209, 103].)

There is an intimate mathematical connection between fixed points, μ^*, of R_η and "stable distributions" in probability theory. It is worthwhile to note that fixed points, μ^*, cannot be strongly mixing. See e.g. [468, 467, 318] and the references given there for a discussion of these probabilistic aspects. We stress, however, that the main concepts of the renormalization group are more general than their probabilistic formulation!

Let $M_{\text{fp}} = M_{\text{fp}}(R_\eta)$ be the manifold of all fixed points of R_η. Now choose some particular fixed point $\mu^* \in M_{\text{fp}}$, and let $M_{\text{fp}}(R_\eta, \mu^*)$ be the connected component of M_{fp} that contains μ^*. Since a certain class of coordinate transformations, like

$$\varphi_j \to \alpha\varphi_j \,, \quad \text{for all } j \in \mathbb{Z}^d \,, \tag{4.15}$$

for some positive α independent of j, commute with R_η, the fixed points of R_η are not isolated, and the linearization of R_η at some fixed point μ^* will generally have an eigenvalue 1 (and possibly further eigenvalues) corresponding to coordinate transformations.

Under suitable hypotheses on R_η and M, one can decompose M in the vicinity of $\mu^* \in M_{\text{fp}}$ into a stable manifold, $M_s(\mu^*)$, and an unstable manifold $M_u(\mu^*)$ [see Figure 4.1]. States on $M_s(\mu^*)$ are driven towards μ^*, states on $M_u(\mu^*)$ are driven away from μ^*, under the action of R_η. The tangent space, $\mathcal{R}$, to $M_u(\mu^*)$ at μ^* is the linear space spanned by eigenvectors of $DR_\eta(\mu^*)$ (the derivative of R_η at μ^*) corresponding to eigenvalues of modulus > 1. It is called the space of "*relevant perturbations*". The space $\mathcal{I}$ of "*irrelevant perturbations*" is the tangent space to $M_s(\mu^*)$ and is spanned by eigenvectors of $DR_\eta(\mu^*)$

corresponding to eigenvalues of modulus < 1. The space $\mathcal{M}$ of "*marginal perturbations*" is spanned by eigenvectors of $DR_\eta(\mu^*)$ corresponding to eigenvalues of modulus 1. Generically $\mathcal{M}$ will be the tangent space, $\mathcal{C}$, to $M_{\text{fp}}(R_\eta, \mu^*)$, and, in a neighborhood of μ^*, each point in $M_{\text{fp}}(R_\eta, \mu^*)$ can be reached by applying a coordinate transformation to μ^*. However, it may happen that the dimension of $\mathcal{M}$ is larger than the one of $\mathcal{C}$. In that case, linear analysis is insufficient. It may happen that one can enlarge M_s, (or M_u, or both) by submanifolds of points which are driven towards (away from) μ^* with "asymptotically vanishing speed". This is precisely what appears to happen in the models introduced in Section 1.4 in four dimensions: $\dim \mathcal{M} = 2 = \dim \mathcal{C} + 1$; (moreover, $\dim \mathcal{R} = 1$). However, all fixed points appear to be scale-invariant Gaussian measures, and M_s can be enlarged by a one dimensional submanifold tangent to a direction in $\mathcal{M}$ at μ^*.

In the situation described here one expects logarithmic corrections to scaling laws.

[Another possibility compatible with $\dim \mathcal{M} > \dim \mathcal{C}$ is the appearance of a stable, periodic cycle. For the transformation R_η defined in (4.8)–(4.11) one should be able to rule out this possibility.]

Suppose now that R_η depends on a continuous parameter, δ, and that δ_0 is some "critical" value of δ such that

$$\begin{aligned} \dim \mathcal{M} &= \dim \mathcal{C} \ , \quad \text{for } \delta > \delta_0 \ , \\ \dim \mathcal{M} &> \dim \mathcal{C} \ , \quad \text{for } \delta = \delta_0 \ . \end{aligned} \tag{4.16}$$

Then δ_0 is a bifurcation point, and one expects the emergence of new fixed points (or periodic cycles) for $\delta < \delta_0$. In the study of the models mentioned above, it was proposed by Wilson and Fisher [519] to identify δ with the *dimension* d and to interpolate analytically in d.[2] The critical dimension, corresponding to δ_0, is 4, and above four dimensions the fixed points governing the critical behavior of those models are Gaussian, and $\eta = 0$. There are partial results towards showing that the "relevant" fixed points in dimension 4 are Gaussian, as well [5, 213, 15, 298]. This will be discussed in Chapters 6 and 15.

Next, we discuss how critical exponents are related to the spectrum of $DR_\eta(\mu^*)$, where R_η is the transformation defined in (4.9)–(4.12). We consider a simple case: In a neighborhood of μ^*, $M_{\text{fp}}(R_\eta, \mu^*)$ is obtained by applying suitable coordinate transformations in spin space to μ^*. By adopting some normalization condition which fixes the choice of coordinates we can project out the marginal directions associated with M_{fp}. We assume that, after this reduction, the tangent space at μ^* splits into a *one-dimensional* space of relevant perturbations and a *codimension-one* space of irrelevant perturbations (in particular, there are no further marginal perturbations). Taking smoothness properties of R_η in some neighborhood of μ^* for granted, we conclude that in some neighborhood of μ^* there exist a one-dimensional unstable and a codimension-one stable manifold passing through μ^*.

[2] Another possibility is to identify δ with the *range of the interaction.*

Next, let $\{\mu_\beta\}_{\beta>0}$ be a family of Gibbs measures of some spin system crossing the stable manifold, $M_s(\mu^*)$, transversally at some value β_c of the parameter β. We assume that, for all $\beta < \beta_c$, μ_β is extremal invariant, and that the inverse correlation length (or "mass", see Section 1.4), $m(\beta)$, is positive and continuous in β, with

$$m(\beta) \searrow 0 , \quad \text{as } \beta \nearrow \beta_c , \tag{4.17}$$

as discussed at the beginning of Chapter 3. (The class of all spin systems whose Gibbs states have these properties, for some given R_η and μ^*, is called a *universality class.*)

Now fix some number $m^* > 0$, and let $M(j, m^*)$ be the manifold of extremal translation-invariant probability measures, μ, on the space of spin configurations, φ, which have the property that

$$\int d\mu(\varphi)\, \varphi_0 \varphi_x \tag{4.18}$$

has exponential decay rate (mass) $L^{-j}m^*$, as $|x| \to \infty$. If the space M of measures on which R_η acts is chosen appropriately, $M(j, m^*)$ will typically be of codimension 1, and $M(\infty, m^*) = M_s(\mu^*)$, in some neighborhood of μ^*. Hence, for j large enough, $M_u(\mu^*)$ will typically cross $M(j, m^*)$ transversally at some point μ_j. We assume that $\{\mu_\beta\}_{\beta<\beta_c}$ crosses $M(j, m^*)$ transversally at a point μ_{β_j}, for large enough j — which is consistent with (4.17). Clearly the sequence $\{\beta_j\}$ converges to β_c, as $j \to \infty$. Furthermore, by the definition of R_η, see (4.8)–(4.11), Section 4.1, and the definition of $M(j, m^*)$,

$$R_\eta M(j, m^*) = M(j-1, m^*) , \tag{4.19}$$

for all j.

Let λ be the unique, simple eigenvalue of $DR_\eta(\mu^*)$ which is larger than 1. In a neighborhood of μ^*, $M_u(\mu^*)$ can be given a metric such that

$$\text{dist}\,(\mu_j, \mu^*)/\text{dist}\,(\mu_{j+1}, \mu^*) \to \lambda , \quad \text{as } j \to \infty , \tag{4.20}$$

as follows from (4.19). Thus if μ_{β_c} is sufficiently "close" to μ^* it follows from our assumptions on $\{\mu_\beta\}_{\beta>0}$ (see Figure 4.2) that

$$|\beta_j - \beta_c| \sim \lambda^{-j} , \quad \text{as } j \to \infty . \tag{4.21}$$

By the definition of $M(j, m^*)$,

$$m(\beta_j) = L^{-j}m^* . \tag{4.22}$$

Thus, if we set $t = \beta_c - \beta$ and $m(t) \equiv m(\beta)$, $\beta < \beta_c$, we obtain from (4.21) and (4.22)

$$m(t) \sim t^{\ln L/\ln\lambda} \qquad \text{as } t \to 0 . \tag{4.23}$$

In terms of critical exponents,

$$\nu = \ln L/\ln\lambda . \tag{4.24}$$

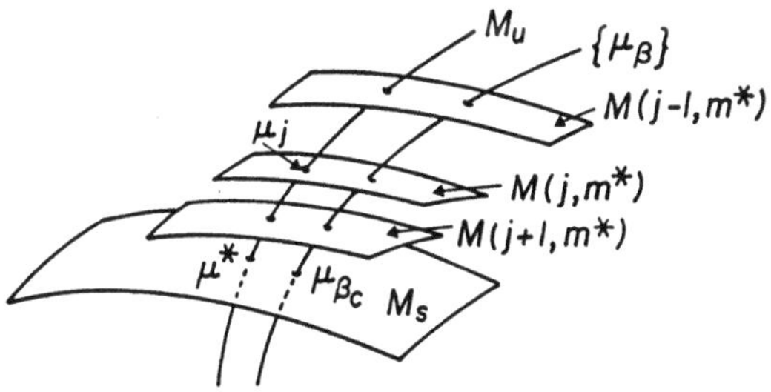

Fig. 4.2. Vicinity of a fixed point μ^*, with stable manifold M_s and unstable manifold M_u. Here $M(j, m^*)$ is the manifold of theories having mass $L^{-j}m^*$, and $\{\mu_\beta\}$ is some one-parameter family of models. The renormalization transformation R_η maps $M(j, m^*)$ onto $M(j-1, m^*)$

Thanks to relation (3.18), the exponent γ of the susceptibility is determined by η and ν.

From what we have said so far, or by inspecting Figures 4.1 and 4.2, it follows that the renormalization group transformation, R_η, drives every family $\{\mu_\beta\}$ of Gibbs states with the properties stated above (transversal crossing of $M(j, m^*)$, for j large enough; see Figure 4.2) towards the unstable manifold, $M_u(\mu^*)$,of the fixed point μ^*. More precisely, given a state $\tilde{\mu} \in M_u(\mu^*)$, there is a sequence of inverse temperatures $(\beta_n = \beta_n(\tilde{\mu}))_{n=1,2,3,\ldots}$ converging to β_c such that

$$\lim_{n\to\infty} R_\eta^n \mu_{\beta_n} = \tilde{\mu} \ . \tag{4.25}$$

Moreover, by (4.20)–(4.24),

$$|\beta_n - \beta_c| \sim \lambda^{-n} \ , \quad \text{as } n \to \infty \ , \tag{4.26}$$

with $\lambda = L^{1/\nu}$. Comparing (4.25) and (4.26) with (4.13), we see that $\tilde{\mu}$ corresponds to a scaling ($\equiv$ continuum) limit of μ_β, as $\beta \nearrow \beta_c$. Hence, $\tilde{\mu}$ corresponds to some "Euclidean field theory." In our example, *all* scaling limits of *all* models whose Gibbs states form a one-parameter family crossing $M_s(\mu^*)$ transversally (i.e. which belong to the universality class of μ^*) form a one-parameter family of "Euclidean field theories," $M_u(\mu^*)$. By (4.20)–(4.24), the exponent ν can be calculated by following the action of R_η along $M_u(\mu^*)$. This explains why critical exponents of lattice field theories and spin systems can be computed from continuum Euclidean field theory.

The fixed point, μ^*, is invariant under R_η and therefore scale-invariant; see (4.11). But scale-invariant Euclidean field theories are (by non-rigorous but plausible arguments) *conformal-invariant theories.* Therefore, fixed points of the renormalization-group transformation R_η, and hence *universality classes* of lattice models, correspond to *conformal field theories.*

The exponents η and ν can be directly calculated from the conformal theory if that theory is known explicitly, see (1.47), (3.27) and (3.29).

This concludes our general discussion of the basic renormalization-group strategy.

Remarks.

1) We find it useful to sketch the conjectural features of the renormalization group applied to three-dimensional, one-component lattice field theories and spin systems with short-range ferromagnetic interactions (e.g. $\lambda\varphi_3^4$, or the three-dimensional Ising model). A massless Gaussian measure, $d\mu_0$, which is specified by the properties

$$\begin{aligned}\int d\mu_0(\varphi)\varphi_x &= 0\,,\\ \int d\mu_0(\varphi)\varphi_0\varphi_x &\underset{|x|\to\infty}{\sim} |x|^{-1}\,,\end{aligned}$$

is a fixed point of R_η, with $\eta = 0$.

We study the action of $R_{\eta=0}$ on the space of measures invariant under $\varphi_x \mapsto -\varphi_x$, for all x. Then the unstable manifold near μ_0 is two-dimensional, its tangent space, $\mathcal{R}$, at μ_0 (relevant perturbations) is spanned by $\{\sum_x \varphi_x^2, \sum_x \varphi_x^4\}$. There is a marginal direction, i.e. $\mathcal{M}$ is one-dimensional, corresponding to the perturbation $\sum_x \varphi_x^6$. It is expected that $\mathcal{M}$ is tangent to $M_s(\mu_0)$ (triviality of $\lambda\varphi_3^6$-theory; see Chapter 5). One believes that there is a non-Gaussian fixed point, μ^*, of a renormalization group transformation R_η, with $\eta > 0$, which has a *stable manifold*, $M_s(\mu^*)$, intersecting $M_u(\mu_0)$ in a curve connecting μ_0 to μ^*. The fixed point μ^* is supposed to describe the universality class of critical $\lambda\varphi_3^4$ and the three-dimensional Ising model. Its unstable manifold is expected to be one-dimensional, and there are no non-trivial marginal perturbations of μ^*.

This picture has been verified, to a considerable extent, in the hierarchical approximation [342, 184]. If correct, this picture shows that μ^* can be constructed from φ^4 perturbations of μ_0.

The picture is believed to be *much simpler* in $d \geq 4$ dimensions: there is a unique fixed point — the Gaussian fixed point — and the unstable manifold is one-dimensional, with tangent space spanned by $\{\sum_x \varphi_x^2\}$ (massive free fields). This situation is studied in detail in Chapters 6 and 15, by using completely different (random-walk) methods.

In two dimensions ($d = 2$), the situation is very different. The unstable manifold of the Gaussian fixed point is known to be *infinite-dimensional*, see [260]. Interactions like $\sum_x P(\varphi_x)$, where P is a polynomial that is bounded from below, belong to the plane tangent to the unstable manifold at the Gaussian fixed point, as do interactions like $\sum_x e^{\alpha\varphi_x}$ or $\sum_x \cos(\alpha\varphi_x + \theta)$ for sufficiently small α. On the other hand, recent progress in two-dimensional conformal field theory has established the existence of infinitely many *non-Gaussian fixed points* (the A-series of minimal models [58, 104, 103, 105, 526]) describing multi-critical behavior More precisely, these investigations suggest that a non-Gaussian fixed point describing the critical behavior of Ising-type models can be constructed by studying perturbations of the Gaussian fixed point by a degree-4 polynomial ("φ^4 theory"), and that non-trivial fixed points describing tri-critical or quadri-critical behavior can be reached by perturbing the Gaussian fixed point by polynomial interactions of degree 6 or 8, respectively; etc. Note, however, that while the existence of these fixed points has been essentially established, the

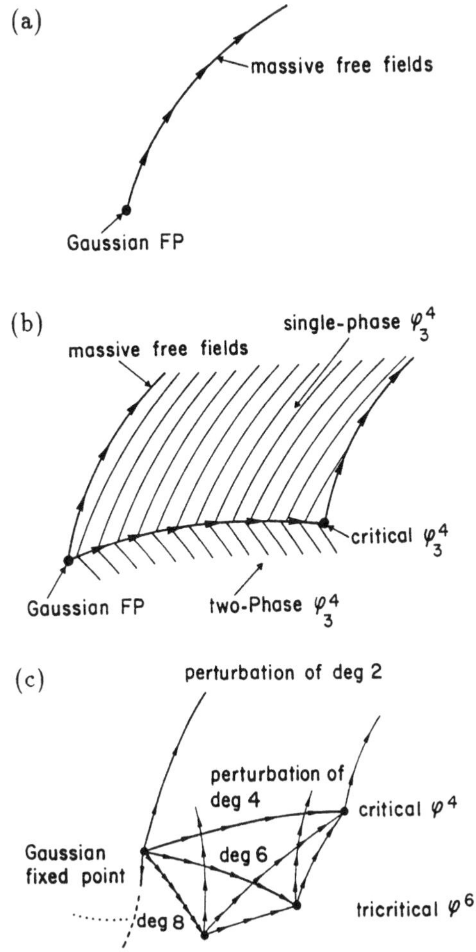

Fig. 4.3. Conjectured structure of the manifold of Euclidean Field Theories: **(a)** $d \geq 4$, **(b)** $d = 3$, **(c)** $d = 2$

precise structure of the renormalization flow between those fixed points (see Figure 4.3) is conjectural.

2) The ideas and concepts discussed here have other interesting applications to relativistic quantum field theory and statistical mechanics: As we have argued in Section 4.1, (4.8) through (4.13), one can use renormalization transformations, R_η, and their fixed points in order to construct the scaling limits, $G^*(x_1, \ldots, x_n)$, of the correlation functions of a spin system which, under general and explicit conditions [224, 225], can be shown to be the Euclidean Green's functions of a relativistic quantum field theory. So far, constructive quantum field theory has — in this language — been mostly concerned with the analysis of *Gaussian fixed points* of the transformations R_η, with $\eta = 0$, and the action of $R_{\eta=0}$ in a small neighborhood of those fixed points. For a study of non-Gaussian fixed points,see however [242, 249, 182, 342].

3) Another application of those ideas concerns the phenomenon of *asymptotic symmetry enhancement.* One example of this phenomenon is found in the fact that in many models the scaling limits, $G^*(x_1, \dots, x_n)$, of the correlation functions on some spin system are invariant under *all* simultaneous *Euclidean motions* of their arguments, although the functions $G_\theta(x_1, \dots, x_n)$ are only invariant under translations by an arbitrary vector a $\in \mathbb{Z}^d_{\theta^{-1}}$. Other examples concern the generation of internal symmetries in the scaling limit. See e.g. [224, 225, 226] for such examples. (Symmetry enhancement arises whenever a fixed point, μ^*, and the *marginal* and *relevant* perturbations of μ^* have a large, "accidental" symmetry group.)

4) Renormalization group methods can also be applied to *dynamics*:Let φ_t denote a smooth flow on a finite dimensional manifold, M. Consider the following mapping on the space of all such flows on M:

$$R_{\theta,\Lambda} : \varphi_t \to (R_{\theta,\Lambda}\varphi)_t \equiv \Lambda^{-1} \circ \varphi_{\theta t} \circ \Lambda \, , \tag{4.27}$$

where Λ is a smooth mapping from M into M, (a coordinate transformation). The mapping $R_{\theta,\Lambda}$ is the analogue of the transformation R_η defined in (4.9)–(4.12). When time is discrete, i.e. $t = n = 1, 2, 3 \dots$, and

$$\varphi_t = \varphi^n \, , \tag{4.28}$$

for some mapping φ from M into M, one would study, for example,

$$R_\Lambda : \varphi \to R_\Lambda \varphi = \Lambda^{-1} \circ \varphi \circ \varphi \circ \Lambda \, . \tag{4.29}$$

This is the Feigenbaum map. It poses very interesting, mathematical problems and serves to understand phenomena like the period doubling bifurcations and the onset of turbulence; see [180, 181, 349, 112]. (This is one among few examples where non-trivial fixed points have been constructed.)

4.3 A brief sketch of the tree expansion

In this section[3] we briefly sketch a version of the renormalization group, originally proposed by Kogut and Wilson [516, 517, 519, 520], which has been developed by Gallavotti and Nicolò [235, 236, 232, 182, 183, 186] and others [188, 189, 191, 192, 187]. It is somewhat different in spirit from that in Sections 4.1 and 4.2. One defines renormalization-group transformations on some space of Hamiltonians (or, in the language of Euclidean field theory, effective actions) rather than on a space of measures. The renormalization-group transformation maps an effective action on a momentum scale $|p| \approx \gamma^k$ to an effective action on a scale γ^{k-1}. This yields an exact recursion relation for the effective actions on arbitrary momentum scales. This is not a particularly convenient

[3]Useful references for this section are Gallavotti [232] and Felder [183].

starting point for a non-perturbative renormalization group analysis of statistical or field theory models, but it is an extremely convenient starting point for a perturbative analysis, in the form of expansions in running coupling constants [188, 189, 191, 192, 187, 242, 249, 244, 245, 243, 246]. It provides a powerful tool to perform explicit calculations and to explore the vicinity of the Gaussian fixed point, μ_0, and thereby construct (at least perturbatively) various "renormalizable" and "non-renormalizable" [32, 33, 182, 186] models of Euclidean field theory.

We consider a free field, φ_x, $x \in \mathbb{R}^d$, with Gaussian distribution $d\mu_0(\varphi)$ specified by its mean and covariance,[4]

$$\begin{aligned} \int d\mu_0(\varphi)\varphi_x &= 0\,, \\ \int d\mu_0(\varphi)\varphi_x\varphi_y &= C(x-y)\,, \end{aligned} \tag{4.30}$$

where

$$C(x-y) = \frac{1}{(2\pi)^d}\int p^{-2}e^{ip(x-y)}d^dp \tag{4.31}$$

Let

$$\begin{aligned} \widehat{C}^j(p) &\equiv \frac{1}{p^2}\{e^{-(p^2/\gamma^{2j})} - e^{-(p^2/\gamma^{2(j-1)})}\} > 0 \\ C^j(x-y) &\equiv \frac{1}{(2\pi)^d}\int \widehat{C}^j(p)e^{ip(x-y)}d^dp\,, \end{aligned} \tag{4.32}$$

where $\gamma > 1$ is some fixed scale factor. We can decompose φ into fields $\varphi^{(j)}$, localized in slices of momentum space, as follows.

$$\varphi_x = \sum_{j=-\infty}^{\infty} \varphi_x^{(j)}\,, \tag{4.33}$$

where $\varphi^{(j)}$ has a Gaussian distribution $d\mu_0^j$ with mean 0 and covariance C^j. It is easy to check that

$$C^j(x-y) = \gamma^{(d-2)j}C^0(\gamma^j(x-y))\,, \tag{4.34}$$

so that $\varphi_x^{(j)}$ and $\gamma^{(d-2)j/2}\varphi^{(0)}_{\gamma^j x}$ have identical distribution. Let $\partial^m = \prod_\mu(\partial/\partial x^\mu)^{m_\mu}$, $|m| = \sum m_\mu$. By (4.32) and (4.34)

$$|\partial^m C^j(x)| \leq c(m)\gamma^{(d-2+|m|)j}e^{-M\gamma^j|x|} \tag{4.35}$$

for some constants $c(m) < \infty$ and $M > 0$.

We now introduce a model of Euclidean field theory with an ultraviolet cutoff on scale $N < \infty$. We set

[4] In this section, we must assume some familiarity with free field theory (e.g. Wick ordering) and constructive field theory, see [260, 462] for background.

$$\varphi^{\leq N} = \sum_{j=-\infty}^{N} \varphi^{(j)} \tag{4.36}$$

and define a self-interaction, $V^{(N)}(\varphi^{\leq N})$, by

$$V^{(N)}(\varphi^{\leq N}) = \sum_{\alpha} \gamma^{-\Delta(\alpha)N} \lambda_\alpha(N) M_\alpha(\varphi^{\leq N}) \ , \tag{4.37}$$

where α ranges over the set $\{2', 2, 4, \ldots, 2t\}$, and $\lambda_{2t}(N) > 0$, $\lambda_{2'}(N) > -\frac{1}{2}$. Moreover,

$$\begin{aligned} M_{2'}(\varphi) &= \int : (\nabla \varphi_x)^2 : d^d x \\ M_2(\varphi) &= \int : \varphi_x^2 : d^d x \\ &\vdots \\ M_{2t}(\varphi) &= \int : \varphi_x^{2t} : d^2 x \ , \end{aligned} \tag{4.38}$$

where $: - :$ indicates the usual Wick order with respect to $d\mu_0$; see [260, 462]. Finally, $\Delta(\alpha)$ denotes the scaling dimension of $M_\alpha(\varphi)$, i.e.

$$\Delta(2') = 0 \ , \quad \Delta(2) = -2 \quad \Delta(4) = d - 4, \ldots, \Delta(2t) = (d-2)t - d \ . \tag{4.39}$$

If $\Delta(\alpha) < 0$ $M_\alpha(\varphi)$ is a relevant perturbation of $d\mu_0$; if $\Delta(\alpha) = 0$ $M_\alpha(\varphi)$ is marginal, and if $\Delta(\alpha) > 0$ $M_\alpha(\varphi)$ is irrelevant. The factors $\gamma^{-\Delta(\alpha)N}$ on the r.h.s. of (4.37) are chosen so that the coefficients $\lambda_\alpha(N)$ are dimensionless.

The goal is to make sense of the measures

$$d\mu^{(N)}(\varphi) = \frac{e^{-V^{(N)}(\varphi^{\leq N})}}{Z^{(N)}} d\mu_0(\varphi) \ , \tag{4.40}$$

with $Z^{(N)}$ chosen such that $\int d\mu^{(N)}(\varphi) = 1$, as N becomes large. As long as $N < \infty$ and $\lambda_{2t}(N) > 0$, $\lambda_{2'}(N) > -\frac{1}{2}$, it is easy to see that (4.40) is well defined; see [260]. In order to study the behavior, as $N \to \infty$, it is convenient to analyze, following Wilson [239], the *effective potentials*. They are recursively defined by

$$V^{(k)}(\varphi^{\leq k}) = -\ln E_{k+1} \exp\left[-V^{(k+1)}(\varphi^{\leq k+1})\right] \ , \tag{4.41}$$

with

$$E_k F(\varphi^{\leq k}) \equiv \int d\mu_0^k(\varphi^{(k)}) F(\varphi^{\leq k-1} + \varphi^{(k)}) \ , \tag{4.42}$$

and $V^{(N)}(\varphi^{\leq N})$ given by (4.37). The knowledge of the effective potentials $V^{(k)}(\varphi^{\leq k})$, $k = 0, 1, 2\ldots$ is equivalent to knowing $d\mu^{(N)}(\varphi)$. See e.g. [232] for details.

One may now expand $V^{(k)}$ in powers of $\varphi^{\leq k}$:

$$V^{(k)}(\varphi^{\leq k}) = \sum_{n=0}^{\infty} \int V_{2n}^{(k)}(x_2, \ldots, x_{2n}) : \varphi_{x_1}^{\leq k} \ldots \varphi_{x_{2n}}^{\leq k} : \tag{4.43}$$

The kernels $V_{2n}^{(k)}$ are Euclidean-invariant distributions whose behavior for different values of n and k we wish to study. The goal is to choose the coefficients $\lambda_\alpha(N)$ (with $\lambda_{2t}(N) > 0$, $\lambda_{2'}(N) > -\frac{1}{2}$) to depend on N in such a way that the kernels $V_{2n}^{(k)}$ remain well defined when the ultraviolet cutoff $N \to \infty$, for all finite n and k. [We shall not analyze the convergence of the series on the r.h.s. of (4.43), although that is necessary if one wants to prove a non-perturbative result.]

A basic idea behind the renormalization group strategy, already sketched in the last two sections, is that the flow

$$\ldots \to V^{(k+1)} \to V^{(k)} \to \ldots \tag{4.44}$$

can be parametrized by finitely many parameters, corresponding to the *relevant* and *marginal operators*. The operator $M_{2'}$ is always marginal ($\Delta(2') = 0$), while $M_2, \ldots, M_{2t}$ are perturbatively relevant, i.e. $\Delta(\alpha) < 0$, for $d > 2(t+1)/t$. M_{2t} becomes marginal ($\Delta(2t) = 0$), when d reaches $2t/t-1$. [An interpretation of these facts in terms of intersection properties of simple random walk can be found in Chapter 5] We thus want to parametrize the flow (4.44), with $V^{(k)}$ given in terms of $V^{(k+1)}$ by (4.41), by a finite number of *running coupling constants*, $\lambda_\alpha(k)$, $\alpha = 2', 2, \ldots, 2t$, of relevant, marginal and possibly a finite number of irrelevant operators. They are defined by the equation

$$\sum_\alpha \gamma^{-\Delta(\alpha)k} \lambda_\alpha(k) M_\alpha(\varphi^{\leq k}) = L_k V^{(k)}(\varphi^{\leq k}) \ , \tag{4.45}$$

where L_k is a projection onto the linear space spanned by $(M_\alpha)_{\alpha=2',2,\ldots,2t}$. More precisely,

$$\begin{aligned} L_k \int V(x_1, \ldots, x_{2n}) : \prod_{i=1}^{2n} \varphi_{x_i}^{\leq k} : \prod_{j=1}^{2n} d^d x_j \\ = \begin{cases} 0 \ , & \text{for } n > t \ , \\ \widehat{V}(0) M_{2n}(\varphi^{\leq k}) \ , & \text{for } 2 < n \leq t \ ; \end{cases} \end{aligned} \tag{4.46}$$

($V \mapsto \widehat{V}$ denotes Fourier transformation), and

$$\begin{aligned} L_k \int V(x_1, x_2) : \varphi_{x_1}^{\leq k} \varphi_{x_2}^{\leq k} : d^d x_1 d^d x_2 \\ = \widehat{C}(0) M_2(\varphi^{\leq k}) + \frac{\partial}{\partial p^2} \widehat{V}(p^2) \Big|_{p=0} M_{2'}(\varphi^{\leq k}) \ . \end{aligned} \tag{4.47}$$

It is sometimes necessary to include some running coupling constants of irrelevant operators, because these can become relevant at a non-Gaussian fixed point of (4.41). Therefore we shall not specify the value of t (as a function of d), yet.

From (4.41) and (4.45)–(4.47) one easily derives a recursion relation for the running coupling constants, at least in the form of a formal power series

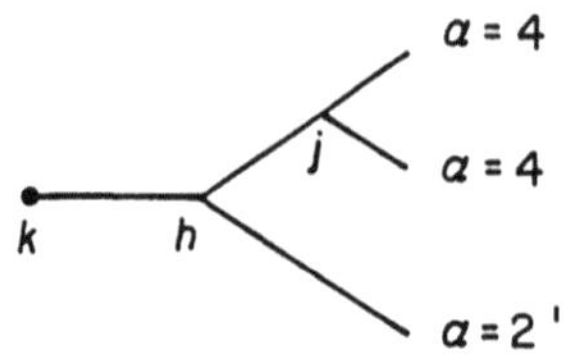

Fig. 4.4. Tree labelling the term (4.50)

expansion. This is one version of the Gallavotti-Nicolò tree expansion which shall now be sketched.

Let E_k^T be the truncated (connected) expectation corresponding to (4.42), i.e.

$$E_k^T F \equiv E_k F \ , E_k^T(F,G) \equiv E_k F \cdot G - (E_k F)(E_k G) \ , \tag{4.48}$$

etc.; see e.g. [260]. From the recursion relation (4.41) and (4.45) we obtain

$$\begin{aligned} V^{(k)} &= \sum_\alpha \gamma^{-\Delta(\alpha)k} \lambda_\alpha(k) M_\alpha(\varphi^{\leq k}) + \\ &+ (1-L_k) \sum_{s=1}^{\infty} \frac{(-1)^s}{s!} E_{k+1}^T(V^{(k+1)}, \dots, V^{(k+1)}) \ . \end{aligned} \tag{4.49}$$

[We have used that

$$\ln E_k e^F = \sum_{s=1}^{\infty} \frac{(-1)^s}{s!} E_k^T(\underbrace{F, \dots, F}_{s \text{ times}}) \ .]$$

In (4.49), the running coupling constants are treated as unknowns. The term $\propto (1-L_k)$ on the r.h.s. is irrelevant. [It will therefore have good "power-counting" properties in the ultraviolet.]

We now analyze (4.49) by iterating it. Iteration yields terms like ($k < h < j$)

$$\begin{aligned} &\gamma^{-2(d-4)j} \lambda_{2'}(j) \lambda_4(j)^2 E_{k+1} E_{k+2} \dots E_{h-1}(1-L_{h-1}) \\ &\times E_h^T \Big(E_{h+1} \dots E_{j-1}(1-L_{j-1}) E_j^T(M_4, M_4), M_{2'} \Big) \ , \end{aligned} \tag{4.50}$$

where we have left out the arguments $\varphi^{\leq j}$ in the M_α's and have used that $E_k F = E_k^T F$, $E_k L_k = L_{k-1} E_k$ and $L_k^2 = L_k$. The term (4.50) can be labelled by the tree of Figure 4.4.

In fact, *all* terms in the iteration of (4.49) can be labelled by trees, whence the name "tree expansion": Iteration of (4.49) yields the following tree expansion.

$$V^{(k)} = L_k V^{(k)} + \sum_{\substack{\theta, \underline{h}, \underline{\alpha} \\ h_{v_0} = k}} \frac{1}{n(\theta)} V(\theta, \underline{h}, \underline{\alpha}) \prod_{i \text{ endpoint of } \theta} \lambda_{\alpha_i}(k_i) \ , \tag{4.51}$$

where θ is a rooted ordered tree, $\underline{h} = \left(h_v\right)_{v \in \mathcal{V}(\theta)}$ are integers associated with the vertices (branching points) of θ, $v \in \mathcal{V}(\theta)$, with $h_{v_1} < h_{v_2}$ if $v_1 < v_2$ in the obvious ordering of $\mathcal{V}(\theta)$. The endpoints, i, of θ bear labels $\alpha_i \in \{2', 2, 4, \ldots, 2t\}$, and $h_i = h_{v(i)}$ is the integer associated with the vertex $v(i)$ to which the endpoint i is connected. The root, v_0, of θ bears the integer k. The sum in (4.51) extends only over *non-trivial, rooted ordered trees*, the trivial tree being the one without branches, assignments of integers $\underline{h}$ to the vertices of a given, non-trivial tree and endpoint labels $\underline{\alpha}$. The combinatorial factor, $n(\theta)$, is given by

$$n(\theta) = \prod_{v \in \mathcal{V}(\theta)} s_v! \;, \tag{4.52}$$

where $s_v + 1$ is the order of the vertex v.

The coefficients $V(\theta, \underline{h}, \underline{d})$ in the expansion (4.51) can be calculated recursively: For the trivial tree, θ_0,

$$V(\theta_0, h_{v_0}, \alpha) = \gamma^{-\Delta(\alpha) h_{v_0}} M_\alpha(\varphi^{\leq h_{v_0}}) \;. \tag{4.53}$$

If the first vertex of θ at which θ branches is v_1, with subtrees $\theta_1, \ldots, \theta_{s_{v_1}}$ growing out of v_1 then

$$V(\theta, \underline{h}, \underline{\alpha}) = E_{h_{v_0}} \cdots E_{h_{v_1}-1} E^T_{h_{v_1}} \left(V(\theta_1, \underline{h}_1, \underline{\alpha}_1), \ldots, V(\theta_{s_{V_1}}, \underline{h}_{s_{v_1}}, \underline{\alpha}_{s_{v_1}})\right) \;, \tag{4.54}$$

where $(\underline{h}_i, \underline{\alpha}_i)$ is the restriction of $(\underline{h}, \underline{\alpha})$ to θ_i.

We observe that (4.51) is an expansion of $V^{(k)}$ in powers of running coupling constants $\lambda_\alpha(h)$, $h \geq k$. $V^{(k)}$ depends on the ultraviolet cutoff N in the way that $h_v \leq N$, for all $v \in \mathcal{V}(\theta)$, and

$$V^{(N)} = L_N V^{(N)} \;. \tag{4.55}$$

The fundamental property of (4.51) is that, order by order in the running coupling constants, $\lambda_\alpha(h)$,the sums over momentum scales $\underline{h}$ on the r.h.s. of (4.51) are convergent. This is proven inductively, using the recursion formula (4.54). The key fact to be used in the proof is that the projection $(1 - L_{h_{v_1}-1})$ on the r.h.s. of equation (4.54) projects the operator

$$E^T_{h_{v_1}} (V(\theta_1, \underline{h}_1, \underline{\alpha}_1), \ldots, V(\theta_{s_{v_1}}, \underline{h}_{s_{v_1}}, \alpha_{s_{v_1}}))$$

onto the linear space of irrelevant operators. The image is therefore an operator with "good ultraviolet power-counting"; (the operators $(1 - L_{h_{v_1}-1})$ play a role analogous to the Taylor subtraction operations in the BPHZ formalism which render Feynman amplitudes convergent). In order to make this more precise, one expands $V(\theta, \underline{h}, \underline{\alpha})$ in powers of the field $\varphi^{\leq h_{v_0}}$:

$$V(\theta, \underline{h}, \underline{\alpha}) = \sum_n \int V(\theta, \underline{h}, \underline{\alpha}; x_1, \ldots, x_{2n}) : \prod_{i=1}^{2n} \varphi_{x_i}^{\leq h_{v_0}} : \prod_{j=1}^{2n} d^d x_j \tag{4.56}$$

The kernels $V(\theta, \underline{h}, \underline{\alpha}; \ldots)$ can be calculated in terms of sums over *Feynman amplitudes:*

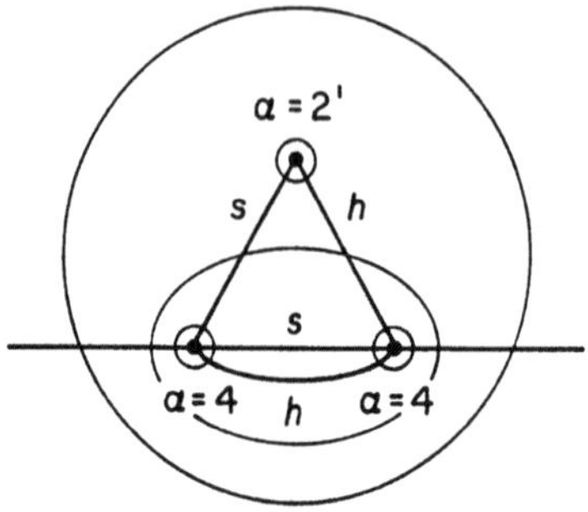

Fig. 4.5. A graph compatible with the tree of Figure 4.4. Each internal line is either hard (h) or soft (s).

$$V(\theta,\underline{h},\underline{\alpha};x_1,\ldots,x_{2n}) = \sum_{\substack{G \text{ compatible} \\ \text{with } \theta,\underline{\alpha}}} V_G(\theta,\underline{h},\underline{\alpha};x_1,\ldots,x_{2n}) \; , \tag{4.57}$$

where the sum extends over all Feynman graphs, G, with $2n$ external lines and l vertices of type $\alpha_1,\ldots,\alpha_l$, where l is the number of endpoints of the tree θ, which are compatible with $\theta,\underline{\alpha}$. A graph G is compatible with $\theta,\underline{\alpha}$ if and only if

(i) for $i=1,\ldots,l$, the ith vertex corresponds to the operator M_α and hence is of order 2, for $\alpha_i = 2', 2$, and of order $2r$, for $\alpha_i = 2r$, with $r \le t$;

(ii) G has two kinds of internal lines: *hard* lines (generated when one evaluates E_h^T-expectations, $h \in \underline{h}$) and *soft* lines (generated by Wick reordering). Hard lines correspond to propagators C^h, defined in (4.32), which are well localized in momentum space and are estimated as in (4.35). Soft lines correspond to propagators $\sum_{j\le k-1} C^j$. If one draws a circle around each set of vertices belonging to a subtree, θ_i, of θ then hard lines connect all such circles contained in an arbitrary larger circle. [This follows inductively from (4.54) by using the definition of connected expectations, $E^T_{h_{v_1}}$.]

A graph compatible with the tree of Figure 4.4 is shown in Figure 4.5 The behavior of Feynman amplitudes corresponding to compatible graphs can be analyzed inductively, using (4.35), and this will prove finiteness of the tree expansion (4.51). For details see [232, 186, 183].

We now return to our basic formulas (4.41) and (4.51): Applying L_k to both sides of the equation

$$V^{(k)}(\varphi^{\le k}) = -\sum_{s=1}^{\infty} \frac{(-1)^s}{s!} E^T_{k+1}(\underbrace{V^{(k+1)},\ldots,V^{(k+1)}}_{s \text{ times}}) \tag{4.58}$$

and then insert (4.51) (with k replaced by $k+1$) on the r.h.s. of (4.58), we obtain a recursion relation for the running coupling constants:

$$\lambda_\alpha(k) = \gamma^{-\Delta(\alpha)}\lambda_\alpha(k+1) - \sum_{\substack{\theta,\underline{h},\underline{\alpha} \\ h_{v_0}=k \\ h_{v_1}=k+1}} \frac{1}{n(\theta)}\beta_\alpha(\theta,\underline{h},\underline{\alpha})\prod_i \lambda_{\alpha_i}(h_i)\ , \tag{4.59}$$

where the sum extends over all non-trivial trees, and the coefficients $\beta_\alpha(\theta, \underline{h}, \underline{\alpha})$ are defined by

$$\sum_\alpha \gamma^{-\Delta(\alpha)k}\beta_\alpha(\theta,\underline{h},\alpha)M_\alpha(\varphi^{\leq k})$$
$$= L_k E_{k+1}\Big(V(\theta_1,\underline{h}_1,\underline{\alpha}_1),\ldots,V(\theta_{s_{v_1}},\underline{h}_{s_{v_1}},\underline{\alpha}_{s_{v_1}})\Big)\ .$$

The product, $\prod_i \lambda_{\alpha_i}(h_i)$, on the r.h.s. of (4.59) is over all endpoints of θ.

The sum on the r.h.s of (4.59) defines, in perturbation theory in the running coupling constants, the *beta functional.* Equation (4.59) reduces the study of the renormalization group recursion formula to the study of the flow of running coupling constant determined by (4.59).

The basic estimate on the coefficients $\beta_\alpha(\theta, \underline{h}, \underline{\alpha})$ required to show that (4.59) makes sense as a termwise well-defined formal power series in the running coupling constants is the following:

$$|\beta(\theta,\underline{h},\underline{\alpha})| \leq C_0^l n(\theta) \prod_{\substack{v\in\mathcal{V}(\theta) \\ v>v_1}} \gamma^{-(h_v-h_v')/2}((t-1)l)!\ , \tag{4.60}$$

where $l \geq 2$ is the number of endpoints of θ, C_0 is some finite constant, v_1 is the earliest branching vertex in θ and, for $v \in \mathcal{V}(\theta)$, v' denotes the branching vertex in $\mathcal{V}(\theta)$ immediately preceding v. Finally, $n(\theta)$ is defined in (4.52).

The estimate (4.60) is valid for renormalizable and non-renormalizable theories $[d \geq 2t/(t-1)]$, but not for super-renormalizable theories.

Estimate (4.60) permits us to resum (4.59) to obtain the termwise well-defined formal expansion

$$\lambda_\alpha(k) = \gamma^{-\Delta(\alpha)}\lambda_\alpha(k+1) + \sum_{\underline{n}:|\underline{n}|\geq 2} \beta_{\alpha,\underline{n}}(k)\underline{\lambda}^{\underline{n}}\ , \tag{4.61}$$

with $\underline{n} = (n_\alpha(h))$, $n_\alpha(h) = 0, 1, 2, \ldots$, for every (α, h), and $\underline{\lambda}^{\underline{n}} = \prod_{\alpha,h} \lambda_\alpha(h)^{n_\alpha(h)}$. One has the bound

$$\Big|\sum_{|\underline{n}|=n} \beta_{\alpha,\underline{n}}(k)\underline{\lambda}^{\underline{n}}\Big| \leq C^n \|\lambda\|_\infty^n((t-1)n)!\ , \tag{4.62}$$

with $\|\lambda\|_\infty = \sup|\lambda_\alpha(h)|$. For details we must refer the reader to [232, 235, 183].

Equations (4.51) and (4.61) are a very elegant starting point for studying renormalization theory and the renormalization group perturbatively and performing explicit calculations. But it yields rigorous, non-perturbative results only for models for which perturbation theory in the running coupling constants converges; see [305, 444, 182, 244, 245, 243, 246, 249, 191, 364]. In other cases, the theory sketched here requires non-trivial modifications: see e.g. the work of

Glimm and Jaffe [254] and Gallavotti and collaborators [230, 231, 59, 60] on φ_3^4, and the work of Balaban [34, 36, 37, 39, 40, 42, 43, 44, 45, 38, 46, 47] and Federbush [171, 51, 178, 172, 175, 176, 173] on Yang-Mills theory.

In the standard examples, like $\lambda\varphi_4^4$-theory, the formalism developed above confirms what earlier approaches predicted. Non-perturbative results for $\lambda\varphi_d^4$, $d \geq 4$ are discussed in Chapters 6 and 15.

In the language of equation (4.61), a fixed point of the renormalization group transformation (4.41) (or (4.51)) corresponds to a fixed point, $\underline{\lambda} \equiv (\lambda_2^*, \lambda_2^*, \lambda_4^*, \dots, \lambda_{2t}^*)$ of the flow equation (4.61). The unstable manifold of that fixed point can be parametrized by solutions,

$$\lambda_\alpha(k) = \lambda_\alpha^* + \delta\lambda_\alpha(k)$$

of (4.61), with $\delta\lambda_\alpha(k)$ small. Knowing such a solution, one can use (4.51) to perturbatively calculate $V^{(k)}$ on all momentum scales k. Critical exponents can be calculated from the behavior of $\delta\lambda_\alpha(k)$ as a function of k. For example, ν can be determined from $\delta\lambda_2(k)$, $k = -1, -2, -3, \dots$.

Renormalized perturbation theory (e.g. for $d = 4$, $t = 2$) is recovered by solving (4.61) under the condition that $\lambda_2(0) = m^2$, $\lambda_4(0) = \lambda_{\text{ren}}$ (renormalized coupling constant). See [235, 188, 189, 191, 192, 187].

4.4 Rigorous uses of block spin transformations

The first mathematically rigorous analysis of a specific example to which the renormalization group strategy outlined in the previous sections can be applied is the one by Bleher and Sinai [67, 68], who analyzed Dyson's hierarchical model. The Hamilton function of this model is chosen in such a way that the renormalization group transformations can be reduced to non-linear transformations acting on some space of densities, f, of the single-spin distribution,

$$d\lambda(\varphi) = f(\varphi)d\varphi\ .$$

The work of Bleher and Sinai was reconsidered and extended in [111, 230, 239, 240, 342, 184] and the references cited there. The study of the hierarchical model had a stimulating influence on the development of the probabilistic approach to the renormalization group, initiated by Jona-Lasinio and his colleagues in Rome [317, 233, 106] and continued by Sinai and Dobrushin (see [467, 468, 318, 468] and the references given there). It was Gallavotti and collaborators [59, 60] who first applied the renormalization group method to (the ultraviolet problem in) *constructive quantum field theory* in a systematic and transparent way, although ideas and techniques related to it — and developed *independently* — can already be found in work of Glimm and Jaffe [254, 253]. These applications concern the construction of the $\lambda\varphi^4$ model in the continuum limit in three dimensions. [This problem is equivalent to the study of a renormalization group transformation analogous to R_η in the vicinity of a *Gaussian* fixed point.] The work in [59, 60] motivated further applications to constructive (gauge) quantum field theory, notably by Balaban [33, 34, 37, 38, 39, 41, 40, 42, 43, 44, 45] and Battle, Federbush and collaborators [49, 50, 48, 171, 174, 51, 178, 172, 175, 176, 173], and to statistical mechanics [239]. These developments have evolved towards a rigorous mathematical theory of renormalization group transformations in the vicinity of Gaussian fixed points. Among systems whose critical behavior is now understood rigorously are the following ones: Dipole gases in $d \geq 2$ dimensions [241, 382, 95]; weakly coupled $\lambda\varphi_d^4$ in $d \geq 4$ dimensions [247, 190, 244, 192, 290, 298]; hierarchical model approximations to large-N $|\varphi|_3^4$ [242, 249], to non-linear σ-models in two dimensions [248, 397], and to the three-dimensional Ising model [342, 184]. A looser interpretation of the renormalization group strategy partially motivated the work in [224, 225, 227, 11].

First applications of renormalization group methods to *dynamics* were made in [180, 181], although the idea to use them in the study of dynamics is certainly older; see e.g. [106].

All the work quoted here involves very intricate analytical and combinatorial methods and can therefore not be sketched here.

In the following chapter we outline another much more special but quite successful approach to critical phenomena which gives rather good results for the models discussed in these notes, near Gaussian fixed points, [5, 213]. It was inspired by a formalism first developed in [494] and made rigorous in [92, 96, 97, 213] relating the theory of classical spin systems to the theory of *random walks*. A related, slightly prior approach, due to Aizenman, may be found in [5].

But mathematically rigorous results on critical phenomena in equilibrium statistical mechanics in dimensions $d < 4$ still do not nearly measure up to the practical successes of the renormalization group. This ought to be a challenge! There has been tremendous progress, though, in understanding critical phenomena in two-dimensional statistical systems thanks to recent developments in two-dimensional conformal field theory [195, 383, 455, 376, 58, 209, 104, 103, 526, 105, 139, 140, 141].

5. Random walks as Euclidean field theory (EFT)

5.1 Definition of the model and statement of results

In this chapter we study a classic problem of probability theory — the intersection properties of simple random walks — using a rigorous blend of perturbation theory and renormalization-group arguments. Aside from its intrinsic mathematical interest, this problem played a key role in motivating the field-theoretic developments described in the remainder of this book, culminating in the triviality theorem for φ^4 field theory and Ising models in dimension $d > 4$. The logic of the field-theoretic arguments (described in more detail in Chapter 6) is roughly the following:

1) An *identity* is derived which represents a lattice field theory in terms of *interacting* (*"non-simple"*) random walks.

2) *Inequalities* are derived which bound certain connected correlation functions of the field theory (in particular, the dimensionless renormalized coupling constant g) in terms of the *intersection properties* of these "field-theoretic" random walks.

3) The intersection properties of *simple* random walks are used as *intuition* to motivate *conjectures* for the intersection properties of "field-theoretic" random walks. (These conjectures must, however, be proven by different methods.)

Thus, while the results of this chapter are not *logically* necessary for field theory, they serve as a valuable "warm-up problem" and as a guide for the more complicated field-theoretic analysis.

In Chapter 3 we learned that passing to the scaling limit is equivalent to approaching a critical point of the lattice theory in such a way that the mass, in physical units (e.g. inverse centimeters), stays fixed. Now the scaling limit of simple random walks on the lattice, obtained by letting the mass m tend to zero in lattice units, is Brownian motion [463, Section 17]. The intersection properties of Brownian paths were determined in the 1940's and 1950's by Dvoretsky, Erdös and Kakutani, using methods of capacity theory. The result is: Two independent Brownian motions in $\mathbb{R}^d$, with distinct starting points, have *zero* probability of intersection if $d \geq 4$ [327, 158] and *nonzero* probability of intersection if $d < 4$ [366, 158].[1] It is therefore reasonable to *conjecture* that a similar result holds for the intersection properties of simple random walks on the lattice, in the limit $m \to 0$. Unfortunately, we do not know any way of deducing such a lattice result as a corollary of the continuum result (or vice versa); but in this chapter we shall give a direct proof of the lattice result.

Let us recall from Section 1.6 that the weight function for simple random walks is

$$\varrho^{[n]}(\omega_1, \ldots, \omega_n) = \prod_{i=1}^{n} \varrho(\omega_i) , \tag{5.1}$$

where

$$\varrho(\omega) = J^{|\omega|} , \quad 0 < J < \frac{1}{2d} . \tag{5.2}$$

The two-point function is

$$K(x, y) = \sum_{\omega:\, x \to y} \varrho(\omega) . \tag{5.3}$$

The parameter playing the role of a mass is given by

$$m^2 = J^{-1} - 2d , \tag{5.4}$$

while the analogue of the susceptibility is

$$\chi = \sum_{y} K(x, y) = (1 - 2dJ)^{-1} = J^{-1} m^{-2} . \tag{5.5}$$

Now let $p_m(x_1, y_1)$ be the *probability that two independent walks, ω_1 and ω_2, starting at points x_1 and y_1, intersect somewhere*,

$$p_m(x_1, y_1) = \chi^{-2} \sum_{x_2, y_2} \sum_{\substack{\omega_1:\, x_1 \to x_2 \\ \omega_2:\, y_1 \to y_2}} \varrho(\omega_1) \varrho(\omega_2)\, \mathrm{I}(\omega_1 \cap \omega_2 \neq \varnothing) ,$$

and let p_m be the "*average intersection probability*"

$$p_m \equiv m^d \sum_{y_1 \in \mathbb{Z}^d} p_m(x_1, y_1) . \tag{5.6}$$

Since $p_m(x_1, y_1)$ is a pure number (it is a *probability*), and m is proportional to the lattice spacing in physical units, p_m is a *dimensionless quantity* which corresponds to the physical coupling constant g introduced in (1.45).[2]

[1] This result can be predicted heuristically by recalling [496] that Brownian paths have Hausdorff dimension $d_H = 2$. Now, two objects of dimension d_H embedded in $\mathbb{R}^d$ should "generically" have nonempty intersection if $2d_H > d$, and empty intersection if $2d_H < d$. The borderline case $2d_H = d$ is more delicate; for Brownian paths it turns out that the Hausdorff dimension is really "infinitesimally less than 2" (see [367, 108] for $d > 2$, and [440, 497] for $d = 2$), so that the intersection is also empty. For further discussion of the (self-)intersection properties of Brownian paths, see [463, pp. 81–87].

[2] An alternate motivation for the m^d factor in (5.6) is to note that the "correlation length" of the simple random walk (5.1)/(5.2) is $\approx m^{-1}$: that is, the two-point function $K(x, y)$ decays exponentially for $|x - y| \gtrsim m^{-1}$. Therefore, it is to be expected (and can in fact be proven) that the intersection probability $p_m(x_1, y_1)$ is negligible for $|x_1 - y_1| \gtrsim m^{-1}$. Therefore, the "average intersection probability" p_m is in essence the average of $p_m(x_1, y_1)$ over the region $|x_1 - y_1| \lesssim m^{-1}$.

The main results on p_m are summarized in the following theorem:

Theorem 5.1 *As* $m \to 0$,

$$p_m \sim \begin{cases} \text{const} > 0 & d < 4 \\ |\log m|^{-1} & d = 4 \\ m^{d-4} & d > 4 \end{cases}$$

The lower bounds on p_m in Theorem 5.1 were first proven by Erdös and Taylor [164]; their proof was further simplified by Sokal [476] and Felder and Fröhlich [185]. The upper bounds on p_m for $d \geq 4$ were also claimed by Erdös and Taylor, but their proof for $d = 4$ was unfortunately incorrect. A correct proof of the upper bound for $d = 4$ was first given by Lawler [351]; simpler proofs were given by Felder and Fröhlich [185], Aizenman [7] and Park [425].

The plan of this chapter is as follows: In Section 5.2 we give a heuristic renormalization-group argument that explains why Theorem 5.1 is expected to hold. In Section 5.3 we give two slightly different proofs of the lower bound on p_m: one is a variant of the Erdös-Taylor "dilution trick", while the other is based on the Schwarz inequality. In Section 5.4.1 we prove the upper bound for $d > 4$ by computing the expected number of intersections; this technique yields a precursor of the Aizenman-Fröhlich inequality which, as discussed in the next chapter, is central to the proof of triviality of the φ^4 field theory. In Sections 5.4.2 and 5.4.3 we give two very different proofs of the (much deeper) upper bound for $d = 4$: the first is a renormalization-group argument due to Felder and Fröhlich [185] and Aizenman [7], while the second is a direct inclusion-exclusion argument due to Park [425]. The latter is closely related to the "passing-to-the-left" trick to be discussed in Section 12.3.1. Unfortunately, neither of these two proofs has yet been generalized to SAWs or "field-theoretic" walks.

5.2 Heuristic renormalization-group argument

We start by giving a heuristic (perturbative) "proof" of Theorem 5.1 and more general results, using a renormalization-group flow equation. In Section 5.4.2 we will show how to make this renormalization-group argument rigorous.

In order to set up the perturbation theory, we first introduce a model of simple random walks with *weak* mutual avoidance, in which intersections between walks ω_1 and ω_2 are penalized by (roughly speaking) a factor $e^{-\lambda}$. As $\lambda \to +\infty$, this reduces to a pair of simple random walks with *strict* mutual avoidance, and a suitable connected four-point function in this model yields the intersection probability $p_m(x_1, y_1)$. On the other hand, by expanding in powers of λ around $\lambda = 0$, we can develop a perturbation expansion.

For each path ω, we first introduce a probability measure $d\tilde{\nu}_\omega$ on $[0, \infty)^{\mathbb{Z}^d}$ by the formula

$$d\widetilde{\nu}_\omega(t) = \prod_z d\widetilde{\nu}_{n_z(\omega)}(t_z) , \tag{5.7}$$

where $n_z(\omega)$ is the number of times that ω visits the site z, and

$$d\widetilde{\nu}_n(s) = \begin{cases} \delta(s)\, ds & \text{if } n = 0 \\ \\ \frac{s^{n-1}}{\Gamma(n)} e^{-s} \chi_{[0,\infty)}(s)\, ds & \text{if } n \geq 1 . \end{cases} \tag{5.8}$$

The variables $t = \{t_z\}$ can be thought of as "waiting times" (or "local times") for a Poisson jump process associated with the walk ω. In particular, $t_z = 0$ with $\widetilde{\nu}_\omega$-probability 1 if $z \notin \omega$, and $t_z > 0$ with $\widetilde{\nu}_\omega$-probability 1 if $z \in \omega$. Since $d\widetilde{\nu}_\omega(t)$ is a probability measure, we obviously have

$$\begin{aligned} K(x,y) &\equiv \sum_{\omega:\, x \to y} J^\omega \\ &= \sum_{\omega:\, x \to y} J^\omega \int d\widetilde{\nu}_\omega(t) . \end{aligned}$$

We next introduce a "connected four-point function"

$$K^c_\lambda(x_1 x_2 \mid y_1 y_2) = \sum_{\substack{\omega_1:\, x_1 \to x_2 \\ \omega_2:\, y_1 \to y_2}} J^{\omega_1} J^{\omega_2} \int d\widetilde{\nu}_{\omega_1}(t^1)\, d\widetilde{\nu}_{\omega_2}(t^2)\, I_\lambda(t^1, t^2) , \tag{5.9}$$

where

$$I_\lambda(t^1, t^2) = \exp(-\lambda \sum_j t^1_j t^2_j) - 1 \tag{5.10}$$

and $0 \leq \lambda < \infty$. Note that

(a) $-1 \leq I_\lambda \leq 0$

(b) I_λ is a decreasing function of λ

(c) $I_0(t^1, t^2) = 0$

(d) With probability 1 (with respect to $d\widetilde{\nu}_{\omega_1} \times d\widetilde{\nu}_{\omega_2}$),

$$I_\lambda(t^1, t^2) = 0 \quad \text{if } \omega_1 \cap \omega_2 = \varnothing \tag{5.11}$$

(e) With probability 1 (with respect to $d\widetilde{\nu}_{\omega_1} \times d\widetilde{\nu}_{\omega_2}$),

$$\lim_{\lambda \to \infty} I_\lambda(t^1, t^2) = \left\{ \begin{array}{ll} -1 & \text{if } \omega_1 \cap \omega_2 \neq \varnothing \\ 0 & \text{if } \omega_1 \cap \omega_2 = \varnothing \end{array} \right\} . \tag{5.12}$$

(f)

$$I^{(2n+1)}_\lambda(t^1, t^2) \leq I_\lambda(t^1, t^2) \leq I^{(2n)}_\lambda(t^1, t^2) , \tag{5.13}$$

where $I^{(k)}_\lambda(t^1, t^2)$ arises by replacing the exponential by its Taylor series expansion up to k^{th} order in λ.

The function I_λ thus interpolates between zero [at $\lambda = 0$] and minus the indicator function $\mathrm{I}(\omega_1 \cap \omega_2 \neq \varnothing)$ [at $\lambda = \infty$]. In particular, we can define finite-λ analogues of $p_m(x_1, y_1)$ and p_m, namely

$$p_{m,\lambda}(x_1, y_1) = -\chi^{-2} \sum_{x_2, y_2} K^c_\lambda(x_1\ x_2 \mid y_1\ y_2)\ . \tag{5.14}$$

and

$$p_{m,\lambda} = m^d \sum_{y_1} p_{m,\lambda}(x_1, y_1)\ . \tag{5.15}$$

As $\lambda \to \infty$ these approach $p_m(x_1, y_1)$ and p_m, respectively.

The first thing to do in perturbation theory is to expand the four-point function $K^c_\lambda(x_1\ x_2 \mid y_1\ y_2)$ in powers of λ, using the following splitting lemma.

Lemma 5.2

$$\begin{aligned} &\sum_{\omega:\, x \longrightarrow y} J^{\omega} \int d\widetilde{\nu}_\omega(t)\, t_z F(t) \\ &= \sum_{\substack{\omega_1:\, x \longrightarrow z \\ \omega_2:\, z \longrightarrow y}} J^{\omega_1+\omega_2} \int d\widetilde{\nu}_{\omega_1}(t^1) d\widetilde{\nu}_{\omega_2}(t^2) F(t^1 + t^2)\ . \end{aligned} \tag{5.16}$$

The proof, which is not difficult, can be found in [97]. This lemma can be used to evaluate $K^c_\lambda(x_1 x_2 \mid y_1 y_2)$ by computing the integral with respect to $d\widetilde{\nu}_{\omega_1}(t^1) d\widetilde{\nu}_{\omega_2}(t^2)$ of

$$I_\lambda(t^1, t^2) = \sum_{n=1}^{\infty} \frac{(-\lambda)^n}{n!} \sum_{z_1, \ldots, z_n} t^1_{z_1} t^2_{z_1} \cdots t^1_{z_n} t^2_{z_n} \tag{5.17}$$

After resumming each term over all resulting random walks, the final outcome can conveniently be described in terms of Feynman diagrams. Using the symbols

○ $\longleftrightarrow$ $p_{m,\lambda}$

● $\longleftrightarrow$ an internal vertex (summed over)

—— $\longleftrightarrow$ a bare propagator $K(x, y)$.

and adopting the convention that all external vertices but one are to be summed over all possible positions, the final result can then be depicted in the form:

○ $= J^2 m^{d-4} \Big[-\lambda$  $+ \lambda^2 b_2$ 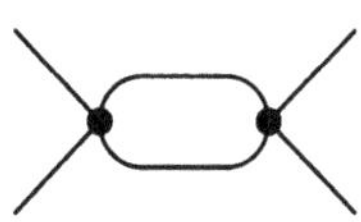

$-\lambda^3 b_3$ 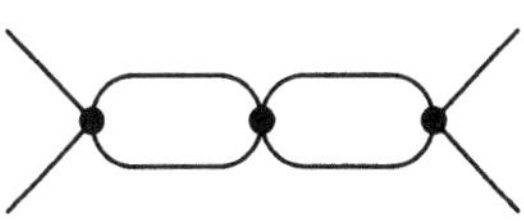 $- \lambda^3 b_3'$ 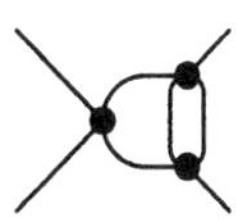 $+ O(\lambda^4) \Big]\ .$

(5.18)

This expansion resembles Feynman perturbation theory for g in lattice $\lambda\varphi^4$ theory, with the difference that *no* self-energy diagrams occur. Note that, by (5.13), this expansion yields an *upper bound* on $p_{m,\lambda}$ if broken off after the $(2n+1)^{st}$ order, while it yields a *lower bound* when broken off after the $(2n)^{th}$ order. A similar result can be proven for ordinary $\lambda\varphi^4$ theory using the methods of Chapter 6; see [97, 74].

It is natural to ask whether the expansion (5.18) comes from some lattice field theory related to $\lambda\varphi^4$. The answer is *yes*! Indeed, consider the theory of p N-component scalar fields $\boldsymbol{\varphi}_1, \ldots, \boldsymbol{\varphi}_p$ with an interaction Hamiltonian

$$H_{\text{int}}(\varphi) = \frac{\lambda_1}{4!}\sum_{j=1}^{p}|\boldsymbol{\varphi}_j|^4 + \frac{\lambda_2}{4!}\sum_{\substack{i,j=1\\ i\neq j}}^{p}|\boldsymbol{\varphi}_i|^2|\boldsymbol{\varphi}_j|^2 . \tag{5.19}$$

Then the intersection properties of simple random walks can be expressed *formally* in terms of the correlation functions of the field theory (5.19) analytically continued to $N=0$:

$$K(x,y) = \lim_{N\to 0}\langle \varphi^{(1)}_{1x}\,\varphi^{(1)}_{1y}\rangle_{\lambda_1=0,\lambda_2=3\lambda,N} \tag{5.20}$$

$$K^c_\lambda(x_1x_2 \mid y_1y_2) = \lim_{N\to 0}\langle \varphi^{(1)}_{1x_1}\,\varphi^{(1)}_{1x_2}\,;\varphi^{(1)}_{2y_1}\,\varphi^{(1)}_{2y_2}\rangle_{\lambda_1=0,\lambda_2=3\lambda,N} \tag{5.21}$$

In fact, the term $\sim\lambda_1|\boldsymbol{\varphi}_j|^4$ in (5.19) generates a weak self-avoidance for the jth walk, while the term $\sim\lambda_2|\boldsymbol{\varphi}_i|^2|\boldsymbol{\varphi}_j|^2$ generates a weak mutual avoidance between the ith and jth walks; so the intersection properties of *simple* random walks are obtained by taking $\lambda_1=0$. (More discussion of analytic continuation in N can be found in Section 9.2.2.) We can therefore apply *formally* the standard methods of perturbative renormalization-group theory [77] to deduce the *conjectured* behavior of $p_{m,\lambda}$ as $m\to 0$. To do this, we define a pair of dimensionless renormalized coupling constants, g and $\tilde g$: here

$$g \equiv 3p_{m,\lambda} \equiv -3\chi^{-2}m^d \sum_{x_2,y_1,y_2}\lim_{N\to 0}\langle \varphi^{(1)}_{1x_1};\varphi^{(1)}_{1x_2};\varphi^{(1)}_{2y_1};\varphi^{(1)}_{2y_2}\rangle , \tag{5.22}$$

while

$$\tilde g \equiv -\chi^{-2}m^d \sum_{x_2,x_3,x_4}\lim_{N\to 0}\langle \varphi^{(1)}_{1x_1};\varphi^{(1)}_{1x_2};\varphi^{(1)}_{1x_3};\varphi^{(1)}_{1x_4}\rangle \tag{5.23}$$

(note the different internal indices 1122 vs. 1111). Define next the Callan-Symanzik β-functions for the flow of g and $\tilde g$ under variations of $\log m$,

$$W \equiv m\frac{dg}{dm}$$

$$\widetilde W \equiv m\frac{d\tilde g}{dm}$$

W and $\widetilde W$ are given initially as functions of λ_1, λ_2 and m; they can be computed formally in an (unrenormalized) perturbation expansion in powers of λ_1 and λ_2. By expanding λ_1 and λ_2 in powers series in g and $\tilde g$ that expansion can be reverted to express W and $\widetilde{W}$ as an expansion in powers of g, $\tilde{g}$ and $\varepsilon = 4 - d$; and the coefficients in this expansion are finite as $m \to 0$. This expansion has been calculated to leading order in g, $\tilde{g}$ and ε by Brézin and Zinn-Justin [83]:

$$W(g, \tilde{g}) = -\varepsilon g + \frac{N+2}{3} g\tilde{g} + \frac{4 + (p-2)N}{6} g^2 + \text{third-order terms} \tag{5.24}$$

$$\widetilde{W}(g, \tilde{g}) = -\varepsilon \tilde{g} + \frac{(p-1)N}{6} g^2 + \frac{N+8}{6} \tilde{g}^2 + \text{third-order terms.} \tag{5.25}$$

In particular, when $N = 0$, the line $\tilde{g} = 0$ is mapped into itself by the RG flow. For $d < 4$, the flow at small g is controlled by a nontrivial fixed point at $g^* = 3\varepsilon/2 + O(\varepsilon^2)$; for $d \geq 4$ it is controlled by the Gaussian fixed point $g^* = 0$. *Assuming* that these fixed points are the ones that control the critical behavior also for $\lambda = \infty$, we can predict the asymptotic behavior of p_m as $m \to 0$:

$$g \equiv 3p_m \sim \begin{cases} 3\varepsilon/2 + O(\varepsilon^2) & d < 4 \\ |\log m|^{-1} & d = 4 \\ m^{d-4} & d > 4 \,, \end{cases} \quad . \tag{5.26}$$

This concludes the non-rigorous argument explaining why Theorem 5.1 is expected to be true.

5.3 Proof of lower bound

In this section we give two proofs of the lower bound on p_m claimed in Theorem 5.1. The first proof uses a variant of the Erdös–Taylor "dilution trick". The second proof is based on the Schwarz inequality.

First proof. Recall that

$$p_{m,\lambda}(x_1, y_1) = -\chi^{-2} \sum_{x_2, y_2} K_\lambda^c(x_1\, x_2 \mid y_1\, y_2)$$

and

$$p_{m,\lambda} = -m^d \chi^{-2} \sum_{x_2, y_1, y_2} K_\lambda^c(x_1\, x_2 \mid y_1\, y_2) \,.$$

These quantities are monotone increasing in λ, and approach $p_m(x_1, y_1)$ and p_m, respectively, as $\lambda \to \infty$. By (5.13) we obtain a *lower bound* on $p_{m,\lambda}(x_1, y_1)$ if we expand I_λ to second order in λ. By (5.9) and the splitting lemma (Lemma 5.2), we get

$$-K_\lambda^c(x_1\, x_2 \mid y_1\, y_2)$$

$$\geq \lambda \sum_z K(x_1, z) K(x_2, z) K(y_1, z) K(y_2, z)$$

$$-\frac{\lambda^2}{2}\sum_{z,w} K(x_1,z)K(y_1,z)K(z,w)^2K(w,x_2)K(w,y_2)$$

$$-\frac{\lambda^2}{2}\sum_{z,w} K(x_1,z)K(y_2,z)K(z,w)^2K(w,x_2)K(w,y_1)$$

$\equiv \lambda$

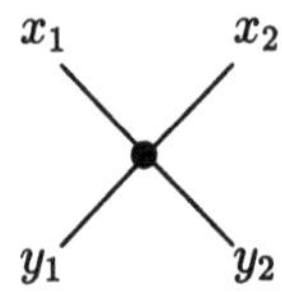

$-\dfrac{\lambda^2}{2}\Bigg[$

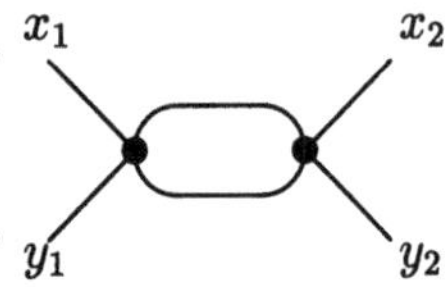

$+$

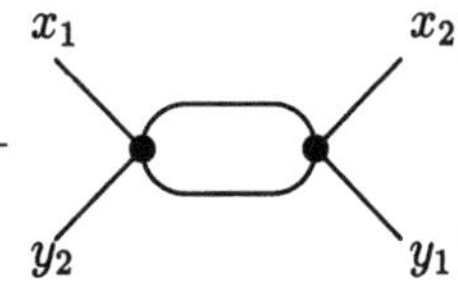

$\Bigg]$. (5.27)

Summing over x_2, y_1 and y_2 we obtain

$$p_{m,\lambda} \geq m^d \chi^2[\lambda - B(m)\lambda^2]$$

where

$$B(m) \equiv \sum_z K(0,z)^2 \tag{5.28}$$

is the "bubble diagram". Since $p_m \geq p_{m,\lambda}$ for all $\lambda \geq 0$, we can obtain a lower bound on p_m by choosing any λ we please, in particular the optimal value $\lambda = [2B(m)]^{-1}$. This gives

$$p_m \geq \frac{J^{-2}}{4B(m)} m^{d-4} \, . \tag{5.29}$$

Recall that $\chi = J^{-1}m^{-2}$. A simple computation yields

$$B(m) \sim \begin{cases} \frac{1}{\varepsilon} m^{-\varepsilon} & d = 4-\varepsilon \;\; (\varepsilon > 0) \\ |\log m| & d = 4 \\ \mathrm{const}(d) < \infty & d > 4 \end{cases} \tag{5.30}$$

We have therefore proven the lower bound in Theorem 5.1. (This argument is an application of the "dilution trick" discussed in full detail in Section 12.3.1.)

Second proof. Let $\mathcal{N}(\omega_1,\omega_2)$ be the number of pairs of "times" at which ω_1 and ω_2 intersect:

$$\mathcal{N}(\omega_1,\omega_2) \equiv \#\{(r,s)\colon \omega_1(r) = \omega_2(s)\} \, . \tag{5.31}$$

Then, for any positive measure E on the space of pairs (ω_1,ω_2), the Schwarz inequality implies that

$$\begin{aligned} E(\mathcal{N}) &= E(\mathcal{N}\,\mathrm{I}(\mathcal{N} \geq 1)) \\ &\leq E(\mathcal{N}^2)^{1/2}\, E(\mathrm{I}(\mathcal{N} \geq 1))^{1/2} \end{aligned}$$

and hence

$$E(\mathrm{I}(\mathcal{N} \geq 1)) \geq \frac{E(\mathcal{N})^2}{E(\mathcal{N}^2)} \,. \tag{5.32}$$

We shall apply this inequality with

$$E(F(\omega_1, \omega_2)) \equiv \sum_{x_2, y_1, y_2} \sum_{\substack{\omega_1 \colon x_1 \to x_2 \\ \omega_2 \colon y_1 \to y_2}} \varrho(\omega_1)\, \varrho(\omega_2)\, F(\omega_1, \omega_2) \tag{5.33}$$

and x_1 fixed. We need only compute $E(\mathcal{N})$ and $E(\mathcal{N}^2)$.

Let us start with $E(\mathcal{N})$:

$$\begin{aligned} E(\mathcal{N}) &= \sum_{r,s \geq 0} \sum_{x_2, y_1, y_2} \sum_{\substack{\omega_1 \colon x_1 \to x_2 \\ \omega_2 \colon y_1 \to y_2}} \varrho(\omega_1)\, \varrho(\omega_2)\, \mathrm{I}(\omega_1(r) = \omega_2(s)) \\ &= \sum_{z} \sum_{r,s \geq 0} \sum_{x_2, y_1, y_2} \sum_{\substack{\omega_1 \colon x_1 \to x_2 \\ \omega_2 \colon y_1 \to y_2}} \varrho(\omega_1)\, \varrho(\omega_2)\, \mathrm{I}(\omega_1(r) = \omega_2(s) = z) \,. \end{aligned} \tag{5.34}$$

For fixed z, r, s, the only terms that contribute are those with $\omega_1(r) = \omega_2(s) = z$. So we can split $\omega_1 \equiv \omega_1' \circ \omega_1''$ at time r, and split $\omega_2 \equiv \omega_2' \circ \omega_2''$ at time s. Therefore

$$\begin{aligned} E(\mathcal{N}) &= \sum_{z} \sum_{r,s \geq 0} \sum_{x_2, y_1, y_2} \sum_{\substack{\omega_1' \colon x_1 \to z \\ \omega_1'' \colon z \to x_2 \\ \omega_2' \colon y_1 \to z \\ \omega_2'' \colon z \to y_2 \\ |\omega_1'| = r \\ |\omega_2'| = s}} \varrho(\omega_1' \circ \omega_1'')\, \varrho(\omega_2' \circ \omega_2'') \\ &= \sum_{z} \sum_{x_2, y_1, y_2} \sum_{\substack{\omega_1' \colon x_1 \to z \\ \omega_1'' \colon z \to x_2 \\ \omega_2' \colon y_1 \to z \\ \omega_2'' \colon z \to y_2}} \varrho(\omega_1' \circ \omega_1'')\, \varrho(\omega_2' \circ \omega_2'') \end{aligned} \tag{5.35}$$

$$= \sum_{z, x_2, y_1, y_2}$$ 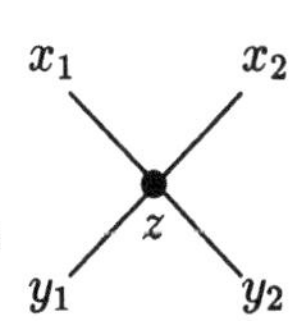 (5.36)

Now we use translation invariance: instead of fixing x_1 and summing over z, x_2, y_1, y_2, we can fix z and sum over x_1, x_2, y_1, y_2. The result is

$$E(\mathcal{N}) = \left(\sum_{x_1,x_2} \sum_{\substack{\omega' :\, x_1 \to z \\ \omega'' :\, z \to x_2}} \varrho(\omega' \circ \omega'') \right)^2 . \tag{5.37}$$

For simple random walks, the weights factorize as

$$\varrho(\omega' \circ \omega'') = \varrho(\omega')\varrho(\omega'') , \tag{5.38}$$

so we have simply

$$E(\mathcal{N}) = \chi^4 . \tag{5.39}$$

Remark. Everything up to (5.37) applies to arbitrary random-walk models. Moreover, we have quite generally

$$\begin{aligned} \sum_{x_1,x_2} \sum_{\substack{\omega' :\, x_1 \to z \\ \omega'' :\, z \to x_2}} \varrho(\omega' \circ \omega'') &= \sum_{z,x_2} \sum_{\substack{\omega' :\, x_1 \to z \\ \omega'' :\, z \to x_2}} \varrho(\omega' \circ \omega'') \\ &= \sum_{x_2} \sum_{\omega :\, x_1 \to x_2} \varrho(\omega) \left(|\omega| + 1 \right) , \end{aligned} \tag{5.40}$$

since $|\omega| + 1$ is the number of ways that a walk ω can be split into $\omega' \circ \omega''$. The difficult part is to compute or bound (5.40) — the "mean length" — in terms of more familiar quantities. For models in which the weight $\varrho(\omega)$ has a trivial J-dependence as in (1.99),

$$\varrho(\omega) = J^{|\omega|} \exp[-U(\omega)] \tag{5.41}$$

(in Section 9.2.1 we call these "polymer-chain models"), it follows immediately that (5.40) equals $\chi + J\partial\chi/\partial J$. (For simple random walks, $\chi + J\partial\chi/\partial J = \chi^2$!) Unfortunately, for "field-theoretic" walks things are more complicated (see Section 11.3.1), and we do not know how to compute (5.40).

Next we compute $E(\mathcal{N}^2)$:

$$\begin{aligned} E(\mathcal{N}^2) &= \sum_{\substack{r,s \geq 0 \\ r',s' \geq 0}} \sum_{x_2,y_1,y_2} \sum_{\substack{\omega_1 :\, x_1 \to x_2 \\ \omega_2 :\, y_1 \to y_2}} \begin{array}{c} \varrho(\omega_1)\, \varrho(\omega_2)\, \mathrm{I}(\omega_1(r) = \omega_2(s)) \\ \times \mathrm{I}(\omega_1(r') = \omega_2(s')) \end{array} \\ &= \sum_{z,z'} \sum_{\substack{r,s \geq 0 \\ r',s' \geq 0}} \sum_{x_2,y_1,y_2} \sum_{\substack{\omega_1 :\, x_1 \to x_2 \\ \omega_2 :\, y_1 \to y_2}} \begin{array}{c} \varrho(\omega_1)\, \varrho(\omega_2)\, \mathrm{I}(\omega_1(r) = \omega_2(s) = z) \\ \times \mathrm{I}(\omega_1(r') = \omega_2(s') = z') . \end{array} \end{aligned} \tag{5.42}$$

We note that

$$\begin{aligned} \sum_{\substack{r,s \geq 0 \\ r',s' \geq 0}} \cdots &\leq 2 \sum_{\substack{r,s,r',s' \geq 0 \\ r \leq r'}} \cdots \\ &\leq 2 \sum_{\substack{r,s,r',s' \geq 0 \\ r \leq r' \\ s \leq s'}} \cdots + 2 \sum_{\substack{r,s,r',s' \geq 0 \\ r \leq r' \\ s \geq s'}} \cdots . \end{aligned}$$

We now split $\omega_1 = \omega_1' \circ \omega_1'' \circ \omega_1'''$ at times r and r', and split $\omega_2 = \omega_2' \circ \omega_2'' \circ \omega_2'''$ at times s and s' (or s' and s). By the same reasoning as in (5.34)–(5.35), we find

$$E(\mathcal{N}^2) \;\le\; 2 \sum_{z,z',x_2,y_1,y_2}$$

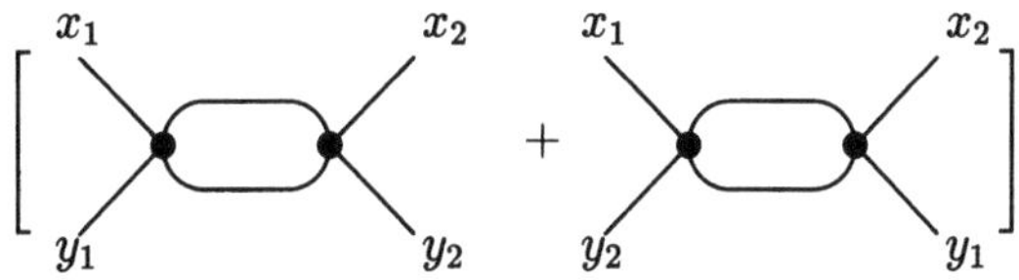

$$= 4\sum_{z'} \Bigg(\sum_{x_1,x_2} \sum_{\substack{\omega' :\, x_1 \to z \\ \omega'' :\, z \to z' \\ \omega''' :\, z' \to x_2}} \varrho(\omega' \circ \omega'' \circ \omega''') \Bigg)^2 \tag{5.43}$$

For simple random walks, the weights again factorize,

$$\varrho(\omega' \circ \omega'' \circ \omega''') \le \varrho(\omega')\varrho(\omega'')\,\varrho(\omega''') \,, \tag{5.44}$$

so we have

$$E(\mathcal{N}^2) \;\le\; 4\chi^4 B(m) \tag{5.45}$$

where $B(m) = \sum_z K(0,z)^2$ is the "bubble diagram" found already in (5.28).

Remark. For self-avoiding walks, we can write

$$\varrho(\omega' \circ \omega'' \circ \omega''') \;\le\; \varrho(\omega')\varrho(\omega'')\,\varrho(\omega''') \,, \tag{5.46}$$

and still deduce (5.45). But this is a bad bound for $d < 4$ (see Section 12.3.3).

Putting together (5.32), (5.39) and (5.45), we get

$$E(\,\mathrm{I}(\mathcal{N} \ge 1)) \;\ge\; \frac{\chi^4}{4B(m)} \tag{5.47}$$

and hence

$$p_m \;\ge\; \frac{m^d \chi^2}{4B(m)} \,, \tag{5.48}$$

exactly as in (5.29).

This Schwarz-inequality argument has been used by Aizenman [7] and by Park [425]. It is an alternative to the "dilution trick", and will be discussed further in Section 12.3.1.

5.4 Proof of upper bound

We now turn to the upper bound on p_m. In Section 5.4.1 we give a simple proof of the upper bound on p_m for $d > 4$. In Sections 5.4.2 and 5.4.3 we give two alternate proofs of the upper bound in the (difficult) borderline case $d = 4$.

5.4.1 Expected number of intersections

In dimension $d > 4$, the upper bound on p_m is very easy: the *probability* that two walks intersect can be bounded above by the *expected number* of intersections $E_m(\#(\omega_1 \cap \omega_2))$.

Lemma 5.3

$$p_m \leq \text{const} \times m^{d-4} . \tag{5.49}$$

Proof.

$$\begin{aligned} p_m &= \chi^{-2} m^d \sum_{y_1, x_2, y_2} \sum_{\substack{\omega_1 :\, x_1 \to x_2 \\ \omega_2 :\, y_1 \to y_2}} \varrho(\omega_1)\, \varrho(\omega_2)\ \mathrm{I}(\omega_1 \cap \omega_2 \neq \varnothing) \\ &\leq \chi^{-2} m^d \sum_{y_1, x_2, y_2} \sum_{z} \sum_{\substack{\omega_1 :\, x_1 \to x_2 \\ \omega_2 :\, y_1 \to y_2}} \varrho(\omega_1)\, \varrho(\omega_2)\ \mathrm{I}(\omega_1 \cap \omega_2 \ni z) \end{aligned} \tag{5.50}$$

Since the weight (5.2) for a simple random walk factorizes under splitting of the walk, we get

$$p_m \leq \chi^{-2} m^d \sum_{y_1, x_2, y_2, z} \sum_{\substack{\omega_1' :\, x_1 \to z \\ \omega_1'' :\, z \to x_2 \\ \omega_2' :\, y_1 \to z \\ \omega_2'' :\, z \to y_2}} \varrho(\omega_1')\, \varrho(\omega_1'') \varrho(\omega_2')\, \varrho(\omega_2'') \tag{5.51}$$

$$\begin{aligned} &= \chi^{-2} m^d \times \chi^4 \\ &= J^{-2} m^{d-4} . \end{aligned} \tag{5.52}$$

In the last two equalities we have used (5.5). The lemma follows from the observation that $J \to J_c > 0$ as $m \to 0$. ∎

The bound provided by this lemma is essentially sharp for dimensions $d > 4$ but very bad for lower dimensions. To explain this fact, we notice that in obtaining the inequality in (5.50) we have bounded the *probability* of intersection by the *mean number* of intersections. This overcounting is severe if walks meet at many places. By comparing (5.52) with the actual result in Theorem 5.1, we see that this is precisely what happens for dimension $d \leq 4$. Indeed, by looking at the ratio $E_m(\#(\omega_1 \cap \omega_2))/p_m \sim m^{d-4}/p_m$, we see that once the walks intercept they do so $\sim |\log m|$ times for $d = 4$ and $\sim m^{d-4}$ times for $d < 4$ — that is, infinitely many times as $m \to 0$. We could in principle eliminate this overcounting by a more careful bookkeeping of the intersections, for instance by considering only the *first* z (in some ordering) at which an intersection takes place. It is not easy to handle the constraints that arise from such a decomposition, but Park [425] has shown how to do it; we present his proof in Section

5.4.3. An alternate proof can be obtained from a rigorous renormalization-group argument to control the *flow* of p_m as m is varied; we present this proof, which is due to Aizenman [7] and Felder and Fröhlich [185], in Section 5.4.2.

Let us mention that the argument in Lemma 5.3 can be generalized to SAWs and to "field-theoretic" walks, yielding a "tree bound" analogous to (5.51); this is the Aizenman-Fröhlich inequality (see Sections 6.2 and 12.2). As in the case of simple random walks, this bound is essentially sharp for dimensions $d > 4$, but is very bad for lower dimensions. Unfortunately, neither of the proofs for $d = 4$ have yet been generalized to SAWs or to φ^4 field theory.

5.4.2 Rigorous renormalization-group argument

Our starting point is the following simple lemma:

Lemma 5.4 *In d dimensions,*

$$m\frac{dp_m}{dm} = (d-4)p_m + 4m^d \sum_x Q(0,x) \ , \tag{5.53}$$

where

$$Q(0,x) = m^6 \sum_{\substack{\omega_1:\, 0 \to \text{any} \\ \omega_2:\, 0 \to \text{any} \\ \omega_3:\, x \to \text{any}}} \left(\prod_{i=1}^{3} J^{|\omega_i|}\right) \mathrm{I}(\omega_1 \cap \omega_3 \neq \varnothing)\ \mathrm{I}(\omega_2 \cap \omega_3 \neq \varnothing) \tag{5.54}$$

Proof. We start from the definition of p_m,

$$p_m = m^{d+4} \sum_{x \in \mathbb{Z}^d} \sum_{\substack{\omega:\, 0 \to \text{any} \\ \omega':\, x \to \text{any}}} J^{|\omega|+|\omega'|}\ \mathrm{I}(\omega \cap \omega' \neq \varnothing) \ , \tag{5.55}$$

with $J = (2d+m^2)^{-1}$. Using $dJ/dm = -2J^2 m$, we calculate $m\, dp_m/dm$ from (5.55) and obtain, using the symmetry in ω and ω',

$$m\frac{dp_m}{dm} = (d+4)p_m - 4m^{d+6} \sum_x \sum_{\substack{\omega:\, 0 \to \text{any} \\ \omega':\, x \to \text{any}}} |\omega| J^{|\omega|+|\omega'|+1}\ \mathrm{I}(\omega \cap \omega' \neq \varnothing) \tag{5.56}$$

Now, note that $|\omega|$ is the number of sites visited by ω (counted with multiplicity). In each of these points we can split ω into two independent walks, ω_1 and ω_2. Setting $\omega_3 \equiv \omega'$, we have

$$\begin{aligned} m\frac{dp_m}{dm} &= (d+4)p_m \\ &\quad -4m^{d+6} \sum_{x,z} \sum_{\substack{\omega_1:\, 0 \to \text{any} \\ \omega_2:\, z \to \text{any} \\ \omega_3:\, x \to \text{any}}} \left(\prod_{i=1}^{3} J^{|\omega_i|}\right) \mathrm{I}(\omega_1 \cap \omega_3 \neq \varnothing \text{ or } \omega_2 \cap \omega_3 \neq \varnothing) \ . \end{aligned} \tag{5.57}$$

But

$$\begin{aligned} \mathrm{I}(\omega_1 \cap \omega_3 \neq \varnothing \text{ or } \omega_2 \cap \omega_3 \neq \varnothing) &= \mathrm{I}(\omega_1 \cap \omega_3 \neq \varnothing) + \mathrm{I}(\omega_2 \cap \omega_3 \neq \varnothing) \\ &\quad - \mathrm{I}(\omega_1 \cap \omega_3 \neq \varnothing \text{ and } \omega_2 \cap \omega_3 \neq \varnothing) . \end{aligned}$$

Inserting this identity into (5.57) and translating z to the origin, we obtain

$$\begin{aligned} m\frac{dp_m}{dm} &= (d-4)p_m \\ &\quad +4m^{d+6} \sum_{x} \sum_{\substack{\omega_1\colon 0 \to \text{any} \\ \omega_2\colon 0 \to \text{any} \\ \omega_3\colon x \to \text{any}}} \left(\prod_{i=1}^{3} J^{|\omega_i|} \right) \mathrm{I}(\omega_1 \cap \omega_2 \neq \varnothing)\, \mathrm{I}(\omega_1 \cap \omega_3 \neq \varnothing) . \end{aligned} \tag{5.58}$$

This concludes the proof of Lemma 5.4. ∎

Next, we prove a lower bound on $Q(0,x)$ in terms of p_m itself:

Lemma 5.5

$$4m^d \sum_x Q(0,x) \geq C_d\, p_m^2$$

where C_d is a finite constant independent of m.

Proof. Let x^α be the coordinate of x with the largest absolute value, (i.e. $|x^\alpha| \geq d^{-1/2}|x|$). Let π_x be the lattice plane perpendicular to the α-axis, intersecting the α-axis in the point $\left[\frac{x^\alpha}{2}\right] e_\alpha$, where e_α is the unit vector in the α direction and $[a]$ is the largest integer $\leq a$. Then by (5.54)

$$\begin{aligned} &Q(0,x) \geq \\ &\quad m^2 \sum_{\omega_3\colon x \to \text{any}} J^{|\omega_3|} \Big[m^2 \sum_{\omega_1\colon 0 \to \text{any}} J^{|\omega_1|}\, \mathrm{I}(\omega_1 \cap \omega_3 \neq \varnothing) \Big]^2 \mathrm{I}(\omega_3 \cap \pi_x \neq \varnothing) . \end{aligned} \tag{5.59}$$

By the Schwarz inequality applied on the r.h.s. of (5.59),

$$\begin{aligned} 4m^d \sum_x Q(0,x) &\geq \Big[m^d \sum_x m^2 \sum_{\omega_3\colon x \to \text{any}} J^{|\omega_3|}\, \mathrm{I}(\omega_3 \cap \pi_x \neq \varnothing) \Big]^{-1} \\ &\quad \times \left\{ m^d \sum_x m^4 \sum_{\substack{\omega_3\colon x \to \text{any} \\ \omega_1\colon 0 \to \text{any}}} \begin{array}{l} J^{|\omega_1|+|\omega_3|}\, \mathrm{I}(\omega_1 \cap \omega_3 \neq \varnothing) \\ \times \frac{1}{2} \Big[\mathrm{I}(\omega_1 \cap \pi_x \neq \varnothing) + \mathrm{I}(\omega_3 \cap \pi_x \neq \varnothing) \Big] \end{array} \right\}^2 \end{aligned} \tag{5.60}$$

by symmetry in ω_1 and ω_3. Now the denominator on the RHS of (5.60) is easily shown to be bounded by a finite constant, *uniformly* in $m > 0$. Moreover,

$$\mathrm{I}(\omega_1 \cap \omega_3 \neq \varnothing)\, \frac{1}{2} \left[\, \mathrm{I}(\omega_1 \cap \pi_x \neq \varnothing) + \mathrm{I}(\omega_3 \cap \pi_x \neq \varnothing)\right] \geq \frac{1}{2}\, \mathrm{I}(\omega_1 \cap \omega_3 \neq \varnothing)\,. \tag{5.61}$$

This proves the lemma. ∎

We are now ready to complete the proof of Theorem 5.1. Putting together Lemmas 5.4 and 5.5, we obtain the RG flow *inequality*

$$m \frac{dp_m}{dm} \geq (d-4) p_m + C_d p_m^2\,. \tag{5.62}$$

The differential inequality (5.62) can easily be integrated, yielding the desired upper bound on p_m. ∎

Let us recapitulate the logic of this argument: It seems difficult to obtain *directly* a sharp upper bound on p_m for $d = 4$. Therefore, we proceed indirectly, by deriving a bound on $m\, dp_m/dm$ in terms of p_m itself. Finally, we integrate this differential inequality. Let us mention that analogous arguments can be used to bound $m\, dp_{m,\lambda}/dm$ in terms of the unknown $p_{m,\lambda}$, with *no* explicit λ-dependence; this generalization is the precise analogue of the heuristic RG flow equation discussed in the preceding section. In essence our arguments prove a non-perturbative lower bound on the RG β-function,

$$W(g) \geq (d-4) g + C_d g^2\,, \tag{5.63}$$

and this bound is valid *for all attainable values of* g, even though we do not know *a priori* which values of g are attainable!

Some final remarks.

1) Similar methods can be used to estimate the intersection probability for *three* simple random walks. Here the critical dimension is $d = 3$.

2) Our methods have been extended in [76] to estimate the intersection probability of non-interacting branched polymers. Here the critical dimension is $d = 8$, in accordance with the belief that the Hausdorff dimension of non-interacting branched polymers is 4.

4) Some of the methods explained in this chapter — in particular Lemma 5.2, inequality (5.13) and inequalities analogous to (5.27) — can be extended to $\lambda\varphi^4$ lattice field theories, and have important applications to the construction of continuum $\lambda\varphi_d^4$ theories for $d = 2, 3$, see Section 6.3 and [97, 96, 74].

5.4.3 Direct inclusion-exclusion argument

In this section we present a beautiful and "elementary" proof of the upper bound in Theorem 5.1, based on recent work by Park [425]. Many of the ideas were also implicit in the original work of Lawler [351]. The method is to use repeatedly the *decoupling identity*

$$1 = \mathrm{I}(\omega \cap \omega' = \varnothing) + \mathrm{I}(\omega \cap \omega' \neq \varnothing)\ , \tag{5.64}$$

manipulating the walks so as to extract as a multiplicative factor the bubble diagram B.

For technical reasons we shall need to consider walks $\omega\colon x \to y$ that *do not return to x after the initial time*; we call such walks "non-return" (NR) walks, and write $\omega\colon x \to y$ (NR). We define the non-return kernel

$$K_{\mathrm{NR}}(x,y) = \sum_{\omega\colon x\to y\,(\mathrm{NR})} \varrho(\omega) \tag{5.65}$$

and the non-return "susceptibilities" (normalization factor)

$$\chi_{\mathrm{NR}} = \sum_y K_{\mathrm{NR}}(x,y)\ . \tag{5.66}$$

These quantities can be related to the ordinary kernel and susceptibility by a "renewal" argument familiar from the theory of Markov chains: Consider an arbitrary walk $\omega\colon x \to y$, and split it at its *last* return to x into $\omega = \omega' \circ \omega''$. Then ω' is an arbitrary (possibly zero-step) walk from $x \to x$, and ω'' is a *non-return* walk from $x \to y$. Moreover, the correspondence between ω and the pair (ω', ω'') is one-to-one and onto. Now, for simple random walks, the weights factorize:

$$\varrho(\omega' \circ \omega'') = \varrho(\omega')\,\varrho(\omega'')\ . \tag{5.67}$$

It follows that

$$K(x,y) = K(x,x) K_{\mathrm{NR}}(x,y)\ . \tag{5.68}$$

To emphasize the translation-invariance, we use the notation

$$L \equiv K(x,x) \tag{5.69}$$

and call L the *loop diagram*; an easy computation in Fourier space yields

$$L \sim \begin{cases} m^{d-2} & d<2 \\ |\log m| & d=2 \\ \mathrm{const}(d)>0 & d>2 \end{cases} \tag{5.70}$$

as $m \to 0$. In summary,

$$\begin{aligned} K_{\mathrm{NR}}(x,y) &= L^{-1}K(x,y) \\ \chi_{\mathrm{NR}} &= L^{-1}\chi \end{aligned}$$

An important role in the argument will be played by the probability of *non-intersection* (except at the origin) of two or three walks starting at the origin. For technical reasons we shall take some of these walks to be non-return walks. (This does not change the behavior as $m \to 0$ in dimension $d > 2$.) We define therefore the unnormalized quantities

$$R_2 = \sum_{\substack{\omega_1\colon 0 \to \text{any (NR)} \\ \omega_2\colon 0 \to \text{any (NR)}}} \varrho(\omega_1)\varrho(\omega_2)\, \mathrm{I}(\omega_1 \cap \omega_2 = \{0\})$$

$$R_3 = \sum_{\substack{\omega_1\colon 0 \to \text{any (NR)} \\ \omega_2\colon 0 \to \text{any} \\ \omega_3\colon 0 \to \text{any (NR)}}} \varrho(\omega_1)\varrho(\omega_2)\varrho(\omega_3)\, \mathrm{I}(\omega_1 \cap (\omega_2 \cup \omega_3) = \{0\})$$

and the corresponding probabilities

$$P_2 = \chi_{\mathrm{NR}}^{-2} R_2 = P\left(\omega_1 \cap \omega_2 = \{0\} \,\middle|\, \begin{matrix} \omega_1\colon 0 \to \text{any (NR)} \\ \omega_2\colon 0 \to \text{any (NR)} \end{matrix}\right)$$

$$P_3 = \chi_{\mathrm{NR}}^{-2} \chi^{-1} R_3 = P\left(\omega_1 \cap (\omega_2 \cup \omega_3) = \{0\} \,\middle|\, \begin{matrix} \omega_1\colon 0 \to \text{any (NR)} \\ \omega_2\colon 0 \to \text{any} \\ \omega_3\colon 0 \to \text{any (NR)} \end{matrix}\right) .$$

(Note that, in the definition of R_3 and P_3, only the intersections ω_1–ω_2 and ω_1–ω_3 are forbidden; the intersection ω_2–ω_3 is permitted.) The main technical result of this section will be the following bound:

Theorem 5.6

$$P_3(m) \lesssim \begin{cases} m^{4-d} & 2 < d < 4 \\ |\log m|^{-1} & d = 4 \\ \mathrm{const}(d) & d > 4 \end{cases} . \tag{5.71}$$

Assuming temporarily the validity of this bound, let us show how to complete the proof of Theorem 5.1:

Proof of the upper bound in Theorem 5.1, assuming Theorem 5.6. By definition,

$$p_m = m^d \chi^{-2} \sum_{x_2, y_1, y_2} \sum_{\substack{\omega_1\colon x_1 \to x_2 \\ \omega_2\colon y_1 \to y_2}} \varrho(\omega_1)\, \varrho(\omega_2)\, \mathrm{I}(\omega_1 \cap \omega_2 \neq \varnothing) . \tag{5.72}$$

Let us fix ω_2, and split $\omega_1 = \omega_1' \circ \omega_1''$ at the *last* time that ω_1 intersects the fixed set ω_2; let z be the intersection point. Now split $\omega_2 = \omega_2' \circ \omega_2''$ at the *last* time that ω_2 visits z. We have

$$p_m = m^d \chi^{-2} \sum_{x_2, y_1, y_2, z} \sum_{\substack{\omega_1'\colon x_1 \to z \\ \omega_1''\colon z \to x_2 \text{ (NR)} \\ \omega_2'\colon y_1 \to z \\ \omega_2''\colon z \to y_2 \text{ (NR)}}} \varrho(\omega_1' \circ \omega_1'')\, \varrho(\omega_2' \circ \omega_2'')\, \mathrm{I}(\omega_1'' \cap (\omega_2' \cup \omega_2'') = \{z\}) . \tag{5.73}$$

Next we use translation invariance to fix z and sum over x_1, x_2, y_1, y_2; we use the factorization property $\varrho(\omega' \circ \omega'') = \varrho(\omega')\varrho(\omega'')$; and we use the invariance of ϱ under reversal of a walk (in this case ω_2'). The unconstrained sum over ω_1' yields χ; the constrained sum over ω_1'', ω_2', ω_2'' yields precisely R_3. Hence

$$p_m = m^d \chi^{-2} \times \chi \times R_3 \; . \tag{5.74}$$

Since $\chi_{\mathrm{NR}} \sim \chi \sim m^{-2}$ (in dimension $d > 2$), the upper bound in Theorem 5.1 follows immediately from Theorem 5.6. ∎

We now begin the proof of Theorem 5.6. The first step is to obtain an upper bound on R_3 in terms of R_2 and an auxiliary quantity that we shall call $\overline{T}$. We start from the definition

$$R_3 \;=\; \sum_{\substack{\omega_1:\, 0 \to \text{any (NR)} \\ \omega_2:\, 0 \to \text{any} \\ \omega_3:\, 0 \to \text{any (NR)}}} \varrho(\omega_1)\varrho(\omega_2)\varrho(\omega_3)\, \mathrm{I}(\omega_1 \cap \omega_2 = \{0\})\, \mathrm{I}(\omega_1 \cap \omega_3 = \{0\}) \, ,$$

and insert the identity

$$\mathrm{I}(\omega_1 \cap \omega_2 = \{0\}) \;=\; 1 - \; \mathrm{I}(\omega_1 \cap \omega_2 \neq \{0\}) \, , \tag{5.75}$$

leaving the other indicator function as is. The term obtained by inserting 1 yields exactly χR_2. The other term is

$$\sum_{\substack{\omega_1:\, 0 \to \text{any (NR)} \\ \omega_2:\, 0 \to \text{any} \\ \omega_3:\, 0 \to \text{any (NR)}}} \varrho(\omega_1)\varrho(\omega_2)\varrho(\omega_3)\;\; \mathrm{I}(\omega_1 \cap \omega_3 = \{0\})\;\; \mathrm{I}(\omega_1 \cap \omega_2 \neq \{0\}) \, . \tag{5.76}$$

If $\omega_1 \cap \omega_2 \neq \{0\}$, then we can fix ω_1 and split $\omega_2 = \omega_2' \circ \omega_2''$ at the *first* time that ω_2 intersects $\omega_1 \setminus \{0\}$; let z be the intersection point. Now split $\omega_1 = \omega_1' \circ \omega_1''$ at the *last* time that it visits z. We get

$$\sum_{z \neq 0} \sum_{\substack{\omega_1':\, 0 \to z \text{ (NR)} \\ \omega_1'':\, z \to \text{any (NR)} \\ \omega_2':\, 0 \to z \text{ (NA)} \\ \omega_2'':\, z \to \text{any} \\ \omega_3:\, 0 \to \text{any (NR)}}} \begin{aligned} &\varrho(\omega_1' \circ \omega_1'')\, \varrho(\omega_2' \circ \omega_2'')\, \varrho(\omega_3)\;\; \mathrm{I}(\omega_1'' \not\ni 0) \\ &\times \mathrm{I}((\omega_1' \cup \omega_1'') \cap \omega_3) = \{0\})\;\; \mathrm{I}(\omega_2' \cap (\omega_1' \cup \omega_1'') \subset \{0, z\}) \end{aligned} \tag{5.77}$$

where NA denotes "non-arrival", i.e. walks that visit their *final* point only once (that is, the reverse walk is NR). Using the factorization property $\varrho(\omega' \circ \omega'') = \varrho(\omega')\varrho(\omega'')$, the sum over ω_2'' decouples; we get $\chi\overline{T}$, where

$$\overline{T} \;\equiv\; \sum_z T(z) \tag{5.78}$$

and

$$T(z) \equiv I(z \neq 0) \sum_{\substack{\omega_1'\colon 0 \to z\,(\mathrm{NR}) \\ \omega_1''\colon z \to \mathrm{any}\,(\mathrm{NR}) \\ \omega_2'\colon z \to 0\,(\mathrm{NR}) \\ \omega_3\colon 0 \to \mathrm{any}\,(\mathrm{NR})}} \begin{aligned}[t] &\varrho(\omega_1')\,\varrho(\omega_1'')\,\varrho(\omega_2')\,\varrho(\omega_3)\;I(\omega_1'' \not\ni 0) \\ &\times I((\omega_1' \cup \omega_1'') \cap \omega_3) = \{0\}) \\ &\times I(\omega_2' \cap (\omega_1' \cup \omega_1'') \subset \{0, z\})\,. \end{aligned} \tag{5.79}$$

(Here we have used the invariance of ϱ under reversal of a walk.)

Summarizing, we have proven:

Lemma 5.7

$$R_3 \leq \chi(R_2 - \overline{T})\,. \tag{5.80}$$

The merit of Lemma 5.7 is that it improves the trivial bound $R_3 \leq \chi R_2$. But at this stage it is far from clear how useful this will be, given the messy structure of $T(z)$. The miracle is that by a very different computation we can obtain the *same* quantity $R_2 - \overline{T}$, and thereby deduce a useful upper bound on R_3. The starting point is the trivial identity

$$\begin{aligned} \chi^2 &\equiv \sum_{\substack{\omega_1\colon 0 \to \mathrm{any} \\ \omega_2\colon 0 \to \mathrm{any}}} \varrho(\omega_1)\varrho(\omega_2) \\ &= \sum_{\substack{\omega_1\colon 0 \to \mathrm{any} \\ \omega_2\colon 0 \to \mathrm{any}}} \varrho(\omega_1)\varrho(\omega_2)\; I(\omega_1 \cap \omega_2 = \{0\}) \\ &\qquad + \sum_{\substack{\omega_1\colon 0 \to \mathrm{any} \\ \omega_2\colon 0 \to \mathrm{any}}} \varrho(\omega_1)\varrho(\omega_2)\; I(\omega_1 \cap \omega_2 \neq \{0\})\,. \end{aligned}$$

We want to obtain a *lower* bound. The first term can be bounded below by R_2 (imposing two non-return constraints). In the second term, we fix ω_2 and split $\omega_1 = \omega_1' \circ \omega_1''$ at the *last* time that ω_1 hits $\omega_2 \setminus \{0\}$; call that point z. We then split $\omega_2 = \omega_2' \circ \omega_2''$ at the *last* time it hits z. This term therefore equals

$$\sum_{z \neq 0} \sum_{\substack{\omega_1'\colon 0 \to z \\ \omega_1''\colon z \to \mathrm{any}\,(\mathrm{NR}) \\ \omega_2'\colon 0 \to z \\ \omega_2''\colon z \to \mathrm{any}}} \varrho(\omega_1' \circ \omega_1'')\varrho(\omega_2' \circ \omega_2'')\; I(\omega_1'' \cap (\omega_2' \cup \omega_2'') \subset \{0, z\})\,. \tag{5.81}$$

We obtain a further lower bound by insisting that $\omega_1'' \cap (\omega_2' \cup \omega_2'') = \{z\}$ and that ω_1' and ω_2' be non-return walks:

$$\sum_{z \neq 0} \sum_{\substack{\omega_1'\colon 0 \to z\,(\mathrm{NR}) \\ \omega_1''\colon z \to \mathrm{any}\,(\mathrm{NR}) \\ \omega_2'\colon 0 \to z\,(\mathrm{NR}) \\ \omega_2''\colon z \to \mathrm{any}}} \varrho(\omega_1')\,\varrho(\omega_1'')\varrho(\omega_2')\,\varrho(\omega_2'')\; I(\omega_1'' \cap \omega_2' = \{z\})\; I(\omega_1'' \cap \omega_2'' = \{z\}) \tag{5.82}$$

Now we insert the identity

$$\mathrm{I}(\omega_1'' \cap \omega_2' = \{z\}) \;=\; 1 - \; \mathrm{I}(\omega_1'' \cap \omega_2' \neq \{z\}) \;, \tag{5.83}$$

leaving the other indicator function as is. The term obtained by inserting 1 yields

$$\left(\sum_{z \neq 0} K_{\mathrm{NR}}(0,z)^2 \right) R_2 \,. \tag{5.84}$$

This combines with the term R_2 found earlier to yield

$$\begin{aligned} \left(1 + \sum_{z \neq 0} K_{\mathrm{NR}}(0,z)^2 \right) R_2 \;&=\; \left(\sum_{z} K_{\mathrm{NR}}(0,z)^2 \right) R_2 \\ &=\; L^{-2} B R_2 \end{aligned}$$

since $K_{\mathrm{NR}}(0,0) = 1$; here L and B are the loop and bubble diagrams defined earlier. Summarizing, we have shown that

$$\chi^2 \geq L^{-2} B R_2 - U \tag{5.85}$$

where

$$U = \sum_{z \neq 0} \sum_{\substack{\omega_1' :\, 0 \to z\,(\mathrm{NR}) \\ \omega_1'' :\, z \to \mathrm{any}\,(\mathrm{NR}) \\ \omega_2' :\, 0 \to z\,(\mathrm{NR}) \\ \omega_2'' :\, z \to \mathrm{any}}} \begin{array}{l} [t]\varrho(\omega_1')\, \varrho(\omega_1'') \varrho(\omega_2')\, \varrho(\omega_2'')\; \mathrm{I}(\omega_1'' \cap \omega_2' \neq \{z\}) \\ \quad \times\, \mathrm{I}(\omega_1'' \cap \omega_2'' = \{z\}) \end{array} \tag{5.86}$$

Now we work on U. Since $\omega_1'' \cap \omega_2' \neq \{z\}$, we can fix ω_1'' and split $\omega_2'' = \omega_2''' \circ \omega_2''''$ at the *last* time that ω_2' hits $\omega_1'' \setminus \{z\}$; call the intersection point z'. Then we split $\omega_1'' = \omega_1''' \circ \omega_1''''$ at the *last* time it hits z'. We obtain

$$U \;=\; \sum_{z \neq 0} \sum_{z' \neq z} \sum_{\substack{\omega_1' :\, 0 \to z\,(\mathrm{NR}) \\ \omega_1''' :\, z \to z'\,(\mathrm{NR}) \\ \omega_1'''' :\, z' \to \mathrm{any}\,(\mathrm{NR}) \\ \omega_2''' :\, 0 \to z'\,(\mathrm{NR}) \\ \omega_2'''' :\, z' \to z\,(\mathrm{NR}) \\ \omega_2'' :\, z \to \mathrm{any}\,(\mathrm{NR})}} \begin{array}{l} \varrho(\omega_1')\, \varrho(\omega_1''')\, \varrho(\omega_1'''')\, \varrho(\omega_2''')\, \varrho(\omega_2'''')\, \varrho(\omega_2'') \\ \quad \times \mathrm{I}(\omega_1'''' \not\ni z)\; \mathrm{I}(\omega_2'''' \not\ni 0) \\ \quad \times \mathrm{I}(\omega_2'''' \cap (\omega_1''' \cup \omega_1'''') \subset \{z, z'\}) \\ \quad \times \mathrm{I}((\omega_1''' \cup \omega_1'''') \cap \omega_2'' = \{z\}) \end{array}$$

Note that ω_1' and ω_2''' are not subjected to any constraints except NR, so these sums can be performed immediately, yielding $K_{\mathrm{NR}}(0,z) K_{\mathrm{NR}}(0,z')$. We now throw away the constraint $\mathrm{I}(\omega_2'''' \not\ni 0)$; the remaining sum is (miraculously!) precisely $T(z'-z)$. Dropping the constraint $z \neq 0$, we obtain

$$\begin{aligned} U \;&\leq\; L^{-2} \sum_{z,z'} K(0,z)\, K(0,z')\, T(z'-z) \\ &=\; L^{-2} \sum_{u} (K * K)(u)\, T(u) \\ &\leq\; L^{-2} \left(\sup_u\, (K*K)(u) \right) \left(\sum_u T(u) \right) \\ &=\; L^{-2} B \overline{T} \,. \end{aligned}$$

The last equality follows from the positivity of the Fourier transform of K, which implies that $K * K$ takes its maximum value at $u = 0$; this value is precisely the bubble diagram B.

Summing up, we have shown:

Lemma 5.8

$$\chi^2 \geq L^{-2} B(R_2 - \overline{T}) . \tag{5.87}$$

Completion of the proof of Theorem 5.6. Combining Lemmas 5.7 and 5.8, we conclude that

$$R_3 \leq \frac{L^2 \chi^3}{B} \tag{5.88}$$

and hence

$$P_3 \leq \frac{L^4}{B} . \tag{5.89}$$

This proves Theorem 5.6. ∎

Remark. In fact, it can be proven that $P_3(m)$ has also a lower bound of the form (5.71). Moreover, with very little extra work, one can prove that

$$\left\{ \begin{array}{ll} m^{4-d} & 2 < d < 4 \\ |\log m|^{-1} & d = 4 \\ \mathrm{const}(d) & d > 4 \end{array} \right\} \lesssim P_2(m) \lesssim \left\{ \begin{array}{ll} m^{(4-d)/2} & 2 < d < 4 \\ |\log m|^{-1/2} & d = 4 \\ \mathrm{const}(d) & d > 4 \end{array} \right\} . \tag{5.90}$$

For details, we refer the reader to Lawler [351] or Park [425]. More precise estimates on $P_2(m)$ for $d \leq 4$ have been proven by Lawler and collaborators [351, 353, 352, 101, 99, 100]. Interesting non-rigorous work on $P_2(m)$ in dimension $d = 2$, exploiting two-dimensional conformal invariance, can be found in [150, 368].

6. EFT as a gas of random walks with hard-core interactions

We are now ready to set the stage for Parts II and III of this book. The method exploited in this book consists in representing spin systems in terms of interacting random walks. We can then use geometrical arguments to derive properties of these random walks, which in turn imply differential inequalities for the corresponding spin systems. These derivations are the subject of Part II of the book. The inequalities have a number of consequences for the critical behavior of the system and the nature of the continuum limit. These consequences are detailed in Part III of the book (which can be read independently of Part II).

We review in this chapter a random-walk representation originally proposed by Symanzik [493, 494] for continuum φ^4 theory, and later transcribed to the lattice (with some modifications to be described later) by Brydges, Fröhlich and Spencer (BFS) [92]. The Symanzik-BFS representation consists in essence of a cleverly partially resummed high-temperature expansion. It has been used to derive many results, old and new, on the critical behavior of φ^4 and Ising models [213, 28, 228, 92, 90, 97, 96]. Some of these results — notably the triviality of φ^4 and Ising models in dimension $d > 4$ — were first proven by Aizenman [5], using different though closely related methods.

In this chapter we first present a simple derivation of the representation and then, to give the reader the flavor of things to come, we discuss in broad terms some inequalities and the implications for the triviality and the continuum limit.

6.1 The Symanzik-BFS random-walk representation

We consider an N-component classical spin system (= lattice EFT) as defined in Section 1.4, with Hamiltonian

$$H(\boldsymbol{\varphi}) = -\sum_{\langle x,y\rangle} J_{xy}\boldsymbol{\varphi}_x \cdot \boldsymbol{\varphi}_y - \sum_x \mathbf{h}_x \cdot \boldsymbol{\varphi}_x \,, \tag{6.1}$$

where for simplicity we absorb β into J_{xy}. The single-spin distribution is assumed isotropic, and given by

$$dP_x(\boldsymbol{\varphi}_x) = g(\boldsymbol{\varphi}_x^2)\, d\boldsymbol{\varphi}_x \,.$$

As motivation for the random-walk representation, let us consider first the special case of the Gaussian model, $g(\varphi^2) = \exp[-\frac{B}{2}\varphi^2]$, at zero magnetic field ($h_z = 0$ for all z). Then the two-point function

$$\langle \varphi_x \varphi_y \rangle = (B - J)^{-1}_{xy}$$

(where $J = \{J_{xy}\}$ is the interaction matrix) can be expanded in a Neumann series

$$\langle \varphi_x \varphi_y \rangle = (B^{-1} + B^{-1}JB^{-1} + B^{-1}JB^{-1}JB^{-1} + \ldots)_{xy}$$

that is convergent whenever $B > ||J||$. The terms in this Neumann series can be interpreted as walks from x to y:

$$\langle \varphi_x \varphi_y \rangle = \sum_{\omega: x \to y} J^\omega B^{-|\omega|-1} \tag{6.2}$$

where

$$J^\omega = \prod_{s=0}^{|\omega|-1} J_{\omega(s),\omega(s+1)}$$

as defined in Section 1.6. Similarly, the higher-point correlation functions can be expressed as sums over families of walks, e.g.

$$\langle \varphi_{x_1} \varphi_{x_2} \varphi_{x_3} \varphi_{x_4} \rangle = \sum_{\substack{\omega: x_1 \to x_2 \\ \omega: x_3 \to x_4}} J^{\omega_1} J^{\omega_2} B^{-|\omega_1|-|\omega_2|-2} \; + \text{ two permutations} .$$

These formulae express the well-known connection between Gaussian field theories and simple random walks. In this section we shall prove an analogous connection between non-Gaussian field theories and certain interacting random walks. The derivation of this random walk expansion will be made for general N-component lattice field-models, but all the remaining applications will for simplicity deal only with the $N = 1$ case. (Many of our results can be generalized to $N = 2$. On the other hand, for $N > 2$ many correlation inequalities are lacking, so the generalization of these applications to spins of three or more components is not trivial, and even questionable.)

The starting point of our random-walk representation is the following sequence of integration-by-parts identities:

$$\varphi \int d\nu_n(t)\, g(\varphi^2 + 2t) = -\frac{\partial}{\partial \varphi} \int d\nu_{n+1}(t)\, g(\varphi^2 + 2t) \tag{6.3}$$

($n = 0, 1, 2, 3, \ldots$), with

$$d\nu_n(t) = \begin{cases} \delta(t)\, dt & \text{if } n = 0 \\ \\ \frac{t^{n-1}}{\Gamma(n)} \chi_{[0,\infty)}(t)\, dt & \text{if } n \geq 1 \end{cases} \tag{6.4}$$

As a shorthand we define

$$g_n(\varphi^2) \equiv \int d\nu_n(t)\, g(\varphi^2 + 2t) \tag{6.5}$$

so that (6.3) reads

$$\varphi g_n(\varphi^2) = -\frac{\partial}{\partial\varphi} g_{n+1}(\varphi^2) \,. \tag{6.6}$$

Moreover, for any well-behaved function $W(\varphi)$, a second integration by parts yields

$$\begin{aligned} \int \varphi g_n(\varphi^2) W(\varphi)\, d\varphi &= \int \Big[-\frac{\partial}{\partial\varphi} g_{n+1}(\varphi^2)\Big] W(\varphi)\, d\varphi \\ &= \int g_{n+1}(\varphi^2) \frac{\partial W}{\partial\varphi}\, d\varphi \end{aligned} \tag{6.7}$$

Now let $\mathbf{n} = \{n_x\}_{x\in\mathbb{Z}^d}$ be a collection of nonnegative integers. We define the unnormalized expectations in "background" $\mathbf{n}$,

$$[F(\varphi)]_{\mathbf{n}} = \int F(\varphi) e^{-\beta H(\varphi)} \prod_{x\in\mathbb{Z}^d} g_{n_x}(\varphi_x^2)\, d\varphi_x \,, \tag{6.8}$$

and in particular the partition function

$$Z(\mathbf{n}) = [1]_{\mathbf{n}} \,. \tag{6.9}$$

The original model corresponds to $\mathbf{n} = 0$, while the general case will appear within the random-walk representation.

Let us consider the expectation $[\varphi_x F(\varphi)]_{\mathbf{n}}$, where F is an arbitrary well-behaved function of the fields $\{\varphi\}$. Applying (6.7) with $W = Fe^{-H}$ to the integral over φ_x, and then integrating over the remaining variables $\{\varphi_z\}_{z\neq x}$, we obtain the *fundamental integration-by-parts identity*

$$\begin{aligned} [\varphi_x F]_{\mathbf{n}} &= \left[\frac{\partial F}{\partial\varphi_x} - F\frac{\partial H}{\partial\varphi_x}\right]_{\mathbf{n}+\delta_x} && (6.10) \\ &= \left[\frac{\partial F}{\partial\varphi_x}\right]_{\mathbf{n}+\delta_x} + \mathbf{h}_x [F]_{\mathbf{n}+\delta_x} + \sum_{x_1} J_{xx_1} [\varphi_{x_1} F]_{\mathbf{n}+\delta_x} \,, && (6.11) \end{aligned}$$

where of course

$$(\delta_x)_z \equiv \delta_{xz} = \begin{cases} 1 & \text{if } x = z \\ 0 & \text{otherwise} \end{cases}$$

Consider first the case $F = 1$. Then (6.11) reduces to

$$[\varphi_x] = \mathbf{h}_x [1]_{\mathbf{n}+\delta_x} + \sum_{x_1} J_{xx_1} [\varphi_{x_1}]_{\mathbf{n}+\delta_x} \,. \tag{6.12}$$

Iterating (6.12) until only terms proportional to $[1]_{\mathbf{n}+\ldots}$ are left[1], we obtain a sum over *walks* which start from x and end at a magnetic field $\mathbf{h}_{x'}$:

---

[1]The reader may wonder about the walks that wander forever without ending at a magnetic field. With some extra work, it can be shown that these contributions vanish.

$$[\varphi_x]_{\mathbf{n}} = \sum_{x'} \mathbf{h}_{x'} \sum_{\omega:\, x \to x'} J^{\omega}\, [1]_{\mathbf{n}}\,, \tag{6.13}$$

where $\mathbf{n}(\omega) = \{n_z(\omega)\}$ and $n_z(\omega)$ is the total number of visits of the walk ω to the site z.

Next, consider the case $F = \varphi_y^{(\beta)}$. Then (6.11) reduces to

$$[\varphi_x^{\alpha}\varphi_y^{\beta}]_{\mathbf{n}} = \delta_{xy}\delta^{\alpha\beta}[1]_{\mathbf{n}+\delta_x} + h_x^{(\alpha)}[\varphi_y^{(\beta)}]_{\mathbf{n}+\delta_x} + \sum_{x_1} J_{x x_1}[\varphi_{x_1}^{(\alpha)}\varphi_y^{(\beta)}]_{\mathbf{n}+\delta_x}\,. \tag{6.14}$$

Iterating (6.14) and (6.12) until only terms proportional to $[1]_{\mathbf{n}+\dots}$ are left, we obtain a sum of two classes of terms: the first class consists of walks ω that start from x and end at y; the second class consists of pairs of walks (ω_1, ω_2), one starting from x and ending at a magnetic field $\mathbf{h}_{x'}$, the other starting from y and ending at a magnetic field $\mathbf{h}_{y'}$[2]:

$$\begin{aligned}
[\varphi_x^{(\alpha)}\varphi_y^{(\beta)}]_{\mathbf{n}} &= \delta^{\alpha\beta} \sum_{\omega:\, x\to y} J^{\omega}\, [1]_{\mathbf{n}+\mathbf{n}(\omega)} \\
&\quad + \sum_{x',y'} h_{x'}^{(\alpha)} h_{y'}^{(\beta)} \sum_{\substack{\omega_1:\, x \to x' \\ \omega_2:\, y \to y'}} J^{\omega_1} J^{\omega_2}\, [1]_{\mathbf{n}+\mathbf{n}(\omega_1)+\mathbf{n}(\omega_2)}\,.
\end{aligned} \tag{6.15}$$

Analogous formulae can be obtained for higher-point correlation functions.

Now let us specialize these formulae to the original model $\mathbf{n} = \mathbf{0}$. We obtain *random-walk representations* for the correlation functions,

$$\langle \varphi_x \rangle = \sum_{x'} \mathbf{h}_{x'} \sum_{\omega: x \to x'} \varrho(\omega) \tag{6.16}$$

$$\langle \varphi_x^{(\alpha)}\varphi_y^{(\beta)} \rangle = \delta^{\alpha\beta} \sum_{\omega: x\to y} \varrho(\omega) + \sum_{x',y'} h_{x'}^{(\alpha)} h_{y'}^{(\beta)} \sum_{\substack{\omega_1:\, x \to x' \\ \omega_2:\, y \to y'}} \varrho(\omega_1, \omega_2) \tag{6.17}$$

and so forth, where we have defined, as in Section 1.6, the weights

$$\varrho(\omega_1, \dots, \omega_k) = J^{\omega_1} \dots J^{\omega_k}\, \frac{Z\left(\sum_{\alpha=1}^{k} \mathbf{n}(\omega_\alpha)\right)}{Z(\mathbf{0})}\,. \tag{6.18}$$

This is the random-walk representation promised at the beginning of this section.[3] It generalizes to arbitrary n-point functions in a straightforward manner. When $\mathbf{h}_z = 0$ for all z, this generalization is particularly elegant and simple:

[2] Again, the contribution of walks that wander forever without ending either at a magnetic field or at the site y can be shown to vanish.

[3] This derivation [97] of the random-walk representation was inspired by a discussion with E. H. Lieb which we gratefully acknowledge. An alternate derivation — which is less physically transparent, but in which the convergence problems are more easily solved — can be found in Sections 9.2.2 and 10.2.3 and in the original papers [92, 97].

$$\langle \varphi_{x_1}^{(\alpha_1)} \dots \varphi_{x_{2k}}^{(\alpha_k)} \rangle = \tag{6.19}$$

$$\sum_{\text{pairings } p} \left(\prod_{j=1}^{k} \delta_{\alpha_{p(2j-1)}\alpha_{p(2j)}} \right) \sum_{\substack{\omega_j : x_{p(2j-1)} \to x_{p(2j)} \\ j=1,\dots,k}} \varrho(\omega_1, \dots, \omega_k) \,. \tag{6.20}$$

It is useful to derive a *local-time representation* for the weights $\varrho(\omega_1, \dots, \omega_k)$: Using the simple identity

$$\nu_{n+m} = \nu_m * \nu_n \tag{6.21}$$

where $*$ denotes convolution, it is straightforward to verify that

$$\varrho(\omega_1, \dots, \omega_k) = J^{\omega_1} \dots J^{\omega_k} \int \dots \int \prod_{j=1}^{k} d\nu_{\omega_j}(t^{(j)}) \, \mathcal{Z}\Big(\sum_{j=1}^{k} t^{(j)}\Big) \,, \tag{6.22}$$

where

$$d\nu_\omega(t) = \prod_z d\nu_{n_z(\omega)}(t_z) \tag{6.23}$$

and

$$\mathcal{Z}(t) = Z^{-1} \int \Big(\prod_z g(\varphi_z^2 + 2t_z) \, d\varphi_z \Big) e^{-H(\boldsymbol{\varphi})} \tag{6.24}$$

The variables $t_z^{(j)}$ have the interpretation of *waiting times* or *local times* at the site z for the walk ω_j.

This concludes the general derivation; for the rest of the chapter we shall consider applications for one-component spins ($N = 1$). In particular, for the connected four-point function u_4 at zero magnetic field, introduced in (1.43), we get

$$u_4(x_1, x_2, x_3, x_4) = \sum_p \sum_{\substack{\omega_1 : x_{p(1)} \to x_{p(2)} \\ \omega_2 : x_{p(3)} \to x_{p(4)}}} [\varrho(\omega_1, \omega_2) - \varrho(\omega_1) \, \varrho(\omega_2)] \,. \tag{6.25}$$

Examples.

1) For $g(\varphi^2) = \exp[-\frac{1}{2}(2d + m^2)\varphi^2]$, $h_z = 0$ for all z, formula (6.14) is the usual random-walk representation of the Green's function $(-\Delta + m^2)^{-1}$, while (6.22) is the local-time representation

$$(-\Delta + m^2)^{-1} = \int_0^\infty e^{-m^2 t} e^{\Delta t} \, dt \,. \tag{6.26}$$

2) For $g(\varphi^2) = \exp\Big[-\frac{\lambda}{4}\varphi^4 - \frac{B}{2}\varphi^2\Big]$, we have $g(\varphi^2 + 2t) = \exp\Big[-\frac{\lambda}{4}\varphi^4 - (\frac{B}{2} + \lambda t)\varphi^2 - (\lambda t^2 + Bt)\Big]$. Therefore, the principal effect of the "background" $t = \{t_z\}$ is to generate a space-dependent mass term $\lambda t_z \varphi_z^2$. The consequences of this term will be analyzed in Part II (Chapter 11). Among the results are correlation inequalities which imply the triviality of φ^4 theory in dimension $d > 4$. The random-walk representation has also been used [97, 74] to develop a "skeleton expansion" and prove "skeleton bounds" for lattice $\lambda\varphi^4$ theory [see (5.13),

(5.27) for similar inequalities]. These results have enabled Brydges, Fröhlich and Sokal to give a simple construction of continuum φ_2^4 and φ_3^4 theories, exhibit a mass gap at weak coupling, and establish the asymptotic nature of perturbation theory [96]; see also [74] for some additional results.

Further remarks.

1) The foregoing random-walk representation, due to Brydges, Fröhlich and Spencer (BFS) [92], is in essence a *partially resummed* high-temperature expansion: the weights $\varrho(\omega_1, \ldots, \omega_k)$ are defined in terms of the lattice field theory itself, cf. (6.18) or (6.22). An alternative approach is to generate a *fully expanded* high-temperature expansion, which rewrites the partition function of lattice field theory as a gas of closed random walks interacting via an explicit soft-core exclusion (contact interaction) [493, 494]. This representation — called *Symanzik's complete polymer representation* — further shows that one may interpret $\varrho(\omega_1, \ldots, \omega_n)$ as a correlation function of n polymer chains $\omega_1, \ldots, \omega_n$ in a gas (or "solvent") of polymer loops. See Section 9.2.2. The advantage of the complete polymer representation over the BFS representation is that the weights are given by *explicit* contact interactions; the disadvantage is that one has to deal with the gas of loops. The BFS representation is therefore combinatorially simpler but physically more complicated. As we shall see in the next section, the complicated weights of the BFS representation can in some cases (e.g. φ^4 theory) be controlled by correlation inequalities; this is the strategy underlying the triviality proofs. To date, few rigorous results have been obtained using the complete polymer representation — but this may simply be for lack of trying!

2) In the above, we have considered a quadratic Hamiltonian $H(\varphi)$, and have placed all higher-order couplings (e.g. φ^4) into $g(\varphi^2)$. An alternate approach is to set $g(\varphi^2) = \exp[-\frac{1}{2}(2d + m^2)\varphi^2]$ and place everything else in $H(\varphi)$, e.g.

$$H(\varphi) = -\sum_{\langle x,y \rangle} J_{xy}\varphi_x\varphi_y + \sum_x \Big(\frac{\lambda}{4}\varphi_x^4 - h\varphi_x\Big) . \tag{6.27}$$

Everything up to (6.10) proceeds as before. Now, however, $\partial H/\partial\varphi_x$ contains a term φ_x^3, which generates *branching random walks*. Regrouping terms in this sum, we can obtain formally the ordinary Feynman perturbation theory. However, in this expansion the convergence problems are insurmountable, as Feynman perturbation theory really is divergent!

3) A modified random-walk representation has turned out to be rather useful to prove Borel summability of the $1/N$-expansion for non-critical $O(N)$ nonlinear σ-models on the lattice [348].

6.2 Consequences for triviality

One of the major applications of the random-walk representation is to prove the triviality of the continuum φ_d^4 theory in $d \gtrsim 4$ dimensions. The basic idea is that certain connected correlation functions of the lattice field theory can be bounded in terms of *intersection properties* of the associated random walks. Direct analysis of the random walks yields bounds on their intersection properties that are analogous to (but somewhat weaker than) the results of Chapter 5 for simple random walks. These bounds then have implications for the triviality of the continuum limit.[4] In this section we give an overview of the arguments; the details can be found in Chapters 11, 12 and 15.

All of the triviality theorems are based on a few simple properties of the weights (6.22). Let

$$g(\varphi^2) = \exp\left[-\frac{\lambda}{4}\varphi^4 - \frac{B}{2}\varphi^2\right] \tag{6.28}$$

with $\lambda > 0$ and B real. The corresponding lattice field theory satisfies the Griffiths inequalities [491], in particular

$$\langle \varphi_{x_1} \cdots \varphi_{x_k} \exp[-\sum_z t_z \varphi_z^2] \rangle \leq \langle \varphi_{x_1} \cdots \varphi_{x_k} \rangle \, \langle \exp[-\sum_z t_z \varphi_z^2] \rangle \tag{6.29}$$

if $t_z \geq 0$ for all z. These inequalities and the local-time representation of $[\cdot]_{\mathbf{n}}$ have the following two consequences:

(I) For all $\mathbf{n}$, we have

$$Z(\mathbf{n})^{-1} [\varphi_{x_1} \cdots \varphi_{x_k}]_{\mathbf{n}} \leq \langle \varphi_{x_1} \cdots \varphi_{x_k} \rangle \,, \tag{6.30}$$

When $h_z = 0$ for all z, we can also formulate this inequality as follows:

$$\begin{aligned} &\sum_p \sum_{\substack{\omega_j:\, x_{p(2j-1)} \to x_{p(2j)} \\ j = 1, \ldots, k}} \varrho(\omega_1, \ldots, \omega_k, \omega_1' \ldots, \omega_l') \\ &\leq \Big(\sum_p \sum_{\substack{\omega_j:\, x_{p(2j-1)} \to x_{p(2j)} \\ j = 1, \ldots, k}} \varrho(\omega_1, \ldots, \omega_k) \Big) \varrho(\omega_1', \ldots, \omega_l') \,. \end{aligned} \tag{6.31}$$

(II) If $\{\omega_1 \ldots \omega_k\} \cap \{\omega_1' \ldots \omega_l'\}$ is empty, then

$$\varrho(\omega_1, \ldots, \omega_k, \omega_1', \ldots, \omega_l') \geq \varrho(\omega_1, \ldots, \omega_k) \varrho(\omega_1', \ldots, \omega_l') \,. \tag{6.32}$$

We observe that (6.31) and (6.32) are inequalities going in opposite directions. Loosely speaking, (6.31) says that the interaction between the walks is, on the average, repulsive, while (6.32) implies that the interaction between

[4]This approach is foreshadowed in the profound original paper of Symanzik [494]. Here he develops a *formal* (non-rigorous) Brownian-path representation of continuum φ^4 field theory, and then argues *heuristically* that the non-intersection of Brownian paths in dimension $d \geq 4$ should imply the triviality of continuum φ^4 field theory in these dimensions. The rigorous triviality proof follows a similar intuition, but works on the lattice rather than in the continuum, and uses the BFS representation instead of the complete polymer representation. Unfortunately, the currently available rigorous results for $d = 4$ fall short of proving Symanzik's conjectures.

nonoverlapping walks is in fact attractive. The existence of these complementary inequalities, resulting from a subtle balance of effects, is in great part responsible for the power and yet relative simplicity of the method. The inequalities (I) and (II) are discussed in more detail in Chapter 11.

A second class of inequalities characterizing the weights ϱ refers to the effect of splitting a path. For example, if we split each path ω into two consecutive pieces ω_1 and ω_2, it is not hard to see (Chapter 11) that

$$\sum_{\substack{\omega:\, x \to y \\ \text{such that } z \in \omega}} \varrho(\omega) = \sum_{z'} J_{zz'} \sum_{\substack{\omega_1:\, x \to z \\ \omega_2:\, z' \to y}} \varrho(\omega_1, \omega_2) + \sum_{\omega_1: x \to y} \varrho(\omega_1)\delta_{yz} \,. \tag{6.33}$$

Let us now see how these properties produce important inequalities for the correlation functions of the underlying spin model. For the time being we set $h_z \equiv 0$. From (6.25) and (6.31) we obtain the Lebowitz inequality

$$u_4(x_1, x_2, x_3, x_4) \leq 0 \,. \tag{6.34}$$

On the other hand, we can classify the terms in (6.25) according to whether the paths ω_1 and ω_2 intersect or not. The contribution of the non-intersecting pairs is positive by (6.32), therefore u_4 can be bounded below by the sum over *intersecting* walks alone:

$$\begin{aligned} u_4(x_1, x_2, x_3, x_4) \;&\geq\; \sum_{\text{pairings } p} \sum_{\substack{\omega_1:\, x_{p(1)} \to x_{p(2)} \\ \omega_2:\, x_{p(3)} \to x_{p(4)} \\ \omega_1 \cap \omega_2 \neq \varnothing}} [\varrho(\omega_1, \omega_2) - \varrho(\omega_1)\varrho(\omega_2)] \\ &\geq\; - \sum_{\text{pairings } p} \sum_{z} \sum_{\substack{\omega_1:\, x_{p(1)} \to x_{p(2)} \\ \omega_2:\, x_{p(3)} \to x_{p(4)} \\ \omega_1 \cap \omega_2 \ni z}} \varrho(\omega_1)\varrho(\omega_2) \end{aligned}$$

(since $\varrho(\omega_1, \omega_2) \geq 0$). If we now split the paths ω_1 and ω_2, we get from (6.33)

$$\begin{aligned} &u_4(x_1, x_2, x_3, x_4) \geq \\ &\quad - \sum_{p} \sum_{z,z',z''} J_{zz'} J_{zz''} \sum_{\substack{\omega_1':\, x_{p(1)} \to z \\ \omega_1':\, z' \to x_{p(2)}}} \sum_{\substack{\omega_2':\, x_{p(3)} \to z \\ \omega_2'':\, z'' \to x_{p(4)}}} \varrho(\omega_1', \omega_2'')\varrho(\omega_2', \omega_2'') + \mathcal{E} \end{aligned}$$

where $\mathcal{E}$ is a correction arising from terms with $z = x_j$ for some $j = 1, 2, 3, 4$. Finally, to make contact with physical observables, we use once more (6.31) and (6.17) to obtain

$$u_4 \geq - \sum_{p} \sum_{z,z',z''} J_{zz'} J_{zz''} \langle \varphi_{x_{p(1)}} \varphi_z \rangle \, \langle \varphi_{z'} \varphi_{x_{p(2)}} \rangle \, \langle \varphi_{x_{p(3)}} \varphi_z \rangle \, \langle \varphi_{z''} \varphi_{x_{p(4)}} \rangle + \mathcal{E} \,. \tag{6.35}$$

Diagrammatically,

$$0 \;\geq\; u_4(x_1, x_2, x_3, x_4) \;\geq\; -\sum_p$$

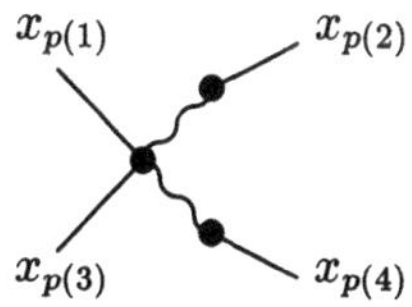

$$+\; \mathcal{E} \qquad (6.36)$$

where x —— y denotes the two-point function ("full propagator") $\langle\varphi_x\varphi_y\rangle$, z ●∿∿● z' denotes $J_{zz'}$, and internal vertices ● are summed over. This is the Aizenman-Fröhlich inequality [5, 213].

To see the implications of this inequality for the continuum limit, we introduce the rescaled four-point Ursell function (see Chapter 3)

$$u_{4,\theta}(x_1, \ldots, x_4) = \alpha(\theta)^4\, u_4(\theta x_1, \ldots, \theta x_4)\;, \qquad (6.37)$$

with parameters $\lambda(\theta)$, $B(\theta)$ and $J(\theta)$ approaching the critical surface as as $\theta \to \infty$ from within the single-phase regime. We also use the *infrared bound*

$$0 \;\leq\; \langle\varphi_x\varphi_y\rangle \;\leq\; \text{const} \times J^{-1}|x-y|^{-(d-2)}\;, \qquad (6.38)$$

which is proven by a clever use of the transfer-matrix formalism [222, 218] together with correlation inequalities [481]. From (6.37), (6.36) and (6.38) we obtain

$$0 \;\geq\; u_{4,\theta}(x_1, x_2, x_3, x_4) \;\geq\; -\,\text{const} \times \theta^{4-d} \qquad (6.39)$$

whenever $|x_i - x_j| \geq \delta$ for $i \neq j$, for some arbitrarily small but positive δ. (The contribution of the correction term $\mathcal{E}$ is of order θ^{2-d}, hence is negligible for $d > 2$.) This proves that, in dimension $d > 4$,

$$S^{\text{cont}}(x_1, \ldots, x_4) \equiv \lim_{\theta\to\infty} u_{4,\theta}(x_1, \ldots, x_4) = 0! \qquad (6.40)$$

It is not hard to extend this result to show that

$$\lim_{\theta\to\infty} u_{n,\theta}(x_1, \ldots, x_n) = 0 \qquad (6.41)$$

for all $n > 2$. Therefore, all theories constructed as a scaling limit of ferromagnetic nearest-neighbor φ^4 lattice theories are *trivial* for $d > 4$.

In *four dimensions*, there is also compelling evidence for triviality, but the rigorous proof has not yet been completed except for small λ [290, 298]. If we assume that, for $|x-y| \geq \delta > 0$ and uniformly in θ,

$$\alpha(\theta)^2\langle\varphi_{\theta x}\varphi_{\theta y}\rangle_{\lambda(\theta),B(\theta),J(\theta)} \;\leq\; k(\varepsilon,\delta)|x-y|^{-\varepsilon}\;, \qquad (6.42)$$

for some constants $\varepsilon > 0$ and $k(\varepsilon,\delta) < \infty$, we obtain the bounds

$$0 \geq u_{4,\theta}(x_1, \ldots, x_4) \geq -\text{const.}\ (\alpha(\theta)^{-1}\theta)^p\;, \qquad (6.43)$$

for some $p > 0$ depending on ε, provided that $|x_i - x_j| \geq \delta > 0$ for $i \neq j$. From this inequality we deduce that the *continuum limit* ($\theta \to \infty$) *in four dimensions is trivial, unless*

$$\alpha(\theta) \sim \theta^{(d-2)/2} = \theta \tag{6.44}$$

i.e. unless the (ultraviolet) dimension of the field φ_{ren} is *canonical.* On the other hand, if φ_{ren} has canonical (free-field) dimensions and the continuum Green's functions $S^{\text{cont}}(x_1, \ldots, x_{2n})$ are scale-invariant, then the theory is a *free-field theory* by a general theorem due to Pohlmeyer [433]. It follows that *there is no non-trivial scale-invariant continuum theory that can be obtained as a scaling limit of ferromagnetic φ^4 lattice theories* compatible with (6.42). (Heuristically this means that the renormalization-group β-function $W(g)$ has no non-trivial roots!) The detailed proofs and interpretation of these results will be discussed in Chapter 15.

Aizenman and Graham [15] have improved inequality (6.36) as follows:

$$0 \geq \bar{u}_4 \geq -\text{const} \times J^2 \chi^2 \frac{\partial \chi}{\partial J} - \mathcal{E}\,. \tag{6.45}$$

When $d = 4$ this inequality implies that in the continuum limit *the theory is trivial unless mean-field theory gives the exact behavior of* $\chi(\beta)$ *as* β *approaches* β_c, that is, unless $\gamma = 1$ *without* multiplicative logarithmic corrections. Moreover, it can be shown that the violations of the mean-field scaling laws in four dimensions are at most logarithmic [15].

In summary, while triviality of φ^4_4, in full generality, is not (yet) a rigorous mathematical theorem, there is strong evidence for that contention.

It is instructive to interpret the above results in terms of the random-walk picture, and in particular to compare them with the results of Chapter 5. The bound (6.36) implies that

$$\begin{aligned} g &\equiv -\chi^{-2} m^d \sum_{x_2,x_3,x_4} u_4(x_1,x_2,x_3,x_4) \\ &\leq 3\chi^{-2} m^d \sum_z \sum_{\omega_1 \cap \omega_2 \ni z} \varrho(\omega_1)\, \varrho(\omega_2) \end{aligned}$$

where $\chi = \sum_y \sum_{\omega: x \to y} \varrho(\omega)$. That is,

$$g \leq 3p$$

where p is the *average intersection probability for two "field-theoretic" walks,* ω_1 and ω_2, with weights $\varrho(\omega_1)$, $\varrho(\omega_2)$. Moreover, by (6.38)

$$0 \leq \langle \varphi_x \varphi_y \rangle = \sum_{\omega: x \to y} \varrho(\omega) \leq \text{const} \times J^{-1} |x-y|^{-(d-2)}\,, \tag{6.46}$$

which says, roughly speaking, that the Hausdorff dimension of a field-theoretic walk ω is at most 2. Hence one expects, by analogy with the arguments in Chapter 5 for *simple* random walks, that $p \to 0$ as the walks become critical ($m \to 0$, $\chi \to \infty$) in dimension $d \geq 4$. For $d > 4$, that is the content of inequalities (6.36)–(6.39); for $d = 4$ it is an open question, which depends on logarithms that have not yet been controlled rigorously.

All these results on triviality refer to the scaling limit taken through systems in the symmetric phase (i.e. zero magnetic field above the critical temperature). The approach through the broken-symmetry phase is more complicated. There are only some partial results based on the random-walk analogue of the Aizenman-Fröhlich inequality (6.36) for $h \neq 0$. We sketch here its derivation. If $h_z \equiv h \geq 0$ for all z then the random walk representations for the one-point and two-point functions are (Chapter 10)

$$\langle \varphi_x \rangle = h \sum_y K(x,y) \tag{6.47}$$

and

$$\langle \varphi_x \varphi_y \rangle = h^2 \sum_{x',y'} \sum_{\substack{\omega_1 \colon x \to x' \\ \omega_2 \colon y \to y'}} \varrho(\omega_1, \omega_2) + K(x,y) \; , \tag{6.48}$$

where

$$K(x,y) \equiv \sum_{\omega \colon x \to y} \varrho(\omega) \tag{6.49}$$

is a rather mysterious kernel that has no obvious physical interpretation.Using a generalization of (6.31),

$$h \sum_{x'} \sum_{\omega_1 : x \to x'} \varrho(\omega_1, \omega_2) \leq \langle \varphi_x \rangle \, \varrho(\omega_2) \tag{6.50}$$

we obtain, by (6.47) and (6.49),

$$\langle \varphi_x \varphi_y \rangle \leq \langle \varphi_x \rangle \langle \varphi_y \rangle + K(x,y) \; . \tag{6.51}$$

This inequality yields an upper bound on χ. Indeed, multiplying by h and summing over y we get, using (6.47),

$$h\chi \equiv h \sum_y \langle \varphi_x ; \varphi_y \rangle \leq \langle \varphi_x \rangle \; . \tag{6.52}$$

This is the weak Griffiths-Hurst-Sherman (GHS) inequality, which should be compared with the Lebowitz inequality (6.34). One simple consequence of this inequality is that, for $h > 0$, the expectation value $\langle \cdot \rangle$ clusters integrably fast.

To obtain an inequality going in the opposite direction, we proceed along the very same steps used to prove the AF inequality (6.36), but using the generalization of (6.25) and (6.33) to the case of nonzero field. The result is (Chapter 12)

$$\langle \varphi_x ; \varphi_y \rangle \geq K(x,y) - |J|^2 \langle \varphi \rangle^2 \sum_z K(x,z) K(y,z) + \mathcal{E}' \; . \tag{6.53}$$

Hence, after multiplying by h and summing over y, we get, using (6.47),

$$h\chi \geq \langle \varphi \rangle - \frac{\beta^2}{h} \langle \varphi \rangle^4 + \mathcal{E}' \; . \tag{6.54}$$

This and more general inequalities are studied in Chapter 12 and applied in Chapter 14 to prove many bounds on critical exponents describing the approach to the critical point as $h \to 0$. Moreover, they are used in Chapter 15 to prove the *nonexistence* of a broken-symmetry phase in $\lambda\varphi^4_d$ $(d > 4)$ with finite renormalized one-point and two-point functions, for a large class of approaches to the critical surface. This result is a precise version of (one form of) the violation of hyperscaling.

6.3 Consequences for nontriviality

A complementary application of the random-walk representation is to prove nontriviality of the continuum φ^4_d theory in dimensions $d = 2, 3$ for weak (bare) couplings. Although these results are not part of the main subject of the present book, we feel it worthwhile to describe them briefly here.

While there is a wealth of methods to study these superrenormalizable field theories (see the references in [96, p. 145]), the one based on random-walk representations is by far the simplest. This method does not rely on any renormalization-group insight (which is what makes it both simpler and less powerful than the other methods); rather, it is based on correlation inequalities, the so-called "skeleton inequalities" [97, 74]. These inequalities provide rigorous upper and lower bounds on the correlation functions in the form of partial power series in the bare couplings with coefficients given by "skeleton" amplitudes. These amplitudes are Feynman diagrams *without self-energy insertions* in which all lines stand for *full* propagators. The skeleton inequalities can be combined with the Schwinger-Dyson equation to derive an inequality bounding the two-point function (full propagator) in terms of itself. By a clever continuity argument (the "forbidden region" argument), the two-point function can be bounded *uniformly in the lattice spacing*. It then follows from the skeleton inequalities that mass-renormalized perturbation theory is asymptotic uniformly in the lattice spacing; in particular, the theory is non-trivial (non-Gaussian) for sufficiently weak coupling. By a refinement of the continuity argument, one can prove the strict positivity of the mass gap. (One can also prove the existence of the continuum limit for *arbitrary* couplings — as long as the theory remains in the single-phase region — but in this case the method says nothing about nontriviality or the mass gap.) The main tool used in the derivation of the skeleton inequalities is the random-walk representation — in particular the "splitting lemma" (Lemma 5.2)— combined with Griffiths inequalities.

The relative simplicity of the method comes at the cost of some obvious limitations. First, the fact that it requires Griffiths inequalities — which at present are known to be valid only in one- and two-component theories — limits its use to such models. (The method also applies to the Edwards model of self-repelling walks [75], which can be considered a "zero-component" model.) Moreover, such inequalities hold only for *real, nonnegative* values of the coupling constants. Hence there is no possibility to obtain Borel summability, which is a consequence of more full-fledged methods [381]. Another aspect that has not yet been proven using skeleton inequalities is the Euclidean (rotation) invariance of the continuum theory.

In the following we describe some of the highlights of the argument proving the nontriviality of the limit. We do not discuss the proof of the actual existence of the continuum limit, which can also be obtained with this method. We only point out that it consists in a compactness argument that proves existence of limits for subsequences. The compactness of the set of Green functions is proven using Griffiths and Gaussian inequalities.

The proof of the nontriviality of the theory rests on the following two ingredients:

1) The first three skeleton inequalities

$$u_4(x_1, x_2, x_3, x_4) \;\leq\; 0 \tag{6.55a}$$

$u_4(x_1, x_2, x_3, x_4) \;\geq\; -6\lambda$ 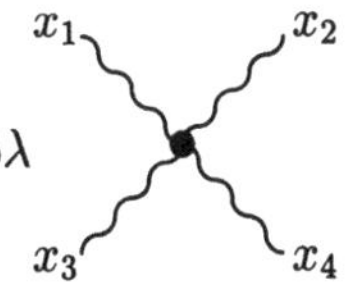

(6.55b)

$u_4(x_1, x_2, x_3, x_4) \;\leq\; -6\lambda$ 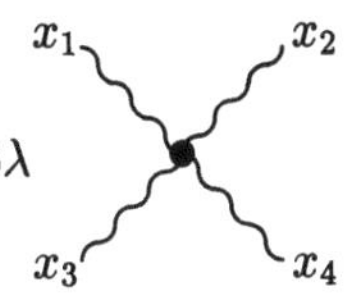

$+18\lambda^2$ [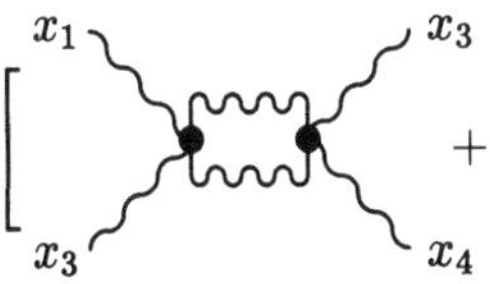

+ two permutations]
(6.55c)

where x  y stands for the full propagator $\langle \varphi_x \varphi_y \rangle$ and integration over internal vertices is understood. The first inequality is just Lebowitz' inequality. The second one is a tree bound that looks very similar to the AF inequality (6.35) but has a very different significance due to the factor λ. Inequality (6.35) is a bound independent of λ, and hence is very useful in high dimensions ($d > 4$) where it implies triviality, but is bad in low dimensions. In contrast, (6.55b), while not terribly useful in $d > 4$, has important consequences for $d = 2, 3$.

We shall not discuss the proof of (6.55b) and (6.55c) [97], but just point out that they result from specific properties of the φ^4 model, rather than from general properties of the weights [contrast (6.32) which implies the AF inequality (6.35)].

2) A uniform bound on the 2-point functions: For a lattice of spacing ε — denoted $\mathbb{Z}_\varepsilon^d$ — let us introduce the norm

$$|||f||| = \sup_{x\in\mathbb{Z}_\varepsilon^d} |f(x)| + \varepsilon^d \sum_{x\in\mathbb{Z}_\varepsilon^d} |f(x)| \tag{6.56}$$

and the notation

$$E^{(\varepsilon)} = \langle\varphi_0\varphi_x\rangle^{(\varepsilon)} - \langle\varphi_0\varphi_x\rangle_G^{(\varepsilon)} \tag{6.57}$$

where $\langle\ \rangle_G$ indicates the expectation in the *Gaussian* theory ($\lambda = 0$): that is, $\langle\varphi_0\varphi_x\rangle_G$ is the *free* propagator. Then, for $m < 1$ there is a positive constant λ_m independent of the lattice spacing ε, such that for $0 \le \lambda \le \lambda_m$ we have

$$|||E^{(\varepsilon)}||| \le c(m)\lambda^2 \tag{6.58}$$

for some constant $c(m)$.

To prove (6.58), the skeleton inequalities are combined with a field equation ("Schwinger-Dyson equation") to obtain bounds on the 2-point function in terms of itself [96]. Estimating full propagators in terms of the $|||\cdot|||$-norm one obtains for $d < 10/3$ a bound of the form

$$|||E^{(\varepsilon)}||| \le \sum_{n=1}^{3} \lambda^n P_n(|||E^{(\varepsilon)}|||) \tag{6.59}$$

for polynomials P_1, P_2 and P_3 with positive coefficients and with P_1 having zero constant term. In addition, with the help of Griffiths and Simon-Lieb inequalities, it is proven that $|||E^{(\varepsilon)}|||$ is continuous in λ for each fixed ε. The bound (6.58) is then proven, via the following "forbidden region" argument: Choose λ_m so small that

$$\sum_{n=1}^{3}(\lambda_m)^n P_n(2) \le 1\ . \tag{6.60}$$

Then, for $\lambda \in [0, \lambda_m]$, the bound $|||E^{(\varepsilon)}||| \le 2$ implies that $|||E^{(\varepsilon)}||| \le 1$ by (6.59). In other words, uniformly in ε, $|||E^{(\varepsilon)}|||$ cannot take values between 1 and 2. Since at $\lambda = 0$ we have $|||E^{(\varepsilon)}||| = 0$ (Gaussian case), it follows, from the continuity of $|||E^{(\varepsilon)}|||$ in λ that $|||E^{(\varepsilon)}||| \le 1$ for $0 \le \lambda \le \lambda_m$ *uniformly in* ε. Two further iterations of (6.59) yield (6.58).

From formulas (6.55c) and (6.58), the nontriviality of φ_d^4 follows, for $d = 2, 3$ and $\lambda > 0$ sufficiently small. Indeed, by (6.58) the difference between the full and Gaussian propagators in the $|||\cdot|||$-norm is of order λ^2. Therefore, the internal integrations in (6.55c) are convergent for $d < 4$ and the term with bracketed coefficient is of order λ^2. Hence, at small λ, this term is dominated by the "tree term", so that $\overline{u}_4 < 0$. Indeed, by (6.55b)–(c) we conclude that u_4 takes precisely the value given by the first-order perturbation expansion — that is, by a tree graph with free propagators — up to an error of order λ^2. Similarly, it can be proven that, up to an error of order λ^2, all the Green functions $\langle \varphi_{x_1} \cdots \varphi_{x_{2n}} \rangle$ take the values given by the first-order perturbation expansion [96].

By additional work along these same lines, the skeleton inequalities can be proven to *all* orders [74]; and this can be employed to prove the asymptoticity of perturbation theory, uniformly in the lattice spacing, to *all* orders [74].

7. Random-surface models

This chapter has been co-authored with Gerhard Hartsleben.

Random-surface (RS) models are fairly natural generalizations of random-walk (RW) models. Although the rest of this book is devoted exclusively to RW models, we present in this chapter a short account of RS models, for the purpose of comparison with RW models and as an introductory review of a subject which is finding an increasing number of applications in condensed-matter physics and quantum field theory.

Examples of problems in condensed-matter physics which involve random surfaces are:

(1) Growth of crystals; properties of crystalline surfaces.

(2) Interfaces and domain walls separating different phases of a physical system such as a fluid or magnet.

(3) Wetting.

(4) Domain-wall wandering in uniaxial commensurate-incommensurate transitions.

(5) Contribution of fluctuations of domain walls bounded by frustration loops to the entropy and free energy of three-dimensional spin-glass models with short-range interactions.

(6) Catalysis of chemical reactions on surfaces of "large" Hausdorff dimension.

(7) Statistical mechanics of branched and membrane-like polymers, of microemulsions, etc.

Some useful references may be [527, 374, 373, 114, 205, 122, 170, 221, 432].

Applications of random surfaces to quantum field theory are:

(1) Random-surface representation of lattice gauge theories [504, 515, 459, 151, 212, 98].

(2) Surface theories as approximations to gauge theories: e.g. large-N lattice gauge theory as a surface theory [123, 332, 344, 214]; surface theories arising as limits of lattice gauge theories, such as self-avoiding surface models and plaquette percolation [506, 161, 153, 14].

(3) Discrete approximations to string theories: Nambu-Goto model [403, 265]; Polyakov model [437, 156, 18]; lattice approximation to string theory [506, 161, 153, 154, 21, 337, 115, 23, 20, 22].

(4) Two-dimensional quantum gravity [216, 338].

This chapter is intended mainly as an illustration of the basic concepts in the theory of random surfaces. We will not attempt an all-encompassing review of the applications; rather we shall concentrate on a particular, yet representative, family of models. We have selected models arising from the regularization of string theory and two-dimensional quantum gravity. These models exhibit many of the interesting problems involved in the theory of random surfaces, and they constitute a natural generalization of the random-walk models (such as simple random walks) which are one of the principal topics of this book.

7.1 Continuum random-surface actions

For the reader's convenience, we give here a short review of the continuum string and quantum-gravity actions referred to in the rest of the chapter. The evolution of strings in d-dimensional space-time is described in terms of world sheets [266, 128], that is, by maps X from a two-dimensional parameter space D_2 into the d-dimensional Minkowski space M^d. The evolution of each world sheet is governed by a suitably chosen action $\mathcal{A}$. Classically, the variation of $\mathcal{A}$ with respect to the world sheets yields the equations of motion; quantum-mechanically one must instead construct functional integrals in which each world sheet is weighted by a factor $\exp(i\mathcal{A}/\hbar)$. The action initially considered in the literature was the Nambu-Goto action [403, 265]

$$\mathcal{A}_{\rm NG}(X) \;=\; \frac{1}{2\pi\alpha'}\int_{D_2} d^2\xi\sqrt{|\det h|}\,, \tag{7.1}$$

where α' is a constant (inverse of the "bare string tension") and $h^{\alpha\beta} = \frac{\partial X}{\partial\xi_\alpha}\frac{\partial X}{\partial\xi_\beta}$ is the metric on D_2 induced by the metric on M^d through the world sheet X. In short, $\mathcal{A}_{\rm NG}$ is proportional to the area of the world sheet measured with the metric h. An alternative action was first proposed by Deser and Zumino [127] and discussed in more detail by Polyakov [437],

$$\begin{aligned}\mathcal{A}_{\rm P}(X,g) \;&=\; \frac{1}{4\pi\alpha'}\int_{D_2} d^2\xi\sqrt{|\det g|}\left[g^{\alpha\beta}\partial_\alpha X_\mu\partial_\beta X_\mu \,+\, \lambda\right]\\ &=\; \mathcal{A}_{\rm P}^0(X,g) + \lambda A(g)\,,\end{aligned} \tag{7.2}$$

where λ is a constant, and the functions $g^{\alpha\beta}$ — components of a Lorentzian metric on D_2 — are extra degrees of freedom. We denote by $A(g)$ the area of the surface D_2 in the metric g. Classically, the action (7.2) is equivalent to (7.1): the variation of $\mathcal{A}_{\rm P}$ with respect to the functions $g^{\alpha\beta}$ yields the equation of motion $g_{\alpha\beta} = \partial_\alpha X\partial_\beta X \equiv h_{\alpha\beta}$ and, moreover, consistency conditions force $\lambda = 0$. Hence, the equations of motion for the fields X are identical for the two actions, (7.1) and (7.2). The extra parameters $g^{\alpha\beta}$ can be interpreted as Lagrange multipliers whose presence simplifies the dependence of the action on the fields X: from a square-root dependence in $\mathcal{A}_{\rm NG}$ to a quadratic dependence in $\mathcal{A}_{\rm P}$. This simpler dependence is crucial in the quantum-mechanical case, where the two actions are no longer equivalent: the quantization of a theory based on (7.2) — that is, the definition of integrals with a weight $\exp(i\mathcal{A}_P/\hbar)$ — seems to be more feasible than for the Nambu-Goto action (7.1). The quantum-mechanical dynamics does not force λ to be zero.

Let us discuss briefly the quantization of the Polyakov action. As for the case of point particles (Section 1.5), it is more convenient to pass to the Euclidean formulation, in which M^d is substituted by the d-dimensional Euclidean space E^d and X_0 is replaced by iX_0. Hence, in the Euclidean formulation of the Polyakov string, one attempts to give a precise mathematical definition of the functional measure

$$e^{-\mathcal{A}_P^0(X,g)/\hbar} e^{-\lambda A(g)/\hbar} \mathcal{D}X\mathcal{D}g \,, \tag{7.3}$$

where g ranges over the Riemannian metrics on a two-dimensional surface D_2 of fixed topological type, $\mathcal{D}X$ is a formal Lebesgue measure, and $\mathcal{D}g$ is the formal volume element on the space of Riemannian metrics.

The measure (7.3) has an alternative — and quite striking — interpretation: it is a theory of a free massless d-component scalar field living on the two-dimensional surface D_2, coupled to two-dimensional quantum gravity. (Note that, in two dimensions, the Einstein-Hilbert action is trivial, i.e. a topological invariant.) The constant λ can then be interpreted as a "bare cosmological constant". Correspondingly, the action $\mathcal{A}_P$ is interpreted as having two parts: the "coupling" $\mathcal{A}_P^0(X,g)$ between the "matter field" X and the "gravitational degrees of freedom" $g^{\alpha\beta}$, and the "pure gravity" part $A(g)$. In particular, models of two-dimensional "pure gravity" are obtained by setting $\mathcal{A}_P^0 = 0$, that is by turning off the couplings between the metric g and the "matter fields" X. This is equivalent to setting $d = 0$.[1]

From the quantum-gravity point of view we can envision possible generalizations of (7.3). For instance, the Polyakov action $\mathcal{A}_P^0$ could be replaced by some other action; in particular, the *free* scalar fields X could be replaced by *interacting* scalar fields (e.g. by adding an X^4 coupling). Moreover, the target space of the maps X need not be Euclidean space-time; it could be a non-Euclidean Riemannian manifold or even a discrete set. In these cases, (7.3) defines a two-dimensional nonlinear σ-model or discrete-spin model (e.g. Ising or Potts model) coupled to quantum gravity. While it may not be clear *a priori* what discrete-spin models on two-dimensional world sheets mean, several years of conformal field theory have taught us how to think of such models. Besides, if the world sheet is discretized they make perfectly good sense. The corresponding lattice models, endowed with an action $\mathcal{A}_{\text{Ising}}^0$ (or $\mathcal{A}_{\text{Potts}}^0$) describing an Ising

[1] A short remark about the word "dimension". When interpreting (7.3) as the quantization of gravity, the reader must be aware that there are two notions of dimension in the theory: the dimension of the Riemannian manifold — two in this chapter — and the dimension d of the matter field X. The latter is the only variable dimension in this chapter, and hence will be the only one referred to in the sequel; "quantum gravity" will always mean "quantum gravity for Riemannian manifolds of dimension two". No such confusion arises when interpreting (7.3) as a string theory, because the word "string" already implies that the parameter space (Riemannian manifold) is two-dimensional.

(or Potts) interaction, are known as Ising (or Potts) models "on a random lattice". They can be considered as toy models for quantum-gravity theories, or as statistical-mechanical models in their own right. Their study has become very popular in recent times, and we shall briefly survey them in Section 7.5.3.

For the Polyakov string (7.3), one might attempt to define string Green functions by considering a parameter space D_2 of fixed genus H, with a number n of boundary components. Such a parameter space is obtained by deleting n small disks from a compact surface of genus H without boundary. One may then try to constrain the maps X: $D_2 \to E^d$ to map the ith boundary component of D_2 to a prescribed loop $\mathcal{L}_i \subset E^d$, for every $i = 1, \ldots, n$. Heuristically, the connected Green functions are then defined by

$$\text{" } G_H(\mathcal{L}_1, \ldots, \mathcal{L}_n) = \int_{\partial X = \mathcal{L}_1 \cup \ldots \cup \mathcal{L}_n} \exp\Big\{-\tfrac{1}{\hbar}[\mathcal{A}^0_{\mathrm{P}}(X,g) + \lambda A(g)]\Big\} \mathcal{D}X\mathcal{D}g \text{ "} . \tag{7.4}$$

The case $n = 0$, that is, integration over genus-H surfaces without boundary, corresponds to the (connected) partition function, and will be denoted E_H. The precise meaning of such formulas in the continuum model is unknown to us. We shall, however, introduce discrete approximations to the Polyakov string model which permit us to define string Green functions analogous to those in (7.4) which do have a precise mathematical meaning. Formulas (7.3)–(7.4) just serve to motivate the quantities we shall consider in the context of discretized models.

Green functions $G_H(\mathcal{L}_1, \ldots, \mathcal{L}_n)$, corresponding to a parameter space with fixed genus H, are then used as building blocks for the Green functions of the putative complete theory that takes into account all possible genera. This is a process analogous to passing from a canonical to a grand canonical ensemble. Formally, to define the full Green function one introduces a "fugacity" N^{-2} (the notation is for future convenience) and weighs each "canonical" function G_H with the factor N^{-2H}:

$$G(\mathcal{L}_1, \ldots, \mathcal{L}_n) = C(n, N) \sum_{H=0}^{\infty} N^{-2H} G_H(\mathcal{L}_1, \ldots, \mathcal{L}_n) , \tag{7.5}$$

The factor N^{-1} is called the bare string coupling constant, and $C(n, N)$ is some constant that depends only on n and N. An expansion like (7.5), whose terms are labelled by the genus, is called a *topological expansion*. Unfortunately, in bosonic string theory the series on the r.h.s. of (7.5) *diverges*, and it is presumably not even Borel summable [281]. It is an *open problem* — and in our opinion one of the most important issues in string theory — to give a rigorous mathematical meaning to the r.h.s. of (7.5). There has been much progress in this direction during the past few years, although the fundamental conceptual issues are still far from resolved; see Section 7.6 below.

In the rest of this chapter we review several attempts to construct regularized versions of the preceding functional integrals, and in particular of the Green functions (7.4). We start, in the next section, with a simple-minded approach based on a lattice regularization for the Nambu-Goto action (7.1). Later (Section 7.4), we discuss regularizations of the Polyakov action in which the integration over metrics g is replaced by sums over random triangulations of a surface. These models can be put in correspondence with matrix field theories (Section 7.5) that can, in many cases, be integrated, yielding explicit solutions for the models at *fixed genus*. In the last section (Section 7.6), we review the use of these matrix models to give meaning to a theory summed over all genera.

7.2 Random-surface models in $\mathbb{Z}^d$

7.2.1 Basic definitions

A random-surface (RS) model in the lattice $\mathbb{Z}^d$ is defined as follows: First one chooses a countable family $\mathcal{E}$ of connected random surfaces in $\mathbb{Z}^d$; this set $\mathcal{E}$ will be the configuration space of the model. A "connected random surface" is, by definition, a connected two-dimensional cell complex, of which each p-cell ($p = 0, 1, 2$) corresponds to a copy of an elementary p-cell in $\mathbb{Z}^d$. The 0-cells in $\mathbb{Z}^d$ are sites of $\mathbb{Z}^d$, the 1-cells are links (i.e. edges joining nearest-neighbor pairs of sites), and the 2-cells are plaquettes. Two 2-cells can be glued along 1-cells only if these 1-cells are copies of a single link in $\mathbb{Z}^d$. Examples of such ensembles are:

(i) All RS contributing to the RS representation of a lattice gauge theory such as the pure $\mathbb{Z}_2$, $U(1)$ or $SU(2)$ models in the confinement phase; see [151, 212, 123, 332, 344, 214, 98].

(ii) *Self-avoiding RS*. (Self-avoiding RS appear in pure $U(n) \times U(m)$ lattice gauge theories, for a fixed value of nm, as $n \to \infty$; in the $\mathbb{Z}_2$ lattice gauge theory; in crystal surface and interface problems; as domain walls in magnets, etc. This is an important ensemble [153].)

(iii) Connected, orientable RS of genus 0 (i.e. without handles). We call these *planar RS* [153].

(iv) Clusters of "occupied plaquettes" (unit squares) in $\mathbb{Z}^d$. They are studied in connection with Bernoulli plaquette percolation [11] and q-state Potts gauge theories in the Fortuin-Kasteleyn representation [14].

(v) Graphs of integer-valued functions on $\mathbb{Z}^2$ which describe the surfaces of the solid-on-solid (SOS) model [357, 400, 503, 501].

Next, one assigns to every random surface $S \in \mathcal{E}$ a statistical weight, $\varrho(S)$. The most conspicuous example of statistical weights, inspired by the Nambu-Goto action defined in (7.1), is

$$\varrho(S) = \begin{cases} e^{-\beta|S|} & S \in \mathcal{E} \\ 0 & S \notin \mathcal{E} \end{cases} \tag{7.6}$$

where $|S|$ indicates the number of plaquettes forming the surface S, and $\mathcal{E}$ is some ensemble of surfaces contributing to the model. We adopt here a statistical-mechanical point of view, hence we use the letter β for the parameter in the exponential and think of it as an inverse temperature. From the quantum-gravity point of view, this parameter is proportional to the bare string tension.

The Green functions of the models are generalizations of the RW Green functions in which points (boundaries of walks) are replaced by loops (boundaries of surfaces). The *n-loop Green function* (or n-loop correlation) is defined as

$$G(\mathcal{L}_1, \ldots, \mathcal{L}_n) \;=\; \sum_{\substack{S \in \mathcal{E} \\ S \text{ connected} \\ \partial S = \mathcal{L}_1 \cup \ldots \cup \mathcal{L}_n}} \varrho(S) \tag{7.7}$$

where $\mathcal{L}_1, \ldots, \mathcal{L}_n$ are lattice loops and ∂S denotes the boundary of S. The expression (7.7) constitutes a regularization of (7.4) or (7.5). All the relevant properties of the model are expressed in terms of these Green functions. For example, if $L \times T$ is a rectangular loop with sides of length L and T, the *string potential* $V(L)$ is defined by

$$V(L) \;=\; \lim_{T \to \infty} -\frac{1}{T} \log G(L \times T) \tag{7.8}$$

and the *string tension* α by

$$\alpha \;=\; \lim_{L \to \infty} \frac{1}{L} V(L) \, . \tag{7.9}$$

The *inverse correlation length* ("glueball mass") is defined by

$$m \;=\; \lim_{a \to \infty} -\frac{1}{a} \log G(\partial p, \partial p_a) \tag{7.10}$$

where p is a plaquette and ∂p_a is a copy of ∂p translated by a lattice units in a lattice direction. A *susceptibility* (or better: specific heat) is defined by

$$\chi \;=\; \sum_{p'} G(\partial p, \partial p') \, . \tag{7.11}$$

Other notions are introduced by analogy to the RW models. For instance, assume that the weights are parametrized by the inverse temperature β in such a way that there exists a value β_c such that the susceptibility diverges for all $\beta < \beta_c$, while it is finite for $\beta > \beta_c$.[2] For the ensembles (ii) and (iii) and for $\varrho(S) = \exp(-\beta|S|)$, β_c exists and, in addition, all Green functions $G(\mathcal{L}_1, \ldots, \mathcal{L}_n)$

[2] We emphasize that the Green functions for surface models are expected to be well behaved at *low* temperature and may not exist at high temperature. This is the opposite of what we expect for RW models (cf. the Gaussian model, which has no low temperature phase, as explained in Section 2.3). This can be understood if one imagines that random-walk expressions roughly represent sums over paths along which spins correlate, and hence they have better convergence properties in the disordered phase where correlations die out fast. On the other hand, random-surface expressions can be associated to sums over Bloch walls, which are more scarce, and hence the sum is better defined, at low temperature. Of course, this just depends on our convention about temperature: β versus $1/\beta$.

are divergent for $\beta < \beta_c$ (If $\mathcal{E}$ is the ensemble of all connected orientable RS with precisely H handles, then β_c is independent of H). In the sequel we shall assume $\beta > \beta_c$ and will manipulate the series (7.7) with the tacit understanding that all the steps can be justified with some more work. We say that β_c is a *critical point* if

$$m(\beta) \searrow 0 \quad \text{as } \beta \searrow \beta_c \, . \tag{7.12}$$

In this case, we may introduce *critical exponents* ν, γ_{str}, η, μ, ... , as follows:

$$m(\beta) \sim (\beta - \beta_c)^{\nu} \tag{7.13}$$

$$\chi(\beta) \sim (\beta - \beta_c)^{-\gamma_{str}} \tag{7.14}$$

$$G_\beta(\partial p, \partial p_a) \sim a^{-(d-2+\eta)} \, , \quad \text{for } 1 \ll a \ll m(\beta)^{-1} \tag{7.15}$$

$$\frac{\partial \alpha}{\partial \beta} \sim (\beta - \beta_c)^{\mu - 1} \, . \tag{7.16}$$

Assuming a weak form of scaling we have that

$$\gamma_{str} = \nu(2 - \eta) \, . \tag{7.17}$$

In addition, for ensembles (ii) and (iii) and for $\varrho(S) = \exp(-\beta|S|)$, assuming a suitable form of scaling, we have:

$$\nu = \frac{1}{2}\mu \, . \tag{7.18}$$

For the planar RS model it is known that if $\chi(\beta) \nearrow \infty$ as $\beta \searrow \beta_c$, then $m(\beta) \searrow 0$ as $\beta \searrow \beta_c$; i.e., β_c corresponds to a critical point [153].

To rigorously define objects with the properties expected of the formal Green functions (7.4), one considers the *continuum limit* of the RS model, defined by rescaling the Green functions in a way similar to the one considered previously for spin models (see Chapter 3). Consider n loops $\mathcal{L}(b_1), \dots, \mathcal{L}(b_n)$ formed, respectively, by two copies of the bonds $b_1, \dots, b_n$. Let $x_1, \dots, x_n$ be the midpoints of $b_1, \dots, b_n$ and let $\mu_1, \dots, \mu_n$ be the directions of $b_1, \dots, b_n$. With the notation $G(x_1, \mu_1, \dots, x_n, \mu_n) \equiv G(\mathcal{L}(b_1), \dots, \mathcal{L}(b_n))$, the rescaled Green functions are

$$G_\theta(y_1, \mu_1, \dots, y_n, \mu_n) \equiv \alpha(\theta)^n G_{\beta(\theta)}(\theta y_1, \mu_1, \dots, \theta y_n, \mu_n) \tag{7.19}$$

where $x_j = \theta y_j$ is the midpoint of some bond in $\mathbb{Z}^d$ with direction μ_j, and $\theta \to \infty$ is a scale parameter. The continuum Green functions are

$$G(y_1, \mu_1, \dots, y_n, \mu_n) = \lim_{\theta \to \infty} G_\theta(y_1, \mu_1, \dots, y_n, \mu_n) \tag{7.20}$$

in the sense of distributions for non-coinciding arguments. Note that it can happen that the continuum Green functions are independent of the vector indices (directions) $\mu_1, \dots, \mu_n$. This will happen if scalar particles (spin 0) are the only ones that survive in the spectrum of the continuum theory. It turns out that this is the case for the planar RS model with weight given in (7.6). The *physical mass* m^* is defined to be

$$m^* \equiv \lim_{\theta\to\infty} \theta m(\beta(\theta)) \tag{7.21}$$

while the *physical string tension* is

$$\alpha^* \equiv \lim_{\theta\to\infty} \theta^2 \alpha(\beta(\theta)) . \tag{7.22}$$

As discussed before, to obtain an interesting continuum limit it is necessary that $\beta(\theta)$ converge to a critical inverse temperature β_c as $\theta \to \infty$. The theory is trivialif α^* is infinite, for instance if $\alpha(\beta)$ remains positive at the critical point. As for the case of random walks, this triviality is associated to some "mean-field behavior": the surfaces of the model collapse into tree-like structures with negligible area but large entropy contribution, and the critical properties are expected to be equal to the properties of *non-interacting branched polymers.* We now turn to a more complete description of such behavior.

7.2.2 A mean-field theory for random-surface models

The basic idea in finding the mean-field theory for random-surface models is that when the dimension, d, is large most of the entropy resides in surfaces which are "tree-like": large collections of filaments of small surface area with a negligible amount of self-intersections. Therefore, in order to construct a mean-field theory we consider the ensemble $\mathcal{E}$ of surfaces made up by gluing together elementary "cylindrical elements" consisting of two overlapping plaquettes glued together along some of the opposite edges (glued along three edges when forming the end of a "branch", along two edges when forming an intermediate segment of a branch, and along one or no edge when forming "T" connections between branches). We shall use the following properties of such surfaces: as $d \to \infty$ no pieces of surface are ever formed which are wider than the minimal width required, i.e. no surfaces enclosing large spheres are formed. This happens for purely combinatorial (entropy) reasons. Moreover, the resulting surfaces are *automatically* self-avoiding, so that distant parts of a surface are statistically independent. All this is easily verified, but see [423, 147, 154] for details.

Let $b_1, \ldots, b_n$ be n bonds in $\mathbb{Z}^d$ viewed as degenerated loops, i.e. each b_i really consists of two copies of a bond glued together at the endpoints. The b_i's are called double bonds. The n-loop Green functions are then

$$G(b_1, \ldots, b_n) = \sum_{S \in \mathbf{T}_{\mathcal{E}}(b_1,\ldots,b_n)} \varrho(S) , \tag{7.23}$$

where $\mathbf{T}_{\mathcal{E}}(b_1, \ldots, b_n)$ is the class of all surfaces, S, in the above ensemble $\mathcal{E}$ for which $\partial S = b_1 \cup \ldots \cup b_n$. This last condition is equivalent to the requirement that the bonds $b_1, \ldots, b_n$ *belong* to S (which is imagined as cut open along such bonds). See [423, 147, 154].

As $d \to \infty$, the lack of self-intersections allows us to decompose each surface unambiguously into maximal pieces joining pairs of bonds (b, b') where b and b' belong to a "fork" or to the boundary. The sum over all possible surfaces factorizes, therefore, into terms corresponding to summing over all possible pieces connecting all pairs (b, b'). Each such term gives precisely a factor $G(b, b')$ (again, we are using the self-avoidance, valid only at $d \to \infty$). This resummation is formalized as follows: We view double bonds, b, as vertices of a graph, $\mathcal{G}$, and pairs of double bonds, (b, b'), as lines of $\mathcal{G}$. Let us fix n double bonds $b_1, \ldots, b_n$, and view them as endpoints of n external lines. Let $\mathcal{T}(b_1, \ldots, b_n)$ denote the family of all connected tree graphs all of whose internal vertices are of order three and whose external lines end in $b_1, \ldots, b_n$. Then

$$G(b_1, \ldots, b_n) \underset{d \text{ large}}{\sim} \sum_{\mathcal{G} \in \mathcal{T}(b_1, \ldots, b_n)} \prod_{(b,b') \in L(G)} G(b, b') \,, \tag{7.24}$$

As $d \to \infty$ (self-avoidance!), every surface $S \in \mathbf{T}_{\mathcal{E}}(b, b')$ can be described as the union of a *unique* walk, ω, made up of the cylindrical elements described above, starting at b and ending at b', and of surfaces in $\mathbf{T}(\tilde{b})$, "$\tilde{b} \in \omega$", i.e., tree-like surfaces rooted at bonds, $\tilde{b}$, belonging to ω. These trees have no intersections with ω, no mutual intersections and no self-intersections. To calculate

$$G(b, b') = \sum_{S \in \mathbf{T}_{\mathcal{E}}(b,b')} \varrho(S) \tag{7.25}$$

in the limit $d \to \infty$, we may first sum over all trees rooted on ω and, subsequently, sum over all walks ω joining b to b'. Each step of the walk has an area 2 (it involves two joined plaquettes) and there are $4|\omega|$ double bonds $\tilde{b}$ where a tree can be rooted (that is, two borders per plaquette). The sum over all trees rooted at $\tilde{b}$ gives a factor $1+G(\tilde{b})$ (the "1" corresponds to the empty tree), which is independent of $\tilde{b}$ by translation-invariance. Thus, for $\varrho(S) = \exp[-\beta|S|]$, the sum over all trees rooted on ω yields

$$\begin{aligned} \varrho(\omega) &= e^{-2\beta|\omega|} \, [1 + G(\tilde{b})]^{4|\omega|} \\ &= e^{-2\beta'|\omega|} \,, \end{aligned} \tag{7.26}$$

as the weight of the walk, where

$$\beta' = \beta'(\beta) = \beta - 2\log(1 + G(b)) \,. \tag{7.27}$$

Thus

$$G(b, b') \underset{d \text{ large}}{\sim} \sum_{\omega : b \to b'} e^{-2\beta'|\omega|} = \left[-\Delta + m^2(2\beta')\right]^{-1}_{b,b'} \,, \tag{7.28}$$

where Δ is the finite-difference Laplacian, and $m^2(\beta)$ is the solution of the equation

$$[2d + m^2(\beta)]^{-1} = e^{-\beta} \,. \tag{7.29}$$

From (7.28) we conclude that the exponent η of $G(b, b')$, introduced in (7.15) has the value

$$\eta = 0\,. \tag{7.30}$$

To compute other exponents, it is useful to notice the following identity for the case $\varrho(S) = e^{-\beta|S|}$:

$$\begin{aligned}
\frac{\partial}{\partial\beta}G(b_1,\ldots,b_n) &= -\sum_{S\in\mathbf{T}_{\mathcal{E}}(b_1,\ldots,b_n)} |S|e^{-\beta|S|} \\
&\simeq -2\sum_{\tilde{b}} \sum_{\substack{S\,\in\,\mathbf{T}_{\mathcal{E}}(b_1,\ldots,b_n)\\ S\,\ni\,\tilde{b}}} e^{-\beta|S|} \\
&= -2\sum_{\tilde{b}} G(b_1,\ldots,b_n,\tilde{b})\,.
\end{aligned} \tag{7.31}$$

(In the intermediate step, the boundary contributions were ignored.) Applying this to

$$\chi(\beta) = \sum_{b'} G(b,b')\,, \tag{7.32}$$

we conclude that

$$\begin{aligned}
\frac{\partial\chi}{\partial\beta} &\underset{d\ \text{large}}{\sim} -\sum_{b',b''} G(b,b',b'') \\
&\underset{d\ \text{large}}{\sim} -2\sum_{b',b''}\sum_{b^*} G(b,b^*)G(b',b^*)G(b'',b^*) \\
&= -2\,\chi^3\,.
\end{aligned} \tag{7.33}$$

We have used the fact that a graph with three external vertices (b,b',b'') must bifurcate at some internal vertex b^*. Integration over β yields

$$\chi(\beta) \sim (\beta-\beta_c)^{-1/2}\,,\ \text{i.e.}\ \gamma_{str} = \frac{1}{2}\,. \tag{7.34}$$

Another consequence of (7.31) is that

$$-\frac{\partial G(b)}{\partial\beta} = \chi(\beta)\,; \tag{7.35}$$

hence (7.34) implies

$$G_{\beta_c}(b) - G_\beta(b) \sim (\beta-\beta_c)^{1/2}\,, \tag{7.36}$$

and this and (7.29) yield

$$m(\beta) \sim (\beta-\beta_c)^{1/4}\,,\ \text{i.e.}\ \nu = \frac{1}{4}\,, \tag{7.37}$$

which also follows from (7.34), (7.30) and the relation $\gamma_{str} = \nu(2-\eta)$.

Finally, we study the behavior of the string tension, $\alpha(\beta)$, near β_c. To be specific, we consider the planar RS model with $\varrho(S) = \exp[-\beta|S|]$, but our final result is more generally valid. We recall that

$$\begin{aligned}\alpha &= \lim_{L\to\infty} -\frac{1}{L^2}\log G(L\times L)\\ &= \lim_{L\to\infty} -\frac{1}{L^2}\log\Big(\sum_{\substack{S\in\mathcal{E}\\ \partial S=L\times L}} e^{-\beta|S|}\Big)\,.\end{aligned} \tag{7.38}$$

In large dimensions, when self-intersections are negligible, every surface $S\in\mathcal{E}$, with $\partial S = L\times L$ can be uniquely decomposed into a "skeleton surface" and trees rooted on its (double) bonds. This skeleton surface is what one gets after "shaving off" all trees; more precisely, we define a skeleton surface R to be any surface which does not decay into two pieces by cutting along any double bond (i.e. by cutting open two bonds of S). This decomposition yields a factorization of the sum inside the log in (7.38). Each skeleton surface R with $\partial R = L\times L$ contains $2|R|+2L$ double bonds and, for each double bound $\tilde{b}$, the sum over all possible trees rooted therein produces a factor $1+G(\tilde{b})$. We therefore get

$$\begin{aligned}\alpha \underset{d\text{ large}}{\sim} &\lim_{L\to\infty} -\frac{1}{L^2}\log\Big\{\sum_{\substack{\text{skeletons, }R\\ \partial R=L\times L}} (1+G(b))^{2L+2|R|}e^{-\beta|R|}\Big\}\\ = &\lim_{L\to\infty} -\frac{1}{L^2}\log\Big\{(1+G(b))^{2L}\sum_{\substack{\text{skeletons, }R\\ \partial R=L\times L}} e^{-\beta'|R|}\Big\}\,,\end{aligned} \tag{7.39}$$

where β' is given by (7.27). Thus

$$\begin{aligned}\alpha(\beta) \underset{d\text{ large}}{\sim} &\lim_{L\to\infty} -\frac{1}{L^2}\log\Big\{\sum_{\substack{\text{skeletons, }R\\ \partial R=L\times L}} e^{-\beta'|R|}\Big\}\,,\\ = &\ \alpha^S(\beta')\end{aligned} \tag{7.40}$$

where α^S is the string tension of the skeleton theory. For large d, $\alpha^S(\beta')\sim\beta'$, and β' remains positive as $\beta\searrow\beta_c$, by (7.29). Hence

$$\alpha(\beta)\searrow\tilde{\alpha}>0\,,\ \text{as } \beta\searrow\beta_c\,, \tag{7.41}$$

with $\tilde{\alpha}=\alpha^S(\beta'(\beta_c))$. However, from (7.27) and (7.35)

$$\begin{aligned}\frac{\partial\alpha}{\partial\beta}(\beta) &\sim \frac{\partial\alpha^S}{\partial\beta'}(\beta'(\beta))\cdot\frac{\partial\beta'}{\partial\beta}(\beta)\\ &\sim \frac{\partial\alpha^S}{\partial\beta'}(\beta'(\beta))\,\chi(\beta)\,.\end{aligned} \tag{7.42}$$

Since the skeleton theory is not yet critical at $\beta'=\beta'(\beta_c)$, $(\partial\alpha^S/\partial\beta')(\beta'(\beta_c))$ is finite. Hence

$$\frac{\partial\alpha}{\partial\beta}(\beta)\sim(\beta-\beta_c)^{-\gamma_{str}}\,, \tag{7.43}$$

i.e.

$$\mu = 1-\gamma_{str} = \frac{1}{2} = 2\nu\,. \tag{7.44}$$

For a more detailed account see [154].

7.2.3 The planar random-surface model

In this section we sketch the analysis of a lattice approximation to the Nambu-Goto model of quantized relativistic strings [403, 265] in the Euclidean (imaginary-time) description. To approximate (7.1), the statistical weights of the surfaces are chosen equal to $\exp[-\beta \times \text{area}]$. This model is a natural random-surface analogue of Brownian motion. As we shall see, its critical behavior is identical to the one of non-interacting branched polymers; and hence its continuum limit is trivial.

As an ensemble of lattice RS we choose the class $\mathcal{E} = \mathcal{E}_{\text{PRS}}$ of all connected complexes, S, constructed by gluing together pairs of adjacent plaquettes of $\mathbb{Z}^d$ along common edges, which are orientable and have the topology of a 2-sphere with an arbitrary, finite number, n, of small discs removed. The boundaries of these discs correspond to connected components of the boundary, ∂S, of a complex $S \in \mathcal{E}_{\text{PRS}}$. We fix ∂S to consist of a union of loops $\mathcal{L}_1, \ldots, \mathcal{L}_n$ of $\mathbb{Z}^d$. As the statistical weights of a "surface" S we choose

$$\varrho(S) = e^{-\beta|S|} . \tag{7.45}$$

The main result [154] for this model is

Theorem 7.1 *Suppose that $\chi(\beta) \nearrow \infty$ as $\beta \searrow \beta_c$, and suppose that the PRS model has a weak self-similarity property defined below; see (7.70), (7.71). Then, mean-field theory provides an exact description of the approach to the critical point, $\beta \searrow \beta_c$. In particular, the critical exponents have the values*

$$\gamma_{str} = \mu = \frac{1}{2}, \quad \nu = \frac{1}{4} \tag{7.46}$$

$$G(b, b') \sim \text{dist}(b, b')^{-(d-2)} \tag{7.47}$$

if $0 \ll \text{dist}(b, b') \ll m(\beta)^{-1}$, *i.e.* $\eta = 0$. *Moreover,*

$$\alpha(\beta) \searrow \tilde{\alpha} > 0 , \text{ as } \beta \searrow \beta_c . \tag{7.48}$$

Remarks.

1) The hypotheses of Theorem 7.1 have been tested numerically in 2, 3 and 5 dimensions [331, 61]. They appear to be valid in all dimensions $d \geq 2$.

2) The same result can be proven for some randomly triangulated models and their associated matrix models described in Sections 7.4 and 7.5 [152].

3) Properties (7.46), (7.47) and (7.48) show that the continuum limit of the two-loop correlation, $G(b, b')$, of the PRS model (i.e. its scaling limit as $\beta \searrow \beta_c$), is the *Euclidean propagator of a single, free scalar field* and that the string tension is infinite.

Sketch of proof [154]. We define a class, $\mathcal{R}^{(2)}$, of *skeleton surfaces* (of order 2) as consisting of all those surfaces $S \in \mathcal{E}_{\text{PRS}}$ which do *not* decay into two disconnected pieces when cut open along any two bonds in S not belonging to the boundary ∂S (a doubly occupied bond of $\mathbb{Z}^d$ counts as *two* bonds of S). More generally, we may define a class, $\mathcal{R}^{(2j)}$, of *skeleton surfaces of order* $2j$ to consist of all those surfaces $S \in \mathcal{E}_{\text{PRS}}$ with the property that if an arbitrary number, i, of discs, $D_1, \ldots, D_i$, with boundaries of length $\geq 2j$ are removed from S then all loops on $S \setminus (D_1 \cup \ldots \cup D_i)$ which are non-contractible with respect to $D_1, \ldots, D_i$ have length $\geq 2j$.

Let $\mathcal{S}(\mathcal{L}_1, \ldots, \mathcal{L}_n)$ be the class of all surfaces $S \in \mathcal{E}_{\text{PRS}}$, with $\partial S = \mathcal{L}_1 \cup \ldots \cup \mathcal{L}_n$, where each $\mathcal{L}_i$ is some loop in $\mathbb{Z}^d$. We define

$$G^{(2j)}(\mathcal{L}_1, \ldots, \mathcal{L}_n) = \sum_{S \in \mathcal{R}^{(2j)} \cap \mathcal{S}(\mathcal{L}_1, \ldots, \mathcal{L}_n)} e^{-\beta |S|} , \tag{7.49}$$

and $m^{(2j)}$, $\alpha^{(2j)}$, $\chi^{(2j)}$ are defined as in (7.9)–(7.11), with G replaced by $G^{(2j)}$. It is immediate that $\chi^{(2j)}(\beta)$ is *convex* in β, and $m^{(2j)}(\beta)$, $\alpha^{(2j)}(\beta)$ are *concave* in β, for all $j = 0, 1, 2, \ldots$. As in [153] (for $j = 0$), one proves that there exists a value β_{2j} independent of $\mathcal{L}_1, \ldots, \mathcal{L}_n$, with $\beta_{2j} \leq \beta_{2j-2} \leq \ldots \leq \beta_0 \equiv \beta_c$, such that $G^{(2j)}(\mathcal{L}_1, \ldots, \mathcal{L}_n)$ converges, for all $\beta > \beta_{2j}$, and diverges, for all $\beta < \beta_{2j}$. Moreover,

$$m^{(2j)} > 0 \text{ and } \alpha^{(2j)} > 0 \quad \text{for all } \beta > \beta_{2j} . \tag{7.50}$$

(The proof is as in [153].) As in the previous section, we define

$$\beta' = \beta'(\beta) = \beta - 2\log(1 + G_\beta(b)) . \tag{7.51}$$

A crucial, but straightforward step in the proof is to prove the equation

$$G_\beta(b, b') = \sum_{n=0}^{\infty} \sum_{b_1, \ldots, b_n} G^{(2)}_{\beta'}(b, b_1) G^{(2)}_{\beta'}(b_1, b_2) \ldots G^{(2)}_{\beta'}(b_n, b') . \tag{7.52}$$

This is proven as follows: We first fix $n+1$ skeleton surfaces, $R_1, \ldots, R_{n+1}$, of order 2, with $\partial R_1 = b \cup b_1$, $\partial R_2 = b_1 \cup b_2, \ldots, \partial R_{n+1} = b_n \cup b'$. The sum over all the surfaces $S \in \mathcal{S}(b, b')$, whose associated skeleton consists of $R_1 \cup \ldots \cup R_{n+1}$, satisfies the identity

$$\sum_{\substack{S \in \mathcal{S}(b,b') \\ \text{skeleton of } S = R_1 \cup \ldots \cup R_{n+1}}} e^{-\beta |S|} = \text{const}\, e^{-\beta'(|R_1| + \ldots + |R_{n+1}|)} \tag{7.53}$$

(each of the $2(|R_1| + \ldots + |R_{n+1}|)$ double bonds contribute with a factor $1 + G(b)$ as in the previous section). Finally, we sum over all possible $R_1, \ldots, R_{n+1}$, all possible $b_1, \ldots, b_n$, and all values of n; see [154] for details. Similarly, we may prove that

$$G_\beta(\mathcal{L}) = G^{(2)}_{\beta'}(\mathcal{L}) = G^{(4)}_{\beta''}(\mathcal{L}) \ldots . \tag{7.54}$$

If we Fourier-transform (7.52) we obtain

$$\widehat{G}_\beta(k) = \frac{\widehat{G}^{(2)}_{\beta'}(k)}{1 - \widehat{G}^{(2)}_{\beta'}(k)} , \tag{7.55}$$

where k is a momentum in the first Brillouin zone. Suppose now that

$$\chi^{(2)}(\beta) = \widehat{G}^{(2)}_{\beta}(k=0) \text{ diverges as } \beta \searrow \beta_2 . \tag{7.56}$$

Then β_c is determined by the equation

$$\chi^{(2)}(\beta'(\beta_c)) = 1 ; \tag{7.57}$$

see (7.55). Since $\chi^{(2)}(\beta)$ is convex in β, it follows that

$$\beta'(\beta_c) > \beta_2 . \tag{7.58}$$

Since by (7.51) $\beta'(\beta_c) < \beta_c$, it follows that

$$\beta_c > \beta_2 . \tag{7.59}$$

For $\beta > \beta_2$, $m^{(2)}(\beta)$ is positive, and hence $\widehat{G}^{(2)}_{\beta}(k)$ is analytic in k^2, for $|k| \ll m^{(2)}(\beta)$. Therefore

$$\widehat{G}^{(2)}_{\beta}(k) = \chi^{(2)}(\beta) - \xi(\beta)k^2 + O(|k|^4) , \tag{7.60}$$

and it follows from the positivity of $G^{(2)}(b, b')$ that $\xi(\beta) > 0$, for all $\beta > \beta_2$. In particular, all this is true for $\beta = \beta'(\beta_c)$, by (7.58). From (7.60) and (7.55) we conclude that

$$\widehat{G}_{\beta}(k) \sim [1 - \chi^{(2)}(\beta') + \xi(\beta')k^2]^{-1} , \tag{7.61}$$

and hence that

$$m(\beta)^2 \sim 1 - \chi^{(2)}(\beta') \sim \chi(\beta)^{-1} \sim (\beta - \beta_c)^{\gamma}_{str} , \tag{7.62}$$

as $\beta \searrow \beta_c$[3]. Thus

$$\nu = \frac{\gamma_{str}}{2} . \tag{7.63}$$

To determine the exponent γ_{str}, we take the derivative of (7.55):

$$\begin{aligned} \frac{\partial \chi}{\partial \beta} &= \frac{\partial \widehat{G}_{\beta}(k=0)}{\partial \beta} \\ &= \left[\frac{1}{1 - \widehat{G}^{(2)}_{\beta'}(0)} + \frac{\widehat{G}^{(2)}_{\beta'}(0)}{(1 - \widehat{G}^{(2)}_{\beta'}(0))^2} \right] \frac{\partial \widehat{G}^{(2)}_{\beta'}(0)}{\partial \beta'} \frac{\partial \beta'}{\partial \beta} . \end{aligned} \tag{7.64}$$

By (7.59), $\widehat{G}^{(2)}_{\beta'}$ and its derivatives are bounded in the vicinity of β_c. On the other hand, by (7.61)–(7.62), $1 - \widehat{G}^{(2)}_{\beta'}(0) \sim \chi(\beta)^{-1}$ and, from definition (7.51), $\partial\beta'/\partial\beta \sim \chi(\beta)$. Thus, we conclude from (7.64) that

$$-\frac{\partial \chi}{\partial \beta} \sim \chi^3 , \text{ i.e. } \gamma_{str} = \frac{1}{2} . \tag{7.65}$$

[3]For related methods see also [94] and references therein.

It also follows directly from (7.61) and the fact that $\xi(\beta'(\beta_c)) > 0$ that $\eta = 0$. Hence

$$\gamma_{str} = \frac{1}{2}\ , \ \nu = \frac{1}{4} \text{ and } \eta = 0 \tag{7.66}$$

which are the mean-field exponents.

Finally, we note that (7.54) implies that

$$\alpha(\beta) \ = \ \alpha^{(2)}(\beta'(\beta))\ . \tag{7.67}$$

Since $\beta'(\beta_c) > \beta_2$, it follows from (7.61) that

$$\alpha(\beta_c) \ = \ \alpha^{(2)}(\beta'(\beta_c)) \ > \ \alpha^{(2)}(\beta_2) \ \geq \ 0\ , \tag{7.68}$$

i.e. $\tilde{\alpha} = \alpha(\beta_c)$ is strictly *positive*. Finally

$$\begin{aligned}\frac{\partial\alpha}{\partial\beta}(\beta) \ &= \ \frac{\partial\alpha^{(2)}}{\partial\beta'}(\beta'(\beta)) \cdot \frac{\partial\beta'}{\partial\beta}(\beta) \\ &= \ \frac{\partial\alpha^{(2)}}{\partial\beta'}(\beta'(\beta)) \ \cdot \chi(\beta) \\ &\sim \ (\beta-\beta_c)^{-\gamma_{str}}\ , \text{ i.e. } \mu = \gamma_{str}\ . \end{aligned} \tag{7.69}$$

Note that $(\partial\alpha^{(2)}/\partial\beta')(\beta'(\beta_c))$ is finite, by concavity and (7.58).

We expect that the behavior of the functions $m^{(2j)}(\beta)$, $\chi^{(2j)}(\beta)$ and $\alpha^{(2j)}(\beta)$ near β_{2j} is qualitatively the same, for all values of j. For example, we expect that if

$$\chi(\beta) \equiv \chi^{(0)}(\beta) \nearrow \infty\ , \text{ as } \beta \searrow \beta_c \tag{7.70}$$

then

$$\chi^{(2j)}(\beta) \nearrow \infty\ , \text{ as } \beta \searrow \beta_{2j}\ , \text{ for all } j > 0\ . \tag{7.71}$$

This is the *self-similarity property of the PRS model* assumed in the theorem. The converse is proven easily: Using an Ornstein-Zernike equation of the form (7.55), for general values of j, one shows that

$$\chi^{(2j)}(\beta) \nearrow \infty\ , \text{ as } \beta \searrow \beta_{2j} \tag{7.72}$$

implies that

$$\chi^{(2j-2)}(\beta) \nearrow \infty\ , \text{ as } \beta \searrow \beta_{2j-2}\ , \tag{7.73}$$

for all j. Thus *if (7.72) holds for* $j = 1$ *all assumptions needed to prove the conclusions of the theorem are valid.*

There are computer experiments suggesting that (7.70) hold in two and three dimensions and that (7.72) holds, for $j = 1$, for some values of $d > 2$ [331, 61]. Moreover, (7.72) holds for all values of j in mean-field theory. ∎

Remarks.

1) Techniques related to (7.55)–(7.61) have also been used in [2].

2) The basic technical results, (7.54)–(7.61) and (7.65) extend to some randomly triangulated models and their associated matrix models described in Sections 7.4 and 7.5, but the basic hypothesis (7.72) (for $j = 1$) has not yet been proven.

7.3 Typical phenomena in random-surface theory

We interrupt, for a moment, our parade of models to describe some typical phenomena encountered in the statistical mechanics of random surfaces, most of which have no analogue in the theory of interacting random walks.

(1) *Surface roughening* [357, 400, 503, 501]. Roughening is exhibited by models of surfaces immersed in an infinite regular lattice, such as $\mathbb{Z}^d$, which have only a *discrete translational symmetry*. Roughening can be viewed as the phenomenon of *enhancement* of the discrete to full *continuous translational symmetry* [224, 225].

Let $P(h|L \times L)$ be the probability that a lattice random surface S with $\partial S = L \times L$ intersects the $(d-2)$-dimensional plane perpendicular to the plane containing $L \times L$ and passing through 0, at a distance h from the latter. For RS models, like the ones defined in (ii)–(v) of Section 7.2.1, one may prove (see [133], [134] and [432] and references therein) that, for $\beta \gg \beta_c$,

$$P(h) \equiv \lim_{L\to\infty} P(h|L \times L) \tag{7.74}$$

exists and decays exponentially in h, for a choice

$$\varrho(S) = e^{-\beta|S|}, \tag{7.75}$$

for example. For the solid-on-solid model (v) (Section 7.2.1), one can prove [224, 225] that there exists some $\beta_R > 0$ such that, for $\beta < \beta_R$, $P(h) = 0$, for all $h < \infty$, and

$$\sum_{h=0}^{\infty} h^2 P(h|L \times L) \sim \log L . \tag{7.76}$$

The existence of a roughening transition characterized by (7.74) and (7.76) is expected for a large class of RS models, including models (i)–(iii) and (v) of Section 7.2.1, with $\varrho(S)$ for example as in (7.75). The roughening transition is expected to be of infinite order. The capillary surface waves present when $\beta < \beta_R$ restore continuous translation invariance broken by the lattice.

Another characterization of the roughening transition is based on the behavior of the string potential $V(L)$, introduced in (7.8). One can prove quite easily that, for large β,

$$V(L) = \alpha(\beta)L + \gamma(\beta) + \delta(\beta)\exp[-L/\xi(\beta)] , \tag{7.77}$$

where α, γ, δ and ξ are finite constants. At β_R, $\xi(\beta)$ *diverges*. (Presumably $\xi(\beta)^{-1}$ has a zero of infinite order at β_R.) On the basis of simple models (e.g. the Gaussian sheet or reformulations of solid-on-solid models in terms of *fermions*) one may argue that, for $\beta < \beta_R$ (where capillary waves are abundant),

$$V(L) = \alpha(\beta)L + \gamma(\beta) + \frac{\kappa(\beta)}{L} , \tag{7.78}$$

where α, γ and κ are finite constants, and $\kappa \simeq \pi(d-2)/24$ for small β [377]. The correction term κ/L reflects the one-dimensional Casimir effect. The string tension, α, is expected to have a singularity at β_R but to remain strictly positive for $\beta \leq \beta_R$.

(2) *Entropic repulsion* [203, 84]. Consider, for example, a solid-on-solid model of an interface in the presence of a wall. The interface, S, is described by assigning a height variable $h_p \in \mathbb{R}$ (or $h_p \in \mathbb{Z}$) to every plaquette p contained in a square $\Lambda \subset \mathbb{Z}^2$ with sides of length L. More precisely, the interface $S = S(h)$ is given by the graph of the height function h, its area by

$$|S(h)| = L^2 + \sum_{\langle pp' \rangle} |h_p - h_{p'}| , \tag{7.79}$$

where $\langle pp' \rangle$ denotes a pair of distinct plaquettes with a common bond, and $h_p = 0$ for all $p \not\subset \Lambda$. The statistical weight of the interface S is given, for example, by

$$\exp[-\beta|S|] . \tag{7.80}$$

The wall is described by the constraint that

$$h_p \geq 0 , \text{ for all } p \subset \Lambda . \tag{7.81}$$

The point is now that, for entropic reasons, the interface will typically settle in a mean position that is quite far away from the wall, because this permits the interface to grow spikes in the direction away from the wall *and towards the wall*. Since such spikes have large entropy, a mean position of the interfaces far away from the wall does not only increase their energy but it also increases their *entropy*. For β small enough, entropy typically wins over energy, and we conclude that

$$\lim_{L\to\infty} \langle h_p \rangle_\Lambda = \infty , \tag{7.82}$$

for any fixed plaquette p. A quantitative estimate,

$$\frac{1}{L^2} \sum_{p \subset \Lambda} \langle h_p \rangle_\Lambda \sim \log L , \text{ as } L \to \infty , \tag{7.83}$$

for the weight of an interface $S(h)$ given by (7.80), has been proven in [84]. Similar phenomena are encountered in systems of two or more interacting surfaces. Entropic repulsion is the mechanism driving *wetting transitions*. This phenomenon of entropic repulsion may also have some bearing on the "breathing transition" described in (4), below.

(3) *Topological complexity* [14]. Random "surfaces", in contrast to random walks, have an intrinsic topology: They may be *non-orientable*, may have *handles*, and may have *pockets* (e.g. in $\mathbb{Z}_n$ and $SU(n)$ lattice gauge theories, $n \geq 3$, pockets arise when n plaquettes are glued together along a common link). When $d \geq 4$, their first (co)homology groups may have a non-trivial torsion subgroup. But even "true" orientable surfaces (as opposed to general 2-dimensional cell complexes) have an intrinsic topology, characterized by the number of holes and handles. "Topological complexity" has some important consequences: it leads to serious difficulties in the $1/n$-expansion and in real or complex interpolations in the parameter n of the n-state Potts model and possibly the pure $SU(n)$ lattice gauge theory [14]. Moreover, it may affect the deconfining transition in RS models (see e.g. Sect. 2 (i) of [11]).

(4) *Collapse to a tree-like surface and "breathing"* [214, 155, 154, 2]. We expect that there are RS models for which there exists a deconfining transition at some value $\beta_{dc} \geq \beta_c$, so that

$$\begin{aligned} \alpha(\beta) &> 0\,, \text{ for all } \beta > \beta_{dc}\,, \\ \alpha(\beta) &\leq 0\,, \text{ for all } \beta < \beta_{dc}\,. \end{aligned} \tag{7.84}$$

This might be the case in some *self-avoiding RS models* in small dimensions. One can argue that, in such a model, for $\beta > \beta_{dc}$, typical random surfaces look like *trees* (in a planar RS model, trees in the graph-theoretic sense), or more complicated networks, made out of *thin, tubular pieces*, with the property that the probability that a closed random surface enclose a sphere of radius R tends rapidly to 0 as $R \nearrow \infty$. In contrast, for $\beta < \beta_{dc}$, surfaces grow fat and wide ("breathe"), and infinite closed random surfaces enclose arbitrarily large spheres.

Mean-field theory predicts that, in large dimension, for any reasonable theory of random surfaces,

$$\alpha(\beta) \searrow \tilde{\alpha} > 0\,, \text{ while } m(\beta) \searrow 0\,, \tag{7.85}$$

as $\beta \searrow \beta_c$ [423, 147, 154]. One might expect that, in models (ii) or (iv) of Section 7.2.1, for example, there exists a dimension $d_c < \infty$ such that (7.85) holds for all $d > d_c$, while, for $d < d_c$,

$$\alpha(\beta) \searrow 0\,, \text{ and } m(\beta) \searrow 0\,, \text{ as } \beta \searrow \beta_c\,. \tag{7.86}$$

Thus, when $d > d_c$, typical random surfaces collapse to tree-like structures when watched from far distances, i.e. in the *scaling limit*, and the critical properties of such a model are described by a model of *branched polymers*.

In plaquette percolation [11], model (iv) of Section 7.2.1, there are several transition points at which an inverse correlation length, m, vanishes *before* the string tension vanishes. It is not known whether the string tension tends to 0 continuously at the transition point. This is also a major unsettled issue in high-dimensional lattice gauge theories.

(5) *Crumpling* [404]. Let us consider a model of random surfaces defined as maps X from a finite rectangular domain Λ in a regular two-dimensional lattice (e.g. triangular, square or honeycomb) into d-dimensional Euclidean space E^d/ The statistical weight of surfaces is given, for example, by the probability measure

$$Z^{-1} \exp\left[-\beta \sum_{\langle i,j \rangle} |X_i - X_j|^2\right] \prod_{i\in\Lambda} d^d X_i \, ,$$

where $\langle i,j \rangle$ denotes a pair of nearest-neighbor sites in the lattice, and $X_j = 0$ for all $j \notin \Lambda$. Then one observes that typical random surfaces in E^d, as described by this model, are *highly crumpled*, in the sense that, for example,

$$\text{diameter(image of } X) \sim \log(\text{diameter of } \Lambda) \, .$$

This phenomenon is believed to be of importance in the statistical mechanics of membranes [404]. The models studied in the next section also have this property.

7.4 Randomly triangulated random-surface models

7.4.1 Definition of the model

In Section 7.2.3 we saw that the lattice regularization of the Nambu-Goto action leads to mean-field results (trivial continuum limit in the sense that the physical string tension is infinite). We now turn to another family of models [216, 115, 333, 21], introduced to regularize a theory based on the Polyakov action, that is, to give meaning to the string Green functions (7.4). These models are based on summing over triangulations, which roughly represent the "gravitational degrees of freedom" of the Polyakov model.

We choose a connected orientable surface R_H of genus H, and delete from it n small discs whose boundaries are denoted by $\gamma_1, \ldots, \gamma_n$. The resulting surface is denoted by $R_H^{(n)}$. We then consider *triangulations* of $R_H^{(n)}$.[4] A triangulation of $R_H^{(n)}$ is, by definition, a 2-dimensional simplicial complex — that is, a collection of 2-cells (triangles), 1-cells (edges) and 0-cells (vertices) with the property that whenever a cell belongs to the complex, all the cells on its boundary also belong to the complex — that is homeomorphic to $R_H^{(n)}$. In particular, this means that #(triangles) − #(edges) + #(vertices) $= 2 - 2H - n$ (Euler formula). In addition, we require that the triangulations be *loop-free*, i.e. no edge can have coinciding endpoints and no two triangles can have more than one common edge. Given a triangulation $\mathcal{T}$, we denote by $F(\mathcal{T})$ the set of faces (triangles) of $\mathcal{T}$, by $E(\mathcal{T})$ the set of edges and by $V(\mathcal{T})$ the set of vertices of $\mathcal{T}$.

We now consider, for each triangulation $\mathcal{T}$ on $R_H^{(n)}$, maps X from $F(\mathcal{T})$ into euclidean space E^d. Two image points $X_i, X_j \in E^d$ are considered to be joined by an edge iff i and j are two adjacent triangles of $\mathcal{T}$. The points $\{X_j \colon \; j \in F(\mathcal{T})\}$

[4] A more detailed discussion of the algebraic-topology background can be found in [401].

are therefore identified with the centers of triangles[5] of a triangulated, piecewise-linear surface $S = S(X;\mathcal{T})$ immersed in E^d. It is assumed that all triangles j adjacent to a component γ_i of the boundary of $R_H^{(n)}$ are mapped to points $X_j \in E^d$ which belong to some fixed loop $\mathcal{L}_i$ in E^d, $i = 1, \ldots, n$. This condition is expressed by the formula

$$\mathrm{bd}_{\mathcal{T}}(S(X;\mathcal{T})) \subset \bigcup_{i=1}^{n} \mathcal{L}_i \,. \tag{7.87}$$

We want the whole construction to remain reparametrization-invariant. Hence, two triangulations of $R_H^{(n)}$ are considered to be equivalent if they determine the same combinatorial structure; metric properties do not play any role in this notion. In consequence, we consider *isomorphism classes of triangulations* or, equivalently, we weight each triangulation $\mathcal{T}$ with a factor $C(\mathcal{T})$ where

$$C(\mathcal{T}) \;=\; \#\{\text{triangulations isomorphic to } \mathcal{T}\} \tag{7.88}$$

is the order of the group of symmetries of the triangulation $\mathcal{T}$.

Given an *action* $\mathcal{A}(S)$ of a surface S, we define the Green functions

$$G_H(\mathcal{L}_1, \ldots, \mathcal{L}_n) \;=\; \sum_{\substack{\mathcal{T} \\ H(\mathcal{T})=H}} \frac{W(\mathcal{T})}{C(\mathcal{T})} \int_{\mathrm{bd}_{\mathcal{T}}(S) \subset \mathcal{L}_1 \cup \ldots \cup \mathcal{L}_n} e^{-\mathcal{A}(S(X;\mathcal{T}))} \prod_{j \in F(\mathcal{T})} d^d X_j \,, \tag{7.89}$$

where the sum ranges over all triangulations of $R_H^{(n)}$. This expression is intended as a regularization of the string Green function (7.4) with the sum over triangulations corresponding to the integration over metrics. Note that we are absorbing the factor $\hbar^{-1}$ into the definition of the discrete action $\mathcal{A}(S)$. The expression (7.4) suggests to choose a weight $W(\mathcal{T})$ of the form

$$W(\mathcal{T}) \sim \exp(-\mu |F(\mathcal{T})|) \tag{7.90}$$

(μ is then interpreted as proportional to the bare cosmological constant); but in principle we could study models with more general weights $W(\mathcal{T})$, as long as the weights only depend on the isomorphism class of $\mathcal{T}$ and decay fast enough so that the sum in (7.89) has adequate convergence properties. The most interesting choice for the action $\mathcal{A}(S)$ is the discretization of Polyakov's action $\mathcal{A}_{\mathrm{P}}^0$ of (7.2), that is,

$$\mathcal{A}_{\mathrm{P}}^0(S(X;\mathcal{T})) \;=\; \frac{\beta}{2} \sum_{(i,j)^* \in E(\mathcal{T})} |X_i - X_j|^2 \,, \tag{7.91}$$

where the sum runs over all pairs (i,j) of adjacent triangles (β is proportional to the bare string tension). Other models recently studied are the "Potts models

[5]Here we depart from the more standard convention of defining X as a map from the *vertices* of triangles into E^d. Our approach, which is equivalent to the more standard one, makes more immediate the connection between the present models and the matrix models introduced in Section 7.5.

on a random lattice". In these models the fields X_i are replaced by *spins* $\sigma_i = 1, \ldots, q$, and the action is the usual Potts Hamiltonian:

$$\mathcal{A}^0_{\text{Potts}}(S(\sigma;\mathcal{T})) = \frac{\beta}{2} \sum_{(i,j)^* \in E(\mathcal{T})} (1 - \delta_{\sigma_i \sigma_j}) + h \sum_i (1 - \delta_{1,\sigma_i}) . \qquad (7.92)$$

Here, as before, the spins live at the *centers* of the triangles. (By duality [57], this is equivalent to a Potts model with spins on the *vertices* of the triangles but at the dual temperature $\overline{\beta} = \log[1 + q/(e^\beta - 1)]$.)

We discuss here the discretization of the Polyakov action; some results for the Potts action will be reviewed in Section 7.5.3. With the action (7.91), the integration over X in (7.89) can be carried out explicitly, because it is given by a Gaussian integral. Indeed, the action (7.91) can be written as

$$\mathcal{A}^0_{\text{P}}(S(X;\mathcal{T})) = \beta \sum_{i,j \in F(\mathcal{T})} X_i D_{ij}(\mathcal{T}) X_j \qquad (7.93)$$

with

$$D_{ij}(\mathcal{T}) = \begin{cases} -1 & \text{if } (i,j)^* \in E(\mathcal{T}) \\ N_i & \text{if } i = j \\ 0 & \text{otherwise} \end{cases} \qquad (7.94)$$

where N_i is the number of triangles adjacent to the i-th triangle. The action (7.93) has a "zero mode" due to translation invariance. For the Green functions this mode is removed by the boundary conditions. For the partition function, however, this "zero mode" must be eliminated "by hand" to avoid the divergence of the integral. This is done, for instance, by fixing the position of one of the triangles, say $X_0 = 0$. The resulting partition function is then:

$$E_H = \sum_{\mathcal{T}:\, H(\mathcal{T}) = H} \frac{W(\mathcal{T})}{C(\mathcal{T})} \int e^{-\mathcal{A}^0_{\text{P}}(S(X;\mathcal{T}))} \, \delta(X_0) \prod_{\substack{j \in F(\mathcal{T}) \\ j \neq 0}} d^d X_j \qquad (7.95)$$

$$= \sum_{\mathcal{T}:\, H(\mathcal{T}) = H} \frac{W(\mathcal{T})}{C(\mathcal{T})} \left[\frac{\pi^{|F(\mathcal{T})|-1}}{\beta^{|F(\mathcal{T})|} \det \widetilde{D}(\mathcal{T})} \right]^{d/2} \qquad (7.96)$$

where $\widetilde{D}$ is the matrix obtained from D by deleting the row and the column corresponding to $i = 0$. Analogous expressions hold for the Green functions [115]. We see that the space dimension d appears here as a parameter, namely as an exponent in (7.96), and hence it can be continued to non-integer values. It is in this sense that one speaks of arbitrary real (or even complex) "dimensions". The case $d = 0$ corresponds to pure gravity, that is, to the theory without the matter fields X. On the other hand, for $d \to +\infty$ the dominant triangulations are those that *minimize* $\det \widetilde{D}$, while for $d \to -\infty$ only the triangulations which *maximize* $\det \widetilde{D}$ survive.

For the action (7.91) with weights (7.90), it can be shown [21] that there exists a β_c such that $G(\mathcal{L}_1, \ldots, \mathcal{L}_n)$ is finite for $\beta > \beta_c$ and infinite for $\beta < \beta_c$, for all n and all loops $\mathcal{L}_1, \ldots, \mathcal{L}_n$, even when the loops degenerate into single points [20]. The critical point β_c is, moreover, independent of the genus H. One may therefore attempt to sum over all genera, following the prescription (7.5), in order to arrive at a non-perturbative version of the theory. However, the sum over all values of H *diverges*, and it is not clear a priori how to arrive at a meaningful definition of that sum. At present we do not know of any direct, non-perturbative definition of RS models that would automatically include a summation over all values of H, except for RS models that are derived from some lattice gauge theory, and the models discussed in Section 7.6.

The physical observables of main interest are the mass gap and the susceptibility. The mass gap is defined through the exponential decay of the two-loop correlation function, $G(\mathcal{L}_1, \mathcal{L}_2)$, in the distance between the loops. It can be checked that, for the weights (7.90), this decay rate is independent of the shape of the loops, so the loops $\mathcal{L}_1, \mathcal{L}_2$ can be contracted to single points x_1, x_2, in which situation the correlation is denoted by $G(x_1, x_2)$. Hence the mass is defined as

$$m \;=\; - \lim_{|x_1 - x_2| \to \infty} \log \frac{G(x_1, x_2)}{|x_1 - x_2|} \,. \tag{7.97}$$

The susceptibility is then $\chi = \int dx\, G(x_0, x)$ and, with the weights (7.90) it is not difficult to see that

$$\chi \;\sim\; - \frac{1}{E_H} \frac{dE_H}{d\mu} \;=\; \langle |F(\mathcal{T})| \rangle \,. \tag{7.98}$$

To define the string tension let $\mathcal{L}_L$ be a rectangular loop with sides of length L and with $4L$ vertices evenly distributed along its perimeter[6]. Then

$$\alpha = - \lim_{L \to \infty} \frac{\log G(\mathcal{L}_L)}{L^2} \,. \tag{7.99}$$

For applications in quantum-gravity or string theory, one is interested in a continuum limit in which finer and finer triangulations converge to the triangulated manifold $R_H^{(n)}$. In particular, the number of triangles $|F(\mathcal{T})|$ must tend to infinity and, moreover, the average $\langle |F(\mathcal{T})| \rangle$ must diverge. By the above formula this implies that in the continuum limit one must reach a *critical model*, that is, $\beta \to \beta_c$; and, moreover, χ must diverge as $\beta \to \beta_c$, which requires

$$\gamma_{str} \;\geq\; 0 \,. \tag{7.100}$$

7.4.2 Properties of the model

We now present results obtained for the discretization of the Poylakov action — that is, weights (7.90) and action (7.91) — in the ***planar*** (i.e. H=0) case. We see from the solution (7.96) that β and μ appear only in the combination $e^{-\mu} \beta^{-d/2}$, or equivalently (for $d \neq 0$)

[6] See [20] for a discussion of why the number of vertices is chosen proportional to L; the value of the tension does not depend on the constant of proportionality.

$$\widetilde{\beta} = \beta e^{2\mu/d} . \qquad (7.101)$$

This is, therefore, the only independent parameter.

To compute the exponent γ_{str} for the discretized Polyakov model (7.91)–(7.96), the starting expression is the Kirchhoff formula (see, for instance, [402])

$$\det \widetilde{D} = \#\{\text{trees spanning } \mathcal{T}^*\} , \qquad (7.102)$$

where $\mathcal{T}^*$ is the graph whose vertices are the triangles in $\mathcal{T}$ and whose lines are given by pairs of adjacent triangles in $\mathcal{T}$. On the basis of this formula, it is expected that the triangulations minimizing $\det \widetilde{D}$ have a "polymer-like" structure, while those maximizing $\det \widetilde{D}$ are formed by regular arrays of triangles. If so, it has been argued [23] that

$$\min_{\mathcal{T}} \det \widetilde{D} \sim |F(\mathcal{T})|\, a^{F(|\mathcal{T}|)} \qquad (7.103)$$

$$\max_{\mathcal{T}} \det \widetilde{D} \sim |F(\mathcal{T})|^{-1/3}\, b^{F(|\mathcal{T}|)} \qquad (7.104)$$

where a, b are some positive constants. Therefore, by (7.96):

$$E_H \sim \sum_{\substack{\mathcal{T} \\ H(\mathcal{T})=H}} \frac{|F(\mathcal{T})|^p}{C(\mathcal{T})} \left(\frac{\beta_c}{\widetilde{\beta}}\right)^{|F(\mathcal{T})|} \quad ; \quad p = \begin{cases} -d/2 & \text{for } d \to +\infty \\ d/6 & \text{for } d \to -\infty . \end{cases} \qquad (7.105)$$

In both limiting regimes, the factor $|F(\mathcal{T})|^p$ obliterates the (d-independent) entropy factor due to the number of triangulations with the same number of triangles. Hence

$$\begin{aligned} E_H &\sim \sum_k k^p \left(\frac{\beta_c}{\beta}\right)^k \\ &\sim (\widetilde{\beta} - \beta_c)^{-p-1} \\ &\sim \begin{cases} (\widetilde{\beta} - \beta_c)^{(d/2)-1} & d \to +\infty \\ (\widetilde{\beta} - \beta_c)^{-(d/6)-1} & d \to -\infty . \end{cases} \end{aligned} \qquad (7.106)$$

Therefore [23]

$$\gamma_{str} \sim \begin{cases} -\frac{d}{2} + \mathrm{O}(1) & d \to +\infty \\ \frac{d}{6} + \mathrm{O}(1) & d \to -\infty . \end{cases} \qquad (7.107)$$

Exact results have also been obtained for $d = -2$, 0 and 1. For $d = -2$, the partition function (7.96) is a linear function of the determinant, that is, of the number of trees spanned by the dual graph, and we get an expansion in terms of tree-like polymers. This permits the application of combinatorial arguments yielding solvable recursion relations [337, 72]. Alternatively, the results for the $d = -2$ model can be deduced from the $d = 0$ model using Parisi-Sourlas dimensional reduction [116]. When $d = 0$, the partition function reduces to a counting problem, which can be solved [78] using the random-matrix models discussed in the next section. Finally, the $d = 1$ model can also be analyzed using random-matrix models [338]: modulo an approximation justified on the basis of universality, the integration reduces to the quantum mechanics of an ideal Fermi gas [78] (see brief review at the end of Section 7.5.4). The results are:

$$\gamma_{str} = \begin{cases} -1 & d = -2 \\ -\frac{1}{2} & d = 0 \\ 0 & d = 1 \end{cases} \tag{7.108}$$

Results (7.107)/(7.108) suggest that there are two critical dimensions, $d_{c_1} = 1$ and $d_{c_2} < \infty$, such that $\gamma_{str} \geq 0$ for $d_{c_1} \leq d \leq d_{c_2}$, and $\gamma_{str} < 0$ otherwise. In this regard, we mention the KPZ prediction [340]

$$\gamma_{str} = \frac{d - 1 - \sqrt{(25-d)(1-d)}}{12}, \tag{7.109}$$

which is believed to be trustworthy for d very small; it agrees with (7.107)/(7.108) at $d = -2, 0, 1$ and $d \to -\infty$, though not at $d \to +\infty$ (where the square root is ambiguous in any case). Numerical data (reviewed, for example, in [338, Section 6] and references therein) indicate that γ_{str} is strictly positive for some values of $d > d_{c_1} = 1$.

It is proven [20] that for the model (7.90)/(7.91)

$$\alpha(\tilde{\beta}) \geq 2\tilde{\beta}. \tag{7.110}$$

Since $\tilde{\beta}_c > 0$, it follows that the string tension at the critical point is nonzero, and hence the physical (continuum-limit) string tension α^* [defined by (7.22)] diverges. This means that the continuum limit is trivial in the sense that the surfaces, in the limit, collapse to very "spiky" objects of negligible area: if $\gamma_{str} > 0$ the system becomes equivalent to a system of branched polymers, and the critical exponents assume their mean-field values. See [215] for some partial results.

The collapse of random surfaces to branched polymers in the continuum limit, for d chosen so that $\gamma_{str} > 0$, is believed to be related to the presence of a tachyon in the particle spectrum of continuum string theories in $d > 1$ dimensions. This, in turn, may be related to the fact that the KPZ formula (7.109) for γ_{str} becomes meaningless when $d > 1$. (See, however [107] for a recent proposal of how to circumvent the $d = 1$ barrier, vaguely related to ideas sketched below.)

In order to obtain a non-trivial continuum limit, the action must be modified so as to suppress the unwanted spikes. A natural suggestion is to introduce terms in the action, depending on the intrinsic or extrinsic curvature, which penalize spike-like objects. The intrinsic-curvature terms [154] may be expected to have little importance in a theory of two-dimensional random surfaces immersed in

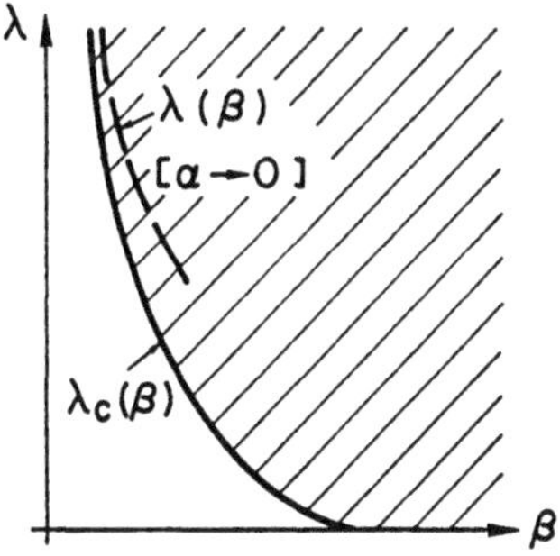

Fig. 7.1. Phase diagram of the model (7.111)

E^d ($d \geq 4$) [22]. Hence only the extrinsic-curvature term is incorporated in the action of the model. The most general form of the new action is [22]

$$\mathcal{A}(S(X;\mathcal{T})) = \beta \sum_{(i,j)^* \in E(\mathcal{T})} h(|X_i - X_j|) + \lambda \sum_{\triangle,\triangle'} f(\theta_{\triangle,\triangle'}) , \tag{7.111}$$

where β and λ are non-negative coupling constants, h and f are non-negative continuous functions, and we denote by $\theta_{\triangle,\triangle'} \in [0,\pi]$ the angles between the embedded triangles $\triangle$ and $\triangle'$ in E^d. To assign a nonzero energy to all triangulations with nonzero curvature, the function f defined on $[0,\pi]$ is assumed to have a unique minimum at zero, with$f(0) = 0$ and $f(\theta) > 0$ if $\theta > 0$.

For this class of models, it is proven [22] that there is a convex curve $(\beta, \lambda_c(\beta))$ in the (β,λ)-plane — the "critical line" (Figure 7.1) — that delimits a convex region $\mathcal{B}$ in whose interior the Green functions are finite. These functions are infinite outside the closure of $\mathcal{B}$. The curve $\lambda_c(\beta)$ is monotonically decreasing and

$$\lambda_c(\beta) \to \infty \quad \text{as} \quad \beta \to 0 . \tag{7.112}$$

In fact, if $h(|x|) = x^2$ and $f(\theta) = 1 - \cos\theta$ ("Gaussian action"), there exist constants c_1, c_2 such that [22]

$$c_1 \beta^{-d/(d-2)} \leq \lambda_c(\beta) \leq c_2 \beta^{-d/(d-2)} . \tag{7.113}$$

Moreover, if $h(|x|) = x^p$, $p \geq 2$, $m(\beta,\lambda)$ and $\alpha(\beta,\lambda)$ are strictly positive in the interior of $\mathcal{B}$. On the critical curve, $\alpha(\beta,\lambda_c(\beta))$ is expected to be positive, but it vanishes as $\beta \to 0$, and $m(\beta,\lambda_c(\beta)) = 0$. More precisely, it is proven that there exist trajectories $(\beta,\lambda(\beta)) \to (0,\infty)$ as $\beta \to 0$, such that $\alpha(\beta,\lambda(\beta)) \to 0$ (Figure 7.1). Similar trajectories $(\beta,\lambda'(\beta))$ are found, where $m(\beta,\lambda'(\beta)) \to 0$ (provided that $p < p_d$, for p_d a certain dimension-dependent constant < 2).

These are encouraging results suggesting the possible non-triviality of the continuum limit. However, the bounds do not yet prove the existence of trajectories $(\beta,\lambda(\beta))$ such that

$$\frac{\alpha(\beta,\lambda(\beta))}{m(\beta,\lambda(\beta))^2} \to \text{const} , \tag{7.114}$$

as required for a continuum theory with finite renormalized string tension and mass.

There are many other surface theories constructed on the same conceptual basis. In applications to condensed-matter physics or fluid dynamics, one may wish to introduce a self-avoidance constraint and/or to restrict the isomorphism classes admitted on the RHS of (7.89) to some small family. This is because a class $\mathcal{T}$ of triangulations of $R_H^{(n)}$ really determines what in condensed-matter physics would be called a distribution of *disclinations* present on the surface $S(X;\mathcal{T})$. Since disclinations may have considerable stress energy, different isomorphism classes may have different a priori probabilities.

Let us finally sketch some partial results for higher genera. Exact solutions [347] show that, as a function of the genus H,

$$\gamma_{str} = \begin{cases} (5H-1)/2 & \text{for } d=0 \\ 2+3(H-1) & \text{for } d=-2 \end{cases} \tag{7.115}$$

which generalizes (7.108). [The first line was obtained only for $H = 0,1,2$, but it is believed to be valid for all genera; the second line was obtained for all genera.] Moreover, Distler and Kawai [131] have argued, on the basis of the consistency of the quantization of the Liouville action, that

$$\gamma_{str} = 2 + (H-1)\frac{(25-d)+\sqrt{(25-d)(1-d)}}{12}, \tag{7.116}$$

an expression that generalizes (7.109). An important aspect of formulas (7.115) and the proposed generalization (7.116) is the *linear dependence on the genus H*. This linear dependence will be seen in Section 7.6.1 to imply that the topological sum — i.e. the sum over H — of (for instance) the free energy is consistent with a scaling relationship. Such a relationship is the basis for nonperturbative theories constructed via the so-called double scaling limit, which is the subject of Section 7.6.

7.5 Random-matrix models

7.5.1 Matrix field theories

The discretized Polyakov model defined by (7.90)–(7.91) can be put in correspondence with an appropriate *matrix* field theory [216, 115]. This correspondence is somewhat analogous to the correspondence between random-walk models and *vector* field theories (Chapter 6), but with the following crucial difference: the random-walk models arise essentially from the *high-temperature* expansion of the vector field theory, which is a *convergent* expansion (on the lattice); while the random-surface models arise from the *coupling-constant* (perturbation-) expansion of the matrix field theory, which is *divergent*. Moreover, for most of the models studied here, the matrix integral is itself divergent, and the coupling-constant expansion is not even Borel summable.

Let us briefly review the main features of a matrix field theory. We consider a field of matrices $\{\varphi(x)\}$ where x ranges over some d-dimensional manifold M, and each $\varphi(x)$ is a $N \times N$ Hermitian matrix. A (Euclidean) matrix field theory is defined by assigning to the space of field configurations $\{\varphi(x)\}$ a (formal) measure of the form

$$e^{-\frac{1}{\hbar}\mathrm{Tr}\, U(\{\varphi(x)\})} \prod_x d\varphi(x) , \tag{7.117}$$

where U is a functional of $\{\varphi(x)\}$ [usually a local functional involving only φ and its first two derivatives], and the measure on the space of $N \times N$ Hermitian matrices is

$$d\varphi(x) \;=\; \prod_{1\le i\le n} d\varphi_{ii}(x) \prod_{1\le i<j\le n} d\mathrm{Re}\,\varphi_{ij}(x) \prod_{1\le i<j\le n} d\mathrm{Im}\,\varphi_{ij}(x) . \tag{7.118}$$

In this section we shall not set $\hbar$ to one, but rather we will consider it a freely adjustable parameter. (We could have denoted the parameter $1/\hbar$ by β, but this would conflict with the notation for the inverse temperature of the statistical-mechanical Potts models of Section 7.5.3.) Below, we shall take the limit $\hbar \to 0$; the theory so obtained is, therefore, often referred to as "semiclassical" (or "low temperature").

As in the standard (vector) $\lambda\varphi^4$ theory, the measure (7.117) is ill-defined due to ultraviolet divergences; one must therefore resort to some limiting procedure to obtain meaningful expressions. One process of regularization consists (as in Section 1.5) in considering lattice approximations and a discretized version of U. On a finite lattice of q sites, the discretized U typically takes the form

$$U(\varphi(1), \ldots, \varphi(q)) \;=\; \sum_{r,s=1}^{q} A_{rs}\varphi(r)\varphi(s) + \sum_{r=1}^{q} V_r(\varphi(r)) , \tag{7.119}$$

where $\{A_{rs}\}$ is positive-definite, and V is a polynomial potential:

$$V_r(\varphi) \;=\; \sum_{p=3}^{p_{\max}} u_p(r)\varphi^p . \tag{7.120}$$

One then wishes to compute integrals with respect to the measure

$$e^{-\frac{1}{\hbar}\mathrm{Tr}\, U(\varphi(1),\ldots,\varphi(q))} \prod_{i=1}^{q} d\varphi(i) . \tag{7.121}$$

Such a theory is called a q-matrix field theory. To obtain a Euclidean field theory, one must first compute the observables at finite q and then let $q \to \infty$, adjusting at the same time $\{A_{rs}\}$ and $\{u_p\}$ as a function of q (infinite-volume and continuum limits for the matrix field theory).

But even for *finite* values of q the matrix theory poses non-trivial problems: To begin with, it is well known that the perturbation expansion in powers of the coefficients u_p ($p \ge 3$) is divergent. This fact alone is perhaps not so serious; it would simply force us, as in ordinary field theory, to take a nonperturbative point of view. The real trouble is that in virtually all situations of interest in random-surface theory, *the integral (7.121) is itself divergent*, because the potential V is unbounded below: either $p_{\max}$ is odd, or else $p_{\max}$ is even but $u_{p_{\max}}(r)$ is negative. Moreover, the perturbation expansion is not Borel summable. (It is not hard to show that, *in certain sectors of the complex coupling-constant space*, the matrix integral is well-defined and the perturbation series is asymptotic or even Borel summable to it. Unfortunately, as we shall see, these sectors are usually *not* the ones where we must know the value of the matrix integral in order to retrieve "physical" information about random surfaces.) At the moment, we regard *formal* matrix integrals as a convenient book-keeping device to handle the *formal* perturbation series, all of whose terms in an arbitrary order of u_p ($p \geq 3$) are *well-defined*.

The correspondence between random-matrix models and models of two-dimensional quantum gravity is made at the level of *formal* power series: The double expansion of a matrix model in powers of $1/N$ (where N is the size of the matrix φ) and u_p ($p \geq 3$), as first studied by 't Hooft [304, 305], is compared to a *formal* series whose terms are indexed by "polygonizations" of two-dimensional compact orientable surfaces of arbitrary genus. It is found that if the power of $1/N$ is equated to the genus of the surface (up to some constant), then the two formal series coincide term-by-term. The point of this observation is that the matrix-model formulation of that series suggests ways of how to "sum it" to a well-defined function. For example, the series expressing the contribution corresponding to a fixed power of $1/N$, i.e. to a fixed genus, are geometrically convergent and can be summed using the method of orthogonal polynomials well known in the theory of matrix integrals [63]. The series expressing the contributions of arbitrary powers of $1/N$, i.e. summing over all genera, can be related, in a certain limit (the so-called double scaling limit discussed in Section 7.6.4), to solutions of certain differential equations (related to KdV flows). These equations can, again, be derived from the method of orthogonal polynomials in the context of matrix integrals. So, the matrix models suggest *non-perturbative* solutions to two-dimensional quantum-gravity models.

For simplicity, let us first consider some 1-matrix models. The computation of the coefficients of the perturbation expansion around the Gaussian model follows the usual Feynman diagrammar based on Wick's formula. The propagator is

$$\langle \varphi_{ij}\varphi_{lk}\rangle_0 = \hbar\, \delta_{ik}\delta_{jl} , \qquad (7.122)$$

where $\langle \cdot \rangle_0$ is the Gaussian measure

$$\langle f(\varphi)\rangle_0 = \frac{\int f(\varphi)\, e^{-\frac{1}{\hbar}\mathrm{Tr}\,\frac{1}{2}\varphi^2}\, d\varphi}{\int e^{-\frac{1}{\hbar}\mathrm{Tr}\,\frac{1}{2}\varphi^2}\, d\varphi} . \qquad (7.123)$$

The double subscripts of φ are taken care of by using double-line propagators in the Feynman diagrams. We see from (7.122) that the contractions are between the first index of one factor and the second index of the other factor. If the position of the index is represented by an arrow (say, outgoing for the second index and ingoing for the first one), this constraint is automatically taken into account by endowing each line with an orientation

$$\langle \varphi_{ij}\overline{\varphi_{kl}}\rangle_0 \;=\; {}^{i}_{j}\!\begin{array}{c}\longrightarrow\\ \longleftarrow\end{array}\!{}^{k}_{l} \;\equiv\; \hbar\,\delta_{ik}\delta_{jl}\,, \tag{7.124}$$

and requiring that, in the graphs, the couplings between lines respect the orientations. Except for this additional convention, the Feynman graphs for a matrix-field theory are constructed in the same way as those for a scalar or vector theory. Many of the graphs so obtained are non-planar: that is, when drawn in a plane, some graphs require propagators to cross at points where there is no vertex. Thus, graphs can be classified according to the minimum genus of a surface on which they can be drawn without self-intersections. This introduces (a little) topology into the game, because the perturbation expansion can be written as a double series in the number of vertices and the genus of the graph. However, the only topological notion needed here is Euler's formula: if we identify the graph with a polyhedron (thinking of propagators as fat edges), so that the loops of the graph corresponds to faces of the polyhedron, we have that

$$\#(\text{vertices}) - \#(\text{edge}) + \#(\text{loops}) \;\equiv\; \text{Euler number} \;=\; 2-2H\,. \tag{7.125}$$

Let us discuss, for concreteness, the perturbation expansion of a 1-matrix φ^3 field theory [$q=1$, $p_{\max}=3$ in (7.119)]. We consider the expansion in powers of u_3 and N of the formal integral

$$\widetilde{Z}(\hbar,u_3,N) \;=\; \int \exp -\tfrac{1}{\hbar}\mathrm{Tr}\left[\frac{1}{2}\varphi^2 + u_3\varphi^3\right] d\varphi\,, \tag{7.126}$$

where φ ranges over all $N\times N$ Hermitian matrices, and $d\varphi$ is given by (7.118). Of course, the integral $\widetilde{Z}$ is defined, naively, only for pure imaginary u_3 (the perturbation series in u_3 being *asymptotic* to $\widetilde{Z}(\hbar,u_3,N)$, for $u_3/\hbar$ on the imaginary axis, with $\hbar$ and N fixed).

We have deliberately considered a redundant number of variables in (7.126): we can set either $\hbar$ or u_3 (but not both) to 1 by a simple rescaling of the matrix fields, at the cost of an unimportant factor in the definition of $\widetilde{Z}$. This redundancy has a pedagogical purpose; it will allow us later to discuss in detail the different parametrizations chosen in the literature.

As a first step, we perform the formal power series expansion of (7.126) in powers of $u_3/\hbar$:

$$\frac{\widetilde{Z}(\hbar,u_3,N)}{\widetilde{Z}(\hbar,0,N)} \underset{\mathrm{f}}{=} \sum_{k=0}^{\infty}\frac{(-u_3)^k}{\hbar^k}\frac{1}{k!}\left\langle\left[\mathrm{Tr}\,\varphi^3\right]^k\right\rangle_0\,, \tag{7.127}$$

where $\langle\,\cdot\,\rangle_0$ is the Gaussian measure (7.123). The symbol "$\underset{\mathrm{f}}{=}$" indicates "equality in the sense of formal power series". The coefficients of this series can be expressed as a collection of Feynman graphs with double-line propagators defined by (7.122) and where each factor $\mathrm{Tr}\,\varphi^3$ in (7.127) corresponds to a vertex with three pairs of lines:

$$\varphi_{ij}\varphi_{jk}\varphi_{ki} \;=\; \tag{7.128}$$

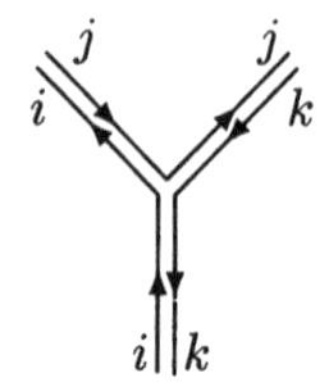

With this convention, each term of order k in (7.127) corresponds to the sum of all Feynman diagrams with k vertices of the form (7.128) where the couplings between pairs of lines respect the orientations of the lines.

To find the contribution of each Feynman diagram, we notice that, by the above formulas, each vertex has a weight $u_3/\hbar$, and each double-line propagator represents a factor of $\hbar$. Moreover, every loop in the space of indices — that is, after contraction of indices, every factor δ_{ii} — yields, upon summation, a factor of N. Thus, each diagram $\mathcal{T}^*$ with k vertices and l propagators contributes a term

$$\frac{1}{C(\mathcal{T}^*)}\,\frac{(-u_3)^k}{\hbar^k}\,\hbar^l N^L \tag{7.129}$$

to the expansion (7.127). Here L is the number of loops (in the space of indices) and $C(\mathcal{T}^*)$ is a combinatorial factor equal to the order of the symmetry group of the diagram $\mathcal{T}^*$.

Up to this point, however, we are obtaining *all* diagrams $\mathcal{T}^*$, connected or not; but, at least for comparison with topological expansions in two-dimensional quantum gravity, we are only interested in *connected* triangulations $\mathcal{T}$. As is well known to the practitioners of Feynman diagrammar, passing to the logarithm of an expansion (in the sense of formal power series) does the trick of retaining only connected diagrams. We therefore define

$$\widetilde{E}(\hbar, u_3, N) \;=\; \log \frac{\widetilde{Z}(\hbar, u_3, N)}{\widetilde{Z}(\hbar, 0, N)}\,, \tag{7.130}$$

and we we obtain from (7.129)

$$\widetilde{E}(\hbar, u_3, N) \underset{\mathrm{f}}{=} \sum_{k=0}^{\infty} \sum_{\substack{\mathcal{T}^*\text{ connected}\\ \text{with } k \text{ vertices}}} \frac{1}{C(\mathcal{T}^*)}(-u_3)^k \hbar^{l-k} N^L\,. \tag{7.131}$$

We now apply Euler's formula (7.125):

$$k - l + L \;=\; 2 - 2H\,, \tag{7.132}$$

with

$$2l \;=\; 3k \tag{7.133}$$

(each propagator connects 2 double entries and each vertex has 3 double entries). As a result, (7.131) can be written as

$$\widetilde{E}(\hbar, u_3, N) \underset{\mathrm{f}}{=} \sum_{k=0}^{\infty} \sum_{\substack{\mathcal{T}^* \text{ connected} \\ \text{with } k \text{ vertices}}} \frac{1}{C(\mathcal{T}^*)} (-u_3)^k (N\hbar)^{k/2} N^{2-2H} . \tag{7.134}$$

If we set

$$X \equiv N\hbar , \tag{7.135}$$

we see that (7.134) is a double series in $u_3\sqrt{X}$ and $1/N$. We reorder it to obtain

$$\widetilde{E}(u_3\sqrt{X}, N) \underset{\mathrm{f}}{=} \sum_{H=0}^{\infty} \frac{1}{N^{2H}} N^2 \widetilde{E}_H(u_3\sqrt{X}) , \tag{7.136}$$

with

$$\widetilde{E}_H(u_3\sqrt{X}) = \sum_{\substack{\mathcal{T}^* \text{ connected} \\ H(\mathcal{T}^*) = H}} \frac{1}{C(\mathcal{T}^*)} (-u_3\sqrt{X})^{|V(\mathcal{T}^*)|} , \tag{7.137}$$

where $V(\mathcal{T}^*)$ denotes the sets of vertices of $\mathcal{T}^*$. (Note that only even powers of $|V(\mathcal{T}^*)|$ show up, because Wick's theorem forces all graphs without external legs to have an even number of vertices, for a φ^3 theory.)

We see that this perturbation expansion has precisely the form of the topological expansion postulated for quantum-gravity models.

The expansion for a general potential $V = \sum_{r=3}^{p_{\max}} u_p \varphi^p$ is a straightforward generalization of the above one. We now have vertices with $3, 4, \ldots, p\mathrm{max}$ pairs of lines; and each vertex with p pairs of lines contributes a factor $(-u_p/\hbar)$. The relationship between the number of edges and number of vertices of the graph is, for the general case,

$$2l = 3|V_3| + 4|V_4| + \ldots + p\mathrm{max}|V_{p\mathrm{max}}| \tag{7.138}$$

where V_p represents the set of vertices with p pairs of lines. Therefore, (7.136)–(7.137) generalize to

$$\widetilde{E}(u_3\sqrt{X}, u_4 X, \ldots, u_{p\mathrm{max}} X^{(p_{\max}/2)-1}, N) \underset{\mathrm{f}}{=} \sum_{H=0}^{\infty} \frac{1}{N^{2H}} N^2 \widetilde{E}_H(u_3\sqrt{X}, u_4 X, \ldots, u_{p\mathrm{max}} X^{(p\mathrm{max}/2)-1}) \tag{7.139}$$

with

$$\widetilde{E}_H\left(\{u_p X^{(p/2)-1}\}\right) = \sum_{\substack{\mathcal{P}^* \text{ connected} \\ H(\mathcal{P}^*) = H}} \frac{1}{C(\mathcal{P}^*)} \prod_{p=3}^{p\mathrm{max}} \left(-u_p X^{(p/2)-1}\right)^{|V_p(\mathcal{P}^*)|} , \tag{7.140}$$

where $\mathcal{P}^*$ (the dual of a "polygonization") ranges over Feynman diagrams with internal vertices of $3, 4, \ldots, p\mathrm{max}$ pairs of lines (and no external vertices).

We notice that in these expressions the parameters u_p and $\hbar$ appear only in the combinations $u_p X^{(p/2)-1}$, where $X = N\hbar$. This is an explicit manifestation of the redundancy of parameters pointed out above. The non-redundant parametrizations used in the literature can be divided into two major groups:

i) Each u_p is chosen to depend on X so that

$$u_p X^{(p/2)-1} = v_p \tag{7.141}$$

where v_p is a constant independent of X. In this case, the change $\varphi \to \varphi/\sqrt{\hbar}$ leads to the form

$$\widetilde{Z}(v_3, \ldots, v_{\mathrm{pmax}}, N) = \int \exp\left\{-\mathrm{Tr}\left[\frac{\varphi}{2} + \sum_{p=3}^{\mathrm{pmax}} \frac{v_p}{N^{(p/2)-1}}\varphi^p\right]\right\} d\varphi \tag{7.142}$$

for the partition function.

ii) All parameters u_p are considered to be independent of X. This is the choice we shall adopt in the sequel: It amounts to consider $\widetilde{Z}$ and $\widetilde{E}$ as functions of $\hbar$ and N only, while $\widetilde{E}_H$ depends on both only through the combination $X = N\hbar$. For this choice, the partition function can be written in the form

$$\widetilde{Z}(\hbar, N) = \int \exp\left\{-\frac{N}{X}\mathrm{Tr}\left[\frac{\varphi}{2} + \sum_{p=3}^{\mathrm{pmax}} u_p\varphi^p\right]\right\} d\varphi \tag{7.143}$$

which suggests the use of steepest-descent techniques for large values of N.

For general q-matrix models with actions (7.119), the Feynman diagrams are the same as for the 1-matrix case, but each vertex is endowed with a label r. A vertex labelled r with p double lines contributes a factor $(-u_p(r)/\hbar)$. The formula for the propagator is now:

$$\langle \varphi_{ij}(r)\varphi_{lk}(s)\rangle_0 = (A^{-1})_{rs}\, \hbar\, \delta_{ik}\delta_{jl} \tag{7.144}$$

where $\langle\cdot\rangle_0$ is the Gaussian measure with density $\exp\left[-\frac{1}{\hbar}\mathrm{Tr}\,\sum_{r,s} A_{rs}\varphi(r)\varphi(s)\right]$. Therefore, the same topological expansion (7.139) is obtained, but depending on the combinations $u_p(r)X^{(p/2)-1}$ ($3 \le p \le \mathrm{pmax}$, $1 \le r \le q$) and with the fixed-genus sums (7.140) including a factor $(A^{-1})_{rs}$ for each (double-line) propagator connecting a vertex labelled r with one labelled s.

Formally, the same expansion remains valid for continuum models, with the discrete label r changed into the variable x.

7.5.2 Random-matrix models and random triangulations: Pure gravity

The relation between random triangulations and matrix models arises from the observation that, for a triangulation $\mathcal{T}$, each vertex of the dual graph $\mathcal{T}^*$ — i.e. the graph whose vertices are the centers i of the triangles and whose edges are the pairs (i,j) where i and j are adjacent triangles — is of order three. That is, every Feynman diagram in the (formal) expansion of a φ^3 field theory can be labelled by the dual, $\mathcal{T}^*$, of a triangulation $\mathcal{T}$. An example of this association is presented in Figure 7.2. Moreover, from (7.96) and (7.90) we have that the

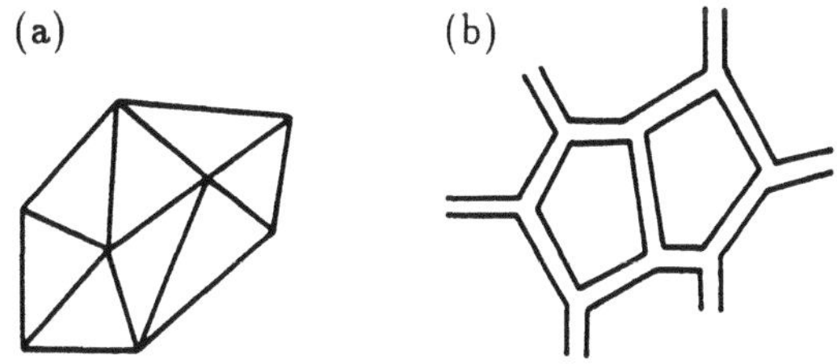

Fig. 7.2. (a) A triangulation; **(b)** the corresponding dual graph

topological expansion for the log of the partition function of a zero-dimensional (pure gravity) theory is

$$E(\mu, N) \underset{\mathrm{f}}{=} \sum_{H=0}^{\infty} \frac{1}{N^{2H}} C(n, N) E_H(\mu) , \tag{7.145}$$

with

$$E_H(\mu) = \sum_{\mathcal{T}:\, H(\mathcal{T})=H} \frac{e^{-\mu|F(\mathcal{T})|}}{C(\mathcal{T})} , \tag{7.146}$$

where $\mathcal{T}$ ranges over the triangulations of a *connected* surface of genus H. As the symmetry group of a graph and that of its dual triangulation are isomorphic:

$$C(\mathcal{T}^*) = C(\mathcal{T}) , \tag{7.147}$$

expansion (7.145)–(7.146) is identical to expansion (7.136)–(7.137) if we take $-u_3\sqrt{X} = e^{-\mu}$. Therefore, random-triangulation pure-gravity models are in formal correspondence with one-*matrix* φ^3 field theories. That is why double-line propagators are used in Figure 7.2. The precise statement of this correspondence is the identity

$$\widetilde{E}(u_3\sqrt{X} = -e^{-\mu}, N) \underset{\mathrm{f}}{=} E(\mu, N) . \tag{7.148}$$

For pure imaginary u_3, the integral (7.126) is well-defined, and the limit

$$\widetilde{E}_{H=0}(u_3\sqrt{X}) = \lim_{\substack{N\to\infty \\ \hbar\to 0 \\ N\hbar = X}} \frac{1}{N^2}\widetilde{E}(u_3\sqrt{X}) \tag{7.149}$$

exists. It is therefore appropriate to call the limit $N \to \infty, \hbar \to 0, N\hbar = X$, the *planar limit* of a matrix theory; it can be thought of as a semiclassical limit. The relation (7.149) is often interpreted by saying that, for large N, the perturbation expansion of an $N \times N$ matrix field theory becomes dominated by planar diagrams. The sum of these diagrams corresponds to planar pure gravity. An exact computation of the r.h.s. of (7.149) was presented in [78] using a steepest-descent argument. The resulting expression is in fact analytic in $u_3\sqrt{X}$ in a disc around the origin, and hence it can be analytically continued to real $u_3\sqrt{X} = -e^{-\mu}$. By (7.148) and (7.145), this yields the convergent expansion

$$E_{H=0}(\mu) = -\frac{1}{2}\sum_{k=1}^{\infty} \frac{[72e^{-2\mu}]^k}{(k+2)!} \frac{\Gamma(3k/2)}{\Gamma(1+k/2)} . \tag{7.150}$$

(As mentioned above, the dependence on $(e^{-\mu})^2 = (u_3\sqrt{X})^2$ is a consequence of Wick's theorem.) This series converges for $e^{-2\mu} \le e^{-2\mu_c} = 1/108\sqrt{3}$, and, for large k, the terms behave as

$$\sim e^{-2(\mu-\mu_c)k} k^{-7/2} . \tag{7.151}$$

Hence,

$$E_{H=0}(\mu) \sim (\mu - \mu_c)^{5/2} . \tag{7.152}$$

In particular,

$$\chi = \frac{\partial^2 E}{\partial \mu^2} \sim (\mu - \mu_c)^{1/2} , \qquad \text{i.e., } \gamma_{str} = -\frac{1}{2} . \tag{7.153}$$

These results represent a solution of a *discretized version* of Polyakov's string in zero dimension, for genus zero. For the motivating physical application — the regularization of (7.4) — we are interested in the behavior in a continuum limit. As discussed above, a continuum limit is reached when the model approaches a critical point. Therefore, we must study the regime $\mu \to \mu_c$ or, in matrix-model language, $X \to X_c$. We postpone the discussion of this limit to Section 7.6. We only mention here that, in discussing such limits, it is customary to choose $u_3 = e^{-\mu_c}$, so $X_c = 1$.

More generally, we can consider quantum-gravity models based on "polygonizations" $\mathcal{P}$, that is, on the use of random p-gons with $p = 3, 4, \ldots, p\text{max}$. Equations (7.139)–(7.140) then show that such models are in correspondence with 1-matrix theories with a polynomial potential of degree $p$max and where the vertex weights are $-u_3\sqrt{X}, -u_4 X, \ldots, -u_{p\text{max}} X^{(p_{\text{max}}/2)-1}$. If all these weights are chosen to be equal to $e^{-\mu}$, we obtain a regularized model with the *same* critical exponents as the model based on triangulations (i.e., we remain in the same universality class). For instance, the model with $V(\varphi) = u_4\varphi^4$ (random quadrangulations), was first analyzed in [78], where the planar limit was computed. With the identification $u_4 X = -e^{-\mu}$, it yields

$$E_{H=0}(\mu) = -\sum_{k=1}^{\infty} (12e^{-\mu})^k \frac{(2k-1)!}{k!(k+2)!} . \tag{7.154}$$

This series has the same behavior (7.152) — and hence leads to the same exponent $\gamma_{str} = -1/2$ — except that the radius of convergence is $e^{-\mu_c} = 1/48$. Expressions for higher genus have also been computed [63]; in particular,

$$E_{H=1}(\mu) = -\frac{1}{24}\sum_{k=1}^{\infty} \frac{(12e^{-\mu})^k}{k} \left[4^k - \frac{(2k)!}{(k!)^2}\right] \tag{7.155}$$

and

$$E_{H=2}(\mu) = \frac{1}{5\cdot 3^3\cdot 2^5}\sum_{k=3}^{\infty}(12e^{-\mu})^k(k-1)\Big[\frac{195\cdot 4^k}{8} - (28k+9)\frac{(2k)!}{(k!)^2}\Big] . \quad (7.156)$$

The radius of convergence is independent of the genus, $e^{-\mu_c} = 1/48$, but the critical exponents are not. From the last two series one obtains $\gamma_{str} = 2$, for $H = 1$; and $\gamma_{str} = 9/2$, for $H = 2$. The above values of γ_{str} constitute, precisely, the first line of (7.115).

A more interesting possibility is to assign different weights to different types of vertices (or, equivalently, to different polygons) and then to "fine-tune" these weights so as to obtain models with different critical behavior. This was explored by Kazakov [336], who found an infinite sequence of multicritical points characterized by exponents $\gamma_{str} = -1/k$, $k = 2, 3, \ldots$ .However, these models require some of the coefficients $u_p X^{(p/2)-1}$ to be *positive*. Such models do not give rise to a *probability theory* of random surfaces, since some surfaces have *negative weights*. See Section 7.6.3.

7.5.3 Random-matrix models for gravity coupled to matter fields: Potts spins

Models of gravity coupled to matter fields can be related to random-matrix models with internal degrees of freedom. An interesting case is obtained for theories of q $N\times N$ matrices with actions of the form (7.119) for the particular choice $A_{rs} = -\delta_{rs} + c(1-\delta_{rs})$; that is, for actions

$$U(\varphi(1),\ldots,\varphi(q)) = -\sum_{r=1}^{q}\varphi^2(r) + c\sum_{\substack{r,s=1\\ r\neq s}}^{q}\varphi(r)\varphi(s) + \sum_{r=1}^{q}\sum_{p=3}^{\text{pmax}}u_p(r)\varphi^p(r) . \quad (7.157)$$

For this action, (7.144) yields the propagators [335]

$$\langle\varphi_{ij}(r)\varphi_{lk}(s)\rangle_0 = \delta_{ik}\delta_{jl}\,\hbar A\Big[\delta_{rs} + \frac{B}{A}(1-\delta_{rs})\Big] \quad (7.158)$$

with

$$A = \frac{[1-c(q-2)]}{(1+c)[1-c(q-1)]} \quad (7.159)$$

$$B = \frac{c}{(1+c)[1-c(q-1)]} . \quad (7.160)$$

Therefore, as discussed in Section 7.5.1, we have the asymptotic expansion

$$\widetilde{E}\Big(\{u_p(r)X^{(p/2)-1}\},c,N\Big) \underset{\mathrm{f}}{=} \sum_{H=0}^{\infty}\frac{1}{N^{2H}}N^2\widetilde{E}\Big(\{u_p(r)X^{(p/2)-1}\},c\Big) , \quad (7.161)$$

where $\widetilde{E}_H$ is the sum over all Feynman graphs with vertices labelled by an index $r = 1, 2, \ldots, q$ and having p double lines. A p-vertex labelled r has a weight

$$- u_p(r)(XA)^{(p/2)-1} , \tag{7.162}$$

and each propagator connecting vertices labelled r and s contributes with a factor

$$\delta_{rs} + \frac{B}{A}(1 - \delta_{rs}) . \tag{7.163}$$

It is not hard to see that this asymptotic series can be identified with the topological expansion for the Potts model on random polygonizations. Indeed, the action (7.92) for the Potts model yields the topological expansion

$$E(\beta, h) \underset{\mathrm{f}}{=} \sum_{H=0}^{\infty} \frac{1}{N^2} C(n, N) E_H(\beta, h) , \tag{7.164}$$

where, by (7.89) (with $n = 0$) and (7.90), E_H corresponds to the sum over all possible polygonizations with 3-gons, 4-gons, ..., $p$max-gons, and where each (center of) polygon is labelled by a variable $\sigma = 1, 2, \ldots, q$. A polygon labelled σ has a weight

$$e^{-\mu} e^{h(\delta_{\sigma,1}-1)} = \begin{cases} e^{-(\mu+h)} & \sigma \neq 1 \\ e^{-\mu} & \sigma = 1 ; \end{cases} \tag{7.165}$$

and for each pair of adjacent polygons with centers labelled σ, σ' there is a factor

$$e^{\beta(\delta_{\sigma\sigma'}-1)} = \delta_{\sigma\sigma'} + e^{-\beta}(1 + \delta_{\sigma\sigma'}) . \tag{7.166}$$

Comparing (7.161)–(7.163) with (7.164)–(7.166) we see that, due to the duality between p-graphs and polygonizations, the topological expansion for the random Potts model at inverse temperature β, magnetic field h and "polygon-fugacity" $e^{-\mu}$, is identical to the perturbation expansion of the q-matrix model with actions (7.157), provided parameters are related by

$$e^{-\beta} = \frac{B}{A} = \frac{c}{1 - c(q-2)} \tag{7.167}$$

and

$$- u_p(r)(XA)^{(p/2)-1} = \begin{cases} e^{-(\mu+h)} & r \neq 1 \\ e^{-\mu} & r = 1 . \end{cases} \tag{7.168}$$

These results pertain to the Potts models with spins placed at the center of each polygon. By the high-temperature–low-temperature duality [57], they also correspond to a model with spins at the vertices of the polygons but at the dual temperature $\overline{\beta} = \log[1 + q/(e^{\beta} - 1)]$.

In the *planar* limit, the 2-matrix model with action (7.157) — which corresponds to the Ising model on a planar random lattice — has been solved for a φ^4 [334, 71] or φ^3 potential [71, 335]. In the former case, the solution is a consequence of the two-matrix integral obtained by Mehta [390]. The model exhibits a first-order phase transition characterized by the critical exponents [71]

$$\alpha = -1, \ \beta = \frac{1}{2}, \ \gamma = 2, \ \delta = 5 \tag{7.169}$$

and

$$\gamma_{str} = \begin{cases} -1/2 & \text{for noncritical temperatures} \\ -1/3 & \text{for the critical temperature}. \end{cases} \qquad (7.170)$$

We recall that γ characterizes the critical behavior of $\partial^2 \log E_{H=0}/\partial h^2$, while γ_{str} (associated to the entropy of random surfaces, see (7.180) below, or to the "entropy of lattices" [71]) corresponds to $\partial^2 \log E_{H=0}/\partial \mu^2$. The critical exponents are different from those of a standard Ising model on a (fixed) two-dimensional lattice. The fluctuations of the lattice alter the critical behavior.

For $q > 2$, the matrix integration reduces, for large N, to a system of integral equations which are amenable to numerical computations but somewhat complicated for analytical treatment [335]. On the other hand, analytic continuation in q yields exact solutions for two interesting limiting cases [335]: $q \to 1$ (random percolation) and $q \to 0$ (random tree percolation; it coincides with the discretized Polyakov string with $d = -2$). For details on such limits, see [335].

7.5.4 Random-matrix models for gravity coupled to matter fields: Polyakov string

The discretized Polyakov model (7.90)–(7.91) in dimensions $d > 0$ can be related to a matrix field theory in d dimensions with propagator e^{Δ}, where Δ is the discrete Laplacian. Indeed, we can write (7.94) purely in terms of the dual graphs $\mathcal{T}^*$:

$$E_H = \sum_{\substack{\mathcal{T}^* \\ H(\mathcal{T}^*)=H}} \int \exp\left\{-\mu|V(\mathcal{T}^*)| - \frac{\beta}{2} \sum_{(i^*,j^*)\in E(\mathcal{T}^*)} [X_{i^*} - X_{j^*}]^2\right\} \prod_{i\neq 0} d^d X_{i^*} . \qquad (7.171)$$

Here (i^*, j^*) denotes the edge dual to the pair of adjacent triangles (i, j) and X_{i^*} is defined, using any well-defined pattern of choices, from the identification $X_{i^*} - X_{j^*} \equiv X_i - X_j$. Changing variables to $Y_j = \sqrt{\beta} X_j$, the expression reduces to a model with $\beta' = 1$ and $\mu' = \mu + (d/2)\log\beta$. The resulting integral can be interpreted as a sum over all φ^3 Feynman diagrams, $\mathcal{T}^*$, in which each vertex has a weight $e^{-\mu'}$ and where the propagator between two points Y_i and Y_j is Gaussian: $\exp[-\frac{1}{2}(Y_i - Y_j)^2]$. A simple computation in Fourier space shows that the exponential is the propagator of a Gaussian theory in $\mathbb{R}^d$ with covariance matrix $(e^{\Delta})(x, y)$. Therefore, following the arguments of the previous section, one concludes that the topological expansion $E(\mu, \beta, N)$ is in correspondence with the (connected) partition function for a d-dimensional $N \times N$-matrix field theory with action given by the trace of

$$U(\{\varphi(x)\}) = \int dx^d \left\{\tfrac{1}{2}\varphi(x)(e^{-\Delta}\varphi)(x) + u_3\varphi(x)^3\right\} \qquad (7.172)$$

with $-u_3\sqrt{X} = e^{-\mu'}$. More generally, we can replace the φ^3 interaction by a general potential $V(\varphi(x)) = \sum_{p=3}^{pmax} u_p \varphi^p(x)$, and obtain a matrix model whose perturbation expansion coincides with the topological expansion of a d-dimensional Polyakov string regularized via random polygons of $3, 4, \ldots, pmax$ sides.

Universality arguments suggest that the inverse propagator $e^{-\Delta}$ in (7.172) can be replaced by $-\Delta + \text{const}$ without changing the long-distance properties of the theory. *Formally*, the resulting theory is then a *unitary* field theory. *In the planar limit*, the theory given by (7.172) is well defined for $d \le 6$ and *is*, in fact, *unitary*. (It is conceivable that it can be defined for larger values of d, and that it remains unitary.) The unitarity of the planar theory permits one to define a *unitary* free-string propagator.

For $d = 1$ the resulting matrix model was solved in [78]. In this case, x can be interpreted as a (Euclidean) time t, and the functions $\varphi(t)$ as trajectories in an $N \times N$-dimensional space. If we accept the approximation

$$e^{-\Delta} = e^{-d^2/dt^2} \approx 1 - \frac{d^2}{dt^2}, \tag{7.173}$$

then (7.172) corresponds to the action obtained for the *quantum mechanics* of N^2 coupled (anharmonic) oscillators, with Hamiltonian

$$\begin{aligned} &-\frac{1}{2}\sum_{i,j}\frac{\partial^2}{\partial\varphi_{ij}^2} + \frac{1}{2}\sum_{i,j}\varphi_{ij}\varphi_{ji} + V(\{\varphi_{ij}\}) \\ &= N\Big[-\frac{1}{2}\sum_{i,j}\frac{\partial^2}{\partial\psi_{ij}^2} + \frac{1}{2}\sum_{i,j}\psi_{ij}\psi_{ji} + V\big(\{\psi_{ij}/\sqrt{N}\}\big)\Big] \end{aligned} \tag{7.174}$$

where the second line is obtained through the change of variables $\psi_{ij} = N^{1/2}\varphi_{ij}$. The factor N in the last expression implies that for large N the logarithm of the partition function tends to the ground state energy of the Hamiltonian. Resorting to the variational principle, an integration over angular variables makes the problem equivalent to the determination of the ground-state energy of an ideal Fermi gas for N particles (corresponding to the eigenvalues of φ) with single-particle Hamiltonian

$$H(\lambda) = -\frac{1}{2}\frac{\partial^2}{\partial\lambda^2} + V(\lambda/\sqrt{N}). \tag{7.175}$$

For large N, the WKB approximation becomes valid and, in the planar limit [78, 338, 345], yields the expression

$$E_{H=0}(\mu') = \varepsilon - \frac{1}{3\pi}\int_{2\varepsilon-\lambda^2-2V(\lambda)\ge 0} P^3(\lambda)\, d\lambda, \tag{7.176}$$

where

$$P(\lambda) = \sqrt{2\varepsilon - \lambda^2 - 2V(\lambda)} \tag{7.177}$$

can be thought of as the momentum of a classical particle of energy ε moving in the potential $\lambda^2/2 + V(\lambda)$ [345]. The value ε is determined by the condition

$$\int_{2\varepsilon-\lambda^2-2V(\lambda)\ge 0} P(\lambda)\, d\lambda = \pi. \tag{7.178}$$

For the action (7.172), the study of the singularity of $E_{H=0}$ yields an exponent [338]

$$\gamma_{str} = 0. \tag{7.179}$$

7.6 The topological expansion: Non-perturbative results for the pure gravity case

7.6.1 Simple vs. double scaling limit

The approach to random-surface models reviewed in the previous section is, from the point of view of string theory, intrinsically perturbative. The results apply only to models of surfaces of *fixed genus*; they are obtained by letting the bare string coupling constant $1/N$ go to zero, and so correspond to a (naive) weak-coupling limit of the string theory. It has long been known, however, that perturbative string theory has many unphysical features (highly degenerate vacuum state, unbroken supersymmetry, etc.). It is expected that a non-perturbative theory will have radically different properties, and it might not exhibit these unwanted features. From this point of view, the divergence and non-Borel summability of the topological expansion can even be taken as encouraging signs.

Of course, the main question is how to construct such a non-perturbative theory — that is, how to obtain a mathematically meaningful expression collecting information from infinitely many genera. Moreover, the theory should be a generalization of the perturbative theory: starting from it, one should be able to reconstruct the topological expansion and the fixed-genus results. One must keep in mind that the goal is to obtain a theory *in the continuum limit*, and hence we have the freedom of choosing the particular way in which this continuum limit is achieved. For instance, in the previous section the continuum was approached in a two-step process: we *first* computed the planar limit, $(N,\hbar) \to (\infty,0)$ with $N\hbar = X$ fixed, and *then* we studied the regime $X \to X_c$ to extract the physically interesting information, for example the exponent γ_{str}. We shall call this process the *simple scaling limit.* The problem with this approach is that it erases all the information stored in higher genera; only surfaces with the topology of the sphere contribute in the end. (By taking derivatives with respect to $1/N$ prior to letting $N \to \infty$, one can extract also information for *fixed but finite* genus H. But this is still essentially perturbative.)

An alternative way to take the continuum limit, with all genera contributing, is based on the discovery [144, 79, 282] of a scaling relationship between the bare string-coupling and cosmological constants. This relationship is based on two facts:

First fact: The exponent γ_{str} is related to the entropy of random surfaces [525, 155, 346]. Indeed, the partition function for triangulations $\mathcal{T}$, or more generally polygonizations $\mathcal{P}$, with fixed area A and genus H is expected to behave as follows [525, 346]:

$$\sum_{\substack{\mathcal{P}\,:\,\text{Area of }\mathcal{P}=A\\ H(\mathcal{P})=H}} \frac{W(\mathcal{P})}{C(\mathcal{P})} \int e^{-\mathcal{A}(S(X;\mathcal{T}))} \prod_{j\in F(\mathcal{T})} d^d X_j$$

$$\sim \; C(H) e^{-(\mu-\mu_c)A} A^{\gamma_{str}(H)-3} \qquad (7.180)$$

with μ_c independent of H. [The area of a polygonization $\mathcal{P}$ is computed by assigning, for example, unit area to each triangle and area $p-2$ (= # triangles forming the polygon) to each p-gon.] Therefore,

$$\begin{aligned} E(\mu, N) &\sim \sum_{H=0}^{\infty} N^{-2H} C(H) \int dA\, e^{-(\mu-\mu_c)A} A^{\gamma_{str}(H)-3} \\ &\sim \sum_{H=0}^{\infty} \frac{C(H)}{N^{2H}(\mu-\mu_c)^{\gamma_{str}-2}} \,. \end{aligned} \tag{7.181}$$

Second fact: The exponent γ_{str} depends *linearly* on the genus H. It is expected that, at least for dimensions $d \leq 1$, there is a relationship of the form [cf. (7.116)]

$$\gamma_{str} - 2 \;=\; (1-H)(\gamma_0 - 2)\,, \tag{7.182}$$

where the critical index γ_0, the *string anomalous dimension*, is expected to be universal. On the basis of the planar ($H = 0$) results, we expect $\gamma_0 \leq 2$.

Combining (7.181) with (7.182) we obtain

$$E(\mu, N) \sim (\mu-\mu_c)^{2-\gamma_0} \sum_{H=0}^{\infty} C(H) \left[\frac{1}{N^2(\mu-\mu_c)^{2-\gamma_0}}\right]^H . \tag{7.183}$$

This shows that the genus-dependent part of the partition function is in fact a power series in the scaling variable

$$\frac{1}{N^2(\mu-\mu_c)^{2-\gamma_0}} \sim \frac{1}{N^2\left[\log(X_c/X)\right]^{2-\gamma_0}} \,. \tag{7.184}$$

On the right side of this expression we have made the transition to matrix-model notation, with $X = N\hbar$ and identifying $u_p X^{(p/2)-1} = -e^{-\mu}$. The contribution from higher genera is therefore preserved if the $N \to \infty$ limit is performed *keeping the scaling variable* (7.184) *fixed.* In particular, under the assumption $\gamma_0 \leq 2$ it follows that we must let $X \to X_c$ as $N \to \infty$, and in this limiting regime we can expand the log in (7.184) to obtain the condition that

$$\frac{\hbar^2}{(X_c - X)^{2-\gamma_0}} \tag{7.185}$$

must be kept (asymptotically) fixed. This variable is referred to as the *true* (or renormalized) *string coupling constant.* Equivalently, we must fix

$$T \;=\; \hbar^{-2/(2-\gamma_0)}(X_c - X)\,. \tag{7.186}$$

The limit $(N, \hbar) \to (\infty, 0)$, $X \to X_c$, with T fixed, is called the *double scaling limit*. We see that it corresponds to approaching the continuum limit *simultaneously for all genera*, as opposed to the simple scaling limit in which only the planar contribution survives the continuum limit.

This double scaling limit, suitably performed, has yielded non-perturbative theories of two-dimensional quantum gravity [144, 79, 282], which we shall briefly discuss in the rest of this chapter. For simplicity, we consider only the pure gravity case (1-matrix models). The proper definition of such limits is based on matrix integration techniques that we proceed to review in the next subsection.

7.6.2 The mathematical toolbox

We consider a 1-matrix model of pure two-dimensional gravity whose partition function is given by

$$Z = \int d\varphi \, \exp\left[-\frac{1}{\hbar}\mathrm{Tr}\, U(\varphi)\right] , \tag{7.187}$$

with

$$U(\varphi) = \tfrac{1}{2}\varphi^2 + \sum_{p=3}^{p\mathrm{max}} u_p \varphi^p . \tag{7.188}$$

We temporarily assume here that $p\mathrm{max}$ is even and $u_{p\mathrm{max}} > 0$. The method to compute the matrix integral defining Z is based on the fact that the measure and the integrand are invariant under unitary transformations $\varphi \to \varphi^{\mathcal{U}} \equiv \mathcal{U}\varphi\mathcal{U}^{-1}$. This allows us to reduce the N^2-dimensional integral to an N-dimensional integral over diagonal matrices.

Lemma 7.2 *Let $f(\varphi)$ be a real-valued function invariant under unitary transformations $\varphi \to \varphi^{\mathcal{U}}$. Then*

$$\int d\varphi \; f(\varphi) = \Omega_N \int \prod_{k=1}^{N} d\lambda_k \; f(\Lambda) \left(\det \|\lambda_i^{j-1}\|_{i,j=1}^{N}\right)^2 , \tag{7.189}$$

where $\Lambda = \mathrm{diag}(\lambda_1, \ldots, \lambda_N)$, $\|\lambda_i^{j-1}\|_{i,j=1}^{N}$ denotes the $N \times N$ matrix whose $(i,j)^{th}$ element is λ_i^{j-1}, and

$$\Omega_N = \frac{\pi^{N(N-1)/2}}{\prod_{k=1}^{N} k!} . \tag{7.190}$$

Proof. The crucial trick is to multiply and divide the integral by $1/\Delta(\varphi) \equiv \int d\mathcal{U} \prod_{1\le i<j\le N} \delta(\varphi_{ij}^{\mathcal{U}})$, where $d\mathcal{U}$ is the Haar measure of the unitary group. It is clear that $\Delta(\varphi)$ is invariant under unitary transformations. Thus

$$\begin{aligned}\int d\varphi \, f(\varphi) &= \int d\mathcal{U} \int d\varphi \; f(\varphi)\, \Delta(\varphi) \prod_{i<j} \delta(\varphi_{ij}^{\mathcal{U}}) \\ &= \Omega_N \int d\varphi \; f(\varphi)\, \Delta(\varphi) \prod_{i<j} \delta(\varphi_{ij}) \\ &= \Omega_N \int d\Lambda \; f(\Lambda)\, \Delta(\Lambda) ,\end{aligned} \tag{7.191}$$

where $\Omega_N = \int d\mathcal{U}$ is the volume of the unitary group, and in the second equality we have used the invariance of $f(\varphi)$, $d\varphi$ and $\Delta(\varphi)$ under the transformation $\varphi \to \varphi^{\mathcal{U}}$. To compute $\Delta(\Lambda)$ we use the parametrization $\mathcal{U} = e^A$ where A is an anti-Hermitian matrix. Since Λ is diagonal, the δ function is satisfied at $\mathcal{U} = I$ (and only there), and the value of $\Delta(\varphi)$ is determined by the behavior in an infinitesimal neighborhood of $\mathcal{U} = I$. Now, asymptotically for $\mathcal{U} \to I$, $d\mathcal{U}$ is given by $d\mathcal{U} = \prod_i d\mathcal{U}_{ii} \prod_{i<j} d(\mathrm{Re}A_{ij}) d(\mathrm{Im}A_{ij})$ where $d\mathcal{U}_{ii}$ is normalized Haar measure on the circle group $U(1)$. Hence, using the fact that $(\mathcal{U}\Lambda\mathcal{U}^{-1})_{ij} = \lambda_i\delta_{ij} + A_{ij}(\lambda_j - \lambda_i) + O(A^2)$, we obtain

$$\begin{aligned} &\Delta(\Lambda)^{-1} \\ &= \int \prod_{i<j} d(\mathrm{Re}A_{ij})\, d(\mathrm{Im}A_{ij})\, \delta\Big(\mathrm{Re}A_{ij}(\lambda_j - \lambda_i) + \ldots\Big)\delta\Big(\mathrm{Im}A_{ij}(\lambda_j - \lambda_i) + \ldots\Big) \\ &= \prod_{i<j} \left(\frac{1}{\lambda_j - \lambda_i}\right)^2 = \left(\det \|\lambda_i^{j-1}\|_{i,j=1}^N\right)^2 , \end{aligned} \tag{7.192}$$

where the last equality is the well-known Vandermonde determinant. This concludes the proof of the lemma, except for the constant Ω_N, which can be obtained by explicit calculation of the Gaussian matrix integral, i.e. by choosing $f(\varphi) = \exp(-\frac{1}{2}\mathrm{Tr}\,\varphi^2)$. ∎

The basic tool in the analysis of matrix integration is the use of orthogonal polynomials with respect to the measure

$$d\mu(\lambda) = \exp\left[-\tfrac{1}{\hbar}U(\lambda)\right] d\lambda . \tag{7.193}$$

We shall denote these polynomials by P_n; they are traditionally chosen with the normalization

$$P_n(\lambda) = \lambda^n + \text{terms of lower order in } \lambda , \tag{7.194}$$

which implies

$$\lambda^n = P_n(\lambda) + \text{terms linear in } P_i \text{ with } i < n . \tag{7.195}$$

If U is even, the polynomials P_n have a well-defined parity equal to $(-1)^n$. Let h_n be the square μ-norm of P_n. Then

$$\int P_n(\lambda) P_m(\lambda)\, d\mu(\lambda) = \delta_{nm} h_n . \tag{7.196}$$

We emphasize that h_n depends on U and $\hbar$, but this dependence will often be suppressed in order to lighten the notation.

We shall denote

$$\langle P_i | f | P_j \rangle \equiv \int d\mu(\lambda)\, P_i(\lambda) P_j(\lambda) f(\lambda) \tag{7.197}$$

$$\langle P_i | P_j \rangle \equiv \langle P_i | 1 | P_j \rangle \tag{7.198}$$

This notation is in fact more than a mere convenience: it embodies a straightforward Hilbert-space interpretation. In such a formalism, (7.197) is the bra-ket notation for the scalar product $(P_i, f[\hat{\lambda}]P_j)$, where $\hat{\lambda}$ is the (Hermitian) operator of "multiplication by λ". In this section, however, we shall take a rather pedestrian approach without making use of the full power of this formalism [282].

In the next lemma we summarize some basic properties of the orthogonal polynomials $\{P_n\}$.

Lemma 7.3

$$\lambda P_n(\lambda) = P_{n+1}(\lambda) + S_n P_n(\lambda) + R_n P_{n-1}(\lambda) , \tag{7.199}$$

where

$$R_n = \frac{h_n}{h_{n-1}} , \tag{7.200}$$

and, for U even,

$$S_n = 0 . \tag{7.201}$$

Proof. Clearly $\lambda P_n(\lambda)$ is a polynomial of degree $n+1$ with leading term λ^{n+1}. Now, by (7.195) and (7.196), we have

$$\int d\mu(\lambda)\, \lambda^i P_n(\lambda) = 0 \qquad \text{for all } i < n , \tag{7.202}$$

which implies that

$$\int d\mu(\lambda)\, [\lambda P_n(\lambda)]\, P_i(\lambda) = 0 \qquad \text{for all } i < n-1 . \tag{7.203}$$

This proves the recursion relation (7.199). The constant R_n is determined by the following calculation:

$$\begin{aligned} h_n &= \langle P_n|P_n\rangle = \langle P_n|\lambda P_{n-1}\rangle = \langle \lambda P_n|P_{n-1}\rangle & (7.204)\\ &= \langle P_{n+1} + S_n P_n + R_n P_{n-1}|P_{n-1}\rangle = R_n\langle P_{n-1}|P_{n-1}\rangle = R_n h_{n-1} . & (7.205) \end{aligned}$$

Finally, for even U the polynomials have a well-defined parity equal to $(-1)^n$, so $S_n = h_n^{-1}\langle \lambda P_n|P_n\rangle$ must vanish. ∎

Lemma 7.4 *For arbitrary $\lambda_1, \dots, \lambda_N$, we have*

$$\det \|\lambda_i^{j-1}\|_{i,j=1}^N = \det \|P_{j-1}(\lambda_i)\|_{i,j=1}^N . \tag{7.206}$$

Proof. This follows from (7.195) plus well-known properties of the determinant. ∎

It is a remarkable fact that the partition function Z can be expressed completely in terms of the recursion parameters R_n and the normalization h_0 of the measure $d\mu$.

Lemma 7.5

$$Z(U,\hbar) = \Omega_N N!\, h_0^N \prod_{i=1}^{N-1} R_i^{N-i}\,, \tag{7.207}$$

where Ω_N is defined in (7.190).

Proof. From (7.189) and (7.206) we have that

$$\begin{aligned} Z(U,\hbar) &= \Omega_N \int \prod_{k=1}^{N} d\lambda_k \left[\det(P_{j-1}(\lambda_i))\right]^2 \exp\left[-\frac{1}{\hbar}\sum_{k=1}^{N} U(\lambda_k)\right] \\ &= \Omega_N \int \prod_{k=1}^{N} d\mu(\lambda_k) \left[\det(P_{j-1}(\lambda_i))\right]^2 \\ &= \Omega_N \sum_{\sigma_1,\sigma_2\in\Pi(N)} \mathrm{sgn}(\sigma_1)\mathrm{sgn}(\sigma_2) \prod_{k=1}^{N} \int d\mu(\lambda_k)\, P_{\sigma_1(k)-1}(\lambda_k)\, P_{\sigma_2(k)-1}(\lambda_k) \\ &= \Omega_N \sum_{\sigma_1,\sigma_2\in\Pi(N)} \mathrm{sgn}(\sigma_1)\mathrm{sgn}(\sigma_2) \prod_{k=1}^{N} h_{\sigma_1(k)-1}\delta_{\sigma_1(k),\sigma_2(k)} \\ &= \Omega_N N! \prod_{k=0}^{N-1} h_k\,. \end{aligned} \tag{7.208}$$

Finally, resorting to (7.200) we obtain

$$\begin{aligned} Z(U,\hbar) &= \Omega_N N! \prod_{k=0}^{N-1}\left(\prod_{i=1}^{k} R_i\right) h_0 \\ &= \Omega_N N!\, h_0^N \prod_{i=1}^{N-1} R_i^{N-i}\,. \ \blacksquare \end{aligned}$$

Using this lemma, the free energy $\widetilde{E} = -\log Z(U,\hbar,N)$ is seen to be given by

$$\widetilde{E} = -\log\left(\Omega_N N!\right) - N\log h_0 - N\sum_{i=1}^{N-1}\left(1-\frac{i}{N}\right)\log R_i\,. \tag{7.209}$$

The first two summands on the r.h.s. constitute an uninteresting normalization; the singular behavior is contained in the term

$$\widetilde{E}_{sing} \equiv -\frac{1}{N}\sum_{i=1}^{N-1}\left(1-\frac{i}{N}\right)\log R_i. \tag{7.210}$$

The factor N^{-2} in the definition of $\widetilde{E}_{sing}$ is introduced due to the quadratic divergence of $\widetilde{E}$ as $N\to\infty$.

Our main goal is to compute this object $\widetilde{E}_{sing}$ which, as we see, is determined by the coefficients $R_1, \ldots, R_{N-1}$. These coefficients are computed using the following fundamental equations.

Lemma 7.6 *(pre-string equations)*

$$\langle P_n|U'|P_n\rangle = 0 \tag{7.211}$$
$$\langle P_n|U'|P_{n-1}\rangle = n\hbar\, h_{n-1} \tag{7.212}$$

Proof. The proof follows from (7.194) and (7.196) by applying integration by parts. As an example we prove (7.212):

$$\begin{aligned}
\langle P_n|U'|P_{n-1}\rangle &= \int d\lambda\, \exp\left[-\frac{1}{\hbar}U(\lambda)\right]\, U'(\lambda)P_n(\lambda)P_{n-1}(\lambda) \\
&= \hbar\int d\lambda\, \exp\left[-\frac{1}{\hbar}U(\lambda)\right]\, [P_n(\lambda)P'_{n-1}(\lambda) + P'_n(\lambda)P_{n-1}(\lambda)] \\
&= \hbar\int d\mu(\lambda)\, [n\lambda^{n-1} + \ldots]P_{n-1}(\lambda) \\
&= n\hbar h_{n-1}
\end{aligned} \tag{7.213}$$

A similar argument shows that $\langle P_n|U'|P_m\rangle = 0$ unless $m = n \pm 1$. ∎

Let us see now how these pre-string equations can be used to generate equations for the coefficients $R_1, \ldots, R_{N-1}$. The "matrix elements" $\langle P_n|U'|P_{n-1}\rangle$ appearing in (7.211)–(7.212) are linear combinations of elements of the form $\langle P_n|\lambda^l|P_{n-1}\rangle = \langle \lambda^l P_n|P_{n-1}\rangle$ which can be calculated using (7.199). This equation shows that, for instance, $\lambda^l P_n(\lambda)$ has three contributions of the form $\lambda^{l-1}P_j$, corresponding to $j = n+1$ (weight $= 1$), $j = n$ (weight $= S_n$) and $j = n-1$ (weight $= R_n$). The corresponding book-keeping can be done beautifully in a graphical way [63]: we draw a horizontal l-axis ("time", decreasing as one proceeds towards the right) and a vertical n-axis, so $\lambda^l P_n$ corresponds to the point $[l, n]$, and the three preceding contributions correspond respectively to upwards, horizontal and downwards [Figure 7.3(a)]. Repeating this argument starting from the each of the points $[l-1, j]$ ($j = n, n \pm 1$) and proceeding all the way down to $l = 0$, one obtains "paths" or "staircases" formed by sequences of these three types of steps. Each path starts at $[l, n]$, ends at some $[0, j]$ with $n - l \le j \le n + l$, and it has associated a weight equal to the product of the weights of each elementary step [see example in Figure 7.3(b)]. In detail,

$$\lambda^l P_n(\lambda) = \sum_{j=n-l}^{n+l}\Bigg[\sum_{\substack{\text{path } p\\ p(0)=n\\ p(l)=j}} W[p]\Bigg]P_j(\lambda)\,, \tag{7.214}$$

where a path is a function p: $\{0, 1, \ldots, l\} \to \mathbb{N}$ with $|p(j+1) - p(j)| \le 1$, the weight $W[p]$ is defined by $W[p] = \prod_{j=0}^{l-1} w[p(j), p(j+1) - p(j)]$ with the "one-step-weight"

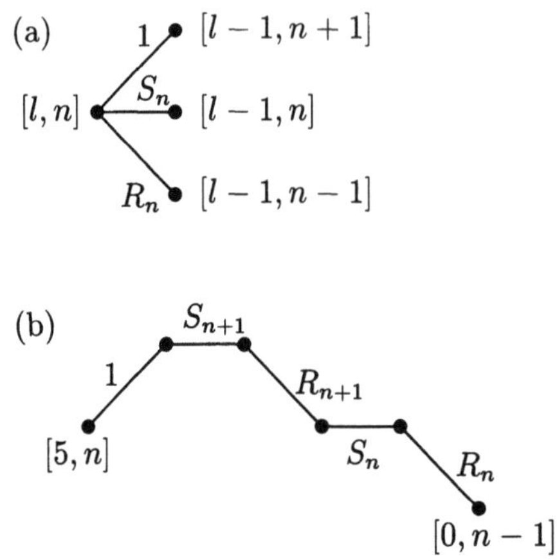

Fig. 7.3. (a) Graphical summary of the recursive relation (7.199). **(b)** Example of a graph contributing to $\langle P_n|\lambda^5|P_{n-1}\rangle$ having a weight of $S_{n+1}R_{n+1}S_nR_n$

$$w[p(j),k] = \begin{cases} 1 & \text{for } k=1 \\ S_{p(j)} & \text{for } k=0 \\ R_{p(j)} & \text{for } k=-1 \end{cases} \tag{7.215}$$

By orthogonality of the polynomials P_n, only some of the terms in (7.214) contribute to the pre-string equations. For instance, the only paths that contribute to (7.212) are staircases starting at $[l,n]$ and ending at $[0,n-1]$ (as in the example of Figure 7.3). As an illustration, let us consider the case of an *even* potential U. Then $S_n = 0$, so the staircases *do not have horizontal steps.* Therefore, the equation $\langle P_n|U'|P_n\rangle = 0$ is identically satisfied (parity!), and the elements $\langle P_n|\lambda^{2l-1}|P_{n-1}\rangle$ of the pre-string equation (7.212) require staircases with $l-1$ steps up and l steps down. Therefore, for

$$U(\lambda) = \sum_{l=0}^{(\deg U)/2} u_{2l}\lambda^{2l}\,, \tag{7.216}$$

the pre-string equations (7.211)–(7.212) reduce to

$$\sum_{l=0}^{(\deg U)/2} 2l\,u_{2l} \sum_{j=n-l-1}^{n+l-1} \sum_{\substack{\text{path } p \\ p(0)=n \\ p(2l-1)=n-1 \\ l-1 \text{ steps up} \\ l \text{ steps down}}} W[p] = n\,\hbar \tag{7.217}$$

(the factor h_{n-1} cancels out since $\langle P_{n-1}|P_{n-1}\rangle = h_{n-1}$). This is a system of algebraic equations for the coefficients $\{R_n\}$, which is complicated to solve. Fortunately, we are only interested in scaling limits which, in particular, require $N \to \infty$. For these limits, the discrete index l can be replaced, at least formally, by a *continuous* variable, so that equation (7.217) becomes either a simpler algebraic equation (simple scaling limit) or a differential equation (double scaling limit).

Note also that equations (7.217) for the coefficients R_n remain well-defined even if U is not a stable potential, e.g. for $U(\varphi) = u_2\varphi^4$ with u_2 *negative*. For such models we may therefore *define* the partition function Z as $e^{-\widetilde{E}_{sing}}$, with $\widetilde{E}_{sing}$ given by (7.210). However, the *meaning* of the equations (7.217) in this case is far from clear, since the connection with orthogonal polynomials is lost [every non-identically-zero polynomial has infinite norm with respect to the measure (7.193).

7.6.3 The simple scaling limit

In the simple scaling limit we let $(N,\hbar) \to (\infty, 0)$ holding the product $X = N\hbar$ fixed. Accordingly, we must also take in the above equations the limit $n \to \infty$ keeping $x \equiv n\hbar$ fixed. Proceeding formally, we *assume* the existence of functions

$$S(x) = \lim_{\substack{n\to\infty\\ \hbar\to 0\\ n\hbar = x}} S_n\,; \quad R(x) = \lim_{\substack{n\to\infty\\ \hbar\to 0\\ n\hbar = x}} R_n\,, \tag{7.218}$$

where S_n and R_n are defined by (7.199). In addition, we *assume* that the functions $S(x), R(x)$ are *smooth* on the interval of interest, $0 \le x \le X$, and that they satisfy the (formal) limit of the pre-string equations. To obtain these limit equations we formally expand

$$\begin{aligned} R_{n+j} &\sim R(x+j\hbar) \\ &\sim R(x) + j\hbar R'(x) + \frac{j^2\hbar^2}{2}R''(x) + \ldots && (7.219)\\ &\to R(x) && (7.220) \end{aligned}$$

and similarly for S. Therefore, in the simple scaling limit, we *drop all derivative terms* and the staircases acquire the weights

$$[R(x)]^{\#\ \text{downwards steps}}\,[S(x)]^{\#\ \text{horizontal steps}}\,. \tag{7.221}$$

For example, for U even, each path with $l-1$ steps up and l steps down has a weight $[R(x)]^l$, and there are

$$\binom{2l-1}{l} = \frac{1}{2lB(l,l+1)} \tag{7.222}$$

such paths (B is the beta function). Therefore, for even (polynomial) potentials the pre-string equation formally becomes

$$x = \sum_{l=0}^{(\deg U)/2} u_{2l}\frac{[R(x)]^l}{B(l,l+1)} \equiv W(R(x))\,. \tag{7.223}$$

More generally, for the weights (7.221), the paths contributing to $\langle P_n|\lambda^l|P_j\rangle$ have a weight given by the coefficient of the z^{j-n} term in a Laurent expansion of $(\frac{R}{z}+S+z)^l$. Therefore, for each finite l we have

$$\lim_{\substack{n\to\infty\\ \hbar\to 0\\ n\hbar=x}} \frac{1}{h_n}\langle P_n|\lambda^l|P_j\rangle \;=\; \oint \frac{dz}{2\pi i}\,\frac{1}{z^{j-n+1}}\left(z+S(x)+\frac{R(x)}{z}\right)^l . \tag{7.224}$$

As a consequence, the pre-string equations (7.211)–(7.212) take the form

$$\oint \frac{dz}{2\pi i}\frac{1}{z}U'(z+S+\frac{R}{z}\Big) \;=\; 0 \tag{7.225}$$

$$\oint \frac{dz}{2\pi i}U'\Big(z+S+\frac{R}{z}\Big) \;=\; x\,. \tag{7.226}$$

This form has the additional advantage that it is meaningful for general analytic potentials U.

Next, we consider the limit

$$\lim_{\substack{N\to\infty\\ \hbar\to 0\\ N\hbar=X}} \widetilde{E}_{N,sing} \;\equiv\; \widetilde{E}_{0,sing}(X)\,, \tag{7.227}$$

which, by (7.210), is also expressed as

$$\widetilde{E}_{0,sing}(X) \;=\; -\frac{1}{X^2}\int_0^X dx\,(X-x)\log R(x)\,, \tag{7.228}$$

Equations (7.225)–(7.226) and (7.228) could be taken as the starting definition of the theory. We remark that (7.228) really *is* the sum of the planar series, even for models in which U is not stable; see [78].

For the theory to be physically interesting, the function $\widetilde{E}_{0,sing}$ must have a singularity at some *real* X_c. The vicinity of this singularity constitutes the simple scaling limit. For simplicity, let us concentrate on *even* potentials, for which (7.225) is automatically satisfied. We need to solve (7.226) for $R(x)$, plug this solution into (7.228), and then study the resulting expression in the limit $X\to X_c$. We shall henceforth take $X_c=1$, which can always be achieved by rescaling the potential. Let us consider the form (7.223). Our goal is to find various potentials giving rise to different singular behavior. In this regard we notice that the nature of the singularity is more directly determined by the choice of the function $W(R)$, regardless of the potential U that produces it. Following Kazakov [336], we select a family of functions $W_k(R)$, $k\in\mathbb{N}$, for which (7.223) is solved easily, and each of which yields a different "k-critical behavior":

$$W_k(R) \;\equiv\; 1-(1-R)^k\,. \tag{7.229}$$

From (7.223) we see that this ansatz yields solutions $R(x)$ with branch-point singularities at $x=1$:

$$R(x) = 1 - (1-x)^{\frac{1}{k}}\,. \tag{7.230}$$

In particular, as $x \to 1$, we have $R \to R_c = 1$. Integrating (7.228), we obtain for $X \to 1$ $(= X_c)$

$$\begin{aligned}\widetilde{E}_{0,sing}(X) &= \int_X^1 dx\,(1-x)\log[1-(1-x)^{\frac{1}{k}}] + \ldots \\ &= -\frac{1}{(2+\frac{1}{k})}(1-X)^{2+\frac{1}{k}} + O((1-X)^{2+\frac{2}{k}}) + \ldots,\end{aligned} \tag{7.231}$$

where the dots stand for terms that are analytic in X. Therefore, the k-critical behavior corresponds to

$$\gamma_{str} = -\frac{1}{k}\,. \tag{7.232}$$

As these models correspond to the planar ($H = 0$) limit, we also have that $\gamma_0 = -1/k$ [see (7.182)].

These are precisely the models mentioned at the end of Section 7.5.2. From (7.223) we see that the functions W_k correspond to potentials

$$U_k(\lambda) = \sum_{l=1}^{k}(-1)^{l+1}\binom{k}{l}B(l,l+1)\lambda^{2l}\,. \tag{7.233}$$

We observe that the vertex weights $u_p X^{(p/2)-1}$ have alternating sign. This has two important consequences: On the one hand, it implies that the matrix integral diverges if k is even (although the perturbation expansion in *fixed* genus, e.g. $H = 0$, converges to a well-defined function for $|X| < X_c$). Therefore, all the above procedure becomes groundless — the polynomials P_n are not properly defined. The equations (7.225), (7.226) and (7.228) remain well-defined, but their ultimate meaning is problematic. On the other hand, the matrix integrals for k odd are absolutely convergent, but they are in correspondence with sums over random surfaces that inevitably contain contributions with *negative weights*.

We emphasize that adding potentials of higher criticality classes to a k-critical potential does not affect the behavior of $\widetilde{E}_{0,sing}$. This means that the potentials U can be subdivided into universality classes where U_k, defined by (7.233), is a representative of the kth universality class. Therefore, the potentials U_k constitute an exhaustive list of representatives of all the universality classes for the planar pure-gravity theory. The potentials U_k, however, do not seem to exhaust the possible universality classes for the nonperturbative theory, that is, in the double scaling limit.

7.6.4 The double scaling limit

As discussed in Section 7.6.1, the double scaling limit is constructed by letting $(N, \hbar) \to (\infty, 0)$ and, at the same time, $X \to 1 (= X_c)$, keeping a certain combination

$$T = \hbar^{-i}(1 - X) \tag{7.234}$$

fixed. The precise value of the exponent i should follow from the theory, and it determines the string anomalous dimension $\gamma_0 = 2(i-1)/i$ [cf. (7.186)]. This T will be the upper limit for a scaling variable

$$t = \hbar^{-i}(1 - x) \,. \tag{7.235}$$

Notice that for fixed $\hbar$ [corresponding to some fixed measure $\mu^\hbar$ in (7.193) and (7.197)], t takes values on a lattice of spacing $\hbar^{1-i}$, extending from $-\infty$ to $\hbar^{-i}$. In the limit $\hbar \to 0$, the t-lattice approaches the real axis, and sums become integrals according to the prescription

$$\hbar^{1-i} \sum_{n=0}^{N} \longrightarrow - \int_T^{\hbar^{-i} \approx +\infty} dt \tag{7.236}$$

Now, for the *planar* (= simple scaling limit) case discussed above, we found the relation

$$x = 1 - (1 - R)^k \,. \tag{7.237}$$

This suggests the existence of a second scaling variable

$$f^\hbar(t) = \hbar^{-j}(1 - R_{n(t)}) \,, \tag{7.238}$$

where $n(t) = \hbar^{-1}(1 - \hbar^{\frac{2k}{2k+1}} t)$ and j is a yet-to-be-determined exponent. [(7.235)/(7.237) suggest that $j = i/k$, and we shall confirm this shortly.] It is *assumed* that $f^\hbar(t)$ is a good scaling variable, in the sense that there exists a (smooth) function f with

$$\lim_{\hbar \to 0} f^\hbar(t) = f(t) \,. \tag{7.239}$$

To construct a meaningful limit, i and j must be chosen so as to satisfy three important constraints:

Constraint 1: The new scaling variables must have a meaningful interpretation. From the discussion of Section 7.6.1 we already know that large values of t correspond to a small renormalized string coupling constant; in particular, the limit $T \to \infty$ corresponds to a weak renormalized-coupling limit. Moreover, f is related to $R(x)$, the key function in the computation of the free energy: by taking the double scaling limit in (7.209), we obtain, using (7.236) and (7.238),

$$\widetilde{E} \simeq -\hbar^{-2+2i} \int_T^{\hbar^{-i}} dt\, (T - t) \log(1 - \hbar^j f) \tag{7.240}$$

$$\simeq c_1(\hbar) + c_2(\hbar) T + \hbar^{2i+j-2} \int_0^T dt\, (T - t) f(t) \,, \tag{7.241}$$

where we have expanded the logarithm using the fact that $j > 0$. [Note that we do *not* divide by N^2 as in passing to (7.210).] This formula shows that, in order for f to be a sensible object, i and j must satisfy the condition

$$2i + j = 2 . \tag{7.242}$$

Then, equation (7.241) shows that, in the limit $\hbar \to 0$, the scaling function f plays the physical role of the specific heat,

$$f(T) = \frac{d^2}{dT^2}\tilde{E}(T) . \tag{7.243}$$

Constraint 2: The theory should contain the planar series as a limiting case. For the planar series, (7.237) implies that

$$\hbar^i t = (\hbar^j f)^k , \tag{7.244}$$

for the k-critical model. In order to obtain a non-trivial equation between the new scaling variables, valid in particular in the planar limit, we must require that

$$i = kj . \tag{7.245}$$

We observe that these two constraints determine the exponents i and j. From (7.242) and (7.245) we obtain:

$$i = \frac{2k}{2k+1} , \qquad j = \frac{2}{2k+1} \tag{7.246}$$

for the k-critical model. We emphasize that this yields exactly the same scaling variables which we deduced earlier from purely physical considerations (7.186). Furthermore, we see that there is yet a further key constraint satisfied by i, j as given by (7.246):

Constraint 3: *In the double scaling limit, surfaces of higher genus should make a non-trivial contribution to* $\tilde{E}_{sing}$ *(i.e., in the double scaling limit,* $\tilde{E}_{sing}$ *should be* different *from the planar series).* The fact that this constraint is automatically satisfied for the above values of i and j shows that the scaling variables t and f are indeed chosen adequately, and it shows that the basic scaling relationship discussed in Section 7.6.1 is indeed correct.

Four different approaches have been used to determine f in such a way that it retains information from higher genera: Douglas and Shenker [144] proceed in a direct way, starting from the pre-string equations (7.211) and (7.212); Gross and Migdal [282] resort to the theory of resolvents of Schrödinger operators; Alvarez and Windey [19] transform the problem into a diffusion problem; and Douglas [142] uses a Heisenberg algebra, represented on the Hilbert space spanned by the polynomials $\{P_n\}$.

We review first the method of Douglas and Shenker [144], following the presentation in [300, 64]. For concreteness, let us consider the 2-critical potential $U_2(\lambda) = \lambda^2 - \frac{1}{12}\lambda^4$ [cf. (7.233)]. We get from (7.217)

$$n\hbar = R_n \left[2 - \tfrac{1}{3}(R_{n+1} + R_n + R_{n-1})\right] . \tag{7.247}$$

We Taylor expand, as in (7.219), $R_{n\pm 1} \approx R(x \pm \hbar) = R(x) \pm \hbar R'(x) + \hbar^2 R''(x)/2 + \ldots$ with $x = n\hbar$, but we do *not* throw away the derivative terms. From (7.235) it follows that $(d/dx) = -\hbar^{-i}(d/dt)$, and hence we obtain

$$\hbar^i t = \hbar^{2j} f^2 - \tfrac{1}{3}\hbar^{2-2i+j} f'' + O(\hbar^{j-4i+4} f'''') + \ldots \qquad (7.248)$$

where $'$ denotes d/dt. We have used the fact that we are working in the regime $x \to 1$, and hence $R(x) \to R_{crit} = 1$.

By comparing the exponents of $\hbar$ in the different terms of (7.248), we see that with the constraint $i = 2j$ discussed above, we re-obtain the equation $t = f^2$ when dropping all the derivatives (planar limit). Furthermore, constraint 3 is satisfied if at least the lowest-order term involving derivatives survives the limit $\hbar \to 0$; i.e. if $i = 2j = 2 - 2i + j$. This implies $i = 4/5$ and $j = 2/5$, which are exactly the values given in (7.246). All terms involving higher-order derivatives contain a higher power of $\hbar$, and thus disappear as $\hbar \to 0$. In this way one gets the *2-critical non-planar string equation*

$$t = f^2 - \tfrac{1}{3} f'' . \qquad (7.249)$$

This differential equation was studied a century ago by Painlevé, and is known as the Painlevé I-equation.

In general, an analogous argument shows that, for a k-critical potential U_k, only derivatives up to order $2k - 2$ contribute to the limit. In fact, by using a Hamiltonian approach and analytic continuation, Gross and Migdal [282] showed that, for the k-critical model (7.233), the double scaling limit yields the equation

$$t = c_k \mathcal{R}_k[f], \qquad (7.250)$$

where $\mathcal{R}_k[f]$ is the kth Korteweg-deVries (KdV) operator.[7] $\mathcal{R}_k[f]$ contains derivatives up to order $2k - 2$, and since $c_k = (-1)^k k! 2^{k+1}/(2k-1)!!$, (7.250) reads

$$t = f^k + \text{derivative terms}, \qquad (7.251)$$

as expected. Since the planar limit dominates for $t \to \infty$, the solution of (7.250) should behave like $f \sim t^{1/k}(1 + o(1))$, as $t \to \infty$, in order for the topological expansion to be asymptotic to the nonperturbative solution.

Let us now try to reconstruct the genus expansion starting from f. Formulas (7.183)–(7.186) and (7.243) suggest the ansatz

$$f = t^{-\gamma_0} \sum_{H=0}^{\infty} \alpha_H t^{-(2-\gamma_0)H} \qquad (7.252)$$

$$= t^{\frac{1}{k}} \sum_{H=0}^{\infty} \alpha_H t^{-(2+\frac{1}{k})H} \qquad (7.253)$$

We notice that as $t \sim \hbar^{-i} = \hbar^{-\frac{2k}{2k+1}}$, we have that the expansion (7.253) is in powers of $\hbar^{2H} \sim N^{-2H}$, as expected. Substituting ansatz (7.253) into (7.250),

[7] The $(k-1)^{st}$ KdV flow is given by

$$\frac{\partial}{\partial \tau} f(t, \tau) = \frac{\partial}{\partial t} \mathcal{R}_k[f],$$

where τ and t are interpreted as time and space variables, respectively.

we obtain recursion relations for the coefficients α_H. For instance, for the $k=2$ case, (7.249) yields:

$$\alpha_{H+1} = \frac{25H^2-1}{24}\alpha_H - \frac{1}{2}\sum_{j=1}^{H}\alpha_j\alpha_{H+1-j}\,, \tag{7.254}$$

from which it follows that $\alpha_H > 0$ for all H, and $\alpha_H \sim (2H)!$ as $H \to \infty$. This demonstrates the divergence of the topological expansion and the difficulties in applying standard summation methods.

Several delicate points must be emphasized. First of all, we observe that, even when detailed expressions have been obtained for the $k=2$ case, the underlying 2-critical theory is not well-defined, since the matrix integral diverges: U is unbounded from below. In our approach this divergent part has been subtracted away, since we only look at the singular part of (7.209), while the divergence of the matrix integral is hidden in the normalization constant h_0. This problem of normalizing the matrix integral occurs in all the even-k theories, whereas the odd-k theories are well-defined.

Second, to select the physically acceptable solution, equation (7.250) must be supplemented with appropriate boundary and regularity conditions. In this regard, the interpretation of f as "specific heat" is crucial. For the case $k=2$, it can be proven [73] that the only singularities of f in C are double poles with principal part $f(t) = 2(t-t_o)^{-2}$, besides a branch point singularity at ∞. The principal part shows that a pole of f corresponds to a (double) zero of the partition function. So we impose the condition that the physical solution have no poles on $\mathbb{R}^+$ (we have defined our approach such that $t>0$). Since f'' is positive on $\mathbb{R}^-$, one sees from the differential equation (7.249) that an infinite number of poles exist on the negative real axis; one can even prove that, as $t \to -\infty$, the distance between two poles goes to zero like $c|t|^{-1/4}$.

Moreover, if one linearizes (7.249) around the asymptotic solution (7.253) by writing $f = t^{\frac{1}{2}}\sum_{H=0}^{\infty}\alpha_H t^{-(2+\frac{1}{2})H} + \delta f$, one obtains the linear equation

$$\left(\frac{d^2}{dt^2} - 6t^{1/2}\right)\delta f(t) = 0\,, \tag{7.255}$$

which yields, to first order, the solutions $\delta f = \exp(\pm\frac{4}{5}\sqrt{6}t^{5/4})$. Clearly, the coefficient of $\exp(+\frac{4}{5}\sqrt{6}t^{5/4})$ has to be zero, while the coefficient of $\exp(-\frac{4}{5}\sqrt{6}t^{5/4})$ cannot be obtained by comparison with the topological expansion. Thus, the solution f of (7.249) depends on a nonperturbative parameter.

A different approach can be taken to choose "physical" boundary conditions for k odd. In these cases, the integral defining the partition function is well-defined — albeit some polygonizations have negative statistical weight [282] — and we can approach the singular point $X_{crit} = 1$ also from $X > 1$. This means that also negative values of t are meaningful, which suggests that the solution should be pole-free for all real t. Moreover, one can impose the behavior as $t \to -\infty$ as a further boundary condition: for k odd, Brézin, Marinari and Parisi [81] propose to choose boundary conditions $f \sim \pm|t|^{1/k}$ as $t \to \pm\infty$. They argue that, by linearizing around the asymptotic solution, they can fix $k-1$ parameters for each limit $t \to \pm\infty$, so that they can find "natural" boundary conditions for the differential equation (7.250) which is of order $2k-2$. Imposing these conditions, they are able to calculate numerically a pole-free solution for the 3-critical case. Moore [399] proved that such a solution exists for all odd values of k.

7.6.5 Other approaches to the double scaling limit

In this section, we sketch some of the main ideas underlying the other approaches, mentioned above, to the double scaling limit. Alvarez and Windey [19] take as their starting point a particular linear combination of the two pre-string equations (7.211) and (7.212), namely

$$\frac{1}{2h_n}\langle P_n|\lambda U'|P_n\rangle = (n+\tfrac{1}{2})\hbar\,. \tag{7.256}$$

Using scaling variables (7.235), the right-hand side of (7.256) can be written as $1-t\hbar^{\frac{2k}{2k+1}}+\frac{1}{2}\hbar$. In order to calculate the left-hand side, they consider a limit of the matrix elements $M_{nm}(i) \equiv (h_n h_m)^{-1/2}\langle P_n|\lambda^i|P_m\rangle$ where n, m and $i \to \infty$. Using (7.199) and (7.200) one gets a recurrence relation for the matrix elements,

$$M_{nm}(i+1) = \sqrt{R_{m+1}}M_{n,m+1}(i) + S_m M_{n,m}(i) + \sqrt{R_m}M_{n,m-1}(i). \tag{7.257}$$

This equation is a discretized diffusion equation which can be analyzed with the help of random walk methods. The main idea is that, for a large number of "steps" i, a typical walk from n to m looks like the trajectory of a particle following one-dimensional Brownian motian, so on physical grounds one expects to get some diffusion problem by letting $i \to \infty$. There are some subtleties in taking this limit which we have not succeeded in clarifying completely; but see [19].

Finally we want to sketch the approach proposed by Douglas [142] which has the great advantage that it can easily be generalized to describe p-matrix models. We consider the Hilbert space $l^2(\mathbb{Z}^+)$ and the orthonormal basis [cf. (7.197) ff.]

$$|n\rangle \equiv \frac{1}{\sqrt{h_n}}|P_n^\hbar\rangle\,, \quad n = 0, 1, 2, \ldots. \tag{7.258}$$

(We recall that, as functions of λ, the polynomials $P_n \equiv P_n^\hbar$ and their normalization constants $h_n \equiv h_n(\hbar)$ depend on $\hbar$, since the integration measure $d\mu(\lambda) = \exp[-\frac{1}{\hbar}U(\lambda)]\,d\lambda$ is $\hbar$-dependent.) The objects of interest are sequences of vectors $|\psi^\hbar\rangle$ with the following scaling behavior:

$$|\psi^\hbar\rangle = \sum_n \hbar^{\frac{1}{4k+2}}\psi_n^\hbar\,|n\rangle, \tag{7.259}$$

where $\psi_n^\hbar$ is given by

$$\psi_n^\hbar \equiv \psi(t(n)), \tag{7.260}$$

with

$$t = t(n) \equiv \hbar^{-\frac{2k}{2k+1}}(1 - n\hbar), \tag{7.261}$$

and where $\psi(t)$ is a smooth function of rapid decay on the real line. The $\hbar$-factor in (7.259) is chosen such that for sequences of the form (7.259)–(7.260)

$$\lim_{\hbar\to 0} \langle \psi_1^\hbar | \psi_2^\hbar \rangle = \int dt\, \overline{\psi_1(t)}\psi_2(t) \,. \tag{7.262}$$

It then follows that, for suitably chosen families of operators, $\widehat{T}^\hbar$, on $l^2(\mathbb{Z}^+)$ and families of vectors $|\psi^\hbar\rangle$, with $\psi_n^\hbar$ as in (7.260),

$$\lim_{\hbar\to 0} \widehat{T}^\hbar |\psi^\hbar\rangle = \widehat{T}\psi(t)\,, \tag{7.263}$$

where $\widehat{T}$ is an operator on $L^2(\mathbb{R})$. (We permit ourselves to be a little vague here, since precise analytical details will not be supplied in the following.)

In particular, let $\widehat{\lambda}$ be the operator of multiplication by λ. (We recall that the pre-string equations (7.211) and (7.212) involve matrix elements of powers of $\widehat{\lambda}$; so this is an important operator.) A central element in the approaches [282] and [142] is the fact that, in the double scaling limit, the action of $\widehat{\lambda}$ on sequences of vectors of the form (7.259)-(7.260) can be approximated by

$$(\widehat{\lambda}\psi^\hbar)_n = [2 - \hbar^{\frac{2}{2k+1}}\widehat{H} + O(\hbar^{\frac{3}{2k+1}})]\, \psi(t)\,, \tag{7.264}$$

if t is given by (7.261) and where

$$\widehat{H} \equiv \frac{d^2}{dt^2} - f\,, \tag{7.265}$$

where f is defined in (7.239). This is seen by considering the recursion relation (7.199),

$$\widehat{\lambda}|n\rangle = \sqrt{R_{n+1}}|n+1\rangle + \sqrt{R_n}|n-1\rangle. \tag{7.266}$$

Then

$$\begin{aligned}\widehat{\lambda}|\psi^\hbar\rangle &= \sum_n \hbar^{\frac{1}{4k+2}}(\sqrt{R_{n+1}}|n+1\rangle + \sqrt{R_n}|n-1\rangle)\psi_n^\hbar \\ &= \sum_n \hbar^{\frac{1}{4k+2}}(\sqrt{R_n}\psi_{n-1}^\hbar + \sqrt{R_{n+1}}\psi_{n+1}^\hbar)|n\rangle.\end{aligned}$$

Therefore, for $n = \hbar^{-1}(1 - \hbar^{\frac{2k}{2k+1}}t)$, see (7.261), it follows from (7.260) and (7.266) that

$$\begin{aligned}(\widehat{\lambda}\psi^\hbar)_n &\simeq \sqrt{R_n}\psi(t+\hbar^{\frac{1}{2k+1}}) + \sqrt{R_{n+1}}\psi(t-\hbar^{\frac{1}{2k+1}}) &(7.267)\\ &= [1 - \frac{1}{2}\hbar^{\frac{2}{2k+1}}f(t)][2\psi(t) + \psi''(t)\hbar^{\frac{2}{2k+1}}] + O(\hbar^{\frac{3}{2k+1}}) \\ &= [2 - \hbar^{\frac{2}{2k+1}}\widehat{H} + O(\hbar^{\frac{3}{2k+1}})]\psi(t)\,. &(7.268)\end{aligned}$$

This completes the proof of (7.264).

An analogous process can be applied to the operator the operator

$$\hat{\hat{P}} = -\frac{1}{2\hbar}U'(\hat{\lambda}) + d/d\lambda \tag{7.269}$$

which, in the double scaling limit, becomes a differential operator with respect to t. Notice that

$$[\hat{\hat{P}}, \hat{\lambda}] = 1 \,. \tag{7.270}$$

To simplify our notation, let us write

$$\widehat{D} = d/dt. \tag{7.271}$$

From the definition of t we see that, as in the Douglas-Shenker approach, derivatives with respect to t are suppressed by a factor $\hbar^{\frac{1}{2k+1}}$. One *defines* a k-critical potential, $U = U_k$, by the requirement that, after approximating the operator $\hat{\hat{P}}$ by a differential operator in the variable t, via steps similar to (7.267)-(7.268), all terms of order $\hbar^{-i}$ with $i > \frac{2}{2k+1}$ must cancel. This condition implies equation (7.233) for U_k, as shown in [82]. So the dominant term in $\hat{\hat{P}}$ is of order $\hbar^{-\frac{2}{2k+1}}$ and is a differential operator in the variable t of order $2k-1$, i.e.

$$\hat{\hat{P}} = c_k \hbar^{-\frac{2}{2k+1}} \hat{P} + O(\hbar^{-\frac{1}{2k+1}}) \,, \tag{7.272}$$

where the constant c_k is given by $c_k = (-1)^{k+1} 2^{k-1} k!/(2k-1)!!$ and is chosen so as to obtain $\hat{P} = \widehat{D}^{2k-1} + \ldots$, as discussed in [82]. From (7.270) and eq. (7.264) we see that

$$[\widehat{H}, \hat{P}] = \frac{1}{c_k}, \tag{7.273}$$

with c_k as defined above. The trick is now to observe that there exists a *unique* anti-selfadjoint differential operator of order $2k-1$, $\hat{P} = \widehat{D}^{2k-1} + \ldots$, whose commutator with $\widehat{H}$ is a function (*not* a differential operator). Then (7.273) determines the string equation, as will be shown explicitly for the 2-critical model. Starting with the ansatz

$$\hat{P} = \widehat{D}^3 + g(t)\widehat{D}^2 + h(t)\widehat{D} + k(t) \,, \tag{7.274}$$

we obtain the equations

$$k(t) = \frac{1}{2}h(t)' \tag{7.275}$$

and

$$g(t) = 0 \tag{7.276}$$

from the anti-selfadjointness condition. Imposing the condition that the commutator $[\widehat{H}, \hat{P}]$ has to be a function, one derives that

$$h' = -\frac{3}{2}f', \tag{7.277}$$

which implies that $h = -\frac{3}{2}f$, (up to an integration constant which can be set to 0). After multiplying (7.273) with $c_2 = \frac{4}{3}$, we then get

$$2f\,f' - \frac{1}{3}f''' = 1\,, \tag{7.278}$$

which is just the derivative of the Painlevé I-equation (7.249). The nice feature of this approach is that by introducing a formal operator $\widehat{D}^{-1}$ such that

$$\widehat{D}^{-1}\widehat{D} = \widehat{D}\widehat{D}^{-1} = \mathrm{I}, \tag{7.279}$$

$\widehat{P}$ can be obtained explicitly:

We use the fact that

$$\widehat{D}^{-1}g = \sum_{j=0}^{\infty}(-1)^j g^{(j)}\widehat{D}^{-1-j}, \tag{7.280}$$

which follows from (7.279), to obtain an expression for $\widehat{H}^{\frac{1}{2}}$ (chosen to be formally anti-selfadjoint),

$$\widehat{H}^{\frac{1}{2}} = \widehat{D} + \sum_{j=1}^{\infty} h_j(t)\widehat{D}^{-j}, \tag{7.281}$$

where the functions $h_j(t)$ can be obtained recursively from the relation $(\widehat{H}^{\frac{1}{2}})^2 = \widehat{H}$. One gets

$$h_1 = -\frac{1}{2}f, h_2 = \frac{1}{4}f', h_3 = -\frac{1}{8}(f^2 - f''), \ldots. \tag{7.282}$$

Now we split $(\widehat{H}^{\frac{1}{2}})^{2k-1} \equiv \widehat{H}^{\frac{2k-1}{2}}$ into two parts, $\widehat{H}^{\frac{2k-1}{2}} = \widehat{H}_+^{\frac{2k-1}{2}} + \widehat{H}_-^{\frac{2k-1}{2}}$, where

$$\widehat{H}_+^{\frac{2k-1}{2}} = \sum_{j=0}^{2k-1} h_{j,k}(t)\widehat{D}^j \tag{7.283}$$

and

$$\widehat{H}_-^{\frac{2k-1}{2}} = \sum_{j=1}^{\infty} h_{-j,k}(t)\widehat{D}^{-j}. \tag{7.284}$$

Since the left hand side of

$$[\widehat{H}_+^{\frac{2k-1}{2}}, \widehat{H}] = -[\widehat{H}_-^{\frac{2k-1}{2}}, \widehat{H}] \tag{7.285}$$

only contains positive powers of $\widehat{D}$, whereas the right hand side of (7.285) only displays negative powers of $\widehat{D}$, the commutator $[\widehat{H}, \widehat{H}_+^{\frac{2k-1}{2}}]$ has to be a function of t. As shown in [142], $\widehat{H}_+^{\frac{2k-1}{2}}$ is anti-selfadjoint, which implies that $\widehat{P} = \widehat{H}_+^{\frac{2k-1}{2}}$. Using the Gel'fand-Dikii result [251]

$$[\widehat{H}, \widehat{H}_+^{\frac{2k-1}{2}}] = \frac{1}{c_k}\widehat{D}\widetilde{\mathcal{R}}_k[f], \tag{7.286}$$

where the constant c_k is the same as in (7.273) and $\widetilde{\mathcal{R}}_k[f] = f^k + \ldots$ is proportional to the kth KdV polynomial, one obtains

$$1 = \widetilde{\mathcal{R}}'_k[f]. \tag{7.287}$$

Equation (7.287) is equivalent to (7.250) since the integration constant can be fixed to zero by comparison with the planar limit $t = f^k$.

The advantage of this method lies in the fact that it can be generalized to p-matrix models. There $\widehat{H}$ is a differential operator of order $p+1$ and one considers $\widehat{P} = \widehat{H}_+^{\frac{l}{p+1}}$, where $l \in \mathbb{N}$. As shown in [142], these models correspond to $(l, p+1)$ minimal conformal models coupled to two-dimensional gravity. This method can in principle be used to calculate some aspects of 1-dimensional matrix models, since they can be obtained (formally) by considering p-matrix models with appropriate couplings (i.e., a linear chain) and letting p (the number of matrices) tend to infinity. (For $N < \infty$, the limit $p \to \infty$ is well known to exist and to reproduce the one-dimensional theory).

7.6.6 Perturbation of the string equation and KdV flow

In this section, we consider perturbations of the string equations (7.223) and (7.250). As can be seen in (7.223), it is useful to consider operators that have a well-defined scaling behavior. This means that we should consider operators which correspond to a change

$$\delta W(R) = \hbar^{\frac{2(k-l)}{2k+1}}(1-R)^l \tag{7.288}$$

in the function $W(R)$ which determines the simple scaling limit [cf. (7.223)]. The operator corresponding to (7.288) will be denoted by Ω_l. As shown in Gross and Migdal [282], these operators are given by

$$\Omega_l = 2\hbar^{-1}\hbar^{\frac{2(k-l)}{2k+1}} B(\frac{1}{2}, -\frac{1}{2} - l)\widehat{H}^{l+\frac{1}{2}}. \tag{7.289}$$

The factors $\hbar^{\frac{2(k-l)}{2k+1}}$ are introduced to obtain the same scaling for all contributions $\mu_l \Omega_l$. This scaling behavior is seen in a very transparent way by studying the string equation corresponding to

$$U = U_k + \sum_{l=0}^{\infty} \mu_l \Omega_l, \tag{7.290}$$

which is given by

$$t = \widetilde{\mathcal{R}}_k[f] - \sum_{l=0}^{\infty} \mu_l \widetilde{\mathcal{R}}_l[f], \tag{7.291}$$

with $\widetilde{\mathcal{R}}_l[f]$ as introduced in (7.286). To simplify the notation, we write $\underline{\mu}$ for $(\mu_0, \mu_1, \ldots)$. We emphasize the fact that U_l is just a superposition of Ω_0 and Ω_l. Using the Hamiltonian approach (7.289), Gross and Migdal [282] first calculate the expectations

$$\langle \Omega_l \rangle = \frac{\partial}{\partial \mu_l} \widetilde{E}_{sing}|_{\underline{\mu}=\underline{0}} \tag{7.292}$$

first for negative l, and then analytically continue the result to positive integer values of l. In this way, they show that in the double scaling limit

$$\langle \Omega_l \rangle = \frac{1}{l+1} \int^T dt \ \widetilde{\mathcal{R}}_{l+1}[f(t)], \tag{7.293}$$

where f solves the string equation (7.250). Due to (7.250) and the recurrence relations for the $\widetilde{\mathcal{R}}_l[f]$, the action of Ω_0 and Ω_k can be obtained as follows: On the one hand, using (7.243) and $\widetilde{\mathcal{R}}_1[f] = f$, one sees that $\langle \Omega_0 \rangle = \frac{d}{dT}\widetilde{E}$, showing that Ω_0 plays the role of a puncture operator. On the other hand, using (7.243) and the recurrence relation [251]

$$\frac{k+1}{2k+1} \ (-\frac{1}{2}\widehat{D}^2 + f + \widehat{D}^{-1} f \widehat{D}) \ \widetilde{\mathcal{R}}_l[f]) = \widetilde{\mathcal{R}}_{l+1}[f], \tag{7.294}$$

one obtains that $\langle \Omega_k \rangle = \frac{1}{2l+1} T \frac{d}{dT} \widetilde{E}$, which shows that the action of Ω_k is to rescale T, as can also be seen directly from (7.291).

The quantities computed up to now are just first order perturbations in μ_l at $\underline{\mu} = \underline{0}$. But eq. (7.291) suggests a natural generalization of the string equation, namely

$$t = \sum_{l=0}^{\infty} (l + \frac{1}{2}) T_l \mathcal{R}_l[f], \tag{7.295}$$

which allows us to consider "flows in theory space". The quantities T_l introduced above are proportional to μ_l, $T_l = (-1)^l \frac{2^{l+2} l!}{(2l+1)!!} \mu_l$. Expectation values $\langle \Omega_l \rangle_{\underline{T}}$ are obtained in the same way as above, namely

$$\langle \Omega_l \rangle_{\underline{T}} = \frac{(-1)^{l+1} 2^{l+2} l!}{(2l+1)!!} \int^T dt \ \mathcal{R}_{l+1}[f(t)], \tag{7.296}$$

where f is a solution of equation (7.295). Every such solution satisfies the partial differential equation

$$\frac{\partial}{\partial T_l} f = \frac{\partial}{\partial t} \mathcal{R}_{l+1}[f], \tag{7.297}$$

which is just the lth KdV-equation. Equation (7.297) follows from (7.296) by differentiating (7.296) twice with respect to T and setting $T \equiv t$. So we see that *the flow of f under a perturbation corresponding to the operator Ω_l is given by the lth KdV flow.*

This should not mislead us to think that it is possible to start from an odd-critical theory and flow to an even-critical theory, because the boundary conditions at infinity chosen for an odd-critical theory are incompatible with those chosen for an even-critical theory; see [399, 143, 81] for details.

Finally we wish to remark that the calculus of τ-functions can be used to construct solutions to the string equation (7.295): The connection between τ-functions and such solutions is given by

$$\tau^2 = e^{-\widetilde{E}}, \tag{7.298}$$

see [458, 129, 399] for detailed explanations. It is worthwhile to note that the τ-functions obey the equations $L_n \ \tau = 0$, for $n \geq -1$, where the operators L_n are generators of a Virasoro algebra as shown in [129].

7.6.7 Epilogue

Let us conclude with some general comments on open problems and perspectives concerning the theories outlined in this chapter.

(1) As the reader has no doubt realized, we have largely limited our attention to discussing the case of pure gravity; we have made no attempt at a comprehensive review of matrix models of two-dimensional gravity coupled to matter. In order to describe models involving matter, we would have to consider models of more than one random matrix. For example, the q-state Potts model coupled to two-dimensional quantum gravity corresponds to a q-matrix model. A string propagating in a one-dimensional space-time corresponds to a one-dimensional matrix field theory. Such multi-matrix models and one-dimensional field theories have been studied in the literature [142, 80], and the formalism of Douglas outlined at the end of Section 7.6.4 is particularly suitable for this purpose [142, 64]. Unfortunately, it turns out that it is virtually impossible to solve models of matrix field theories in more than one dimension. This is presumably related to the appearance of tachyons in theories of bosonic strings propagating in space-times of dimension larger than one. This barrier at $d = 1$ is also manifest in the KPZ formula quoted in (7.109). Recently, however, a model of two-dimensional quantum gravity in the continuum limit has been proposed, which has features suggesting that it avoids the $d = 1$ barrier [107]. But the tachyon appears to remain among the excitations of that model in more than one dimension. We feel that it is difficult to find a matrix-model formulation of bosonic string theory in dimension $d > 1$.

(2) As we have repeatedly pointed out, there is a basic difference between the k-critical models with k even and those with k odd. While the matrix integrals defining the odd-k models at finite values of N are absolutely convergent, the matrix integrals formally corresponding to the even-k models are divergent, because the potentials U of such models are unstable (the coefficient of the term of highest degree, $\mathrm{Tr}\,(\phi^{2k})$, is negative). In the simple scaling limit, the instability of the even-k models is not felt, since only planar diagrams contribute to this limit. If the coefficients R_n are *defined* as solutions of the algebraic equations (7.217), they provide the correct solution of the k-critical models in the simple scaling limit, *irrespective* of whether k is odd or even. The situation is more problematic in the double scaling limit, to which surfaces of arbitrary genus contribute. There appear to exist several inequivalent definitions of k-critical models in the double scaling limit when k is even:

(a) The first possibility is to regularize a k-critical model with k even by adding a term $u_{2k+2}\phi^{2k+2}$ to the potential of the k-critical model, where $u_{2k+2} > 0$. One might then attempt to define the k-critical model by taking the limit $u_{2k+2} \downarrow 0$. This procedure leads to difficulties in imposing the desired boundary conditions at infinity on the solutions of the string equation [399].

(b) A more pragmatic point of view, adopted by most workers in the field, is to *define* the k-critical theory as a solution of the string equation (7.250) with appropriate boundary and regularity conditions, chosen so as to obtain "consistency" between this string-equation formulation and the random-matrix approach. As for regularity, the solution $f(t)$ is required not to exhibit poles on the real axis; such poles would correspond to (unphysical) zeroes of the partition function of the random-matrix model. The boundary conditions are imposed at infinity. At the very least, $f(t)$ must reproduce the planar theory when t tends to infinity. However, these requirements do not yet completely determine the solution. For instance, for the 2-critical model — equation (7.249) — one still has infinitely many solutions differing by exponentially small terms [see the discussion after formula (7.255)]. Different strategies have been proposed to single out a unique "consistent" or "physical" solution: For k odd, a numerical approach has been formulated [81] that seems to produce at most a discrete number of pole-free solutions. It is based on the integration of (7.250) on a finite region $|t| < L$, subject to the boundary condition $f(t) = t^{1/k}$ for $|t| > L$, and letting $L \to \infty$. Alternatively, a solution can be found by minimizing the functional $\int (t - c_k \mathcal{R}_k[f])^2 dt$.

For the more complicated k-even models, an appealing proposal [117] is to select the solutions $f(t)$ satisfying Schwinger-Dyson (SD) equations derived for the (loop) correlations of the matrix models. However, it is proven in [117] that, at least for the $k = 2$ model, *no real solution of the string equation satisfies the SD equations of the matrix model.* This feature appears to be a purely non-perturbative phenomenon, since the SD equations appear to be satisfied to all orders in the perturbation expansion. Moreover, by going to complex potentials (complex coefficients u_p), a complex solution to the string equation can be uniquely determined [118] by requiring that the region (in the complex t plane) where $f(t)$ has well-defined large-t asymptotic behavior — that is, at most finitely many poles — coincide with the region (in the complex $u_{p\max}$ plane) where the planar limit of the random-matrix model exists. Such solutions $f(t)$ have always a nonvanishing, but exponentially small, (physically undesirable) imaginary part. A natural question is whether both criteria — SD equations and simultaneous existence of asymptotic behavior — lead to the same solution. See also [399] for a possibly related formulation.

(c) A third strategy towards defining the k-critical models for k even is to use stochastic quantization to regularize the divergent matrix integrals [267]. Stochastic quantization of a one-matrix model leads to a one-dimensional, stable supersymmetric matrix field theory which can be "solved" with the help of WKB methods [78, 385, 384, 328].

There are strong indications that strategies (a), (b) and (c) lead to inequivalent definitions of the k-critical models with k even. Moreover, it is far from clear that *any* of these three strategies leads to a theory with reasonable physical properties. (Part of the problem is that *we do not know* what are the relevant physical properties.) Finally, on a purely technical level, if one insists that a reasonable definition should have the property that the planar limit is recovered in the limit where t tends to infinity, one meets the problem that these models cannot be reached from the stable k-critical models (i.e. those with k odd) along some renormalization-group trajectory. Thus, in a sense the "universe" of k-critical models with k even is inaccessible from the "universe" of k-critical models with k odd [322, 118, 399].

(3) From a *mathematical* point of view, one would attempt to define matrix models in such a way that they provide solutions to interesting mathematical problems, in particular to those concerning moduli spaces. It is likely that a definition of the 2-critical model of two-dimensional quantum gravity based on strategy (b) in remark (2) above would allow the computation of quantities related to topological invariants of the moduli spaces of Riemann surfaces; see [130, 521, 429, 430, 311]. So perhaps mathematical principles permit one to select the most natural definitions of matrix models when they are not given by absolutely convergent matrix integrals.

(4) We regard the difficulties in finding an unambiguous definition of a model of two-dimensional pure quantum gravity (or, more generally, of k-critical matrix models with k even) in the double scaling limit, and of crossing the $d = 1$ barrier in the construction of matrix field theories of non-critical strings, as two further manifestations of the intrinsic diseases plaguing bosonic string theory. In our opinion it will be fruitful to study the matrix models and matrix field theories corresponding to *supersymmetric* string theories and to theories of strings propagating in *complex* manifolds. Hopefully, the situation in such theories will look more promising.

(5) On a more fundamental level, we feel that the technical difficulties encountered in string theory and two-dimensional quantum gravity may be an indication that we have not yet found the right general *conceptual framework* for expressing a "theory of random surfaces". In such a situation, it is quite reasonable to study simplified toy models of random surfaces, in order to gain some experience and test new ideas. The matrix models of Sections 7.5 and 7.6 are such toy models. Unfortunately, in this process of simplification and idealization, the *physics* often gets lost; and technical successes can easily blind us to fundamental unresolved *conceptual* issues. In particular, since the matrix models do not describe (as far as we know) any system actually occurring in nature, the physical principles that would tell us which solution (if any) is the most natural one are not available.

(6) In spite of the somewhat pessimistic remarks made above, the reader should realize that there are many problems in statistical physics which are naturally formulated in terms of random surfaces and which can be studied with the help of the techniques reviewed in Sections 7.2 to 7.5 (see also the references quoted in Section 7.3). Indeed, it is our experience that random-surface theory has been most successful where it has been applied to a well-posed, concrete physical problem, such as the roughening and wetting transitions, the critical properties of membranes, etc. In areas that are physically more speculative as well as mathematically more sophisticated — such as string theory and two-dimensional quantum gravity — we suspect that the theory has not yet found its final formulation.

Part II

Random-walk models and random-walk representations of classical lattice spin systems

© Quino (Joaquín Lavado). Reprinted with permission.

8. Introduction

Random-walk expansions have played a key role in recent advances in our understanding of critical phenomena in statistical mechanics and of the continuum limit in quantum field theory. These advances include the proof of triviality of the continuum limit and mean-field critical behavior for φ^4 and Ising models in dimensions $d > 4$ [5, 15, 8, 12, 213, 90, 28, 292, 223], and an extremely simple construction of continuum φ^4 quantum field theories in dimensions $d < 4$ [97, 96, 74, 75, 292]. Our goal in Parts II and III is to present several random-walk expansions from a unified point of view, and to explain the physical results which can (and cannot) be derived from them.

Three random-walk expansions are of particular interest:

- The Brydges-Fröhlich-Spencer (BFS) representation for classical lattice spin systems [92, 213, 90, 28, 292, 97, 74, 292, 223] (which is a variant on Symanzik's [493, 494] representation of continuum field theory);
- The Aizenman random-walk (ARW) representation for Ising spin systems [5, 15, 8, 12]; and
- The random-walk models employed in polymer physics, such as the self-avoiding walk (SAW) and its generalizations.

While the resemblance between these expansions has often been noted, the exact relationship between them has remained unclear. In this monograph we propose a unified framework within which to study these (and other) random-walk expansions. We show that the usefulness of all these expansions derives from a few important properties which we systematize here. To facilitate the comparison among the different expansions, we adopt a unifying notation which may in some cases disagree with (and be more cumbersome than) the notations in use in the respective literatures; we trust that this will not cause undue confusion.

All random-walk models have essentially the same *combinatoric* structure. This structure, which is familiar to anyone who has ever computed the correlation functions of the Gaussian model (free field), is the subject of Chapters 9 and 10. What distinguishes one random-walk model from another are the *weights* attached to particular families of walks. For general random-walk models little of interest can be said (except at "high temperature"), but if the weights satisfy suitable inequalities then many powerful results can be derived. In Chapter 11 we isolate these inequalities, which express two key physical properties:

- repulsiveness (or repulsiveness "on the average") between walks; and
- attractiveness (or non-interaction) between non-overlapping (or "compatible") walks.

These two properties give rise to inequalities going in opposite directions, and it is their combination that is responsible for the powerful results of Aizenman [5, 15, 8, 12], Fröhlich [213, 28] and others.

Not everything can be done with random walks — some of the profound results of Aizenman and collaborators [5, 15, 9, 6, 12, 10, 13, 194] make essential use of the random-current representation for the Ising model, and apparently cannot be derived by the simpler random-walk methods. The random-walk methods appear to have two main limitations:

- The random-walk formalism seems to work best for models that are in some sense "not too far from Gaussian". For example, random-walk methods can be used to construct weakly-coupled superrenormalizable φ^4 quantum field theories [97, 96, 74] and Edwards models [75] in dimensions $d < 4$, and to prove triviality (Gaussianness) of the continuum limit in dimension $d > 4$ [5, 15, 8, 213, 90, 28, 292, 223]. On the other hand, accurate bounds for strongly non-Gaussian models, such as the Ising model in dimension $d < 4$, seem to be obtainable only through the random-current formalism.
- All of the results derivable by random-walk methods seem to involve truncated weights of order at most 2; spin-model results involving truncated weights of order ≥ 3 [12, 460] seem to require random currents. (However, some of the results derived for Ising models using the random-current formalism can also be derived for self-avoiding walks by direct combinatorial arguments — see Sections 12.2.4, 12.6 and 12.7.)

Our goal in this work is thus not only to explain in detail the results which *can* be obtained by random-walk methods, but also to give a brief explanation of those results which *cannot* be obtained by random-walk methods and to indicate briefly why.

The basic idea underlying many of the inequalities in this work (notably the Aizenman-Fröhlich inequalities) is the *inclusion-exclusion principle*: roughly speaking, it bounds the *probability* that two random walks intersect by the *mean number of times* that they intersect. In Sections 12.2 and 12.3 we show how this principle is used, and try to make clear what its limitations are, i.e. in what situations it overcounts. We believe that significant improvements on the Aizenman-Fröhlich inequalities — sufficient, for example, to prove the triviality of φ_4^4 — will have to take account of what the inclusion-exclusion principle throws away. We discuss here two tricks for going beyond naive inclusion-exclusion:

- the "passing to the left" trick (Sections 12.2.3 and 12.3.1), and
- the "dilution" trick (Sections 12.3.1–12.3.4).

These tricks yield inequalities which sum "infinitely many orders" in the expansion parameter; they improve the first-order and second-order inclusion-exclusion inequalities, respectively. Unfortunately, these two tricks are not universally applicable. But we think they will have applications in the future, and are thus worth emphasizing.

In all of the work we discuss, the application of the random-walk (or random-current) formalism is a two-stage process:

1. The random-walk or random-current formalism is employed (usually on a finite-volume system) to derive one or more correlation inequalities. These inequalities ordinarily carry over immediately to the infinite-volume limit.
2. The correlation inequalities are analyzed and their physical consequences extracted.

These stages of the analysis are described in Chapters 12 and 14–15, respectively. (We remark that more direct applications of the random-walk formalism may well be possible, and would surely be of interest; unfortunately we are unable to make any suggestions about how this might be done. Readers interested in this approach might find inspiration in the profound original paper of Symanzik [494].)

The physical consequences of the correlation inequalities fall into two main categories:

- *Inequalities for critical exponents.* Sometimes a correlation inequality leads directly, upon translation from correlation functions into thermodynamic functions, to a bound of immediate physical interest. More often, however, the correlation inequality leads to a *differential inequality* for the thermodynamic functions, which can then be *integrated* to yield a result of physical interest. We illustrate this in Chapter 14.
- *Bounds on the continuum limit.* In some cases a complementary pair of correlation inequalities for the lattice spin system leads directly, upon passage to the continuum limit, to the *triviality* (Gaussianness) of the continuum-limit theory. This is the principal topic of Chapter 15. (In other cases a complementary pair of correlation inequalities can be used to prove *nontriviality*, as was sketched in Section 6.3.)

In deriving these physical consequences, we take care to distinguish between *universal* and *non-universal* bounds, as explained at the beginning of Chapter 14.

One of the main results of the random-walk formalism is a detailed picture of the behavior of spin systems and random-walk models above their upper critical dimension d_c (for short-range interactions d_c equals 4): this behavior includes mean-field critical exponents, the failure of hyperscaling (in various forms), and the triviality of continuum limits. All these results are rigorous versions of the idea that two random walks in dimension $d > 4$ never intersect. For example, by rather straightforward random-walk arguments one can prove the Aizenman-Fröhlich inequality [5, 8, 213, 90, 292]

$$0 \geq u_4(x_1, x_2, x_3, x_4) \gtrsim -\sum_z S(x_1, z)S(x_2, z)S(x_3, z)S(x_4, z) \tag{8.1}$$

and its corollary for thermodynamic (zero-momentum) quantities

$$0 \geq \overline{u}_4 \equiv \frac{\partial^2 \chi}{\partial h^2} \gtrsim -\chi^4 . \tag{8.2}$$

(For a precise statement and discussion, see Section 12.2.) Among the consequences of this inequality are the triviality of the continuum limit, and the failure of the hyperscaling relation $d\nu - 2\Delta_4 + \gamma = 0$, for φ^4 and Ising models in the *single-phase* (symmetric) regime in dimension $d > 4$ [5, 8, 213, 90, 292], and a partial result for $d = 4$ [213]; see Sections 14.1 and 15.2. By an almost identical random-walk argument in nonzero magnetic field one can prove the Fröhlich-Sokal inequality [223]

$$0 \leq K(x_1, x_2) \lesssim S_2^T(x_1, x_2) + \sum_z K(x_1, z)K(x_2, z)S_1(z)^2 \tag{8.3}$$

and its corollary for thermodynamic quantities

$$0 \leq \frac{M}{h} \lesssim \chi + \frac{M^4}{h^2} , \tag{8.4}$$

where K is a rather mysterious kernel which satisfies

$$\sum_y K(x, y)h_y \approx S_1(x) . \tag{8.5}$$

These results are discussed in detail in Section 12.4. Unfortunately, (8.3) does *not* shed light on the presumed triviality of the continuum limit for φ^4 and Ising models in the *two-phase* (non-symmetric) regime in dimension $d > 4$. But it does imply the failure of the hyperscaling relation $d\nu' = \gamma' + 2\beta$ for these models (see Section 14.3), a result with profound physical consequences [509, 323, 198, 201, 510, 200, 288]. Refinements of (8.1) and (8.3) have been proven, using the random-current formalism, by Aizenman and Graham [15, 194] and by Aizenman, Barsky and Fernández [10], respectively. From these refined inequalities one can deduce the mean-field critical behavior $\gamma = 1$ [5, 15]

for Ising and φ^4 models in dimension $d > 4$, $\beta = 1/2$ [12, 10] for Ising models in $d > 4$, and some further partial results on triviality for Ising and φ^4 models in $d = 4$ [15]. We discuss these refinements in Sections 12.2.3 and 12.4.6, and their applications in Chapters 14 and 15.

For strongly-coupled systems in dimension $d < d_c$ our picture is much less complete, but some results can be obtained using the random-current formalism. For example, the Aizenman-Graham inequality implies ([484], [12, Appendix A]) the "spherical-model upper bound" $\gamma \leq 2/(d-2)$ for Ising and φ^4 models in dimension $2 < d < 4$, which is complementary to the "mean-field lower bound" $\gamma \geq 1$ due to Glimm and Jaffe [256, 30, 257]. Similarly, the Aizenman-Barsky-Fernández inequality implies [10] the "mean-field upper bound" $\beta \leq 1/2$, as well as the absence of an intermediate phase with zero magnetization and infinite susceptibility, for Ising and φ^4 models in any dimension. (These latter results had been proven earlier by Aizenman [7, 9] under a more restrictive hypothesis.) The main open question for Ising models in dimension $d < 4$ is the nontriviality of the continuum limit, which is equivalent to the hyperscaling relation $d\nu - 2\Delta_4 + \gamma = 0$. We can offer only very limited progress in this direction: a proof of hyperscaling for the intersection properties of ordinary random walks (IPORW) model in $d < 4$ (Section 12.3.2); and some very weak lower bounds on the renormalized coupling constant $g \equiv -\overline{u}_4/\chi^2\xi^d$ for the self-avoiding walk (Section 12.3.3) and the Ising model (Section 12.3.4).

In all this analysis, a central role is played by the "bubble diagram", which is finite at the critical point in $d > d_c$ and presumably infinite at the critical point in $d \leq d_c$. Indeed, the bubble diagram is so important to our analysis that we introduce new critical exponents b, b', b_c to describe its critical behavior. We remark that in percolation models an analogous role is played by the "triangle diagram" [16, 416, 295, 291], and in branched-polymer (lattice-animal) models by the "square diagram" [76, 299, 296].

Finally, we would like to comment on the relevance of our results for polymer physics.As first noted by deGennes, there is an isomorphism between self-avoiding walks and the $N \to 0$ limit of N-component spin models (see [121, 507] and references therein). This isomorphism has a nice explanation from the point of view of the Symanzik-BFS polymer representation [28, 75], as we discuss in Sections 9.2.2 and 10.2.3. Unfortunately, this isomorphism seems to be useless for the purposes of rigorous analysis, since it is based on an analytic continuation in N (which destroys inequalities). However, it is a very valuable heuristic tool, since it suggests how results (and sometimes proofs) for the Ising model may be adapted to the SAW or vice versa. There is, however, a clash of terminology between the two fields of physics which must be borne in mind. In Table 8.1 we present a brief "dictionary".

Unfortunately, our results in this work for polymer models concern almost entirely the case of zero "magnetic field" (Section 9.2.1), or what is essentially equivalent, the "baby" contact-interacting-walk (CIW) model in nonzero field (Section 10.2.2). From the point of view of polymer physics this means that we treat only systems with a fixed finite number of polymers. A polymer system

Table 8.1. Correspondence between the terminology used for spin and polymer models

Spin models (or field theory)		*Polymer models*
canonical ensemble ($h = 0$)	$\Longleftrightarrow$	canonical ensemble[1]
canonical ensemble ($h \neq 0$)	$\Longleftrightarrow$	grand canonical ensemble[1]
inverse temperature (or ferromagnetic pair interaction)	$\Longleftrightarrow$	monomer activity
magnetic field	$\Longleftrightarrow$	chain end activity
$2n$-point correlation function (at zero magnetic field)	$\Longleftrightarrow$	correlation function of n polymer chains

[1]There is some lack of standardization in the terminology for polymer-model ensembles. Here microcanonical $\equiv$ a fixed number (usually one) of polymer chains of fixed length; canonical $\equiv$ a fixed number (usually one) of polymer chains of variable length; grand canonical $\equiv$ a variable number of polymer chains of variable length.

at nonzero density would correspond to our "mature" contact-interacting-walk model (Section 10.2.1), for which we have no significant results.

In summary, this monograph is part original, part review — many of the results discussed here are contained in work already published [5, 15, 8, 9, 7, 10, 12, 28, 75, 92, 90, 97, 96, 185, 213, 478, 481] by the authors and their collaborators. But even when we review "well-known" material, we hope to provide new ways of looking at some key ideas. At the very least, the process of writing this monograph has proven beneficial for the authors; we hope that the result will prove beneficial for some readers as well.

For the "experts", we mention some of the new results contained in this work:

- A "twice-improved Aizenman-Fröhlich inequality" for the SAW, and more generally for "repulsive simple CIW models" (Section 12.2.2).
- The inequality $(-1)^n u_{2n} \leq 0$ for the SAW (Sections 12.6 and 12.7).
- An extrapolation principle derived from the ABF inequality (Section 14.2.3).
- Mean-field upper and lower bounds for the magnetization, and partial results for the susceptibility, for Ising and φ^4 models in dimension $d > 4$, in a full neighborhood of the critical point in the (β, h)-plane (Section 14.4).
- The critical-exponent inequality $\delta \geq 2\gamma + 1$ (Section 14.4.1).

The first of these results has a very nice physical interpretation, which makes rigorous one half of an appealing intuitive argument due originally to des Cloizeaux [125]. The second of these results is an analogue for the SAW of a recent result of Shlosman [460] for the Ising model. The other results follow from our focus on studying a full neighborhood of the critical point in the (β, h)-plane, going beyond the three "traditional" paths of approach to the critical point.

This monograph can be read on two levels. Readers interested primarily in the physical consequences can jump directly to Part III; we have tried hard to keep this part self-contained, and only occasional references to Chapter 9 (to check notation) should be necessary. Readers interested in the random-walk formalism can read Chapters 9–12. Readers who are not yet familiar with the random-walk methods might find it advisable, on a first reading, to consider only the case of zero magnetic field: they should read Chapter 9 and Sections 11.1–11.2, 11.3.1, 12.1–12.3 and 12.6, ignoring all references to the subscripts j on the weights $\varrho_j(\omega)$.

9. Random-walk models in the absence of magnetic field

9.1 General definitions

Spin systems (such as the Ising model) and polymer models (such as the self-avoiding walk) have long played an important role in the theory of critical phenomena. The analogy between spin systems and random-walk models has intrigued physicists since the 1950's [498, 208, 206, 197, 135], but the precise relations between these two types of models have emerged only gradually [493, 494, 120, 126, 92, 28]. In recent years, several "artificial" random-walk representations have been introduced as tools with which to study spin systems [92, 5, 8]. The purpose of this chapter (and the next one) is to exhibit an underlying mathematical structure which is common to all random-walk models.

We consider a fixed finite set $\mathbb{L}$ called the *lattice*, whose elements are called *sites*[1]. A *bond* is an unoriented pair $\{x, y\}$ of sites $x, y \in \mathbb{L}$, $x \neq y$. An oriented pair (x, y) is called a *step* with initial site x and final site y. A *walk* (or *path*) ω in the lattice is a sequence of sites $\omega(0), \omega(1), \ldots, \omega(N)$ $(N \geq 0)$; we call $|\omega| \equiv N$ the *length* of the walk, $b(\omega) \equiv \omega(0)$ its initial point, and $e(\omega) \equiv \omega(N)$ its final point. If $\omega(0) = x$ and $\omega(N) = y$, we will write $\omega\colon x \to y$. We emphasize that a walk is, for us, an oriented object. The set of walks on $\mathbb{L}$ will be denoted Ω.

To each walk ω there is associated a sequence (perhaps empty) of steps

$$\underline{\omega} \;=\; (\omega(0), \omega(1)),\ (\omega(1), \omega(2)),\ \ldots,\ (\omega(N-1), \omega(N)) \quad ; \tag{9.1}$$

this correspondence is one-to-one *except* that the empty sequence of steps is associated to every zero-length walk irrespective of its initial point.

The *support* of a walk ω is the set of sites visited by the walk:

$$\operatorname{supp}(\omega) \;=\; \{x \in \mathbb{L}\colon\ \omega(i) = x \text{ for some } i\}\ . \tag{9.2}$$

The support of a family of walks is the union of the supports of the individual walks. Two families of walks are said to be *nonoverlapping* if their supports are disjoint.

[1] Of course, the most interesting physics is for infinite-volume systems. But in the present work the random-walk expansions are primarily tools for deriving correlation inequalities, which then carry over immediately to the infinite-volume limit. From this point of view the limitation to finite lattices is natural. Some of our random-walk models can be formulated directly on infinite lattices, with a little extra work.

If ω_1, ω_2 are two walks with $e(\omega_1) = b(\omega_2)$ we denote by $\omega_1 \circ \omega_2$ the *concatenation* of ω_1 and ω_2, i.e. the walk of length $|\omega_1| + |\omega_2|$ obtained by traversing first the steps of ω_1 and then those of ω_2. Likewise we may define the concatenation $\underline{\omega}_1 \circ \underline{\omega}_2$ of any two sequences of steps $\underline{\omega}_1$ and $\underline{\omega}_2$, irrespective of their initial and final points; of course, the sequence of steps $\underline{\omega}_1 \circ \underline{\omega}_2$ will be associated with a walk only if $\underline{\omega}_1$ and $\underline{\omega}_2$ themselves are associated with walks and $e(\underline{\omega}_1) = b(\underline{\omega}_2)$ [or either $\underline{\omega}_1$ or $\underline{\omega}_2$ is empty].

A *random-walk model* is defined by a system of weights $(\varrho^{[n]})_{n \geq 0}$, where each $\varrho^{[n]}$ is a function $\varrho^{[n]}: \Omega^n \to \mathbb{R}$, and $\varrho^{[0]} \equiv 1$. The weights are used to construct the fundamental objects of the models, which we will call "kernels". For $x_1, \ldots, x_{2n} \in \mathbb{L}$,

$$K(x_1 x_2 | \ldots | x_{2n-1} x_{2n}) \equiv \sum_{\substack{\omega_1 : x_1 \to x_2 \\ \vdots \\ \omega_n : x_{2n-1} \to x_{2n}}} \varrho^{[n]}(\omega_1, \ldots, \omega_n) . \tag{9.3}$$

In principle, any system of weights defines a random-walk model, but for the models of physical interest the weights are found to have certain important additional properties which can be grouped into three main categories:

1. *Positivity.* In this work we consider only models with $\varrho^{[n]} \geq 0$.

2. *Factorization properties of the walks.* These are inequalities relating a given weight with products of weights of smaller or equal order. Two kinds of inequalities are of interest:

   a) Inequalities involving the partition of an (ordered) family of walks into (ordered) subfamilies.

   b) Inequalities involving the splitting of a path.

   These properties will be discussed in Chapter 11.

3. *Symmetries of the weights.* For weights describing a system of indistinguishable unoriented polymers, it is natural to assume that $\varrho^{[n]}$ is symmetric under permutation of walks and under inversion of a walk. Here we do *not* assume such symmetries. This generality is forced by the existence of an important random-walk model which lacks such symmetries: the ARW representation for the Ising model. As a consequence, we do not assume that the kernel (9.3) is symmetric under reordering of the pairs or under reordering within one pair.

In the sequel we will omit the superscript $[n]$ for the weights, as the number of arguments will make this clear.

Next we define the $2n$-point *Green functions*, which are the quantities of direct physical relevance for spin systems. For generality, we allow for the presence of internal indices α_i, which may be arbitrary positive integers. We thus define Green functions $S_{2n}^{(\alpha_1 \ldots \alpha_{2n})}(x_1, \ldots, x_{2n})$ by summing the kernels (9.3) over all pairings between the sites $x_1, \ldots, x_{2n}$ which are compatible with the specified internal indices $\alpha_1, \ldots, \alpha_{2n}$: only sites having the *same* value of the internal index may be paired. Each such pairing is in correspondence with a partition of the set $\{1, \ldots, 2n\}$ into sets of two elements, i.e. with a set

$$\Big\{\{\pi(1), \pi(2)\}, \ldots, \{\pi(2n-1), \pi(2n)\}\Big\} \tag{9.4}$$

where π is some permutation of the set $\{1, \ldots, 2n\}$. As we are not assuming any kind of symmetry for the weights, we must specify an order for the pairs in (9.4) and for the elements inside each pair. So we must determine which special family of permutations we consider in (9.4).

If P_{2n} is the group of permutations of $\{1, \ldots, 2n\}$, then the set of pairings of $\{1, \ldots, 2n\}$ is in one-to-one correspondence with P_{2n}/C_{2n}, where C_{2n} is the subgroup of P_{2n} generated by the following operations:

i) the identity,

ii) the n permutations which interchange the elements within one pair, and

iii) the $n(n-1)/2$ permutations which interchange one pair with another.

Each equivalence class of P_{2n}/C_{2n} is formed by $2^n n!$ permutations, each defining the same pairing. To define the Green function S_{2n} we must choose a particular representative of each class. We choose the one which produces the elements within each pair in (9.4) to be ordered in the natural order, and the pairs themselves to be ordered according to their first elements. That is, we define Q_{2n} as the set of permutations $\pi \in P_{2n}$ such that

$$\begin{array}{lcll} \pi(2k-1) & < & \pi(2k) & k = 1, \ldots, n \\ \pi(2k-1) & < & \pi(2k+1) & k = 1, \ldots, n-1 \end{array} \tag{9.5}$$

(The permutations in Q_{2n} can be thought of as constructed in the following way: first pair the integer 1 with one of the remaining $2n-1$ integers, then pick the smallest of the integers not yet paired and pair it with another integer, etc.)

Now we can define the Green functions for the model:

$$\begin{aligned} S_{2n}^{(\alpha_1 \ldots \alpha_{2n})}(x_1, \ldots, x_{2n}) &= \sum_{\pi \in Q_{2n}} \delta_{\alpha_{\pi(1)}\alpha_{\pi(2)}} \cdots \delta_{\alpha_{\pi(2n-1)}\alpha_{\pi(2n)}} \\ &\quad \times \; K(x_{\pi(1)} x_{\pi(2)} | \ldots | x_{\pi(2n-1)} x_{\pi(2n)}) \end{aligned} \tag{9.6a}$$

$$S_{2n+1}^{(\alpha_1 \ldots \alpha_{2n+1})}(x_1, \ldots, x_{2n+1}) = 0\,. \tag{9.6b}$$

In particular, if the internal indices $\alpha_1 \ldots \alpha_{2n}$ are all equal,

$$\begin{aligned} S_{2n}(x_1, \ldots, x_{2n}) &\equiv S_{2n}^{(\alpha \ldots \alpha)}(x_1, \ldots, x_{2n}) \\ &= \sum_{\pi \in Q_{2n}} K(x_{\pi(1)} x_{\pi(2)} | \ldots | x_{\pi(2n-1)} x_{\pi(2n)})\,. \end{aligned} \tag{9.7}$$

We observe that in the generality with which we are stating our definitions, the Green functions S_{2n} may depend on the choice of the set of representatives Q_{2n}, and may turn out not to be symmetric under permutation of sites. It is a nontrivial fact that our major example with non-symmetric weights — the ARW model discussed below — does not exhibit this last pathology. For models with symmetric weights any choice of representatives is equivalent, and in fact we could replace the sum over Q_{2n} by the sum over the whole P_{2n} divided by $2^n n!$.

9.2 Examples

For the following models we consider a symmetric matrix $J = (J_{xy})_{x,y \in \mathrm{L}}$ with nonnegative elements. For a family of bonds $A = \{\{x_i, y_i\}_i\}$ and a function $f\colon \mathbb{R} \to \mathbb{R}$ we will denote

$$[f(J)]^A \equiv \prod_i f(J_{x_i y_i}) \tag{9.8}$$

with the convention $[f(J)]^{\varnothing} = 1$. Analogously, if ω is a walk, we write

$$[f(J)]^\omega \equiv \prod_i f(J_{\omega(i)\omega(i+1)}) \tag{9.9}$$

with the convention that if $|\omega| = 0$, then $[f(J)]^\omega \equiv 1$. If more than one walk is involved we will abbreviate

$$[f(J)]^{\omega_1 + \ldots + \omega_n} \equiv \prod_k [f(J)]^{\omega_k} \tag{9.10}$$

9.2.1 Polymer-chain models

These will be our simplest family of models. They are defined by weights of the form

$$\varrho(\omega_1, \ldots, \omega_k) = J^{\omega_1 + \ldots + \omega_k} \exp[-U_k(\omega_1, \ldots, \omega_k)] \tag{9.11}$$

and are intended to model a polydisperse ensemble of interacting polymer chains $\omega_1, \ldots, \omega_k$. Thus, J is to be interpreted as a monomer activity (or fugacity), and U is to be interpreted as the interaction energy of the walks (which may be $+\infty$ for certain configurations); U is assumed to be a symmetric function of its arguments. Note that U has the important feature that it is independent of J. Therefore, these models have a *trivial J-dependence*. By definition $U_0 = 0$ so that $\varrho^{[0]} \equiv 1$.

These models are the natural discretization of the models of continuum polymer chains introduced by Edwards [160], de Gennes [120] and des Cloizeaux [126], among others. We warn the reader, however, that other models introduced in the polymer-physics literature may employ conventions that disagree with ours. For example, our walks are oriented objects, but polymer molecules are usually considered to be unoriented; and we allow zero-length walks (i.e. walks occupying a single lattice site), while some authors (e.g. [284]) explicitly forbid them.

We call a polymer-chain model *repulsive* if

$$U_{j+k}(\omega_1, \ldots, \omega_{j+k}) \;\geq\; U_j(\omega_1, \ldots, \omega_j) \;+\; U_k(\omega_{j+1}, \ldots, \omega_{j+k}) \tag{9.12}$$

for all j, k and all $\omega_1, \ldots, \omega_{j+k}$. This will be the key condition needed for proving "Gaussian upper bound" correlation inequalities (see Theorems 11.1 and 12.1).

We now define two major classes of polymer-chain models, which we call *pair-interacting walks* (PIW) and *contact-interacting walks* (CIW), respectively.

Pair-interacting walks (PIW). In these models the interaction energy U_k takes the form

$$U_k(\omega_1, \ldots, \omega_k) \;=\; \sum_{i,j=1}^{k} \sum_{r=0}^{|\omega_i|} \sum_{s=0}^{|\omega_j|} V(\omega_i(r), \omega_j(s)) \; [1 - \delta_{i,j}\delta_{r,s}]. \tag{9.13}$$

Clearly this represents a pair interaction V between the "atoms" of the walks $\omega_1, \ldots, \omega_k$; the square bracket indicates the absence of self-interactions of a single "atom". If $V \geq 0$, the model is repulsive in the sense defined above.

Contact-interacting walks (CIW). These are models in which walks interact only when they touch; that is, the interaction energy between nonoverlapping families of walks is zero. More precisely, a *(generalized) CIW model* is one in which

$$U_{j+k}(\omega_1, \ldots, \omega_{j+k}) \;=\; U_j(\omega_1, \ldots, \omega_j) \;+\; U_k(\omega_{j+1}, \ldots, \omega_{j+k}) \tag{9.14}$$

whenever the families $\{\omega_1, \ldots, \omega_j\}$ and $\{\omega_{j+1}, \ldots, \omega_{j+k}\}$ are nonoverlapping. However, we are mainly interested CIW models that arise in the following way: Given a walk ω, define its "visitation" function $n_x(\omega)$ as the number of times ω "hits" (or "visits") the site x, i.e.

$$n_x(\omega) \;=\; \#\{i\colon\; \omega(i) = x\}. \tag{9.15}$$

For a family $\omega_1, \ldots, \omega_k$ of walks, we define the total visitation function

$$n_x(\omega_1, \ldots, \omega_k) \;=\; \sum_{j=1}^{k} n_x(\omega_j). \tag{9.16}$$

We shall define *simple CIW models* as those in which the interaction energy takes the form

$$\exp[-U_k(\omega_1, \ldots, \omega_k)] \;=\; \prod_{x \in \mathbb{L}} c_x(n_x(\omega_1, \ldots, \omega_k)) \tag{9.17}$$

for suitable "weights" $c_x(n)$, where $c_x(0) \equiv 1$ and $c_x(1), c_x(2), \ldots \geq 0$. We note that a simple CIW model is repulsive if it satisfies the condition

$$c_x(n+m) \;\leq\; c_x(n)c_x(m)\,. \tag{9.18}$$

The triviality of the J-dependence together with the properties of the visitation function (see (11.34) and (11.36) below) imply straightforward formulas for the derivatives with respect to J of the weights and Green functions of simple CIW models (see Sections 11.3 and 12.2.4).

Some examples of simple CIW models are:

a) Ordinary random walks.

$$c_x(n) \;=\; \tau^{-n}\,. \tag{9.19}$$

(The usual normalization is $\tau = 1$, but for reasons to be seen in the next section we wish to allow arbitrary $\tau > 0$.)

b) Self-avoiding walk (SAW).

$$c_x(n) = \begin{cases} 1 & \text{if } n = 0, 1 \\ 0 & \text{if } n \geq 2 \end{cases} \tag{9.20}$$

c) The Domb-Joyce model [138, 93].

$$c_x(n) \;=\; e^{-vn(n-1)/2} \qquad (v \geq 0)\,. \tag{9.21}$$

This is equivalent to a PIW model with $V(x,y) \;=\; v\delta_{x,y}$: each self-intersection costs an energy v. If $v = 0$ this is the ordinary random walk (with $\tau = 1$); as $v \to +\infty$ it tends to the SAW.

d) The Edwards model [160, 28, 97, 75].

$$c_x(n) \;=\; \int d\nu_n(t)\; \mathcal{Z}(t) \tag{9.22}$$

where

$$d\nu_n(t) = \begin{cases} \delta(t)\, dt & \text{if } n = 0 \\ \\ \frac{t^{n-1}}{\Gamma(n)}\chi_{[0,\infty)}(t)\, dt & \text{if } n \geq 1 \end{cases} \tag{9.23}$$

and

$$\mathcal{Z}(t) \;=\; \exp[-\lambda t^2 - \tau t] \qquad (\lambda \geq 0)\,. \tag{9.24}$$

If $\lambda = 0$, the Edwards model reduces to the ordinary random walk. If $\lambda \to +\infty$ while $\tau = -(2\lambda \log(\lambda/\pi))^{1/2}$ one obtains the SAW [75, Appendix]. (The weights (9.22)–(9.24) may appear strange, but they arise naturally from the continuous-"time" Edwards model [160] by passing to the imbedded discrete-"time" walk. The variable t has the interpretation of a *waiting time* (or *local time*) in the continuous-time Edwards model.)

Numerous other special cases of the CIW have been introduced in the recent literature; see [157] and references cited therein.

We note that all four of these examples satisfy the condition (9.18), hence are repulsive. This is immediate for the first three examples; for the Edwards model it is a consequence of the fact that [28]

$$d\nu_n * d\nu_m \;=\; d\nu_{n+m} \tag{9.25}$$

where * indicates convolution, together with the inequality

$$\mathcal{Z}(t+s) \;\leq\; \mathcal{Z}(t)\mathcal{Z}(s) \tag{9.26}$$

which follows from (9.24).

We remark that both the PIW and the CIW models can be generalized by allowing the walks ω_i to carry internal indices α_i (expressing the "color", chemical composition, etc. of the walk) and allowing the interaction energy to depend on these internal indices. In the simple CIW models this is most easily expressed by making the visitation numbers *vector-valued*, i.e. $\mathbf{n}_x(\omega_1, \ldots, \omega_k)$ is a vector specifying the number of times that the site x has been visited *by walks of each color*. One interesting case is the "intersection properties of ordinary random walks" (IPORW) model, in which intersections between walks of *different* colors are forbidden (or are penalized with an Edwards-like weight), but intersections between walks of the *same* color (in particular, self-intersections) are allowed. (This model has been studied by Felder and Fröhlich [185]; similar results were obtained by Aizenman [7] in a slightly different formalism.) Many of the correlation inequalities for repulsive CIW models carry over to this "multi-colored" case, but we shall leave this extension to the reader.

9.2.2 BFS representation for continuous spin systems

Consider an N-component isotropic spin system defined by a Hamiltonian

$$H \;=\; \frac{1}{2}\sum_{x,y} J_{xy}\,\boldsymbol{\varphi}_x\boldsymbol{\varphi}_y \tag{9.27}$$

and single-spin measures on $\mathbb{R}^N$

$$dP_x(\boldsymbol{\varphi}_x) \;=\; g_x(\boldsymbol{\varphi}_x^2)\,d\boldsymbol{\varphi}_x\;, \tag{9.28}$$

where $d\boldsymbol{\varphi}_x$ is the Lebesgue measure on $\mathbb{R}^N$, and the functions g_x are assumed to be smooth, strictly positive and decaying faster than exponentially at infinity. (These conditions on g_x are much stronger than necessary, and are imposed solely to avoid uninteresting technical problems. More general models (e.g. Ising) can be handled by taking limits in the final formulae.) The states of such systems are completely described by the correlation functions

$$\langle\varphi_{x_1}^{(\alpha_1)}\cdots\varphi_{x_m}^{(\alpha_m)}\rangle \;=\; \frac{1}{Z}\int\varphi_{x_1}^{(\alpha_1)}\cdots\varphi_{x_m}^{(\alpha_m)}\;e^{\frac{1}{2}\sum_{x,y}J_{xy}\boldsymbol{\varphi}_x\cdot\boldsymbol{\varphi}_y}\prod_{x\in\mathbb{L}}g_x(\boldsymbol{\varphi}_x^2)\,d\boldsymbol{\varphi}_x \tag{9.29}$$

for $x_1, \ldots, x_m \in \mathbb{L}$ and $\alpha_1, \ldots, \alpha_m \in \{1, \ldots, N\}$. Here Z is the partition function

$$Z = \int e^{\frac{1}{2}\sum_{x,y} J_{xy}\varphi_x\cdot\varphi_y} \prod_{x\in\mathbb{L}} g_x(\varphi_x^2)d\varphi_x \,. \tag{9.30}$$

On the other hand, the BFS random-walk model for the interaction J and measures dP_x is defined by the weights

$$\varrho(\omega_1, \ldots, \omega_k) = J^{\omega_1+\ldots+\omega_k} \int \mathcal{Z}(\underline{t}_1 + \ldots + \underline{t}_k) \prod_{i=1}^{k} d\nu_{\omega_i}(\underline{t}_i) \tag{9.31}$$

with $\underline{t} = (t_x)_{x\in\mathbb{L}}$,

$$\mathcal{Z}(\underline{t}) \equiv \frac{1}{Z}\int e^{\frac{1}{2}\sum_{x,y} J_{xy}\varphi_x\cdot\varphi_y} \prod_{x\in\mathbb{L}} g_x(\varphi_x^2 + 2t_x)\, d\varphi_x \tag{9.32}$$

$$= \left\langle \prod_{x\in\mathbb{L}} \frac{g_x(\varphi_x^2 + 2t_x)}{g_x(\varphi_x^2)} \right\rangle \tag{9.33}$$

and the measure $d\nu_\omega(\underline{t})$ on $\mathbb{R}^{\mathbb{L}}$ is defined by

$$d\nu_\omega(\underline{t}) = \prod_{x} d\nu_{n_x(\omega)}(t_x) \tag{9.34}$$

where $n_x(\omega)$ is the visitation function (9.15) and $d\nu_n$ is given by (9.23).

The weights (9.31) of the BFS model can in principle be written in the form (9.11) of the weights of polymer-chain models. However, the corresponding interaction U — given implicitly in terms of the correlations (9.33), or explicitly by equations (9.37)–(9.39) below — would depend on J. That is, the weights (9.31) have, besides the trivial J-dependence analogous to that of the polymer-chain models, another "deep" J-dependence in $\mathcal{Z}(t)$. This deep dependence is made explicit in the "loop" expansion (9.37)–(9.39) below. Both types of J-dependence play in a sense a different role in the properties of the systems. The trivial dependence is related with combinatorial aspects that remain basically the same for all choices of N and of the single-spin measure dP_x. On the other hand, the deep dependence is responsible for more subtle and specific properties of each model (choice of N and dP_x), such as the existence (or not) of correlation inequalities. In this regard the polymer-chain models are useful for understanding the basic combinatorial properties of random-walk models, without the burden of the deep J-dependence.

Another important difference between BFS and polymer-chain models is that the former do not satisfy a repulsiveness condition of the sort of (9.12). In fact, for some families of walks inequalities resembling (9.12) are satisfied (Theorem 11.2 below), but for others the inequality is reversed (Theorem 11.7).

The *fundamental identity* relating an N-component spin system to the corresponding BFS model is [92, 97]:

$$\langle \varphi_{x_1}^{(\alpha_1)} \cdots \varphi_{x_{2m}}^{(\alpha_{2m})} \rangle = S_{2m}^{(\alpha_1\ldots\alpha_{2m})}(x_1, \ldots, x_{2m}) \tag{9.35}$$

and in particular if the internal indices $\alpha_1, \ldots, \alpha_n$ are all equal,

$$\langle \varphi_{x_1}^{(\alpha)} \cdots \varphi_{x_{2m}}^{(\alpha)} \rangle = S_{2m}(x_1, \ldots, x_{2m}) \tag{9.36}$$

for any $\alpha = 1, \ldots, N$. It is this identity which allows us to use spin-system results (like the Griffiths inequalities) in studying the BFS model, and conversely, to transfer results proved for the BFS model back to the spin system. For the convenience of the reader, we give a proof of (9.35) at the end of this section.

For some purposes, it is more suggestive to write the weights as a "loop" expansion ("Symanzik's complete polymer representation"). Indeed, the partition function (9.30) can be written as (see [92] with the correction stated in the last paragraph of [97])

$$Z = \sum_{n=0}^{\infty} \frac{1}{n!} \left(\frac{N}{2}\right)^n \times \sum_{v_1, \ldots, v_n \in \mathbb{L}} \sum_{\substack{\underset{\sim}{\omega}_1 : v_1 \to v_1 \\ \vdots \\ \underset{\sim}{\omega}_n : v_n \to v_n}} \frac{J^{\underset{\sim}{\omega}_1 + \ldots + \underset{\sim}{\omega}_n}}{|\underset{\sim}{\omega}_1| \cdots |\underset{\sim}{\omega}_n|} e^{-U_k^N(\underset{\sim}{\omega}_1, \ldots, \underset{\sim}{\omega}_n)} \tag{9.37}$$

where the interaction U_k^N is a simple contact interaction independent of J, defined by (9.17) with

$$c_x(n) = \frac{\int d\nu_{n+N/2}(t)\, g_x(2t)}{\int d\nu_{N/2}(t)\, g_x(2t)}\,. \tag{9.38}$$

Correspondingly, the weights take the form

$$\begin{aligned} \varrho(\omega_1, \ldots, \omega_k) = {} & J^{\omega_1 + \ldots + \omega_k} \frac{1}{Z} \sum_{n=0}^{\infty} \frac{1}{n!} \left(\frac{N}{2}\right)^n \\ & \times \sum_{v_1, \ldots, v_n \in \mathbb{L}} \sum_{\substack{\underset{\sim}{\omega}_1 : v_1 \to v_1 \\ \vdots \\ \underset{\sim}{\omega}_n : v_n \to v_n}} \frac{J^{\underset{\sim}{\omega}_1 + \ldots + \underset{\sim}{\omega}_n}}{|\underset{\sim}{\omega}_1| \cdots |\underset{\sim}{\omega}_n|} e^{-U_{n+k}^N(\omega_1, \ldots, \omega_k, \underset{\sim}{\omega}_1, \ldots, \underset{\sim}{\omega}_n)} \end{aligned} \tag{9.39}$$

On the one hand, these expressions show that in the BFS models the walks not only exhibit contact interactions among themselves, but they interact also with a "solvent" of closed walks. This interaction with the solvent can be interpreted as the cause for the fact that for nonoverlapping walks the inequality (9.12) is reversed (Theorem 11.7): the mediation of the "loops" produces a net attraction between nonoverlapping walks.

On the other hand, (9.37)-(9.39) make sense for N not necessarily integer, and define an analytic continuation in N of the BFS model (and thus, in view of the fundamental identity (9.35)/(9.36), of the N-component isotropic spin model); in particular, the partition and correlation functions are analytic in a neighborhood of $N = 0$. When the limit $N \to 0$ is performed in (9.37) and (9.39), only the summands with $n = 0$ survive, and we obtain

$$\lim_{N\to 0} Z = 1 \tag{9.40}$$

$$\lim_{N\to 0} \varrho(\omega_1,\ldots,\omega_k) = J^{\omega_1+\ldots+\omega_k} \exp[-U_k(\omega_1,\ldots,\omega_k)] \,. \tag{9.41}$$

The interaction $U_k = \lim_{N\to 0} U_k^N$ is a simple contact interaction with [28]

$$c_x(n) = \frac{\int d\nu_n(t)\, g_x(2t)}{\int d\nu_n(t)\, g_x(0)} \,. \tag{9.42}$$

This is the general version of deGennes' [120] relation between spin models and polymer-chain (simple CIW) models.

We now present some important examples of BFS random-walk models:

a) Gaussian model.

$$g_x(\varphi_x^2) = \exp\left[-\frac{\tau}{2}\varphi_x^2\right] \tag{9.43}$$

where $\tau > 0$. The weights (9.31) coincide with (9.19); the Gaussian model thus corresponds to the ordinary random walk.

b) $|\varphi|^4$ model.

$$g_x(\varphi_x^2) = \exp\left[-\frac{\lambda}{4}\varphi_x^4 - \frac{\tau}{2}\varphi_x^2\right] \tag{9.44}$$

with $\lambda \geq 0$. The case $\lambda = 0$ is the Gaussian model; in the opposite limit $\lambda \to +\infty$ with $\tau = -\lambda N$, it yields the N-vector model (non-linear σ-model) with single-spin measure

$$dP_x(\boldsymbol{\varphi}_x) = \text{const} \times \delta(\boldsymbol{\varphi}_x^2 - N)\, d\boldsymbol{\varphi}_x \quad . \tag{9.45}$$

From (9.42)/(9.44) and (9.22)/(9.24), we see that the $N \to 0$ limit of the $|\varphi|^4$ model is the Edwards model. It can be shown [28, Appendix A] that the $N \to 0$ limit of the N-vector model is the SAW model.

c) Single-spin measures of the BFS class, namely [97, Sections 5 and 6] N=1 or 2 and

$$g_x(\boldsymbol{\varphi}_x^2) = e^{-f_x(\varphi_x^2)} \tag{9.46}$$

with $s \mapsto f_x(s)$ convex and growing at least linearly for s large; and limits of such measures. For $N = 1, 2$ this class includes the previous examples and (for $N = 1$) is more general than the Ellis-Monroe-Newman [162, 163] class. We shall see in Chapters 11 and 12 that the BFS class is the natural class of single-spin measures for deriving correlation inequalities in the BFS random-walk model. Thus, while the definition (9.31)–(9.34) and the fundamental identity (9.35) hold for any N and any single-spin measure of the form (9.28), we are able to prove correlation inequalities only when the single-spin measure is of BFS class. Rather than adopt a cumbersome nomenclature to denote this special case, *we adopt henceforth the convention that "BFS random-walk model" means "BFS random-walk model with 1 or 2 components and single-spin measure of BFS class"*.

PROOF OF THE FUNDAMENTAL IDENTITY (9.35). We first derive an integration-by-parts formula for Gaussian integrals:

Lemma 9.1 *For any matrix M with positive-definite real part and any well-behaved function F,*

$$\int \varphi_x F(\varphi)\, e^{-\frac{1}{2}(\varphi, M\varphi)}\, d\varphi \;=\; \sum_y (M^{-1})_{xy} \int \frac{\partial F}{\partial \varphi_y}\, e^{-\frac{1}{2}(\varphi, M\varphi)}\, d\varphi\,. \tag{9.47}$$

Proof. By the fundamental theorem of calculus,

$$\int \frac{\partial}{\partial \varphi_y} \left[F(\varphi)\, e^{-\frac{1}{2}(\varphi, M\varphi)}\right] d\varphi \;=\; 0 \tag{9.48}$$

(provided that F and its first derivative have sub-Gaussian growth at infinity). Writing this out, we get

$$\int \left[\frac{\partial F}{\partial \varphi_y} - \sum_z M_{yz}\varphi_z F(\varphi)\right] e^{-\frac{1}{2}(\varphi, M\varphi)}\, d\varphi \;=\; 0\,. \tag{9.49}$$

Now multiply both sides by $(M^{-1})_{xy}$ and sum over y. ∎

Note that by substituting successively $F(\varphi) = \varphi_{x_2}$, $F(\varphi) = \varphi_{x_2}\varphi_{x_3}\varphi_{x_4}$, etc. we can derive the usual formulae for the moments of a Gaussian measure.

We now wish to derive an analogous integration-by-parts formula for the non-Gaussian measure (9.29). Consider the quantity

$$\langle \varphi_x^{(\alpha)} F(\boldsymbol{\varphi})\rangle \;=\; Z^{-1} \int \varphi_x^{(\alpha)} F(\boldsymbol{\varphi})\, e^{\frac{1}{2}(\boldsymbol{\varphi}, J\boldsymbol{\varphi})} \prod_z g_z(\boldsymbol{\varphi}_z^2)\, d\boldsymbol{\varphi}_z\,. \tag{9.50}$$

We use the Fourier representation

$$g_z(\boldsymbol{\varphi}_z^2) \;=\; \int_{-\infty}^{\infty} e^{-ia_z\boldsymbol{\varphi}_z^2}\, \widehat{g}_z(a_z)\, da_z\,. \tag{9.51}$$

Note that the assumptions on g_z imply that $\widehat{g}_z$ is an entire analytic function which decays faster than any inverse power as $|\mathrm{Re}\, a_z| \to \infty$. Therefore, we can move the path of integration from the real axis to the line $\mathrm{Im}\, a_z = -C < 0$. Having done so, we insert (9.51) into (9.50). Since, for C sufficiently large, the multiple integration is absolutely convergent, we can interchange the order of integration and apply (for each fixed $a = \{a_z\}$) the Gaussian integration-by-parts formula of the previous Lemma, with $M_{uv}^{(\alpha\beta)} = (2ia_u\delta_{uv} - J_{uv})\delta^{\alpha\beta}$. We obtain

$$\langle \varphi_x^{(\alpha)} F(\boldsymbol{\varphi})\rangle \;=\; Z^{-1} \sum_y \int (2ia - J)_{xy}^{-1} \frac{\partial F}{\partial \varphi_y^{(\alpha)}}\, e^{\frac{1}{2}(\boldsymbol{\varphi}, (J-2ia)\boldsymbol{\varphi})} \prod_z d\boldsymbol{\varphi}_z \prod_z \widehat{g}_z(a_z)\, da_z\,. \tag{9.52}$$

Next we expand $(2ia - J)^{-1}$ in a Neumann series

$$(2ia - J)^{-1} = (2ia)^{-1} + (2ia)^{-1}J(2ia)^{-1} + (2ia)^{-1}J(2ia)^{-1}J(2ia)^{-1} + \ldots \quad (9.53)$$

(which converges for $C = -\operatorname{Im} a_z$ sufficiently large). The sums over matrix indices implicit in (9.53) can be combined into one sum over a random walk; doing this and inserting into (9.52), we get

$$\langle \varphi_x^{(\alpha)} F(\varphi) \rangle = Z^{-1} \sum_y \sum_{\omega:\, x \to y} J^\omega \int \left(\prod_z (2ia_z)^{-n_z(\omega)} \right) \frac{\partial F}{\partial \varphi_y^{(\alpha)}} e^{\frac{1}{2}(\varphi,(J-2ia)\varphi)} \prod_z d\varphi_z \prod_z \widehat{g}_z(a_z)\, da_z\,. \quad (9.54)$$

where $n_z(\omega)$ is the visitation function (9.15). Now use, for each site z, the identity

$$b^{-n} = \begin{cases} \int_0^\infty \frac{e^{-bt}t^{n-1}}{(n-1)!}\, dt & \text{for } n \ge 1 \\ \int_0^\infty e^{-bt}\, \delta(t)\, dt & \text{for } n = 0 \end{cases} \equiv \int e^{-bt}\, d\nu_n(t) \quad (9.55)$$

valid for $\operatorname{Re} b > 0$. Inserting this into (9.54) [with $b = 2ia_z$] we get

$$\begin{aligned} &\langle \varphi_x^{(\alpha)} F(\varphi) \rangle \\ &= Z^{-1} \sum_y \sum_{\omega:\, x \to y} J^\omega \int d\nu_\omega(\underline{t})\, \frac{\partial F}{\partial \varphi_y^{(\alpha)}} e^{\frac{1}{2}(\varphi,J\varphi)} \prod_z e^{-ia_z(\varphi_z^2 + 2t_z)}\, \widehat{g}_z(a_z)\, da_z\, d\varphi_z \\ &= Z^{-1} \sum_y \sum_{\omega:\, x \to y} J^\omega \int d\nu_\omega(\underline{t})\, \frac{\partial F}{\partial \varphi_y^{(\alpha)}} e^{\frac{1}{2}(\varphi,J\varphi)} \prod_z g_z(\varphi_z^2 + 2t_z)\, d\varphi_z \end{aligned} \quad (9.56)$$

where $d\nu_\omega(\underline{t})$ is defined in (9.34). Summarizing, we have proven the following non-Gaussian analogue of Lemma 9.1:

Proposition 9.2

$$\int \varphi_x^{(\alpha)} F(\varphi)\, e^{\frac{1}{2}(\varphi,J\varphi)} \prod_z g_z(\varphi_z^2)\, d\varphi_z = \sum_y \sum_{\omega:\, x \to y} J^\omega \int d\nu_\omega(\underline{t})\, \frac{\partial F}{\partial \varphi_y^{(\alpha)}} e^{\frac{1}{2}(\varphi,J\varphi)} \prod_z g_z(\varphi_z^2 + 2t_z)\, d\varphi_z\,. \quad (9.57)$$

Taking $F(\varphi) = \varphi_{x_2}^{(\alpha_2)}$ in Proposition 9.2, we obtain the fundamental identity for the 2-point function

$$\begin{aligned} \langle \varphi_{x_1}^{(\alpha_1)} \varphi_{x_2}^{(\alpha_2)} \rangle &= \delta^{\alpha_1\alpha_2} \sum_{\omega:\, x_1 \to x_2} J^\omega \int d\nu_\omega(\underline{t})\, \mathcal{Z}(\underline{t}) \\ &= \delta^{\alpha_1\alpha_2} \sum_{\omega:\, x_1 \to x_2} \varrho(\omega)\,, \end{aligned} \quad (9.58)$$

where $\mathcal{Z}$ and ϱ are defined as in (9.31)–(9.33). Similarly, taking $F(\varphi) = \varphi_{x_2}^{(\alpha_2)} \varphi_{x_3}^{(\alpha_3)} \varphi_{x_4}^{(\alpha_4)}$ and using Proposition 9.2 twice, we get

$$\langle \varphi_{x_1}^{(\alpha_1)} \varphi_{x_2}^{(\alpha_2)} \varphi_{x_3}^{(\alpha_3)} \varphi_{x_4}^{(\alpha_4)} \rangle \;=\; \delta^{\alpha_1\alpha_2}\delta^{\alpha_3\alpha_4} \sum_{\substack{\omega_1:\; x_1 \to x_2 \\ \omega_2:\; x_3 \to x_4}} \varrho(\omega_1, \omega_2)$$

$$+ \text{ two permutations} . \tag{9.59}$$

The details, and the generalization to an arbitrary $2n$-point function, are left to the reader. ∎

9.2.3 The ARW representation for the Ising model

The correlation functions of the Ising model are

$$\langle \sigma_{x_1} \ldots \sigma_{x_m} \rangle \;=\; \frac{1}{Z} \operatorname{trace} \left[\sigma_{x_1} \ldots \sigma_{x_m} \, e^{\frac{1}{2}\sum_{x,y} J_{xy}\sigma_x\sigma_y} \right] \tag{9.60}$$

where trace $= \prod_{x\in\Lambda}[\frac{1}{2}\sum_{\sigma_x=\pm 1}]$ and

$$Z \;=\; \operatorname{trace} \left[e^{\frac{1}{2}\sum_{x,y} J_{xy}\sigma_x\sigma_y} \right] . \tag{9.61}$$

The ARW representation of the Ising model [5, 8] is obtained as a resummation of a random-current expansion. The existence of this underlying current representation is responsible for additional flexibility and certain supporting results not shared by the BFS representation, but also for the lack of symmetry of the weights and some degree of nonuniqueness in the definition of the model. Indeed, the random-walk representation is obtained by specifying a precise rule to associate to each random current configuration a unique family of walks. This can be done in many ways, each one defining a different representation. The one that seems geometrically most convenient is the one adopted in [8], which restricts the allowed walks to those satisfying certain consistency conditions that we proceed to discuss.

For each $x \in \mathbb{L}$ we fix an order for the set of steps having x as initial site. Each step (x, y) has associated a set of *cancelled* bonds formed by $\{x, y\}$ itself and all the bonds $\{x, z\}$ such that $(x, z) < (x, y)$ in the above order. A sequence of steps is said to be *consistent* if no step of the sequence uses a bond cancelled by a previous step. A walk is said to be consistent if the associated sequence of steps is consistent. Two consistent walks ω_1, ω_2 are *compatible* (in the given order) if $\omega_1 \circ \omega_2$ is consistent. Walks may be compatible in one order but not in another. Note, however, that if they are nonoverlapping they are compatible in any order. Two additional pieces of notation are needed: for a sequence of steps $\underline{\omega}$, we will denote $\widetilde{\omega}$ the set of bonds cancelled by $\underline{\omega}$; and if A is a set of bonds, Z_A will denote the partition function for the less ferromagnetic system obtained by setting to zero all the coupling constants for bonds not in A .

The weights for the ARW model are:

$$\varrho(\omega_1, \ldots, \omega_k) \;=\; \widetilde{\varrho}(\underline{\omega}_1 \circ \ldots \circ \underline{\omega}_k) \prod_{i=1}^{k} \mathrm{I}[\omega_i \text{ visits only once its final site}] \,, \tag{9.62}$$

where for a sequence of steps $\underline{\omega}$

$$\tilde{\varrho}(\underline{\omega}) \;=\; (\tanh J)^{\underline{\omega}}\ \mathrm{I}[\underline{\omega} \text{ is consistent}]\ (\cosh J)^{\widetilde{\omega}}\ \frac{Z_{\underset{\sim}{\omega}}}{Z}\ . \qquad (9.63)$$

Here I$[Q]$ is the indicator function that takes the value 1 if the condition Q is satisfied and 0 otherwise. We see that the factor involving $\tanh J$ in (9.63) is analogous to the "trivial" J-dependence of the previous random-walk models, but the rest of the weight is more complicated. In particular the requirement of consistency makes the weight not symmetric with respect to permutations of walks or reversals of a walk. We see from (9.62) that zero-step walks are irrelevant for the model. Indeed, if $|\omega_i| = 0$, then

$$\varrho(\omega_1, \ldots, \omega_{i-1}, \omega_i, \omega_{i+1}, \ldots, \omega_k) \;=\; \varrho(\omega_1, \ldots, \omega_{i-1}, \omega_{i+1}, \ldots, \omega_k). \qquad (9.64)$$

In this regard, sequences of steps rather than walks are the fundamental objects of the ARW expansion for the zero-field case.

It may also be instructive to remark that in terms of random currents, the factor $(\cosh J)^{\widetilde{\omega}} Z_{\underset{\sim}{\omega}}/Z$ in (9.63) becomes the probability of having current configurations with even flux numbers on $\widetilde{\omega}$ [8, 12], and hence it is smaller than one. Therefore, we have the simple upper bound:

$$\tilde{\varrho}(\underline{\omega}) \;\leq\; (\tanh J)^{\underline{\omega}}\ \mathrm{I}[\underline{\omega} \text{ is consistent}]\ . \qquad (9.65)$$

The *fundamental identity* relating an Ising modelto the corresponding ARW model is [5, 8]:

$$\langle \sigma_{x_1} \cdots \sigma_{x_{2m}} \rangle \;=\; S_{2m}(x_1, \ldots, x_{2m}) \qquad (9.66)$$

provided that the sites $x_1, \ldots, x_{2m}$ are all distinct. When some sites coincide the Green functions S_{2m} are definitely larger than the Ising correlations — except for the 2-point Green function (in zero magnetic field), which is equal to the Ising 2-point correlation even for coincident sites, i.e. $S_2(x,x) = \langle \sigma_x \sigma_x \rangle = 1$.

It is certainly possible to modify the weights so that the fundamental identity (9.66) holds even if some sites are the same. This can be done for instance, by adding some extra restrictions in the weights that prohibit pairs of walks of *non-zero* length with some repeated endpoint. However, this would notoriously complicate the expression for the weights and destroy the simplicity of the path-splitting inequalities for the ARW model (Section 11.2 below). The most important side effect of having the fundamental identity only for different sites appears when studying the truncated correlations (Chapter 12 below). The expressions obtained for the truncated Green functions S_{2n}^T translate into statements for the truncated Ising correlations only if the sites are noncoincident. To obtain general expressions for the Ising correlations one must in addition analyze on a case-by-case basis the situations in which several spin sites are the same. As all the results reviewed in this work involve only the truncated correlations of low order, this situation is not much of a burden. On the other hand, the random-walk formalism is not an appropriate tool to study the truncated correlations of higher order; one must resort to the more detailed random-current expansion (see Sections 12.5 and 12.6, and [460]).

Let us finally remark that the identity (9.66) proves that the Green functions for the ARW model with noncoincident arguments are symmetric under permutation of arguments — a fact far from obvious given the lack of symmetry of the weights.

10. Random-walk models in the presence of a magnetic field

We now consider a possibly site-dependent "magnetic field", i.e. a vector $\underline{h} = (h_x)_{x \in \mathbb{L}}$. In the presence of a magnetic field the random-walk expansion of $S_n(x_1, \ldots, x_n)$ undergoes a double alteration. On the one hand we must also include paths connecting one x_i with an "external magnetic field". This means that the walks may end up not only at the points $x_1, \ldots, x_n$ ("lattice sources"), but also at any other point in the lattice, where they may be thought to reach an "h-source". This is nothing more than the familiar combinatorics of the Gaussian model (free field) in nonzero magnetic field. On the other hand the expression for the weight depends on the kind of source each path has as endpoint, and it may happen that the same walk must be counted twice with different weights. We are forced to complicate the notation to distinguish the two cases.

For simplicity we write the formulae for the one-component case; the generalization to the N-component case is a simple matter of inserting internal indices and their corresponding Kronecker deltas. In order that the weights $\varrho^{[n]}$ be nonnegative, we assume henceforth that the magnetic fields h_x are all nonnegative.

10.1 General definitions

We will associate to each path ω_i a subscript j_i which takes the value 1 if the path ends up at an h-source and 0 otherwise. Hence now the weights are functions: $\varrho^{[k]}\colon [\Omega \times \{0,1\}]^k \to \mathbb{R}$, with $\varrho^{[0]} \equiv 1$. We shall denote

$$\varrho_{j_1 \ldots j_k}(\omega_1, \ldots, \omega_k) \equiv \varrho^{[k]}[(\omega_1, j_1), \ldots, (\omega_k, j_k)] \tag{10.1}$$

The weights are understood to depend on h in such a way that

$$\varrho_{j_1 \ldots j_k}(\omega_1, \ldots, \omega_k) = 0 \quad \text{whenever } j_m = 1 \text{ and } h_{e(\omega_m)} = 0$$

for one or more m $(1 \le m \le k)$.

There are now many more kernels than in the zero-field case. For instance, for two sites $x_1, x_2 \in \mathbb{L}$ we have the kernel

$$K(x_1, x_2) = \sum_{\omega: x_1 \to x_2} \varrho_0(\omega) \tag{10.2}$$

analogous to (9.3); but also we will consider

$$\sum_{z_1, z_2 \in \mathbb{L}} \sum_{\substack{\omega_1: x_1 \to z_1 \\ \omega_2: x_2 \to z_2}} \varrho_{11}(\omega_1, \omega_2) , \tag{10.3}$$

which can be interpreted as the kernel obtained by pairing each of the sites x_1, x_2 with all possible h-sources. For the sake of compactness we shall adopt a somewhat symbolic notation, denoting with a "$\bullet$" a dummy site that must be summed over the whole lattice. For example

$$\sum_{x \to \bullet} \equiv \sum_{z \in \mathbb{L}} \sum_{\omega: x \to z} \tag{10.4}$$

Correspondingly, we shall denote the kernel (10.3) as $K(x_1 \bullet | x_2 \bullet)$; so (10.3) takes the form

$$K(x_1 \bullet | x_2 \bullet) = \sum_{\substack{\omega: x_1 \to \bullet \\ \omega: x_2 \to \bullet}} \varrho_{11}(\omega_1, \omega_2) . \tag{10.5}$$

In general, given sites $x_1, \ldots, x_n$ one must consider all kernels obtained by pairing these sites and a number $2k - n$ of "$\bullet$"; with $0 \leq 2k - n \leq n$. To write the expression of such objects consider the set $N_n^* = \{1, \ldots, n\} \cup \{\infty\}$, with the total order defined by the standard order of natural numbers plus the relation $\infty > i \ \forall i \in N$. For each $2k \geq n$ let $P_{2k,n}$ be the set of "arrangements of a set with n distinguishable and $2k - n$ indistinguishable elements", i.e. the set of surjective maps π: $\{1, \ldots, 2k\} \to N_n^*$ such that $\#[\pi^{-1}(\infty)] = 2k - n$. If we adopt the convention

$$x_\infty \equiv \bullet \tag{10.6}$$

each $\pi \in P_{2k,n}$ defines a pairing analogous to (9.4) of the set formed by the sites $x_1, \ldots, x_n$ and $2k - n$ bullets. Not all these pairings go as arguments of the kernels K, because we are not interested in h-source–h-source pairing. To define the Green functions, we must again decide which representative of each class of maps π we choose. We do a choice analogous to the one for $h = 0$. Denote $Q_{2k,n}$ the set of $\pi \in P_{2k,n}$ satisfying:

$$\begin{aligned} \pi(2k-1) &< \pi(2k) & k &= 1, \ldots, n \\ \pi(2k-1) &< \pi(2k+1) & k &= 1, \ldots, n-1 \end{aligned} \tag{10.7}$$

Then, for each $\pi \in Q_{2k,n}$ $(2k \leq n)$:

$$K(x_{\pi(1)} x_{\pi(2)} | \ldots | x_{\pi(2k-1)} x_{\pi(2k)}) = \sum_{\substack{\omega_1: x_{\pi(1)} \to x_{\pi(2)} \\ \vdots \\ \omega_k: x_{\pi(2k-1)} \to x_{\pi(2k)}}} \varrho_{j_1 \ldots j_k}(\omega_1, \ldots, \omega_k) \tag{10.8}$$

where

$$j_i = \begin{cases} 1 & \text{if } \pi(2i) = \infty \\ 0 & \text{otherwise} \end{cases} \tag{10.9}$$

The Green functions are:

$$S_n(x_1, \ldots, x_n) = \sum_{k=[(n+1)/2]}^{n} \sum_{\pi \in Q_{2k,n}} K(x_{\pi(1)} x_{\pi(2)} | \ldots | x_{\pi(2k-1)} x_{\pi(2k)}) \tag{10.10}$$

where $[m]$ is the integer part of m. In the absence of a magnetic field we see from the comment following (10.1) that (10.10) reduces to (9.7) with $\varrho(\omega_1, \ldots, \omega_n) = \varrho_{0\ldots0}(\omega_1, \ldots, \omega_n)$.

10.2 Examples

We remind the reader that $b(\omega)$ and $e(\omega)$ denote, respectively, the initial and final points of the walk ω.

10.2.1 Polymer-chain models

We define the weights for the polymer-chain models with a magnetic field so as to preserve the useful identity between simple CIW models and the $N \to 0$ limit of n-component spin models. The weights are:

$$\varrho_{j_1\ldots j_k}(\omega_1, \ldots, \omega_k) = \left(\prod_{i=1}^{k} h_{e(\omega_i)}^{j_i} \right) \varrho^{(h)}(\omega_1, \ldots, \omega_k) \tag{10.11}$$

with

$$\begin{aligned} \varrho^{(h)}(\omega_1, \ldots, \omega_k) &= \frac{1}{Z} J^{\omega_1 + \ldots + \omega_k} \sum_{m=0}^{\infty} \frac{1}{m!} \left(\frac{1}{2}\right)^m \sum_{\omega'_1, \ldots, \omega'_m \in \Omega} J^{\omega'_1 + \ldots + \omega'_m} \\ &\times e^{-U_{m+k}(\omega'_1, \ldots, \omega'_m, \omega_1, \ldots, \omega_k)} \prod_{i=1}^{m} h_{b(\omega'_i)} h_{e(\omega'_i)} \end{aligned} \tag{10.12}$$

Here U is the same interaction energy as for the corresponding zero-field model (e.g. (9.17) for simple CIW models) and Z is the normalization factor:

$$Z = \sum_{m=0}^{\infty} \frac{1}{m!} \left(\frac{1}{2}\right)^m \sum_{\omega'_1, \ldots, \omega'_m \in \Omega} J^{\omega'_1 + \ldots + \omega'_m} e^{-U_m(\omega'_1, \ldots, \omega'_m)} \prod_{i=1}^{m} h_{b(\omega'_i)} h_{e(\omega'_i)} . \tag{10.13}$$

We notice that now the models have acquired a "deep" J-dependence.

The partition function (10.13) can be interpreted as the grand partition function for a gas of open walks with interaction energy U, bond activity J_{xy} and chain end activity $2^{-1/2} h_x$.

10.2.2 Baby polymer-chain models

We have written (10.11) in a way that stresses the existence of two different kinds of h-dependence. The first factor is the trivial h-dependence while the second factor includes a "deep" h-dependence. Again, each dependence is related to different features: the deep dependence is specific to each model (choice of U), while the trivial one embodies the new combinatorial aspects, common to all models, brought about by the existence of a magnetic field.

To study the latter type of properties, we have found it convenient to define some (unphysical) models with only the trivial h-dependence. We call such models "baby" models. In particular we shall consider the baby polymer-chain models which are defined by the weights

$$\varrho_{j_1\ldots j_k}(\omega_1,\ldots,\omega_k) = \left[\prod_{i=1}^{k} h_{e(\omega_i)}^{j_i}\right] \varrho(\omega_1,\ldots,\omega_k) \tag{10.14}$$

where $\varrho(\omega_1,\ldots,\omega_k)$ are the zero-field weights (9.11). We shall use the terminology inherited from the zero-field models. Hence, baby (generalized) CIW models are those whose interaction energy satisfies (9.14), and baby simple CIW models those satisfying (9.17). A baby model is repulsive if (9.12) holds.

The most important simplifying property of these baby polymer-chain models is that their weights conserve the factorization properties of the weights of the corresponding zero-magnetic-field models (Chapter 11).

10.2.3 BFS random-walk models

Recall that for simplicity we are considering one-component models (the generalization to N-component models is easy). The Hamiltonian of the spin system is therefore

$$H = -\frac{1}{2}\sum_{x,y} J_{xy}\varphi_x\varphi_y - \sum_x h_x\varphi_x . \tag{10.15}$$

The corresponding BFS random-walk model [223] has weights of the form (10.11) where $\varrho^{(h)}$ is given by the analogue of (9.31)–(9.33) but with the new Hamiltonian (10.15). That is,

$$\varrho^{(h)}(\omega_1,\ldots,\omega_k) = J^{\omega_1+\ldots+\omega_k} \int \mathcal{Z}(\underline{t}_1+\ldots+\underline{t}_k) \prod_{i=1}^{k} d\nu_{\omega_i}(\underline{t}_i) \tag{10.16}$$

with $\underline{t} = (t_x)_{x\in\mathrm{L}}$,

$$\mathcal{Z}(\underline{t}) \equiv \frac{1}{Z}\int \exp\left[\frac{1}{2}\sum_{x,y} J_{xy}\varphi_x\varphi_y + \sum_x h_x\varphi_x\right] \prod_{x\in\mathrm{L}} g_x(\varphi_x^2+2t_x)\,d\varphi_x \tag{10.17}$$

$$= \left\langle \prod_{x\in\mathrm{L}} \frac{g_x(\varphi_x^2+2t_x)}{g_x(\varphi_x^2)} \right\rangle \tag{10.18}$$

and $d\nu_\omega(\underline{t})$ is as before. The *fundamental identity* relating the spin system to the corresponding BFS random-walk model, now valid for both even and odd n, is

$$\langle \varphi_{x_1} \cdots \varphi_{x_n} \rangle = S_n(x_1, \ldots, x_n) . \tag{10.19}$$

For the convenience of the reader, we sketch the proof of (10.19) at the end of this section.

There is also a "loop" expansion analogous to (9.39): for an N-component model, it reads

$$\varrho^{(h)}(\omega_1, \ldots, \omega_k) =$$

$$J^{\omega_1 + \ldots + \omega_k} \Bigg\{ \frac{1}{Z} \sum_{n,m=0}^{\infty} \frac{1}{n!} \frac{1}{m!} \left(\frac{N}{2}\right)^n \left(\frac{1}{2}\right)^m$$

$$\sum_{\substack{v_1, \ldots, v_n \in \mathbb{L} \\ x_1, y_1 \in \mathbb{L} \\ \vdots \\ x_m, y_m \in \mathbb{L}}} \quad \sum_{\substack{\underset{\sim}{\omega}_1 :\, v_1 \to v_1 \\ \vdots \\ \underset{\sim}{\omega}_n :\, v_n \to v_n \\ \omega'_1 :\, x_1 \to y_1 \\ \vdots \\ \omega'_m :\, x_m \to y_m}} \frac{J^{\omega'_1 + \ldots + \omega'_m + \underset{\sim}{\omega}_1 + \ldots + \underset{\sim}{\omega}_n}}{|\underset{\sim}{\omega}_1| \cdots |\underset{\sim}{\omega}_n|}$$

$$\times \prod_{i=1}^{m} [h_{b(\omega'_i)} h_{e(\omega'_i)}] \, e^{-U^N_{m+n+k}(\omega_1, \ldots, \omega_k, \omega'_1 + \ldots + \omega'_m, \underset{\sim}{\omega}_1, \ldots, \underset{\sim}{\omega}_n)} \Bigg\} \tag{10.20}$$

with

$$Z = \varrho(\varnothing) , \tag{10.21}$$

while the interaction U is the same as in the zero-field case (9.39).As we intended, the "mature" CIW models (10.11) are recovered in the $N \to 0$ limit.

For the same reason as in the zero-field case, *we adopt henceforth the convention that "BFS random-walk model" means "BFS random-walk model with single-spin measure of BFS class".*

PROOF OF THE FUNDAMENTAL IDENTITY (10.19). It is instructive to consider first the Gaussian case. Setting $F(\varphi) = G(\varphi) e^{(h,\varphi)}$ in Lemma 9.1, we obtain

$$\int \varphi_x G(\varphi) \, e^{-\frac{1}{2}(\varphi, M\varphi) + (h,\varphi)} \, d\varphi$$

$$= \sum_y (M^{-1})_{xy} \int \left[\frac{\partial G}{\partial \varphi_y} + h_y G(\varphi) \right] e^{-\frac{1}{2}(\varphi, M\varphi) + (h,\varphi)} \, d\varphi . \tag{10.22}$$

Setting successively $G(\varphi) = 1$, $G(\varphi) = \varphi_{x_2}$, $G(\varphi) = \varphi_{x_2}\varphi_{x_3}$, ... we obtain the usual formulae for the moments of a (normalized) Gaussian measure with nonzero magnetic field:

$$\langle \varphi_x \rangle = (M^{-1}h)_x \tag{10.23a}$$

$$\langle \varphi_{x_1} \varphi_{x_2} \rangle = (M^{-1})_{x_1 x_2} + (M^{-1}h)_{x_1} \langle \varphi_{x_2} \rangle$$

$$= (M^{-1})_{x_1x_2} + (M^{-1}h)_{x_1}(M^{-1}h)_{x_2} \tag{10.23b}$$

$$\begin{aligned}\langle\varphi_{x_1}\varphi_{x_2}\varphi_{x_3}\rangle &= (M^{-1})_{x_1x_2}\langle\varphi_{x_3}\rangle + (M^{-1})_{x_1x_3}\langle\varphi_{x_2}\rangle + (M^{-1}h)_{x_1}\langle\varphi_{x_2}\varphi_{x_3}\rangle \\ &= (M^{-1})_{x_1x_2}(M^{-1}h)_{x_3} + (M^{-1})_{x_1x_3}(M^{-1}h)_{x_2} \\ &\quad + (M^{-1})_{x_2x_3}(M^{-1}h)_{x_1} + (M^{-1}h)_{x_1}(M^{-1}h)_{x_2}(M^{-1}h)_{x_3}\end{aligned} \tag{10.23c}$$

Clearly the combinatorics is exactly what was described in Section 10.1: each site x_i is connected by a "propagator" M^{-1} either to another site x_j or else to a magnetic field.

Exactly the same combinatorics is obtained when we substitute $F(\varphi) = G(\varphi)e^{(h,\varphi)}$ into the non-Gaussian integration-by-parts formula, Proposition 9.2. Instead of "propagators" M^{-1} we obtain sums over walks ω_i with weights $\varrho^{(h)}(\omega_1, \ldots, \omega_k)$.

10.2.4 ARW model

In the setting of the random-current expansion used to define the ARW model, the magnetic field can be incorporated by introducing a layer of "ghost" spins [12]. Then, the random-walk expansion is obtained by a resummation of the currents on this enhanced lattice. However, in order to facilitate the comparison with the previous random-walk models, we will not introduce here the enhanced lattice. Instead, in the expression for the weights we will have to include an extra factor [$T_{j_1\cdots j_k}$ in (10.27) below] that somewhat obscures the similarities with the h=0 case. In addition to this factor, the main difference from the zero-field case is that walks ω ending in an h-source (i.e. with $j = 1$) are not subjected to any constraint regarding the number of visits to the final site. (In the enhanced-lattice picture, the walk ω is completed by a single step from $e(\omega)$ to the corresponding ghost spin.) To write the expression of the weights we need some extra notation. For $\omega \in \Omega$, $j \in \{0,1\}$, let $\pi(\omega, j)$ denote the set

$$\pi(\omega, j) = \begin{cases} \mathrm{supp}(\omega) & \text{if } j = 1 \\ \mathrm{supp}(\omega \setminus \{e(\omega)\}) & \text{if } j = 0 \end{cases} \tag{10.24}$$

If A is a set of bonds and Λ a set of sites, a subscript A, Λ will indicate that the coupling constants of bonds not in A and the magnetic field for sites not in Λ are set equal to zero.

For an Ising model with Hamiltonian

$$H = -\sum_{x,y} J_{xy}\sigma_x\sigma_y - \sum_x h_x\sigma_x , \tag{10.25}$$

the *fundamental identity*

$$\langle\sigma_{x_1}\cdots\sigma_{x_n}\rangle = S_n(x_1, \ldots, x_n) \tag{10.26}$$

holds with [12]

$$\varrho_{j_1 \dots j_k}(\omega_1, \dots, \omega_k) = \prod_{i=1}^{k} (\mathrm{I}[\omega_i \text{ visits only site}])^{1-j_i} \times T^{(h)}_{j_1 \dots j_k}(\omega_1, \dots, \omega_k) \widetilde{\varrho}^{(h)}(\underline{\omega}_1 \circ \dots \circ \underline{\omega}_k) \quad (10.27)$$

with

$$T^{(h)}_{j_1 \dots j_k}(\omega_1, \dots, \omega_k) = \prod_{i=1}^{k} \Big\{ [\tanh h_{e(\omega_i)}]^{j_i} \prod_{x \in \pi(\omega_i, j_i)} \cosh h_x \Big\} \times \frac{Z^{(h)}_{[(\omega_1 \circ \dots \circ \omega_k)^{\sim}]^c, [\pi(\omega_1, j_1) \cup \dots \cup \pi(\omega_k, j_k)]^c}}{Z^{(h)}_{[(\omega_1 \circ \dots \circ \omega_k)^{\sim}]^c}} \quad (10.28)$$

and $\widetilde{\varrho}^{(h)}$ given by (9.63) but including the magnetic field in the partition functions. An important difference with previous models is the very complicated dependence of the weights on the values of j_i. We also note that the analog of property (9.64) is valid only if the zero-step walk ω_i has an associated $j_i = 0$. (Again, this distinction is overcome in the enhanced-lattice picture in which the ARW expansion takes the form of an expansion in sequences of steps as for the zero-field case.)

In the random-current formalism the RHS of (10.28) is equal to the product $\prod_{i=1}^{k} [\tanh h_{e(\omega_i)}]^{j_i}$ times the probability that the flux numbers of the currents for the h-bonds "on top" of $(\pi(\omega_i, j_i))_{1 \le i \le k}$ be even, for a system deprived of the bonds in $\omega_1 \circ \dots \circ \omega_k$. Bounding this probability by one we have (c.f. (9.65))

$$T^{(h)}_{j_1 \dots j_k}(\omega_1, \dots, \omega_k) \le \prod_{i=1}^{k} \Big[\tanh h_{e(\omega_i)} \Big]^{j_i} \quad (10.29)$$

11. Factorization and differentiation of the weights

We now discuss the behavior of the weights of the previous models with regard to the partition of a family of walks into subfamilies and to the splitting of a walk. We also discuss some identities and inequalities on the derivatives of the weights with respect to J or h. Most of the results of this chapter hold for arbitrary $h \geq 0$ (but the only polymer-chain models we consider are the *baby* ones). Many proofs are not given here, but can be found in the cited references.

11.1 Inequalities involving the partition of a family of walks

There are two types of inequalities of interest here: those expressing the *repulsiveness* (or repulsiveness "on the average") of the interaction between walks, and those expressing the *attractiveness* (or vanishing) of the interaction between nonoverlapping (or compatible) walks.

We remind the reader that *for the BFS model, the spins are assumed to have 1 or 2 components and to have single-spin measures in the BFS class* [cf. (9.46)].

Let us shorten our notation for sequences. If $I = (i_1, \ldots, i_k)$ is an ordered k-tuple of indices (from some arbitrary index set), then $\omega_I \equiv (\omega_{i_1}, \ldots, \omega_{i_k})$, $j_I \equiv (j_{i_1}, \ldots, j_{i_k})$ and so forth. Moreover, if $I = (i_1, \ldots, i_k)$ and $I' = (i'_1, \ldots, i'_k)$, then $(\omega_I, \omega_{I'}) \equiv (\omega_{i_1}, \ldots, \omega_{i_k}, \omega_{i'_1}, \ldots, \omega_{i'_k})$, and so forth. If $I = (i_1, \ldots, i_k)$, we define $\mathrm{supp}(\omega_I) = \bigcup_{i=1}^{k} \mathrm{supp}(\omega_{i_k})$. We always allow our sequences to be empty ($k = 0$).

11.1.1 Repulsiveness (or repulsiveness "on the average")

For any repulsive polymer-chain model in zero magnetic field, it follows immediately from (9.11) and (9.12) that

$$\varrho(\omega_1, \ldots, \omega_{k+l}) \;\leq\; \varrho(\omega_1, \ldots, \omega_k)\; \varrho(\omega_{k+1}, \ldots, \omega_{k+l}) \,. \tag{11.1}$$

The same (with arbitrary subscripts $j_1, \ldots, j_{k+l}$) holds in nonzero magnetic field for any repulsive *baby* polymer-chain model.

We have therefore proven:

Theorem 11.1 *For repulsive baby polymer-chain models:*

$$\varrho_{j_{I_1} j_{I_2}}(\omega_{I_1}, \omega_{I_2}) \;\leq\; \varrho_{j_{I_1}}(\omega_{I_1}) \varrho_{j_{I_2}}(\omega_{I_2}) \,. \tag{11.2}$$

Remark. Although on the left-hand sides of (11.1)–(11.2) the walks are specified in a particular order — these of I_1 followed by those of I_2 — this order is obviously irrelevant, since the weights ϱ for a polymer-chain model are symmetric.

For the BFS and ARW models the situation is more complicated. A general repulsiveness inequality of the form (11.1) is quite simply *false*: for certain families of walks the inequality goes in the other direction (Theorem 11.7 and Corollary 11.9), and examples can easily be constructed in which the reverse inequality is strict. However, it does turn out [92, 213, 28, 223, 5, 12] that under suitable conditions the walks are repulsive "on the average". More precisely, an inequality resembling (11.1) can be obtained when one *sums* over combinations of weights that add up to a Green function S_n, since in this case the second Griffiths inequality ($N = 1$) [491, 252, 407, 462] or the Ginibre inequality ($N = 2$) [252, 148] can be applied to the corresponding spin-model correlation function via the fundamental identity (9.36), (9.66), (10.19) or (10.26).

To see what is going on, let us consider the weight $\varrho(\omega_1, \omega_2)$ in the one-component BFS models at zero magnetic field:

$$\varrho(\omega_1, \omega_2) \;=\; J^{\omega_1+\omega_2} \int \mathcal{Z}(\underline{t}_1 + \underline{t}_2)\; d\nu_{\omega_1}(\underline{t}_1)\, d\nu_{\omega_2}(\underline{t}_2) \tag{11.3}$$

where

$$\mathcal{Z}(\underline{t}) \;=\; \frac{1}{Z} \int e^{\frac{1}{2}(\varphi, J\varphi)} \prod_x g_x(\varphi_x^2 + 2t_x)\, d\varphi_x \,. \tag{11.4}$$

We now rewrite the integrand in (11.3) as

$$\begin{aligned} \varrho(\omega_1, \omega_2) \;&=\; J^{\omega_1} \int d\nu_{\omega_1}(\underline{t}_1)\, \mathcal{Z}(\underline{t}_1) \left[J^{\omega_2} \int d\nu_{\omega_2}(\underline{t}_2) \frac{\mathcal{Z}(\underline{t}_1 + \underline{t}_2)}{\mathcal{Z}(\underline{t}_1)} \right] \\ &=\; J^{\omega_1} \int d\nu_{\omega_1}(\underline{t}_1)\, \mathcal{Z}(\underline{t}_1)\, \varrho(\omega_2)_{\underline{t}_1} \,, \end{aligned} \tag{11.5}$$

where $\varrho_{\underline{t}}$ is defined in the same way as ϱ but with each $g_x(\varphi_x^2)$ replaced by $g_x(\varphi_x^2 + 2t_x)$. Unfortunately, we know very little about the monotonicity of $\varrho(\omega_2)_{\underline{t}_1}$ for *fixed* ω_2 as a function of $\underline{t}_1$. However, if we *sum* over $\omega_2 \colon x \to y$, we can use the fundamental identity (9.35), which holds also in a "background $\underline{t}$ field":

$$\sum_{\omega_2 : x \to y} \varrho(\omega_1, \omega_2) \;=\; J^{\omega_1} \int d\nu_{\omega_1}(\underline{t}_1)\, \mathcal{Z}(\underline{t}_1)\, \langle \varphi_x \varphi_y \rangle_{\underline{t}_1} \,, \tag{11.6}$$

where $\langle \cdot \rangle_{\underline{t}}$ denotes normalized expectation in a model whose single-spin measures are $g_x(\varphi_x^2 + 2t_x)\, d\varphi_x$. As we shall prove shortly (Lemma 11.3), the correlation functions $\langle \varphi \ldots \varphi \rangle_{\underline{t}}$ are monotone decreasing in $\underline{t}$, for single-spin measures of BFS class; in particular, they can be bounded above by the corresponding correlations without a background $\underline{t}$ field. Therefore, we can write

$$\begin{aligned}\sum_{\omega_2:x\to y} \varrho(\omega_1,\omega_2) &\leq J^{\omega_1} \int d\nu_{\omega_1}(\underline{t}_1)\, \mathcal{Z}(\underline{t}_1)\, \langle\varphi_x\varphi_y\rangle \\ &= \varrho(\omega_1) \sum_{\omega_2:x\to y} \varrho(\omega_2)\,. \end{aligned} \tag{11.7}$$

This inequality expresses the repulsiveness "on the average" between a pair of BFS walks; the point is that ω_2 must be summed over a class of walks that produces a correlation function via the fundamental identity (in this case $\omega_2 \colon x \to y$).

For a BFS model in nonzero magnetic field, the situation is slightly more complicated. The inequality (11.7) is no longer provable by the above argument, and may not even be true! This is because the sum over walks $\omega_2 \colon x \to y$ does *not* reconstruct a spin-model correlation function $\langle\varphi_x\varphi_y\rangle$; the relevant fundamental identities (10.19) are more complicated:

$$\langle\varphi_x\rangle = \sum_{\omega:x\to\bullet} \varrho_1(\omega) \tag{11.8}$$

$$\langle\varphi_x\varphi_y\rangle = \sum_{\omega:x\to y} \varrho_0(\omega) + \sum_{\substack{\omega:\, x\,\to\,\bullet \\ \omega':\, y\,\to\,\bullet}} \varrho_{11}(\omega,\omega') \tag{11.9}$$

and so forth. These identities determine the classes of walks that must be summed in order to obtain "repulsiveness on the average". For example, one has

$$\sum_{\omega_2:x\to\bullet} \varrho_{j1}(\omega_1,\omega_2) \leq \varrho_j(\omega_1) \sum_{\omega_2:x\to\bullet} \varrho_1(\omega_2)\,, \tag{11.10}$$

and

$$\begin{aligned}&\sum_{\omega_2:x\to y} \varrho_{j0}(\omega_1,\omega_2) + \sum_{\substack{\omega_2:\, x\,\to\,\bullet \\ \omega_3:\, y\,\to\,\bullet}} \varrho_{j11}(\omega_1,\omega_2,\omega_3) \\ &\leq \varrho_j(\omega_1)\Big[\sum_{\omega_2:x\to y} \varrho_0(\omega_2) + \sum_{\substack{\omega_2:\, x\,\to\,\bullet \\ \omega_3:\, y\,\to\,\bullet}} \varrho_{11}(\omega_2,\omega_3)\Big]\,. \end{aligned} \tag{11.11}$$

The general result is the following [92, 213, 28, 223]:

Theorem 11.2 *In the BFS model, let Γ be a family of pairs (ω_I, j_I) such that the fundamental identity (9.36)/(10.19) asserts that*

$$\sum_{(\omega_I,j_I)\in\Gamma} \varrho_{j_I}(\omega_I) = \langle\varphi_{x_1}^{(1)}\cdots\varphi_{x_n}^{(1)}\rangle \tag{11.12}$$

for some sites $x_1,\ldots,x_n$. Then, for any pair $(\omega_{I'}, j_{I'})$ we have

$$\sum_{(\omega_I, j_I) \in \Gamma} \varrho_{j_{I'} j_I}(\omega_{I'}, \omega_I) \;\leq\; \varrho_{j_{I'}}(\omega_{I'}) \sum_{(\omega_I, j_I) \in \Gamma} \varrho_{j_I}(\omega_I)\,. \tag{11.13}$$

Proof. The combinatorial structure of the proof is an obvious generalization of the steps from (11.3) to (11.7). It remains only to prove the claimed monotonicity of $\langle \varphi \ldots \varphi \rangle_{\underline{t}}$ in $\underline{t}$. We consider first the one-component case:

Lemma 11.3 [92] *Consider a one-component ferromagnetic spin system ($J_{xy} \geq 0$, $h_x \geq 0$) with single-spin measure of BFS class. For each site x, let F_x: $\mathbb{R} \to \mathbb{R}$ be a function that is either even or odd, and is nonnegative and increasing on $[0, \infty)$. Then, the expectation $\langle \prod_x F_x(\varphi_x) \rangle_{\underline{t}}$ is nonnegative and is a decreasing function of each t_y. In particular,*

$$0 \;\leq\; \left\langle \prod_x F_x(\varphi_x) \right\rangle_{\underline{t}} \;\leq\; \left\langle \prod_x F_x(\varphi_x) \right\rangle\,. \tag{11.14}$$

Proof. The nonnegativity of $\langle \prod F_x \rangle_{\underline{t}}$ is just Nelson's generalized form of Griffiths' first inequality [252, 407]. To prove the monotonicity, we assume that the single-spin distribution is of the form

$$g_x(\varphi_x^2) \;=\; e^{-f_x(\varphi_x^2)} \tag{11.15}$$

with $s \mapsto f_x(s)$ convex and growing at least linearly for s large. (Limits of such measures can be handled taking limits in the final formulae.) Then an easy calculation using the definition of $\langle \,\cdot\, \rangle_{\underline{t}}$ yields

$$\frac{\partial}{\partial t_y} \left\langle \prod_x F_x(\varphi_x) \right\rangle_{\underline{t}} \;=\; -2 \left\langle \prod_x F_x(\varphi_x)\,;\, f_y'(\varphi_y^2 + 2t_y) \right\rangle_{\underline{t}}\,. \tag{11.16}$$

Now, the quantity $f_y'(\varphi_y^2 + 2t_y)$ is an even function of φ_y, increasing for $\varphi_y \geq 0$. Therefore, Nelson's generalized form of the second Griffiths inequality ([252, 407] and [462, Theorem VIII.14A]) applies and shows that

$$\left\langle \prod_x F_x(\varphi_x)\,;\, f_y'(\varphi_y^2 + 2t_y) \right\rangle_{\underline{t}} \;\geq\; 0\,. \tag{11.17}$$

∎

For the 2-component case, we use the notation $\boldsymbol{\varphi}_x = (|\boldsymbol{\varphi}_x| \cos\theta_x, |\boldsymbol{\varphi}_x| \sin\theta_x)$. The result analogous to Lemma 11.3 is the following:

Lemma 11.4 *Consider a two-component ferromagnetic spin system ($J_{xy} \geq 0$, $h_x^{(1)} \geq 0$, $h_x^{(2)} = 0$) with single-spin measure of BFS class. For each site x, let F_x: $[0, \infty) \to [0, \infty)$ be an increasing function, and let m_x be an integer. Then, the expectation $\langle \prod_x F_x(|\boldsymbol{\varphi}|) \cos m_x\theta_x \rangle_{\underline{t}}$ is nonnegative and is a decreasing function of each t_y. In particular,*

$$0 \leq \left\langle \prod_x F_x(|\varphi|) \cos m_x \theta_x \right\rangle_{\underline{t}} \leq \left\langle \prod_x F_x(|\varphi|) \cos m_x \theta_x \right\rangle . \tag{11.18}$$

Proof. The proof is identical to that of Lemma 11.3, using Dunlop-Newman's form of the Ginibre inequality [148, Theorem 12] in place of Nelson's form of the Griffiths inequality. ∎

Remark. These proofs show clearly why the BFS class is the "natural" class of single-spin measures for proving correlation inequalities in the BFS random-walk model. Unfortunately, we are unable at present to extend these arguments to spin systems of more than two components, since the appropriate Griffiths inequality has not yet been proven [492].

For ARW models the situation is very similar, but there are two complicating issues: the non-symmetry of the weights ϱ and the need for non-coinciding arguments in the fundamental identity. Because of the non-symmetry of the weights, we must require that the walks to be summed over form a *contiguous block* within the list of arguments of ϱ. The restriction to non-coinciding arguments in the fundamental identity imposes a like restriction on the family Γ. The result is the following:

Theorem 11.5 *In the ARW model, let Γ be a family of pairs (ω_I, j_I) such that the fundamental identity (9.66)/(10.26) asserts that*

$$\sum_{(\omega_I, j_I)\in\Gamma} \varrho_{j_I}(\omega_I) = \langle \sigma_{x_1} \dots \sigma_{x_n} \rangle \tag{11.19}$$

for some sites $x_1, \dots, x_n$. (Recall that, in particular, the sites $x_1, \dots, x_n$ must be all different, *except for the case $n = 2$ in zero magnetic field.) Then, for any pairs (ω_{I_1}, j_{I_1}) and (ω_{I_2}, j_{I_2}), we have*

$$\sum_{(\omega_I, j_I)\in\Gamma} \varrho_{j_{I_1} j_I j_{I_2}}(\omega_{I_1}, \omega_I, \omega_{I_2}) \leq \varrho_{j_{I_1} j_{I_2}}(\omega_{I_1}, \omega_{I_2}) \sum_{(\omega_I, j_I)\in\Gamma} \varrho_{j_I}(\omega_I) . \tag{11.20}$$

The proof of a special case of Theorem 11.5 can be found in [12]; the general case is a straightforward extension.

No result analogous to Theorems 11.1, 11.2 and 11.5 exists so far for the "mature" CIW in the presence of magnetic field. One may expect that at least those simple CIW models that are the $N \to 0$ limit of BFS models may inherit an inequality like (11.13). However, there is at present no proof that the Griffiths inequalities are preserved in the $N = 0$ limit, and there are even suspicions that it may not be so [278].

The repulsiveness property stated in Theorems 11.1, 11.2 and 11.5 will lead to "Gaussian upper bound" correlation inequalities (Theorem 12.1).

11.1.2 Attractiveness (or noninteraction) between nonoverlapping (or compatible) walks

Attractiveness (or noninteraction) between nonoverlapping
(or compatible) walks

By definition of contact interaction, the interaction between nonoverlapping families of walks vanishes, so the weights factorize:

Theorem 11.6 *For baby (generalized) CIW models*

$$\mathrm{supp}(\omega_{I_1})\cap\mathrm{supp}(\omega_{I_2}) = \varnothing \implies \varrho_{j_{I_1}j_{I_2}}(\omega_{I_1},\omega_{I_2}) = \varrho_{j_{I_1}}(\omega_{I_1})\varrho_{j_{I_2}}(\omega_{I_2}) . \quad (11.21)$$

For the BFS and ARW models, the interaction between nonoverlapping families of walks is *attractive*:

Theorem 11.7

a) For BFS models

$$\mathrm{supp}(\omega_{I_1})\cap\mathrm{supp}(\omega_{I_2}) = \varnothing \implies \varrho_{j_{I_1}j_{I_2}}(\omega_{I_1},\omega_{I_2}) \geq \varrho_{j_{I_1}}(\omega_{I_1})\varrho_{j_{I_2}}(\omega_{I_2}) . \quad (11.22)$$

b) For ARW models

$$\begin{aligned} \omega_1\circ\ldots\circ\omega_{k+l} \text{ consistent} \implies \varrho_{j_1\ldots j_{k+l}}(\omega_1,\ldots,\omega_{k+l}) \geq \\ \varrho_{j_1\ldots j_k}(\omega_1,\ldots,\omega_k)\ \varrho_{j_{k+1}\ldots j_{k+l}}(\omega_{k+1},\ldots,\omega_{k+l}) \end{aligned} \quad (11.23)$$

Theorem 11.7(a) was proven in [213, 28, 223] for zero magnetic field, and in [223] for nonzero magnetic field. Theorem 11.7(b) was proven in [8] for zero field, and in [12] for nonzero field. Since [223] is (and forever will remain) unpublished, we give here the proof of Theorem 11.7(a):

Proof of Theorem 11.7(a). Note first that if x is a site *not* visited by ω, then the measure $d\nu_\omega(\underline{t})$ forces t_x to be 0. Theorem 11.7(a) is then an immediate consequence of the following lemma:

Lemma 11.8 *In a BFS model with 1 or 2 components and single-spin measure on BFS class,*

$$\mathcal{Z}(\underline{t}+\underline{t}') \geq \mathcal{Z}(\underline{t})\mathcal{Z}(\underline{t}') \quad (11.24)$$

where $\underline{t} = \{t_x\}$ *and* $\underline{t}' = \{t'_x\}$ *are nonnegative numbers satisfying* $t_x t'_x = 0$ *for all* $x \in \mathbb{L}$.

Proof. Assume that the single-spin distribution is of the form

$$g_x(\varphi_x^2) = e^{-f_x(\varphi_x^2)} \quad (11.25)$$

with $s \mapsto f_x(s)$ convex and growing at least linearly for s large. (Limits of such measures can be handled taking limits in the final formulae.) By the fundamental theorem of calculus,

$$\log \mathcal{Z}(\underline{t}+\underline{t}') = \log \mathcal{Z}(\underline{t}) + \int_0^1 d\alpha \frac{d}{d\alpha} \log \mathcal{Z}(\underline{t}+\alpha\underline{t}') . \tag{11.26}$$

We then have

$$\begin{aligned} \frac{d}{d\alpha} \log \mathcal{Z}(\underline{t}+\alpha\underline{t}') &= -2\sum_x t'_x \left\langle f'_x(\varphi_x^2 + 2t_x + 2\alpha t'_x) \right\rangle_{\underline{t}+\alpha\underline{t}'} \\ &= -2\sum_x t'_x \left\langle f'_x(\varphi_x^2 + 2\alpha t'_x) \right\rangle_{\underline{t}+\alpha\underline{t}'} \\ &\geq -2\sum_x t'_x \left\langle f'_x(\varphi_x^2 + 2\alpha t'_x) \right\rangle_{\alpha\underline{t}'} \\ &= \frac{d}{d\alpha} \log \mathcal{Z}(\alpha\underline{t}') . \end{aligned} \tag{11.27}$$

Here the first equality is an easy calculation using the definition of $\mathcal{Z}(\underline{t})$; the second inequality holds because by hypothesis $t_x = 0$ whenever $t'_x \neq 0$; the inequality is a consequence of Lemmas 11.3 and 11.4, since $f'_x(\varphi_x^2 + 2\alpha t'_x)$ is an increasing function of φ_x^2; and the final equality is just the first equality run backwards. Inserting (11.27) into (11.26) and integrating, we get

$$\log \mathcal{Z}(\underline{t}+\underline{t}') \geq \log \mathcal{Z}(\underline{t}) + \log \mathcal{Z}(\underline{t}') \tag{11.28}$$

■

Remark. In the φ^4 case

$$g_x(\varphi_x^2) = \exp\left[-\frac{\lambda}{4}\varphi_x^4 - \frac{\tau}{2}\varphi_x^2\right] , \tag{11.29}$$

the inequality (11.24) can be made more precise:

$$\mathcal{Z}(\underline{t}+\underline{t}') \geq \mathcal{Z}(\underline{t})\mathcal{Z}(\underline{t}') \exp\left[-2\lambda \sum_x t_x t'_x\right] \tag{11.30}$$

for *any* choice of nonnegative numbers $\{t_x\}$, $\{t'_x\}$. See [213, 28, 97] for a proof.

We remark that (11.22) is satisfied also by the ARW models because nonoverlapping families are compatible. Hence, we can more simply state:

Corollary 11.9 *For the baby (generalized) CIW, BFS and ARW models:*

$$\mathrm{supp}(\omega_{I_1}) \cap \mathrm{supp}(\omega_{I_2}) = \varnothing \Longrightarrow \varrho_{j_{I_1} j_{I_2}}(\omega_{I_1}, \omega_{I_2}) \geq \varrho_{j_{I_1}}(\omega_{I_1}) \varrho_{j_{I_2}}(\omega_{I_2}) . \tag{11.31}$$

with equality for the baby (generalized) CIW models.

Theorem 11.7 thus gives a bound for the BFS and ARW models which is opposite to that of Theorems 11.1, 11.2 and 11.5. These two results, taken together, are somewhat surprising: for example, (11.13) and (11.20) state that the interaction between two families of walks (summing over both overlapping and nonoverlapping cases) is "on the average" repulsive, while (11.31) states that in the nonoverlapping case the interaction is attractive. It is the interplay between these two bounds which leads to the profound results of [5, 213, 223, 12]; see Theorems 12.4 and 12.11 below.

11.2 Inequalities involving the splitting of a walk

The deepest applications of the random-walk formalism involve splitting a walk into two pieces (often at the point of intersection with a second walk). In this section we study the identities and inequalities for the weights ϱ which result from such a splitting.

We consider two ways in which a walk ω may be split in two pieces ω_1 and ω_2, according to whether an "intermediate site" or "intermediate step" is used.

1) *Intermediate site.* Let ω_1 and ω_2 be walks with $e(\omega_1) = b(\omega_2)$; then for the ARW model it follows immediately from the definitions (9.62) or (10.27) that

$$\mathrm{I}[\omega_1 \text{ visits } e(\omega_1) \text{ only once}]\ \varrho_j(\omega_1 \circ \omega_2) \;=\; \varrho_{0j}(\omega_1, \omega_2)\,. \tag{11.32}$$

In particular, if we are given the walk ω_1 and wish to sum over all walks ω_2 which run from $e(\omega_1)$ to a certain site y, we obtain

$$\mathrm{I}[n_{e(\omega_1)}(\omega_1) = 1] \sum_{\omega_2 : e(\omega_1) \to y} \varrho_j(\omega_1 \circ \omega_2) \;=\; \sum_{\omega_2 : e(\omega_1) \to y} \varrho_{0j}(\omega_1, \omega_2)\,. \tag{11.33}$$

The identities (11.32) and (11.33) indicate that intermediate-site splitting is very natural for the ARW model. But it is less natural for the simple CIW and BFS models: Notice first that

$$n_x(\omega_1 \circ \omega_2) = \begin{cases} n_x(\omega_1) + n_x(\omega_2) & \text{if } x \neq e(\omega_1) \\ n_x(\omega_1) + n_x(\omega_2) - 1 & \text{if } x = e(\omega_1) \end{cases} \tag{11.34}$$

The last line is due to the fact that when ω_1 and ω_2 are considered separately, the site $e(\omega_1)$ is visited one extra time as the initial point of ω_2. This discrepancy between $n_x(\omega_1 \circ \omega_2)$ and $n_x(\omega_1) + n_x(\omega_2)$ makes intermediate-site splitting inconvenient for the CIW and BFS models (except for the purpose of deriving λ-dependent "skeleton inequalities", where it is exactly what is needed [97, 75, 74, 292]). However, for the baby SAW a direct argument yields

$$\varrho_j(\omega_1 \circ \omega_2) \;=\; \varrho_0(\omega_1)\varrho_j(\omega_2)\, \mathrm{I}[\omega_1 \cap \omega_2 = \{e(\omega_1)\}] \leq \varrho_0(\omega_1)\varrho_j(\omega_2)\,, \tag{11.35}$$

and so intermediate-site splitting can be employed in this case.

2) *Intermediate step.* Let ω_1 and ω_2 be walks; we consider concatenating ω_1 and ω_2 with a single step $(e(\omega_1), b(\omega_2))$ in-between. We get

$$n_x(\omega_1 \circ (e(\omega_1), b(\omega_2)) \circ \omega_2) \;=\; n_x(\omega_1) + n_x(\omega_2) \tag{11.36}$$

for *all* x [in contrast to (11.34)], and thus for the simple baby CIW and BFS models we find

$$\varrho_j(\omega_1 \circ (e(\omega_1), b(\omega_2)) \circ \omega_2) = J_{e(\omega_1),b(\omega_2)}\, \varrho_{0j}(\omega_1, \omega_2) \tag{11.37}$$

In particular, if we are given a walk ω_1 and wish to sum over all possible ways of completing ω_1 beyond its endpoint, we find

$$\sum_{\omega' : e(\omega_1) \to y} \varrho_j(\omega_1 \circ \omega') = \varrho_j(\omega_1)\delta_{e(\omega_1),y} \;+\; \sum_z \sum_{\omega_2 : z \to y} J_{e(\omega_1),z}\, \varrho_{0j}(\omega_1, \omega_2)\;. \tag{11.38}$$

The first term on the right hand side comes from a zero-length walk ω'; the second term comes from all other walks $\omega' \equiv (e(\omega_1), z) \circ \omega_2$.

It should be noted that an intermediate-step splitting expression analogous to (11.37) holds also for the ARW model, except that it is an inequality rather than an identity. Indeed, using the probabilistic interpretation of some of the factors of the weights (10.27)/(9.63) [see remark preceding (9.65)], it is not hard to see that

$$\mathrm{I}[n_{e(\omega_1)}(\omega_1) = 1]\, \varrho_j(\omega_1 \circ (e(\omega_1), b(\omega_2)) \circ \omega_2) \;\le\; \tanh(J_{e(\omega_1),b(\omega_2)})\, \varrho_{0j}(\omega_1, \omega_2) \tag{11.39}$$

and hence an expression analogous to (11.38) holds for the ARW model but replacing "=" by "$\le$" and $J_{e(\omega_1),z}$ by $\tanh(J_{e(\omega_1),z})$.

The general form of the path-splitting results is:

Proposition 11.10

a) For the simple baby CIW and BFS random-walk models

$$\sum_{\omega' :\, e(\omega) \to y} \varrho_{jj_I}(\omega \circ \omega', \omega_I) = \varrho_{jj_I}(\omega, \omega_I)\delta_{e(\omega),y} \;+\; \sum_z \sum_{\omega' : z \to y} J_{e(\omega),z}\, \varrho_{0jj_I}(\omega, \omega', \omega_I) \tag{11.40}$$

[intermediate-step splitting].

b) For the ARW model

$$\mathrm{I}[n_{e(\omega)}(\omega) = 1] \sum_{\omega' : e(\omega) \to y} \varrho_{j_{I_1} j j_{I_2}}(\omega_{I_1}, \omega \circ \omega', \omega_{I_2}) = \sum_{\omega' : e(\omega) \to y} \varrho_{j_{I_1} 0 j j_{I_2}}(\omega_{I_1}, \omega, \omega', \omega_{I_2}) \tag{11.41}$$

[intermediate-site splitting]. Moreover, we also have the following intermediate-step splitting inequality

$$\mathrm{I}[n_{e(\omega)}(\omega) = 1] \sum_{\omega' : e(\omega) \to y} \varrho_{j_{I_1} j j_{I_2}}(\omega_{I_1}, \omega \circ \omega', \omega_{I_2}) \;\le\; \varrho_{j_{I_1} j j_{I_2}}(\omega_{I_1}, \omega, \omega_{I_2})\delta_{e(\omega),y}$$

$$+ \sum_z \sum_{\omega' : z \to y} \tanh(J_{e(\omega),z})\, \varrho_{j_{I_1} 0 j j_{I_2}}(\omega_{I_1}, \omega, \omega', \omega_{I_2}) \tag{11.42}$$

c) For the baby SAW model, if $e(\omega_1) = b(\omega_2)$ *then*

$$\varrho_{jjI}(\omega_1 \circ \omega_2, \omega_I) = \varrho_0(\omega_1)\, \varrho_{jjI}(\omega_2, \omega_I)\, \mathrm{I}[\omega_1 \cap \omega_2 = \{e(\omega_1)\}] \,. \tag{11.43}$$

[intermediate-site splitting].

These expressions imply several path-splitting inequalities. We shall concentrate in a very useful particular case which involves sums of the form

$$\sum_{\substack{\omega:\, x \to y \\ \mathrm{supp}(\omega) \ni z}} \varrho_j(\omega) \tag{11.44}$$

for some fixed site z. In such situation we can split the walk ω at z. In order to have a well-defined splitting, we have to specify at *which* visit to z (in case there is more than one) the walk is to be split. For example, we could split ω at its first visit to z, or at its last visit to z, etc. In order to avoid problems with the ARW model, we choose to split the path the *first* time it hits z (so the indicator function in the LHS of (11.32)/(11.41) becomes unity). In such a fashion we obtain from the previous proposition that for the baby CIW and BFS models ($h = 0$):

$$\sum_{\substack{\omega:\, x \to y \\ \mathrm{supp}(\omega) \ni z}} \varrho_j(\omega) = \sum_{\omega':x \to z} \mathrm{I}[n_z(\omega') = 1] \left\{\varrho_j(\omega')\delta_{z,y} + \sum_{z'} J_{zz'} \sum_{\omega'':z' \to y} \varrho_{0j}(\omega', \omega'')\right\} . \tag{11.45}$$

For the ARW model intermediate-step splitting (11.42) yields an analogous *inequality*. But better, we also have an *identity* through intermediate-site splitting (11.41):

$$\begin{aligned} \sum_{\substack{\omega:\, x \to y \\ \mathrm{supp}(\omega) \ni z}} \varrho_j(\omega) &= \sum_{\substack{\omega':\, x \to z \\ \omega'':\, z \to y}} \mathrm{I}[n_z(\omega') = 1]\ \varrho_{0j}(\omega', \omega'') \\ &= \sum_{\substack{\omega':\, x \to z \\ \omega'':\, z \to y}} \varrho_{0j}(\omega', \omega'') \,, \end{aligned} \tag{11.46}$$

since the indicator function is built into the weight ϱ (cf. (9.62)).

We can summarize these formulae in terms of kernels. In (11.45) we bound the indicator function by unity, thereby obtaining an *inequality* (which is an equality for the SAW, because the indicator function *equals* unity by definition of such walks). In (11.46) we can obtain directly an *identity* in terms of kernels. The results are:

Proposition 11.11

a) For the baby simple CIW, BFS and ARW models:

$$\sum_{\substack{\omega:\, x \to y \\ \mathrm{supp}(\omega) \ni z}} \varrho_0(\omega) \le K(xz)\delta_{zy} + \sum_{z'} J_{zz'} K(xz|z'y) \tag{11.47}$$

and

$$\sum_{\substack{\omega: x \to \bullet \\ \mathrm{supp}(\omega) \ni z}} \varrho_1(\omega) \;\leq\; K(xz)h_z \;+\; \sum_{z'} J_{zz'} K(xz|z'\bullet) \qquad (11.48)$$

with equality in both cases for the baby SAW. For the ARW model $J_{zz'}$ and h_z can be replaced by $\tanh(J_{zz'})$ *and* $\tanh(h_z)$*, respectively.*

b) For the ARW model:

$$\sum_{\substack{\omega: x \to y \\ \mathrm{supp}(\omega) \ni z}} \varrho_0(\omega) \;=\; K(xz|zy) \qquad (11.49)$$

and

$$\sum_{\substack{\omega: x \to y \\ \mathrm{supp}(\omega) \ni z}} \varrho_1(\omega) \;=\; K(xz|z\bullet) \qquad (11.50)$$

11.3 Differentiation of the weights with respect to J or h

11.3.1 Differentiation with respect to J

As discussed above, we can distinguish between two types of J-dependence of the weights: the "trivial" and the "deep" dependence. It turns out that these dependences have opposite monotonicity properties: the trivial J-dependence is an increasing function of each J_{yz}, while the deep dependence is — for the models analyzed here — decreasing in the couplings. Therefore the derivative of the weights with respect to the coupling constants will have two contributions — T_1 and T_2 below — of opposite sign.

We shall discuss in some detail the simplest differentiation formula – namely that for $\frac{\partial \varrho_j(\omega)}{\partial J_{yz}}$ – which already exhibits all the relevant features. The general results stated in Theorem 11.12 below are just straightforward generalizations.

The warm-up case is that of polymer-chain models in zero magnetic field, which only exhibit the trivial J-dependence. In particular, for simple CIW models the properties (11.34) and (11.36) of the visitation function imply simple formulas for the derivatives of the weights with respect to the coupling constants. For instance:

$$\frac{\partial \varrho(\omega)}{\partial J_{yz}} \;=\; \sum_{\omega', \omega''} \varrho(\omega', \omega'')\, \mathrm{I}[\omega = \omega' \circ (y,z) \circ \omega''] + [y \Longleftrightarrow z] \;\equiv\; T_1 \qquad (11.51)$$

This identity will play an important role in Section 12.2.4. Analogously, for the *baby* simple CIW model in a nonzero magnetic field

$$\frac{\partial \varrho_j(\omega)}{\partial J_{yz}} \;=\; \sum_{\omega', \omega''} \varrho_{0j}(\omega', \omega'')\, \mathrm{I}[\omega = \omega' \circ (y,z) \circ \omega''] + [y \Longleftrightarrow z] \;\equiv\; T_1' \qquad (11.52)$$

For BFS models we have in addition the deep J-dependence. Instead of (11.51), we have now for the zero-magnetic-field case:

$$\frac{\partial \varrho(\omega)}{\partial J_{yz}} = T_1 + T_2 \tag{11.53}$$

where

$$T_1 = \sum_{\omega',\omega''} \varrho(\omega',\omega'')\, \mathrm{I}[\omega = \omega' \circ (y,z) \circ \omega''] + +[y \Longleftrightarrow z] \tag{11.54}$$

[as in the simple CIW models], and the deep J-dependence yields the additional term

$$\begin{aligned} T_2 &= J^\omega \int d\nu_\omega(\underline{t}) \frac{\partial \mathcal{Z}(\underline{t})}{\partial J_{yz}} \\ &= \sum_{\omega':y\to z} [\varrho(\omega,\omega') - \varrho(\omega)\varrho(\omega')] \end{aligned} \tag{11.55}$$

(see [97, 223] for details), which is nonpositive by Theorem 11.2. In nonzero magnetic field:

$$\frac{\partial \varrho_j(\omega)}{\partial J_{yz}} = T_1' + T_2' \tag{11.56}$$

with the same T_1' as the baby simple CIW models:

$$T_1' = \sum_{\omega',\omega''} \varrho_{0j}(\omega',\omega'')\ ,\ \mathrm{I}[\omega = \omega' \circ (y,z) \circ \omega''] + [y \Longleftrightarrow z] \tag{11.57}$$

and

$$\begin{aligned} T_2' = & \sum_{\omega':y\to z} [\varrho_{j0}(\omega,\omega') - \varrho_j(\omega)\varrho_0(\omega')] \\ & + \sum_{\substack{\omega':y\to\bullet \\ \omega'':z\to\bullet}} [\varrho_{j11}(\omega,\omega',\omega'') - \varrho_j(\omega)\varrho_{11}(\omega',\omega'')]\ , \end{aligned} \tag{11.58}$$

which is nonpositive by Theorem 11.2.

For the ARW model in zero magnetic field we have, from (9.63),

$$\varrho(\omega) = (\tanh J)^{\underline{\omega}}\, \mathrm{I}[n_{e(\omega)}(\omega) = 1] \frac{(\cosh J)^{\widetilde{\omega}}}{\langle \exp(\sum_{\{x,y\}\in\widetilde{\omega}} J_{xy}\sigma_x\sigma_y)\rangle_{\underline{\omega}}}\ . \tag{11.59}$$

Hence,

$$\frac{\partial \varrho(\omega)}{\partial J_{yz}} = T_1 + T_2 \tag{11.60}$$

with

$$\begin{aligned} T_1 &= \left[\frac{\partial}{\partial J_{yz}}(\tanh J)^{\underline{\omega}}\right] \mathrm{I}[n_{e(\omega)}(\omega) = 1] \frac{(\cosh J)^{\widetilde{\omega}}}{\langle \exp(\sum_{\{x,y\}\in\widetilde{\omega}} J_{xy}\sigma_x\sigma_y)\rangle_{\underset{\sim}{\omega}}} \\ &\leq \operatorname{cosech}^2(J_{yz}) \sum_{\omega',\omega''} \varrho(\omega',\omega'')\,\mathrm{I}[\omega = \omega' \circ (y,z) \circ \omega'']\,\mathrm{I}[n_{e(\omega')}(\omega') = 1] \end{aligned} \tag{11.61}$$

and

$$\begin{aligned} T_2 &= (\tanh J)^{\underline{\omega}}\, \mathrm{I}[n_{e(\omega)}(\omega) = 1] \frac{\partial}{\partial J_{yz}} \left[\frac{(\cosh J)^{\widetilde{\omega}}}{\langle \exp(\sum_{\{x,y\}\in\widetilde{\omega}} J_{xy}\sigma_x\sigma_y)\rangle_{\underset{\sim}{\omega}}}\right] \\ &= \varrho(\omega)\left[\tanh(J_{yz})\,\mathrm{I}[\{y,z\}\in\widetilde{\omega}\right] - \langle\sigma_y\sigma_z\rangle] \,. \end{aligned} \tag{11.62}$$

By the second Griffiths inequality, $T_2 \leq 0$.

The results for nonzero field are similarly obtained by differentiating:

$$\begin{aligned} \varrho_j(\omega) &= (\tanh J)^{\underline{\omega}}\, [\tanh h_{e(\omega)}]^j\, \mathrm{I}[n_{e(\omega)}(\omega) = 1]^{1-j} \\ &\quad\times \frac{(\cosh J)^{\widetilde{\omega}} \prod_{x\in\pi(\omega,j)} \cosh h_x}{\langle \exp(\sum_{\{x,y\}\in\widetilde{\omega}} J_{xy}\sigma_x\sigma_y \;+\; \sum_{x\in\pi(\omega,j)} h_x\sigma_x)\rangle^{(h)}_{\underset{\sim}{\omega},\pi(\omega,j)^c}} \end{aligned} \tag{11.63}$$

The reader may now convince himself/herself that all these results generalize to weights of higher order in a straightforward manner. The bottom line is the following:

Theorem 11.12 *For baby polymer-chain, BFS and ARW models:*

$$\frac{\partial \varrho_{j_1\ldots j_k}(\omega_1,\ldots,\omega_k)}{\partial J_{yz}} = T_1 + T_2 \tag{11.64}$$

with $T_1 \geq 0$ and $T_2 \leq 0$. Moreover, $T_1 = 0$ if and only if the bond $\{y,z\}$ is not used by any of the walks $\omega_1,\ldots,\omega_k$. Explicitly:

a) For baby simple CIW and BFS models

$$\begin{aligned} T_1 = \sum_{i=1}^{k} \sum_{\omega_i',\omega_i''} & \varrho_{j_1\ldots j_{i-1}0 j_i j_{i+1}\ldots j_k}(\omega_1,\ldots,\omega_{i-1},\omega_i',\omega_i'',\omega_{i+1},\ldots,\omega_k) \\ & \times \mathrm{I}[\omega_i = \omega_i' \circ (y,z) \circ \omega_i''] \end{aligned} \tag{11.65}$$

while for ARW models

$$\begin{aligned} & T_1 \leq \\ & \operatorname{cosech}(J_{yz}) \sum_{i=1}^{k} \sum_{\omega_i',\omega_i''} \varrho_{j_1\ldots j_{i-1}0 j_i j_{i+1}\ldots j_k}(\omega_1,\ldots,\omega_{i-1},\omega_i',\omega_i'',\omega_{i+1},\ldots,\omega_k) \\ & \qquad \times \mathrm{I}[\omega_i = \omega_i' \circ (y,z) \circ \omega_i'']\, \mathrm{I}[n_{e(\omega_i')}(\omega_i') = 1] \,. \end{aligned} \tag{11.66}$$

b) For baby polymer-chain models

$$T_2 \;=\; 0 \,, \tag{11.67}$$

for BFS models

$$\begin{aligned} T_2 \;=\; & \sum_{\omega':y\to z} \Big[\varrho_{j_1\ldots j_k 0}(\omega_1,\ldots,\omega_k,\omega') \;-\; \varrho_{j_1\ldots j_k}(\omega_1,\ldots,\omega_k)\varrho_0(\omega')\Big] \\ & + \sum_{\substack{\omega':y\to\bullet \\ \omega'':z\to\bullet}} \Big[\varrho_{j_1\ldots j_k 11}(\omega_1,\ldots,\omega_k,\omega',\omega'') \\ & \qquad\qquad - \varrho_{j_1\ldots j_k}(\omega_1,\ldots,\omega_k)\varrho_{11}(\omega',\omega'')\Big] \end{aligned} \tag{11.68}$$

and for ARW models

$$T_2 \;=\; \varrho_{j_1\ldots j_k}(\omega_1,\ldots,\omega_k)\Big\{\tanh(J_{yz})\mathrm{I}[\{y,z\}\in(\omega_1\circ\ldots\circ\omega_k)\;]-\langle\sigma_y\sigma_z\rangle\Big\}\,. \tag{11.69}$$

As a consequence:

Corollary 11.13 *For baby polymer-chain, BFS and ARW models*

$$\frac{\partial\varrho_{j_1\ldots j_k}(\omega_1,\ldots,\omega_k)}{\partial J_{yz}} \;\leq\; 0 \tag{11.70}$$

whenever

$$\{y,z\}\notin \underline{\omega}_1\cup\ldots\cup\underline{\omega}_k\,. \tag{11.71}$$

11.3.2 Differentiation with respect to *h*

In a similar fashion, the derivatives of the weights with respect to a magnetic field are seen to be — for the *baby* polymer-chain, BFS and ARW models — the sum of two terms of different sign: a nonnegative contribution due to the "trivial" h-dependence and a nonpositive contribution caused by the "deep" h-dependence.

Theorem 11.14 *For baby polymer-chain, BFS and ARW models*

$$\frac{\partial\varrho_{j_1\ldots j_k}(\omega_1,\ldots,\omega_k)}{\partial h_z} \;=\; T_1'+T_2' \tag{11.72}$$

with $T_1'\geq 0$ and $T_2'\leq 0$. Moreover, $T_1'=0$ if and only if the site z is not the endpoint of any walk ω_i with $j_i=1$. Explicitly:

a) For baby polymer-chain and BFS models

$$T_1' \;=\; \sum_{i=1}^{k}\delta_{z\,e(\omega_i)}\;j_i\;\varrho_{j_1\ldots j_{i-1}0j_{i+1}\ldots j_k}(\omega_1,\ldots,\omega_k)\,, \tag{11.73}$$

while for ARW models

$$T_1' \;\le\; \operatorname{cosech}(h_z) \sum_{i=1}^{k} \delta_{z\, e(\omega_i)}\; j_i\; \varrho_{j_1 \cdots j_{i-1} 0 j_{i+1} \cdots j_k}(\omega_1, \ldots, \omega_k) \;, \tag{11.74}$$

b) For baby polymer-chain models

$$T_2' \;=\; 0 \;, \tag{11.75}$$

for BFS models

$$T_2' = \sum_{\omega' : z \to \bullet} \Big[\varrho_{j_1 \cdots j_k 1}(\omega_1, \ldots, \omega_k, \omega') - \varrho_{j_1 \cdots j_k}(\omega_1, \ldots, \omega_k) \varrho_1(\omega') \Big] \tag{11.76}$$

and for ARW models

$$T_2' = \varrho_{j_1 \cdots j_k}(\omega_1, \ldots, \omega_k) \Big\{ \tanh(h_z)\; \mathrm{I}[z \in \pi(\omega_1, j_1) \cup \ldots \cup \pi(\omega_k, j_k)] - \langle \sigma_z \rangle \Big\} \;. \tag{11.77}$$

Corollary 11.15 *For baby polymer-chain, BFS and ARW models*

$$\frac{\partial \varrho_{j_1 \cdots j_k}(\omega_1, \ldots, \omega_k)}{\partial h_z} \;\le\; 0 \tag{11.78}$$

whenever

$$z \notin \{ e(\omega_i) \colon\; j_i = 1;\; i = 1, \ldots, k \} \;. \tag{11.79}$$

In particular,

$$\frac{\partial \varrho_{0 \ldots 0}(\omega_1, \ldots, \omega_k)}{\partial h_z} \le 0 \;. \tag{11.80}$$

12. Correlation inequalities: A survey of results

In this chapter we give a survey of the correlation inequalities which can be obtained using the random-walk formalism. We also discuss some profound inequalities which *cannot* be obtained within the random-walk formalism, but which need instead the full power of the random-current formalism. Some of the results in this chapter are given without proof; full details can be found in the cited references, especially [223] and [12].

We remind the reader that *BFS models are assumed to have 1 or 2 components and to have single-spin measures of BFS class.*

We distinguish two general types of correlation inequalities, which we call "pointwise" and "summed". The "pointwise" inequalities [e.g. (12.23)] are bounds on the correlation functions evaluated at *individual* lattice sites; they make no reference to any geometric structure of the lattice. The "summed" [e.g. (12.31)] inequalities are bounds on *sums* of correlation functions over the lattice, and assume that the model is translation-invariant. [If the lattice has a group structure (e.g. $\mathbb{Z}^d$ or a torus in $\mathbb{Z}^d$), we call a model *translation-invariant* if $J_{x+a,y+a} = J_{x,y}$ and $h_x = h \; \forall x, y, a$. For such models we denote $|J| = \sum_y J_{xy}$.]

As mentioned previously, all correlation inequalities are derived initially for a "finite-volume" system. In most cases we prove first a "pointwise" inequality. Such an inequality carries over immediately to the infinite-volume limit[1] provided that the boundary conditions employed in taking this limit respect the hypotheses of the inequality (e.g. zero or periodic boundary conditions for zero-magnetic-field inequalities; zero, periodic or plus boundary conditions for nonzero-magnetic-field inequalities). For a translation-invariant model, the corresponding "summed" inequality can then be derived either in finite volume (with periodic boundary conditions) or in infinite volume (for a suitable translation-invariant Gibbs state) by taking the "pointwise" inequality and summing it. However, in a few cases [e.g. (12.166)] no suitable "pointwise" inequality is known, and the "summed" inequality must be proven directly on a finite-volume system with periodic boundary conditions. Such an inequality can be carried over to the infinite-volume limit modulo technical problems of interchanging limit and summation.

[1] Pointwise inequalities involving a sum over sites z [e.g. the Aizenman-Fröhlich inequality (12.23)] require a slight additional argument, e.g. invoking the dominated convergence theorem or the monotone convergence theorem.

12.1 Gaussian upper bounds

These are inequalities bounding $2n$-point Green functions by sums of products of Green functions of lower order, which become equalities for the Gaussian model (9.43)/(9.19). They are an almost immediate consequence of Theorems 11.1, 11.2 11.5. For simplicity, we consider only the zero-field case.

The simplest case is that of repulsive polymer chains, for which (11.2) implies ($h = 0$)

$$\begin{aligned} &K(x_1x_2|\dots|x_{2(j+k)-1}x_{2(j+k)}) \;\le \\ &\qquad K(x_1x_2|\dots|x_{2j-1}x_{2j})\, K(x_{2j+1}x_{2j+2}|\dots|x_{2(j+k)-1}x_{2(j+k)}) \end{aligned} \tag{12.1}$$

for any $1 \le j \le k$. Similar inequalities hold in nonzero magnetic field for repulsive *baby* polymer-chain models.

For the BFS and ARW models, however, the general repulsiveness inequality (11.2) does not hold; we have only the special cases (11.13) and (11.20) in which one factor in the partition is *summed* over a class of walks leading to some Green function S_n (cf. the hypotheses (11.12) and (11.19)). Let us analyze the simplest case, in which this factor consists of a single walk ω . The conditions (11.12) and (11.19) then read

$$\sum_{\omega:x_1\to x_2} \varrho(\omega) \;=\; S_2(x_1,x_2)\;, \tag{12.2}$$

and (11.13) and (11.20) immediately imply that

$$K(x_1x_2|\dots|x_{2k-1}x_{2k}) \;\le\; S_2(x_1,x_2)K(x_3x_4|\dots|x_{2k-1}x_{2k})\;. \tag{12.3}$$

Analogous (but notationally more involved) expressions hold when other than the first pair of sites is singled out. In particular, if (12.3) is summed over all $\pi \in Q_{2k}$, we obtain:

Theorem 12.1 *For repulsive polymer-chain, BFS and ARW models with no magnetic field, the strong Gaussian inequality [97, Section 4 and references therein] holds:*

$$S_{2k}(x_1,\dots,x_{2k}) \;\le\; \sum_{\pi\in Q_{2k}} S_2(x_{\pi(1)},x_{\pi(2)})S_{2(k-1)}(x_{\pi(3)},\dots,x_{\pi(2k)})\;. \tag{12.4}$$

By iteration of (12.4) one obtains the ordinary Gaussian inequality:

$$S_{2k}(x_1,\dots,x_{2k}) \;\le\; \sum_{\pi\in Q_{2k}} \prod_{i=1}^{k} S_2(x_{\pi(2i-1)},x_{\pi(2i)}) \tag{12.5}$$

If we apply 11.13 for families ω_I of more than one walk, we obtain, for the case h=0, expressions analogous to (12.3) but in which more than one pair of sites is factorized. In this fashion one obtains generalizations of (12.4) to other partitions of the sequence $1,\dots,2k$ (Theorem 5.2 in [97]).

12.2 Truncated four-point function in zero magnetic field: Lebowitz and Aizenman-Fröhlich inequalities

The truncated four-point function in zero magnetic field is defined as

$$\begin{aligned} S_4^T(x_1,x_2,x_3,x_4) &\equiv S_4(x_1,x_2,x_3,x_4) - S_2(x_1,x_2)S_2(x_3,x_4) \\ &\quad -S_2(x_1,x_3)S_2(x_2,x_4) - S_2(x_1,x_4)S_2(x_2,x_3) . \end{aligned} \tag{12.6}$$

In terms of kernels,

$$\begin{aligned} S_4^T(x_1,x_2,x_3,x_4) &= \sum_{\pi\in Q_4} K^T(x_{\pi(1)}x_{\pi(2)}|x_{\pi(3)}x_{\pi(4)}) \\ &= K^T(x_1x_2|x_3x_4) + K^T(x_1x_3|x_2x_4) + K^T(x_1x_4|x_2x_3) , \end{aligned} \tag{12.7}$$

where the truncated kernels are defined as

$$K^T(x_1x_2|x_3x_4) \equiv K(x_1x_2|x_3x_4) - K(x_1x_2)K(x_3x_4) . \tag{12.8}$$

In terms of weights,

$$K^T(x_1x_2|x_3x_4) = \sum_{\substack{\omega_1:x_1\to x_2 \\ \omega_2:x_3\to x_4}} \varrho^T(\omega_1,\omega_2) , \tag{12.9}$$

where the truncated weights are

$$\varrho^T(\omega_1,\omega_2) \equiv \varrho(\omega_1,\omega_2) - \varrho(\omega_1)\varrho(\omega_2) . \tag{12.10}$$

Considerable effort has been devoted to obtaining bounds on the truncated four-point functions. In this chapter we review such bounds using the unifying framework of the random-walk formalism.

12.2.1 Upper bound (Lebowitz inequality)

Theorems 11.1, 11.2 and 11.5 immediately imply upper bounds for the above-defined objects. Indeed, from (11.2) we have that for repulsive polymer-chain models:

$$\varrho^T(\omega_1,\omega_2) \le 0 , \tag{12.11}$$

while from (11.13) and (11.20) we have that for the BFS and ARW models

$$\sum_{\omega_2:x\to y} \varrho^T(\omega_1,\omega_2) \le 0 . \tag{12.12}$$

As an immediate consequence of (12.11) and (12.12) we have:

Theorem 12.2 *For the repulsive polymer-chain, BFS and ARW models (h=0),*

$$K^T(x_1x_2|x_3x_4) \;\leq\; 0 \tag{12.13}$$

and hence

$$S_4^T(x_1,x_2,x_3,x_4) \;\leq\; 0\,. \tag{12.14}$$

(12.14) is the Lebowitz inequality [279, 362, 491, 162, 92, 97, 5], which has had numerous applications in quantum field theory and statistical mechanics. It is, of course, just the Gaussian inequality (12.3)–(12.5) specialized to $k = 2$.

12.2.2 Nontrivial lower bounds (Aizenman-Fröhlich inequalities)

Obviously, we can obtain a trivial lower bound on the truncated four-point function by neglecting the nonnegative term $S_4(x_1,x_2,x_3,x_4)$ in the RHS of (12.6). However, the random-walk formalism allows us to account for some of the cancellations among the terms on the RHS of (12.6), and so to obtain sharper lower bounds. The key step is to use Theorem 11.7 together with path-splitting inequalities to obtain lower bounds on the truncated weights.

The procedure is as follows. From Corollary 11.9 we have that for the (generalized) CIW, BFS and ARW models,

$$\varrho^T(\omega_1,\omega_2) \geq \begin{cases} 0 & \text{if } \mathrm{supp}(\omega_1)\cap\mathrm{supp}(\omega_2) = \varnothing \\ -\varrho(\omega_1)\varrho(\omega_2) & \text{otherwise} \end{cases} \tag{12.15}$$

with equality for the SAW. Therefore,

$$K^T(x_1x_2|x_3x_4) \;\geq\; -\sum_{\substack{\omega_1: x_1\to x_2 \\ \omega_2: x_3\to x_4}} \varrho(\omega_1)\varrho(\omega_2)\ \mathrm{I}[\mathrm{supp}(\omega_1)\cap\mathrm{supp}(\omega_2)\neq\varnothing] \tag{12.16}$$

with equality for the SAW. Using the trivial inequality[2]

$$\begin{aligned} \mathrm{I}[\mathrm{supp}(\omega_1)\cap\mathrm{supp}(\omega_2)\neq\varnothing] \;&\leq\; \sum_z \mathrm{I}[\mathrm{supp}(\omega_1)\cap\mathrm{supp}(\omega_2)\ni z] \\ &=\; \sum_z \mathrm{I}[\mathrm{supp}(\omega_1)\ni z]\mathrm{I}[\mathrm{supp}(\omega_2)\ni z] \end{aligned} \tag{12.17}$$

and then using Proposition 11.11 to split both ω_1 and ω_2 through z, we obtain:

Proposition 12.3
a) For the simple CIW, BFS and ARW models

[2] This is the first-order inclusion-exclusion inequality; see Section 12.3 for a fuller discussion.

$$K^T(x_1x_2|x_3x_4) \geq -\sum_z \Big[K(x_1z)\delta_{z,x_2} + \sum_{z_1} J_{zz_1} K(x_1z|z_1x_2)\Big] \times \Big[K(x_3z)\delta_{z,x_4} + \sum_{z_2} J_{zz_2} K(x_3z|z_2x_4)\Big] \quad (12.18)$$

For the ARW model J_{zz_i} *can be replaced by* $\tanh(J_{zz_i})$.

b) For the ARW model

$$K^T(x_1x_2|x_3x_4) \geq -\sum_z K(x_1z|zx_2)K(x_3z|zx_4) \quad (12.19)$$

Let us emphasize what was thrown away in the passage from (12.16) to (12.18)/(12.19). There is a large overcounting in (12.17): loosely speaking, the *probability* of intersection has been replaced by the *mean number of sites* at which the intersection takes place. This is a very bad bound in the critical region for dimension $d \leq 4$: *precisely because* ω_1 and ω_2 are very likely to meet, they are also likely to meet at *very many places.* This is the major overcounting (and it is in fact the only one for the ARW model with intermediate-site splitting).

An additional error arises in the case of intermediate-step splitting, from the use of Proposition 11.11. For the CIW and BFS models, we used the inequality (11.47) instead of the identity (11.45), thereby throwing away the indicator functions $\mathrm{I}[n_z(\omega_1') = 1]$ and $\mathrm{I}[n_z(\omega_2') = 1]$. Loosely speaking, the mean number of *sites* at which intersection takes place has been replaced by the mean number of *times* at which intersections take place. We expect that this causes an overcounting by a factor of order $\langle\varphi_z^2\rangle/\langle\varphi_z^2\rangle_0$, where $\langle\,\cdot\,\rangle_0$ denotes expectation in the single-spin distribution (9.28). Such an overcounting is serious only for near-Gaussian models in dimension $d \leq 2$. An analogous error arises for the ARW model if intermediate-step splitting is used [i.e. (12.18)], in this case not from neglect of the indicator functions (they are superfluous) but rather because (11.42) and hence (11.47) are only *inequalities.*

The foregoing "errors" can in principle be remedied by specifying a rule for defining uniquely a "privileged" intersection point $z(\omega_1, \omega_2)$, thereby eliminating the overcounting. For instance, we can order the sites visited by ω_1 with the natural order induced by the direction of travel of ω_1, and define $z(\omega_1, \omega_2)$ as the last site of ω_1 where an intersection with ω_2 takes place. For the benefit of the ARW case we split the walks the *first* time they hit $z(\omega_1, \omega_2)$. With this definition we have an *identity* for the right-hand side of (12.16): for example, for the ARW model with intermediate-site splitting, we obtain

$$-\sum_{\substack{\omega_1: x_1 \to x_2 \\ \omega_2: x_3 \to x_4}} \varrho(\omega_1)\varrho(\omega_2)\, \mathrm{I}[\mathrm{supp}(\omega_1) \cap \mathrm{supp}(\omega_2) \neq \varnothing] =$$

$$-\sum_z \sum_{\substack{\omega_1': x_1 \to z \\ \omega_1'': z \to x_2 \\ \omega_2': x_3 \to z \\ \omega_2'': z \to x_4}} \varrho(\omega_1', \omega_1'')\varrho(\omega_2', \omega_2'')\, \mathrm{I}\Big[\mathrm{supp}(\omega_1'') \cap (\mathrm{supp}(\omega_2') \cup \mathrm{supp}(\omega_2'')) = \{z\}\Big]\,. \quad (12.20)$$

For the BFS model, we have an analogous identity using intermediate-step splitting and the indicator functions

$$\mathrm{I}\Big[\mathrm{supp}(\omega_1'') \cap (\mathrm{supp}(\omega_2') \cup \mathrm{supp}(\omega_2'')) = \varnothing \text{ or } \{z\}\Big]\ \mathrm{I}[n_z(\omega_1') = 1]\ \mathrm{I}[n_z(\omega_2') = 1]\ . \tag{12.21}$$

However, we have no idea what to do with these indicator functions, other than to throw them away! A more careful treatment of these constraints could well lead to important new results, including the triviality of the φ_4^4 quantum field theory.

The Aizenman-Fröhlich inequalities are obtained by making a further "error" in Proposition 12.3: by the Gaussian bound (12.3), we can replace the factors $K(\cdot\,|\,\cdot)$ by S_2S_2. [In Sections 12.2.3 and 12.2.4 we analyze this "error" in more detail and present some improved inequalities.] We thus have:

Theorem 12.4 *In the absence of magnetic field:*

$$K^T(x_1x_2|x_3,x_4) \;\geq\; -V(x_1x_2|x_3x_4) \tag{12.22}$$

and

$$S_4^T(x_1,x_2,x_3,x_4) \;\geq\; -[V(x_1x_2|x_3x_4) \;+\; 2 \text{ permutations}] \tag{12.23}$$

with

a) For repulsive simple CIW, BFS and ARW models

$$\begin{aligned} V(x_1x_2|x_3x_4) \;&=\; \sum_z S_2(x_1,z)S_2(x_3,z)\Big[\delta_{z,x_2} + \sum_{z_1} J_{zz_1}S_2(z_1,x_2)\Big] \\ &\qquad \Big[\delta_{z,x_4} + \sum_{z_2} J_{zz_2}S_2(z_2,x_4)\Big]\ . \end{aligned} \tag{12.24}$$

For the ARW model J_{zz_i} can be replaced by $\tanh(J_{zz_i})$.

b) For ARW models

$$V(x_1x_2|x_3x_4) \;=\; \sum_z S_2(x_1,z)S_2(x_3,z)S_2(z,x_2)S_2(z,x_4)\ . \tag{12.25}$$

(12.24)/(12.25) is the Aizenman-Fröhlich inequality [5, 213, 301] in intermediate-step and intermediate-site forms, respectively.

Let us remark that (12.23)/(12.25) implies the analogous expression for the Ising-model correlations,

$$\langle \sigma_{x_1};\sigma_{x_2};\sigma_{x_3};\sigma_{x_4}\rangle \;\geq\; -3\sum_z \langle\sigma_{x_1}\sigma_z\rangle\langle\sigma_{x_3}\sigma_z\rangle\langle\sigma_z\sigma_{x_2}\rangle\langle\sigma_z\sigma_{x_4}\rangle\ , \tag{12.26}$$

in principle only if the sites $x_1,\ldots,x_4$ are all different. However, a simple check of the cases with 2, 3 or 4 coincident sites shows that (12.26) is valid for all $x_1,\ldots,x_4$. (In fact, the factor 3 in (12.26) is not optimal, and with the aid of the random-current formalism it can be reduced to 2; see Ref. [5].) In what follows we shall use the notation $u_4(x_1, x_2, x_3, x_4) \equiv \langle \varphi_{x_1}; \varphi_{x_2}; \varphi_{x_3}; \varphi_{x_4} \rangle$ for spin models (including Ising), in order to distinguish (where necessary) from S_4^T. We have $u_4(x_1, x_2, x_3, x_4) = S_4^T(x_1, x_2, x_3, x_4)$ in all cases *except* for the ARW model when some of the sites coincide. On the other hand $S_2(x_1, x_2) = \langle \varphi_{x_1} \varphi_{x_2} \rangle$ in *all* cases (provided $h = 0$).

Let us also note that for the Ising model the inequality with intermediate-site splitting [(12.23)/(12.25)] is sharper than the one with intermediate-step splitting [(12.23)/(12.24)] even if in the latter J_{zz_i} is replaced by $\tanh(J_{zz_i})$. Indeed, by Griffiths' third inequality [271] (which is a special case of the Simon-Lieb inequality [464, 372]) we have

$$\langle \sigma_z \sigma_x \rangle \;\leq\; \sum_{z'} (\tanh J_{zz'}) \langle \sigma_{z'} \sigma_x \rangle \tag{12.27}$$

— valid only if $x \neq z$ — which implies that (12.24) is an upper bound for (12.25).

The Lebowitz and Aizenman-Fröhlich inequalities were used in [5, 213] to prove the triviality of the continuum limit for φ_d^4 field theories in dimension $d > 4$. These inequalities are usually represented pictorially as follows: For the Ising model

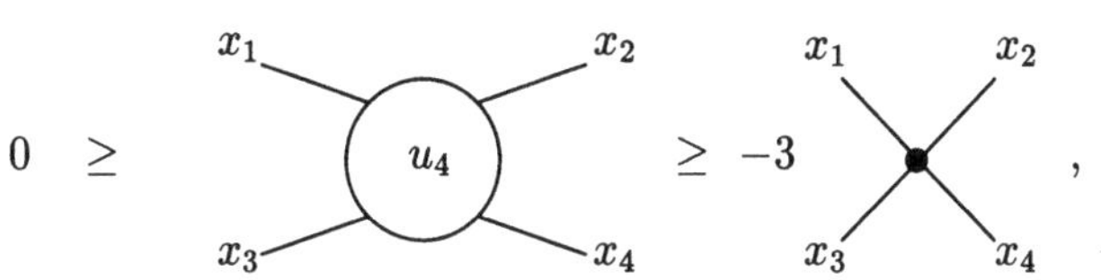

and for the repulsive simple CIW, BFS and ARW models

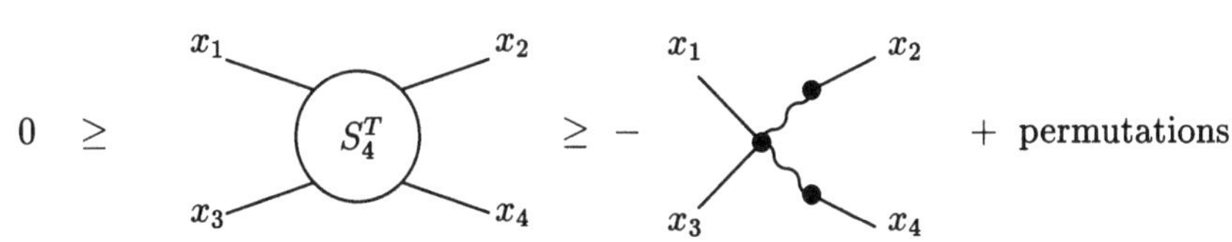

where x —— z denotes the two-point function $S_2(x, z)$, z ●∿∿● z_1 denotes J_{zz_1}, and a sum over internal vertices ● is understood. For the intermediate-step cases, the pictorial representation (12.28) ignores the terms involving δ_{z,x_2} and/or δ_{z,x_4}, but those terms are indeed insignificant in the critical regime.

For translation-invariant models, we have the following summed versions of (12.23)/(12.25). If we adopt the standard magnetic notation

$$\chi \;=\; \sum_x S_2(0, x) \tag{12.28}$$

$$\overline{u}_4 \;=\; \sum_{x_2, x_3, x_4} u_4(x_1, x_2, x_3, x_4) \tag{12.29}$$

then, summing (12.23)/(12.25) over x_2, x_3 and x_4 we obtain

$$\overline{u}_4 \geq -3|J|^2\chi^4(1 + \frac{1}{|J|\chi})^2 \tag{12.30}$$

for repulsive simple CIW, BFS and ARW models, and

$$\overline{u}_4 \geq -3\chi^4 \tag{12.31}$$

for ARW models.

12.2.3 Once-improved Aizenman-Fröhlich inequality (Aizenman-Graham inequality)

The above inequalities are probably the best that can be obtained for spin models using only the random-walk expansion. However, by resorting to the more powerful random-current representation, Aizenman and Graham [15] were able to improve the lower bounds (12.25) and (for some one-component spin models) (12.24). Their result is the following:

Theorem 12.5 (Aizenman and Graham) [15]
a) For ARW models:

$$u_4(x_1, x_2, x_3, x_4) \geq -\sum_{z,z_1} S_2(x_1, z)S_2(x_2, z)\tanh(J_{zz_1})\frac{\partial}{\partial J_{zz_1}}S_2(x_3, x_4) - \varepsilon(x_1, x_2, x_3, x_4) \,. \tag{12.32}$$

b) For ferromagnetic 1-component spin models with single-spin measures in the Griffiths-Simon class [466] (in short "models in the GS class"; this class of models includes the Ising and 1-component φ^4 models, among others):

$$u_4(x_1, x_2, x_3, x_4) \geq -\sum_{z,z_1,z_2} S_2(x_1, z)J_{zz_1}S_2(z_1, x_2)J_{z_1z_2}\frac{\partial}{\partial J_{z_1z_2}}S_2(x_3, x_4) - \varepsilon(x_1, x_2, x_3, x_4) \,. \tag{12.33}$$

The "error term" $\varepsilon(x_1, x_2, x_3, x_4)$ is a term that becomes negligible in the critical region. For (12.32) *the error term is*

$$\varepsilon(x_1, x_2, x_3, x_4) = S_2(x_1, x_4)S_2(x_2, x_4)S_2(x_3, x_4) + [x_3 \Longleftrightarrow x_4] \,. \tag{12.34}$$

For (12.33) *it is*[3]

[3]This "error term" is a corrected version of the slightly mistaken [17] "error term" of equation 4.1 of reference [15]. The mistake in the latter is due to the fact that the restriction "$z \neq x$" in Proposition 7.1 of [15] was not properly taken into account during the proof of Proposition 4.1 of [15]. In particular, the bound (7.13) of [15] is valid only if $v \neq x_3$; hence one must add a contribution due to spins at sites v with $v = x_3$ — we are using the notation of [15] — which produces an extra error term in equation (4.1) of [15] of the form

$$\begin{aligned}\varepsilon(x_1,x_2,x_3,x_4) &= \Big[\sum_z S_2(x_1,z)J_{zx_4}S_2(x_4,x_3)S_2(x_4,x_2) \\ &\quad + [x_1 \Longleftrightarrow x_3] + S_2(x_4,x_2)S_2(x_1,x_4)\delta_{x_3,x_4}\Big] \\ &\quad + \Big[x_4 \Longleftrightarrow x_3\Big] .\end{aligned} \tag{12.37}$$

Diagrammatically, (12.33) can be represented in the form

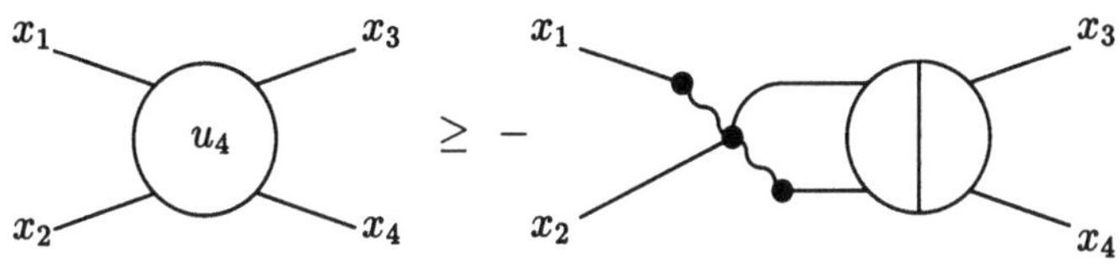

(12.38)

where

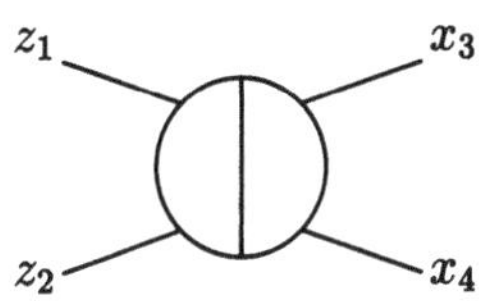

$$\equiv \frac{\partial}{\partial J_{z_1 z_2}} S_2(x_3,x_4) , \tag{12.39}$$

and the "error term" is omitted.

These are "once-improved" versions of the Aizenman-Fröhlich inequality (12.24): *one* of the products S_2S_2 has been replaced by $\partial S_2/\partial J$, which by Lebowitz inequality (12.14) is an improvement:

$$\begin{aligned}\frac{\partial}{\partial J_{z_1 z_2}} S_2(x_3,x_4) &= \langle \varphi_{z_1}\varphi_{z_2}\varphi_{x_3}\varphi_{x_4}\rangle - S_2(z_1,z_2)S_2(x_3,x_4) \\ &= S_2(z_1,x_3)S_2(z_2,x_4) + [z_1 \Longleftrightarrow z_2] + u_4(z_1,z_2,x_3,x_4) \\ &\leq S_2(z_1,x_3)S_2(z_2,x_4) + [z_1 \Longleftrightarrow z_2]\end{aligned} \tag{12.40}$$

Diagrammatically, the identity (12.40) is

$$\sum_w S_2(x_1,w)J_{wx_3}S_2(x_3,x_4)S_2(x_2,x_3) + [x_1 \Longleftrightarrow x_3] \tag{12.35}$$

(the permutation is brought about by the (x_1,x_3) symmetrization of the term). Analogously, in the last term of the upper bound in (7.11) of [15] one must consider separately the case $x_1 = x_3$ which yields an extra contribution

$$S_2(x_1,x_2)S_2(x_1,x_4)\delta_{x_1,x_3} + [x_1 \Longleftrightarrow x_3] . \tag{12.36}$$

In addition, the formulas (12.32)-(12.37) differ from those of Proposition 4.1 of [15] by a $(1 \Longleftrightarrow 4)$ permutation of the subscripts.

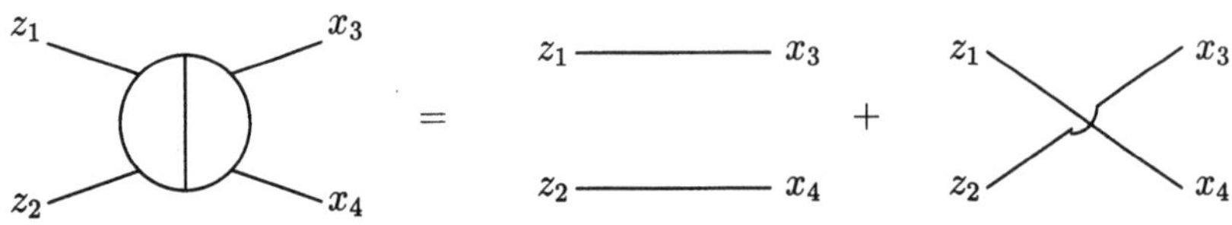

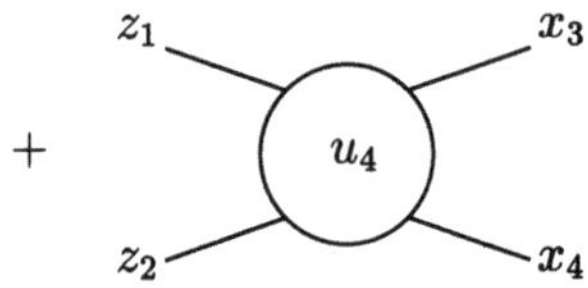

(12.41)

The inequalities (12.32) and (12.33) are rather deep results. Their proofs require a detailed account of the delicate correlations involved in the truncated functions. This level of detail can be achieved via the extremely flexible random-current representation of the Ising model but seems to be beyond the reach of the simpler but less informative random-walk representations — except in the case of simple CIW models (see below). Consequently, such results have not been proven for general BFS models, but only for models that can be approximated by sequences of Ising models (GS class of models [466]).

For translation-invariant models we can obtain "summed" versions of (12.32) and (12.33). For instance, summing (12.33) over x_2, x_3 and x_4 we obtain for models in the GS class:

$$\begin{aligned} \overline{u}_4 &\geq -2\beta|J|\chi^2\frac{\partial\chi}{\partial\beta} - 2|J|\chi^3 - 2\chi^2 \\ &\equiv -2\beta|J|\chi^2\frac{\partial\chi}{\partial\beta} - \varepsilon\,, \end{aligned} \tag{12.42}$$

where we have denoted

$$\beta\frac{\partial\chi}{\partial\beta} = \frac{1}{2}\sum_{x_2,x_3,x_4} J_{x_3x_4}\frac{\partial}{\partial J_{x_3x_4}}S_2(x_1,x_2)\,. \tag{12.43}$$

The important inequality (12.42) implies that *if* in $d=4$ there is a logarithmic correction to the mean-field behavior of χ — the belief in the existence of such correction is widespread but a rigorous proof is lacking — then a massive φ^4 continuum field theory in $d=4$ is necessarily trivial [15].

Moreover, if the Aizenman-Graham inequality (12.33) is combined with the identity (12.40), we obtain

$$\begin{aligned} &J_{x_1x_2}\frac{\partial}{\partial J_{x_1x_2}}S_2(x_3,x_4) \geq \\ &\quad J_{x_1x_2}\Big[S_2(x_1,x_3)S_2(x_2,x_4) + S_2(x_1,x_4)S_2(x_2,x_3)\Big] \\ &\quad - J_{x_1x_2}\Big[\sum_{z,z_1,z_2} S_2(x_1,z)J_{zz_1}S_2(z_1,x_2)J_{z_1z_2}\frac{\partial}{\partial J_{z_1z_2}}S_2(x_3,x_4) \\ &\qquad - \varepsilon(x_1,x_2,x_3,x_4)\Big] \end{aligned} \tag{12.44}$$

or, diagrammatically,

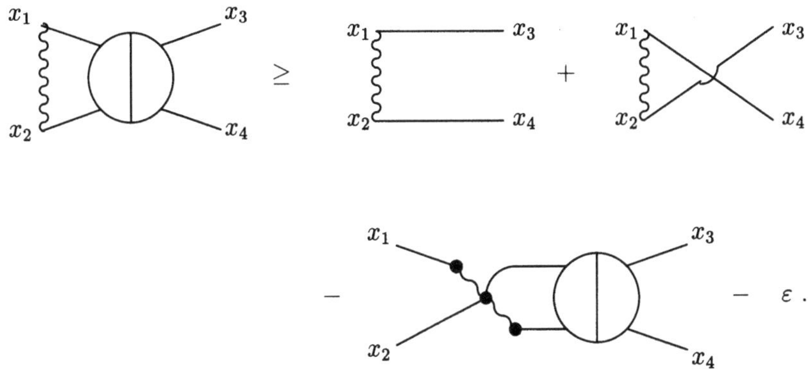

(12.45)

We now sum over x_2, x_3, x_4 and use translation invariance; we obtain

$$\beta\frac{\partial\chi}{\partial\beta} \geq |J|\chi^2 - B'_J\beta\,\frac{\partial\chi}{\partial\beta} - \bar{\varepsilon} \tag{12.46}$$

where B'_J is the modified "bubble" diagram

$$B'_J = \sum_{z,z_1,x_2} S_2(x_1,z)J_{x_1x_2}J_{zz_1}S_2(z_1,x_2)$$

$=$ 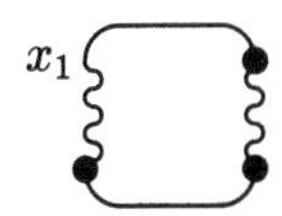

(12.47)

and $\bar{\varepsilon} = \sum_{x_2,x_3,x_4} J_{x_1x_2}\varepsilon(x_1,x_2,x_3,x_4)$. The key fact is that the third term in the right side contains as a factor $\beta\partial\chi/\partial\beta$ which appears on the left side; so we can bring it to the left side and solve for $\beta\partial\chi/\partial\beta$:

$$\beta\frac{\partial\chi}{\partial\beta} \geq \frac{|J|\chi^2 - \bar{\varepsilon}}{1+B'_J}\ . \tag{12.48}$$

Finally, the modified bubble diagram B'_J can be bounded by the Schwarz inequality in terms of the usual bubble diagram

$$B_0 = \sum_x [S_2(0,x)]^2\ , \tag{12.49}$$

to obtain

$$\beta\frac{\partial\chi}{\partial\beta} \geq \frac{|J|\chi^2}{1+|J|^2B_0}[1 - \frac{2|J|B_0}{\chi} - \frac{B_0}{\chi^2}] \tag{12.50}$$

This trick of finding an inequality for a quantity in terms of itself, and then "passing it to the left", has the net result of yielding rigorous inequalities which sum infinitely many orders in the "expansion parameter" B_0. We shall discuss another technique of this type in Section 12.3.

The bound (12.50), together with the bounds on B_0 obtained from reflection positivity, implies that the critical exponent γ for *nearest-neighbor* ferromagnetic models of the GS class obeys [15]:

a) for $d > 4$ a mean-field upper bound $\gamma \leq 1$,

b) for $d = 4$ a mean-field upper bound with a correction of at most a logarithm to the first power,

c) for $2 < d < 4$ a "spherical-model upper bound" $\gamma \leq 2/(d-2)$.

These results will be more fully discussed in Section 14.1.

We remark that an analogue of the Aizenman-Graham inequality was proven for the SAW@ in [28]; however, the same method in fact yields a stronger inequality — see the next section.

12.2.4 Twice-improved Aizenman-Fröhlich inequality

We note that the Aizenman-Graham factor $\frac{\partial}{\partial J_{zz_1}} S_2(x_1, x_3)$ is not only smaller than the factor $S_2(x_1, z)S_2(x_3, z)$ appearing in (12.24), but is also smaller than the factor $K(x_1z|z_1x_3) + (x_1 \Longleftrightarrow x_3)$ appearing in the bound for S_4^T obtained from (12.18). Indeed, for BFS and ARW models (when x_1, x_3, z, z_1 are all different sites):

$$\begin{aligned} \frac{\partial}{\partial J_{zz_1}} S_2(x_1, x_3) &= S_4(x_1, x_3, z, z_1) - S_2(x_1, x_3)S_2(z, z_1) \\ &= K(x_1z|z_1x_3) + K(x_1z_1|zx_3) + K^T(x_1x_3|zz_1) \end{aligned} \tag{12.51}$$

So the Aizenman-Graham inequality can definitely *not* be derived from Proposition 12.3. On the other hand, for simple CIW models the formula for $\partial S_2/\partial J$ is

$$\frac{\partial}{\partial J_{yz}} S_2(x_1, x_2) = K(x_1y|zx_2) + K(x_1z|yx_2) \; ; \tag{12.52}$$

this is an immediate consequence of (11.51). Note that there is no term K^T! For these models, therefore, one can prove not only an "Aizenman-Graham" inequality ("once-improved Aizenman-Fröhlich inequality") but in fact a "twice-improved Aizenman-Fröhlich inequality" [483]. Indeed, from (12.18) and (12.52),

$$\begin{aligned} K^T(x_1x_2|x_3x_4) \geq & -\sum_z \left[S_2(x_1, z)\delta_{z,x_2} + (1/2) \sum_{z_1} J_{zz_1} \frac{\partial}{\partial J_{zz_1}} S_2(x_1, x_2) \right] \\ & \times \left[S_2(x_3, z)\delta_{z,x_4} + (1/2) \sum_{z_2} J_{zz_2} \frac{\partial}{\partial J_{zz_2}} S_2(x_3, x_4) \right] . \end{aligned} \tag{12.53}$$

Diagrammatically,

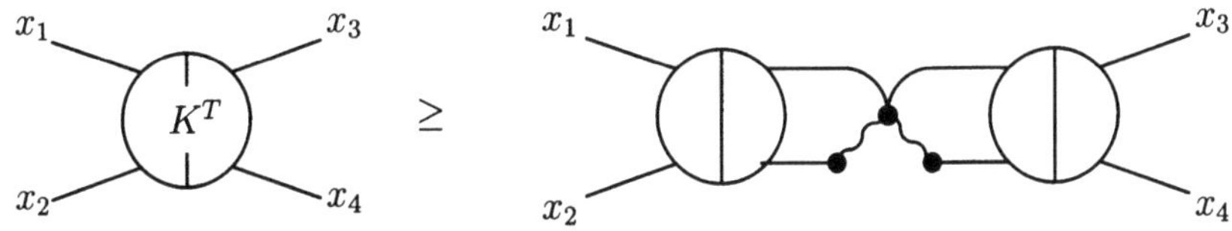

(12.54)

For the special case of the SAW model, it is natural to write this as

(12.55)

where the dashed lines indicate nonintersection between the pair of walks which they separate.

Hence, for translation-invariant simple CIW models, we have instead of (12.42) the bound[4]

$$\overline{u}_4 \;\geq\; -3\left(\beta\frac{\partial\chi}{\partial\beta} \;+\; \chi\right)^2 , \tag{12.56}$$

which implies the triviality of the (massive) continuum limit whenever $d\nu > 2$ (here ν is the critical exponent for the correlation length). See Section 15.2 for details.

Similarly, we could hope to combine (12.53) with

$$\beta\frac{\partial\chi}{\partial\beta} \;=\; \sum_{x_2,x_3,x_4} J_{x_3x_4}[S_2(x_1,x_3)S_2(x_2,x_4) \;+\; K^T(x_1x_3|x_2x_4)] \tag{12.57}$$

in order to obtain a lower bound on $\beta\partial\chi/\partial\beta$ which refines (12.50). However, we are unable to deal with the resulting sums (the "channels" get mixed up). In any case, (12.53) implies for the simple CIW models a *weaker* bound similar to (12.33), from which one can derive a bound on $\beta\partial\chi/\partial\beta$ similar to (12.50).

12.3 Inequalities involving infinitely many orders in the expansion parameter (non-Gaussian upper bounds)

We shall discuss in this section the kind of situations one has to deal with when attempting to use the random-walk formalism to prove non-Gaussian upper bounds, i.e., upper bounds that are *not* saturated for the Gaussian model. Such bounds are of relevance, for instance, in proving non-mean-field behavior at

[4]This bound is implicit in the inequality $c_{N_1,N_2} \leq (N_1+1)(N_2+1)c_{N_1}c_{N_2}$ of Domb and McKenzie [389]. See Section 13.2 for definitions.

the critical point or the nontriviality of continuum limits. The methods used to obtain such bounds fall into two distinct categories. For the near-Gaussian cases (e.g. $\lambda\varphi^4$ BFS model or Edwards model for λ *small*) the required upper bounds are obtained as a polynomial in the bare coupling constant λ , in the form of an "skeleton" inequality [97, 74, 292, 75]. The resulting bounds are powerful enough to control the continuum limit for weakly-coupled *superrenormalizable* φ^4 or Edwards models (i.e. $d < 4$) [96, 74, 75], but are useless in the renormalizable and nonrenormalizable cases. On the other hand, for the strongly-coupled cases (e.g. ARW and SAW models) the expansion parameter is instead the "bubble" diagram B_0, which can be extremely *large* in the critical regime (below or at the upper critical dimension). Therefore, techniques different from the one used for weakly-coupled models are needed — ones which "sum infinitely many orders in B_0". We have already seen one example of such a technique in the preceding section, namely the trick of "passing to the left" used in obtaining the non-Gaussian bound (12.48). This example shows the typical features expected in such a technique: it allowed us to pass from a factor $1 - B'_J$ roughly present in (12.46) to the factor $(1 + B'_J)^{-1}$ obtained in (12.48). This transformed a useless inequality of first order in B'_J into a useful one involving all orders in B'_J. In this section we introduce another example of such a technique which we call the "dilution trick". This technique was originally introduced by Erdös and Taylor [164, Theorem 7]; the simplified form we present here is due to Sokal, Felder and Fröhlich [476], and was used within the random-current formalism for Ising models by Aizenman and Fernández [12].

12.3.1 General setup for the dilution trick

The basic problem we are confronted with is to obtain bounds for an indicator function

$$\mathrm{I}[A_1 \cup \ldots \cup A_n] \tag{12.58}$$

where $A_1, \ldots, A_n$ is a *very large* family of measurable sets (events). An expression specifically designed for our purposes is the inclusion-exclusion principle

$$\mathrm{I}[A_1 \cup \ldots \cup A_n] \;=\; \sum_{i=1}^{n} (-1)^{i-1} \mathrm{I}_i \tag{12.59}$$

with

$$\mathrm{I}_i \;=\; \sum_{1 \le j_1 < \ldots < j_i \le n} \mathrm{I}[A_{j_1} \cap \ldots \cap A_{j_i}] \,. \tag{12.60}$$

Moreover, the truncations of the RHS of (12.59) constitute a sequence of upper and lower bounds:

$$\mathrm{I}[A_1 \cup \ldots \cup A_n] \;\le\; \sum_{i=1}^{k} (-1)^{i-1} \mathrm{I}_i \quad , \qquad k \text{ odd} \tag{12.61}$$

and

$$\mathrm{I}[A_1 \cup \ldots \cup A_n] \;\geq\; \sum_{i=1}^{k}(-1)^{i-1}\mathrm{I}_i \quad , \qquad k \text{ even} \tag{12.62}$$

Broadly speaking, the "passing to the left trick" is a trick used to obtain a meaningful result out of the simplest of the upper bounds (12.61), namely the one with $k = 1$

$$\mathrm{I}[A_1 \cup \ldots \cup A_n] \;\leq\; \sum_{i=1}^{n}\mathrm{I}[A_i] \;; \tag{12.63}$$

while the "dilution trick" was developed to obtain results out of the simplest ($k = 2$) of the lower bounds (12.62):

$$\begin{aligned}\mathrm{I}[A_1 \cup \ldots \cup A_n] \;&\geq\; \sum_{i=1}^{n}\mathrm{I}[A_i] \;-\; \sum_{1\leq j_1<j_2\leq n} \mathrm{I}[A_{j_1} \cap A_{j_2}] \\ &=\; \sum_{i=1}^{n}\mathrm{I}[A_i] \;-\; (1/2) \sum_{\substack{j_1, j_2 = 1 \\ j_1 \neq j_2}}^{n} \mathrm{I}[A_{j_1} \cap A_{j_2}] \,. \end{aligned} \tag{12.64}$$

For the applications of this work, we shall integrate these inequalities with respect to some finite measure E:

$$E(A_1 \cup \ldots \cup A_n) \;\leq\; \sum_{i=1}^{n} E(A_i) \tag{12.65}$$

$$E(A_1 \cup \ldots \cup A_n) \;\geq\; \sum_{i=1}^{n} E(A_i) \;-\; (1/2) \sum_{\substack{i,j = 1 \\ i \neq j}}^{n} E(A_{j_1} \cap A_{j_2}) \,. \tag{12.66}$$

We notice that for n large both (12.65) and (12.66) are *very bad bounds* unless the occurrence of several events at the same time is very rare. In fact, when $\sum_j E(A_i \cap A_j) \gtrsim E(A_i)$ the bound (12.66) is worse than the trivial bound $E(A_1 \cup \ldots \cup A_n) \geq 0$. This kind of pathology occurs in the study of truncated Green functions when the model is strongly coupled.

The events of interest in this work are of the form

$$A_z \;=\; \{\mathrm{supp}(\omega_1) \cap \mathrm{supp}(\omega_2) \ni z\} \tag{12.67}$$

or, especially for the ARW model,

$$A_{(x,y)} \;=\; \{C \cap \underline{\omega} \ni (x,y)\} \tag{12.68}$$

where C is a random set of steps (associated to a random set of bonds). In the critical regime, walks are very likely to meet — except for the weakly-coupled systems — and the probability of multiple intersection is comparable with that of a single intersection. Therefore the bounds (12.65) and (12.66) are too generous to be useful.

Let us see how the dilution trick can improve the lower bound (12.66). First, let us introduce an "expansion parameter" that determines the regime in which (12.66) needs to be improved. To do this, it is convenient to write

$$E(A_i \cap A_j) = E(A_j|A_i)\, E(A_i)\,, \tag{12.69}$$

hence, in (12.66)

$$E(A_1 \cup \ldots \cup A_n) \geq \sum_i \Big[1 - (1/2) \sum_{j:j \neq i} E(A_j|A_i)\Big] E(A_i)\,. \tag{12.70}$$

If B is a quantity such that

$$(1/2) \sum_{j:\, j \neq i} E(A_j|A_i) \leq B \tag{12.71}$$

then

$$E(A_1 \cup \ldots \cup A_n) \geq (1 - B) \sum_{i=1}^{n} E(A_i)\,. \tag{12.72}$$

We see that this bound is useless whenever $B \geq 1$. In particular this occurs in the critical region for strongly-coupled systems, for which B is typically huge. One possible way to obtain a better bound would be to resort to higher-order inequalities of the family (12.62). However, for $i << n$ each term I (i) has approximately $i!$ summands, so the series may involve a delicate balance among large terms of alternating sign, which could be very difficult to control. In the limit $n \to \infty$ the series is quite possibly divergent. On the other hand, one might suspect that the true behavior in (12.72) is not $(1 - B)$ but rather $1/(1 + B)$, similarly to what was found in (12.48). The "dilution trick" provides a rigorous way to obtain a bound of this sort without having to deal explicitly with all the subsequent terms in (12.59). The method is based on the following trivial observation: if S is *any* subset of the "lattice" $\{1, \ldots, n\}$, then

$$\mathrm{I}[A_1 \cup \ldots \cup A_n] \geq \mathrm{I}[\cup_{i \in S} A_i] \tag{12.73}$$

and hence

$$\mathrm{I}[A_1 \cup \ldots \cup A_n] \geq \sum_{i \in S} \mathrm{I}[A_i] - (1/2) \sum_{\substack{i,j \in S \\ i \neq j}} \mathrm{I}[A_i \cap A_j]\,. \tag{12.74}$$

The idea is to take S to be a "diluted" or "thin" (but not *too* thin) subset, so that $\sum_j E(A_i \cap A_j)$ is sufficiently smaller than $E(A_i)$, but $\sum_i E(A_i)$ is not *too* small. The "thinness" of S must be adjusted *a posteriori* in an optimal way, which of course depends on the measure E. In the original version of Erdös and Taylor [164] the events were of the form (12.67) and S was taken to be a periodic sublattice of the lattice $\mathbb{Z}^d$, with spacing optimized as a function of J (distance to criticality). Sokal [476] observed that the computations can be simplified slightly if S is taken to be a random subset of the lattice, in which each point y belongs to S with probability $p_y \in [0, 1]$, independently for each y. Averaging over the randomness, (12.74) becomes

$$\mathrm{I}[A_1 \cup \ldots \cup A_n] \geq \sum_{i=1}^{n} p_i \mathrm{I}[A_i] - (1/2) \sum_{\substack{i,j = 1 \\ i \neq j}}^{n} p_i p_j \mathrm{I}[A_i \cap A_j]\,. \tag{12.75}$$

In fact, the same procedure can be applied to any of the bounds (12.62). In this fashion we obtain the following generalization of (12.62), which could be of independent interest.

Proposition 12.6 (Diluted inclusion-exclusion) *For any family of numbers $p_1, \ldots, p_n \in [0,1]$, and for any even number k,*

$$\mathrm{I}[A_1 \cup \ldots \cup A_n] \ \geq \ \sum_{i=1}^{k} (-1)^{i-1} \sum_{1 \leq j_1 < \ldots < j_i \leq n} p_{j_1} \cdots p_{j_i} \, \mathrm{I}[A_{j_1} \cap \ldots \cap A_{j_i}] \, . \tag{12.76}$$

It would be interesting to find a purely combinatorial/analytic proof of these inequalities.

Let us see how (12.75) can be used to improve (12.72). We take $p_i = p$ for all i and assume (12.71). Repeating the steps leading to (12.72) we obtain

$$E(A_1 \cup \ldots \cup A_n) \ \geq \ [p - p^2 B] \sum_i E(A_i) \, . \tag{12.77}$$

We can now optimize the choice of p. In general, for any $a, b \geq 0$, we have

$$\max_{0 \leq p \leq 1} [ap - bp^2] = \left\{ \begin{array}{ll} a^2/(4b) & \text{if } 0 \leq a < 2b \\ a - b & \text{if } a \geq 2b \end{array} \right\} \geq \frac{a^2}{2a + 4b} \, , \tag{12.78}$$

and hence,

$$E(A_1 \cup \ldots \cup A_n) \ \geq \ \frac{1}{2 + 4B} \sum_i E(A_i) \, . \tag{12.79}$$

This is exactly the bound we expected. Note the importance of the constraint $i \neq j$ in the last term of (12.75): any term $i = j$ would pick up only a factor p, *not* p^2, and hence compete with the first term of (12.75). Note also that the optimal value of p is

$$p \ = \ \min[1, 1/2B] \tag{12.80}$$

which decreases with B, hence amounting to an effective "dilution" in those cases in which the starting bound (12.64) does not work.

Michael Aizenman (private communication) has pointed out to us the following simple proof of a slightly strengthened version of (12.79). Note first that if N is any nonnegative integer-valued random variable and E is any positive measure, we have, by the Schwarz inequality,

$$\begin{aligned} E(N) \ &= \ E(N \cdot I[N \geq 1]) \\ &\leq \ E(I[N \geq 1])^{1/2} \, E(N^2)^{1/2} \end{aligned}$$

and hence

$$E(I(N \geq 1)) \ \geq \ \frac{E(N)^2}{E(N^2)} \tag{12.81}$$

provided that $E(N^2) < \infty$. So take $N = \sum_{i=1}^n I(A_i)$; then

$$\begin{aligned} E(A_1 \cup \ldots \cup A_n) &= E(I[N \geq 1]) \\ &\geq \frac{\left(\sum_{i=1}^n E(A_i)\right)^2}{\sum_{i,j=1}^n E(A_i \cap A_j)} \\ &\geq \frac{\sum_{i=1}^n E(A_i)}{1+2B} \, . \end{aligned}$$

The "passing to the left trick" does not admit a general probabilistic formulation, but it relies on some particular properties of the model, chiefly the path-splitting inequalities. With them one can prove in some cases that for events of the form (12.67)-(12.68)

$$\sum_i E(A_i) \leq BE(A_1^c \cap \ldots \cap A_n^c) \, . \tag{12.82}$$

Such bound is useful to obtain a lower bound for the *complement* of the event $A_1 \cup \ldots \cup A_n$:

$$E(A_1^c \cap \ldots \cap A_n^c) = E(U) - E(A_1 \cup \ldots \cup A_n) \tag{12.83}$$

where U is the universe. Using the first order inclusion-exclusion inequality (12.65) together with (12.82) and passing to the left one gets:

$$E(A_1^c \cap \ldots \cap A_n^c) \geq \frac{E(U)}{1+B} \, . \tag{12.84}$$

An example of such a bound is (12.50). Indeed, using the random-current formalism $\partial\chi/\partial\beta$ can be written as a sum over pairs of currents subject to the constraint that two of the sources are *not* connected by the total current. Therefore, a lower bound on $\beta\partial\chi/\partial\beta$ corresponds to a lower bound of an expression of the form $E(\bigcap_{(x,y)} A^c_{(x,y)})$ for events of the type (12.68). One can obtain (12.50) directly by using the inclusion-exclusion bound (12.63) and the steps (12.82)-(12.84) without resorting explicitly to the Aizenman-Graham bound on u_4 [194] (see also Theorem 5.6 of [12]).

12.3.2 Upper bound on the truncated four-point function for the intersection properties of ordinary random walks (IPORW model)

The analysis of the truncated four-point function in zero magnetic field of the "intersection properties of ordinary random walks" (IPORW) model [185] provides a simple example for the application of the "dilution trick". (This is in fact, the original application of Erdös and Taylor [164].) The truncated four-point kernel for this model is, by definition,

$$K^T(x_1x_2|x_3x_4) = - \sum_{\substack{\omega_1: x_1 \to x_2 \\ \omega_2: x_3 \to x_4}} J^{\omega_1+\omega_2} \, \mathrm{I}[\mathrm{supp}(\omega_1) \cap \mathrm{supp}(\omega_2) \neq \varnothing] \, . \tag{12.85}$$

We are interested in

$$\overline{K}^T \equiv \sum_{x_2,x_3,x_4} K^T(x_1x_2|x_3x_4) . \tag{12.86}$$

The indicator function in (12.85) has the form (12.58) for events (12.67). The starting point is the use of the lower bound (12.74)

$$\begin{aligned} &\mathrm{I}[\mathrm{supp}(\omega_1) \cap \mathrm{supp}(\omega_2) \neq \varnothing] \\ &\geq \mathrm{I}[\mathrm{supp}(\omega_1) \cap \mathrm{supp}(\omega_2) \cap S \neq \varnothing] \\ &\geq \sum_{z \in S} \mathrm{I}[\mathrm{supp}(\omega_1) \cap \mathrm{supp}(\omega_2) \ni z] \\ &\qquad - (1/2) \sum_{\substack{z_1, z_2 \in S \\ z_1 \neq z_2}} \mathrm{I}[\mathrm{supp}(\omega_1) \cap \mathrm{supp}(\omega_2) \ni z_1, z_2] . \end{aligned} \tag{12.87}$$

where S is an arbitrary ("diluting") subset of the lattice L. Taking S to be a random subset of L such that each site y belongs to S with independent probability p and averaging over the randomness we obtain a particular case of (12.75):

$$\begin{aligned} &\mathrm{I}[\mathrm{supp}(\omega_1) \cap \mathrm{supp}(\omega_2) \neq \varnothing] \\ &\geq p \sum_{z} \mathrm{I}[\mathrm{supp}(\omega_1) \cap \mathrm{supp}(\omega_2) \ni z] \\ &\qquad - (p^2/2) \sum_{\substack{z_1, z_2 \\ z_1 \neq z_2}} \mathrm{I}[\mathrm{supp}(\omega_1) \cap \mathrm{supp}(\omega_2) \ni z_1, z_2] . \end{aligned} \tag{12.88}$$

Inserting this into (12.85) we obtain

$$-\overline{K}^T \geq p \sum_{z} E(A_z) - (p^2/2) \sum_{\substack{z_1, z_2 \\ z_1 \neq z_2}} E(A_{z_1} \cap A_{z_2}) \tag{12.89}$$

with:

$$E(A_z) = \sum_{x_2,x_3,x_4} \sum_{\substack{\omega_1 : x_1 \to x_2 \\ \omega_2 : x_3 \to x_4}} J^{\omega_1+\omega_2} \mathrm{I}[\mathrm{supp}(\omega_1) \cap \mathrm{supp}(\omega_2) \ni z] \tag{12.90}$$

and

$$E(A_{z_1} \cap A_{z_2}) = \sum_{x_2,x_3,x_4} \sum_{\substack{\omega_1 : x_1 \to x_2 \\ \omega_2 : x_3 \to x_4}} J^{\omega_1+\omega_2} \mathrm{I}[\mathrm{supp}(\omega_1) \cap \mathrm{supp}(\omega_2) \ni z_1, z_2] \tag{12.91}$$

If we use the identities[5]

[5] These are trivial special cases of the "splitting lemma" of [97].

$$\sum_{\omega:x\to y} J^\omega \, \mathrm{I}[\mathrm{supp}(\omega) \ni z] \;=\; \sum_{\substack{\omega':x\to z\\ \omega'':z\to y}} J^{\omega'+\omega''} \, \mathrm{I}[n_z(\omega')=1] \tag{12.92}$$

and (for $z_1 \neq z_2$)

$$\begin{aligned} &\sum_{\omega:x\to y} J^\omega \, \mathrm{I}[\mathrm{supp}(\omega) \ni z_1, z_2] \\ &\quad= \sum_{\substack{\omega':x\to z_1\\ \omega'':z_1\to z_2\\ \omega''':z_2\to y}} J^{\omega'+\omega''+\omega'''} \, \mathrm{I}[n_{z_1}(\omega')=1] \\ &\qquad\qquad \times \mathrm{I}[n_{z_2}(\omega')=0] \, \mathrm{I}[n_{z_2}(\omega'')=1] \\ &\qquad + \Big[z_1 \Longleftrightarrow z_2\Big] , \end{aligned} \tag{12.93}$$

which are obtained by splitting the path ω the *first* time it hits z (resp. hits z_1 and z_2); we obtain

$$E(A_z) = \sum_{x_2,x_3,x_4} F(x_1,z)F(x_3,z)S_2(z,x_2)S_2(z,x_4) \tag{12.94}$$

and

$$\begin{aligned} E(A_{z_1}\cap A_{z_2}) \;\le\; &\sum_{x_2,x_3,x_4} \Big\{ \Big[F(x_1,z_1)F(x_3,z_1)F(z_1,z_2)^2 S_2(z_2,x_2)S_2(z_2,x_4)\Big] \\ &+ \Big[x_3 \Longleftrightarrow x_4\Big]\Big\} + \Big\{x_1 \Longleftrightarrow x_2\Big\} \end{aligned} \tag{12.95}$$

where we have written

$$F(x,y) \;=\; \sum_{\omega:x\to y} J^\omega \, \mathrm{I}[n_y(\omega)=1] \, . \tag{12.96}$$

We see that for a translation-invariant model:

$$(1/2)\sum_{z_2} E(A_{z_1}\cap A_{z_2}) \;\le\; B\, E(A_{z_1}) \tag{12.97}$$

where

$$B \;\equiv\; 2\sum_{x\neq 0}[F(x,0)]^2 \, . \tag{12.98}$$

Therefore, (12.89) yields the following particular case of (12.77):

$$\begin{aligned} -\overline{u}^T \;&\ge\; (p - p^2 B)\sum_z E(A_z) \\ &\ge\; \frac{1}{2+4B} \sum_{z,x_2,x_3,x_4} F(x_1,z)F(x_3,z)S_2(z,x_2)S_2(z,x_4) \, , \end{aligned} \tag{12.99}$$

where the last inequality was obtained by choosing the optimal value of p according to (12.78).

Finally, we remark that, because of the Markovian nature of the ordinary random walk, we have

$$\begin{aligned} F(x,y) &= \sum_{\omega:x\to y} J^{\omega}\,\{1 \;-\; \mathrm{I}[n_y(\omega)\geq 2]\} \\ &= S_2(x,y) \;-\; \sum_{\omega:x\to y} J^{\omega}\; \mathrm{I}[n_y(\omega)\geq 2] \\ &= S_2(x,y) \;-\; \sum_{\substack{\omega':x\to y\\ \omega'':y\to y}} J^{\omega'+\omega''}\; \mathrm{I}[n_y(\omega')=1]\; \mathrm{I}[n_y(\omega'')\geq 2] \\ &= S_2(x,y) \;-\; F(x,y)[S_2(y,y)-1] \end{aligned}$$

and hence

$$F(x,y) \;=\; S_2(x,y)/S_2(y,y)\;. \tag{12.100}$$

Therefore, (12.99) can be written more compactly as

$$\overline{K}^T \;\geq\; -\frac{\chi^4}{2[S_2(0,0)]^2+\;8B_0'}\quad, \tag{12.101}$$

where B_0' denotes the restricted bubble diagram

$$B_0' \;=\; \sum_{x\neq 0}[S_2(0,x)]^2\quad. \tag{12.102}$$

We notice that for the IPORW model

$$\beta\frac{\partial\chi}{\partial\beta} \;=\; |J|\chi^2\;. \tag{12.103}$$

hence, from (12.57) Therefore, (12.101) can also be written

$$\overline{K}^T \;\geq\; -\frac{|J|}{2[S_2(0,0)]^2+\;8B_0'}\;(\beta\frac{\partial\chi}{\partial\beta})^2\;. \tag{12.104}$$

For SAW and Ising models, by contrast, (12.103) is replaced by an inequality, and (12.104) would be strictly weaker than (12.101). We emphasize that this subsection is just a rephrasing of the proof of Erdös and Taylor [164, Theorem 7] (see also [476]).

We remark that a result like (12.101) can alternatively be proven [185] by using an "skeleton inequality" in the Edwards-like generalization of the IPORW model together with the monotonicity of K^T in λ.

12.3.3 Upper bound on the truncated four-point function for the SAW

The study of the SAW would be the natural next step after the IPORW model. However, in this case we are not able to complete the analysis successfully and obtain a bound analogous to (12.101) or even to (12.104). The analysis of S_4^T for the SAW proceeds along the same lines as for the IPORW. The truncated four-point kernel for the SAW is, by definition,

$$K^T(x_1x_2|x_3x_4) = - \sum_{\substack{\omega_1 : x_1 \to x_2 \\ \omega_2 : x_3 \to x_4}} \varrho(\omega_1)\varrho(\omega_2) \, \mathrm{I}[\mathrm{supp}(\omega_1) \cap \mathrm{supp}(\omega_2) \neq \varnothing] \, . \tag{12.105}$$

where $\varrho(\omega) = 1$ if ω is self-avoiding and 0 otherwise. Now we use the "diluted" inclusion-exclusion inequality (12.88) to obtain

$$\begin{aligned} -K^T(x_1x_2|x_3x_4) \geq \; & p \sum_z F(x_1, z, x_2) F(x_3, z, x_4) \\ & -p^2 \sum_{\substack{z_1, z_2 \\ z_1 \neq z_2}} \Big\{ F(x_1, z_1, z_2, x_2) F(x_3, z_1, z_2, x_4) + [x_3 \Longleftrightarrow x_4] \Big\} \end{aligned} \tag{12.106}$$

where

$$\begin{aligned} F(x, y, z) \; & \equiv \; \sum_{\omega : x \to y} \varrho(\omega) \, \mathrm{I}[\mathrm{supp}\omega \ni z] \\ & = \; K(x, z)\delta_{z,y} + \sum_{z'} J_{zz'} K(xz|z'y) \end{aligned} \tag{12.107}$$

and

$$\begin{aligned} F(x, z_1, z_2, y) \; & \equiv \; \sum_{\omega : x \to y} \varrho(\omega) \, \mathrm{I}[\omega \text{ visits } z_1 \text{ and } \textit{subsequently} \text{ visits } z_2] \\ & = \; \sum_{z_1'} J_{z_1 z_1'} K(xz_1|z_1'z_2)\delta_{z_2,y} + \sum_{z_1', z_2'} J_{z_1 z_1'} J_{z_2 z_2'} K(xz_1|z_1'z_2|z_2'y) \end{aligned} \tag{12.108}$$

where the equality in (12.107) is the path-splitting identity (11.47) for the SAW, and the equality in (12.108) is proved similarly. Pictorially:

$-K^T(x_1x_2|x_3x_4) \;\geq\; p$ 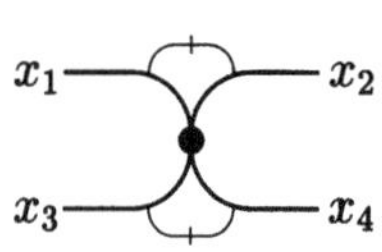

$-p^2 \Big[$ 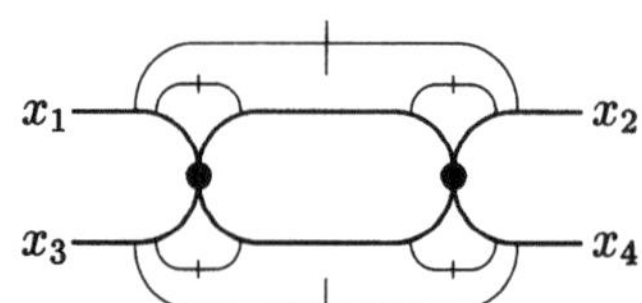

$+ \; (x_3 \Longleftrightarrow x_4) \Big]$

(12.109)

where each symbol stands for the requirement that the walks connected by the symbol do not intersect. If we sum over x_2, x_3, x_4 we obtain

$$-\overline{K}_T \ \geq \ p(\frac{\partial \chi}{\partial \beta} \ + \ \chi)^2 \ - \ 2p^2 T \tag{12.110}$$

where

$$T \ = \ \sum_{x_2, x_3, x_4} \ \sum_{\substack{z_1, z_2 \\ z_1 \neq z_2}} \ F(x_1, z_1, z_2, x_2) F(x_3, z_1, z_2, x_4) \ . \tag{12.111}$$

At this point we find ourselves unable to bound T in an optimal way. The problem can be understood from the diagrammatic formulation (12.109) where T corresponds to the last diagram. In order to apply the optimization procedure (12.78), we would like to show that the last diagram in (12.109) can be bounded by a "bubble diagram" times the first diagram, i.e.,

$$T \ \leq \ \text{const} \times B(\beta \frac{\partial \chi}{\partial \beta} + \chi)^2 \qquad (?). \tag{12.112}$$

Looking at the diagram for T we see that if we neglect the four "little" constraints we may be able to pull a bubble factor from the part of the diagram. Unfortunately, what is left is not quite the same as the first diagram because in the diagram for T, z_1 and z_2 *are not necessarily "neighbors"*[6]. It is, of course, possible to make an even cruder estimate on T by neglecting *all* the constraints ; we then get

$$T \ \leq \ \text{const} \times B\chi^4 \tag{12.113}$$

and hence, by (12.78),

$$-\overline{K}_T \ \geq \ \text{const} \times \frac{(\beta \frac{\partial \chi}{\partial \beta} + \chi)^4}{B\chi^4} \quad . \tag{12.114}$$

This is a terrible bound in dimension $d < 4$, but it is not so bad in dimension $d \geq 4$ (see Section 14.1).

Even though the standard dilution trick seems to fail for the SAW, we do believe that some bound similar to (12.104) may be true. We are encouraged by the fact that the other example of a strongly-coupled model — the Ising model — does exhibit such a bound, whose proof, however, requires more than the random-walk representation (see next).

---

[6]Indeed, a heuristic scaling computation shows that if $d < 4$ (more precisely, if $d\nu < 3\gamma - 1$), then the dominant contribution to the sum comes from $z_1 - z_2$ of order ξ — *not* of order 1 — and one recovers the crude bound (12.113). Thus, it appears that in order to do better than (12.113) it is essential to keep the four "little" constraints.

12.3.4 Upper bound on the truncated four-point function for the ARW (Ising) model

The "dilution trick" — involving events of the form (12.68) — can be successfully applied to the ARW (Ising) model, but working on inequalities derived from the more detailed random-current formalism. Indeed, by mimicking the proof of Theorem 5.7 of reference [12] one can prove [194]

$$
\begin{aligned}
&\langle \sigma_0; \sigma_x; \sigma_y; \sigma_z \rangle \leq \\
&\quad -(1/2) \sum_{u,v} p_{\{u,v\}} J_{u,v} \langle \sigma_u \sigma_x \rangle \langle \sigma_u \sigma_0 \rangle \langle \sigma_y \sigma_z ; \sigma_u \sigma_v \rangle \\
&\quad + (1/4) \sum_{u,v,s,t} p_{\{u,v\}} p_{\{s,t\}} J_{u,v} J_{s,t} [(\langle \sigma_x \sigma_s \rangle \langle \sigma_u \sigma_s \rangle \langle \sigma_u \sigma_0 \rangle) + (s \Longleftrightarrow t)] \\
&\qquad \times \Big[\big\{ [(2 \langle \sigma_z \sigma_u \rangle \langle \sigma_v \sigma_s \rangle \langle \sigma_y \sigma_t \rangle) + (s \Longleftrightarrow t)] + [(u,v) \Longleftrightarrow (s,t)] \big\} \\
&\qquad\quad + (1/2) \big\{ u \Longleftrightarrow v \big\} \Big]
\end{aligned}
\tag{12.115}
$$

(each permutation refers to the preceding factor with the same delimiter), for any family $\{p_{\{x,y\}}\}$ of the numbers interval $[0,1]$. With the choice $p_{\{x,y\}} = p$ for all bonds, (12.115) yields the summed version

$$
\overline{u}_4 \;\leq\; -p\chi^2 (\beta \frac{\partial \chi}{\partial \beta}) \;+\; 12 p^2 |J|^2 \chi^4 B_0 \,, \tag{12.116}
$$

for any $p \in [0,1]$. Choosing p as in (12.78), namely

$$
p \quad = \quad \min \left(1, \; \frac{\partial \chi / \partial \beta}{24 |J|^2 \chi^2 B_0} \right) \tag{12.117}
$$

we therefore obtain

$$
\overline{u}_4 \;\leq\; - \frac{\left(\beta \frac{\partial \chi}{\partial \beta} \right)^2}{48 |J|^2 B_0 \;+\; 2\chi^{-2} \left(\beta \frac{\partial \chi}{\partial \beta} \right)} \tag{12.118}
$$

(the second term in the denominator is insignificantly small in the vicinity of the critical point). This result is analogous to (12.104). However, we cannot obtain a result analogous to (12.101) because for the Ising model χ^2 and $\beta \partial \chi / \partial \beta$ are not proportional as they are for the IPORW model. Instead of (12.103) we now have (12.50) (whose proof involves the "passing to the left trick"), which incorporates an extra factor B_0^{-2}. Inserting (12.50) in (12.118) we obtain:

$$
\overline{u}_4 \;\leq\; - \frac{\chi^4}{48 B_0 (1 + 2|J| B_0)^2} \;+\; \varepsilon \tag{12.119}
$$

where ε becomes negligible in the critical region. The two extra powers of the bubble brought by (12.50) can be associated, loosely speaking, with the strong coupling present in the *self-interaction* of each of the two walks of the expression for $\partial\chi/\partial\beta$. In the IPORW model these self-interactions vanish — only the interaction *between* walks is strongly coupled — hence the two extra factors B_0^{-1} are missing.

12.4 The kernel $K(x,y)$ and the truncated two-point function in nonzero magnetic field

12.4.1 The kernel $K(x,y)$ and the magnetization

For the nonzero-field case, the simplest kernels (10.8) are $S_1(x) = K(x\bullet)$ and

$$K(x,y) \;=\; \sum_{\omega:x\to y} \varrho_0(\omega)\;. \tag{12.120}$$

In the zero-magnetic-field situation we have $K(x,y) = S_2(x,y)$, which is monotonic in the coupling constants J (by the second Griffiths inequality). However, in the presence of magnetic field, $K(x,y)$ becomes a rather mysterious object with no immediate physical meaning, no known monotonicity in J (except for the baby CIW models), and whose definition in the ARW models depends on the arbitrary choices associated with the notion of consistency. Nevertheless, $K(x,y)$ appears quite naturally when handling the random-walk expansions for the Green functions of low order. For instance, from (10.10) and (10.11) we have for the CIW and BFS models:

$$S_1(x) \;=\; \sum_{y\in L} K(x,y)h_y\;. \tag{12.121}$$

The analogous relation for the ARW models is more involved. In fact, we do not have an equality but rather inequalities:

$$\sum_{y\in L} K(x,y)\,\frac{\tanh(h_y)}{1+S_1(y)\tanh(h_y)} \;\le\; S_1(x) \;\le\; \sum_{y\in L} K(x,y)\tanh(h_y) \tag{12.122}$$

The proof of (12.122), which uses random currents, can be found in [12].

The two previous expressions look more suggestive for translation invariant models. For such models let us adopt the standard magnetic notation:

$$M \;\equiv\; S_1(x) \tag{12.123}$$

and

$$|J| \;=\; \sum_{y} J_{xy} \tag{12.124}$$

Proposition 12.7 *Let us assume the systems to be translation-invariant, then:*

a) For CIW and BFS models

$$\sum_{y \in L} K(x,y) \;=\; \frac{M}{h} \,. \tag{12.125}$$

b) For the ARW models

$$\frac{M}{\tanh(h)} \;\leq\; \sum_{y} K(x,y) \;\leq\; \frac{M}{\tanh(h)}[1 + M\tanh(h)] \,. \tag{12.126}$$

For h small (12.126) is just as good as (12.125).

12.4.2 The kernel $K(x,y)$ and the two-point function

The kernel $K(x,y)$ is also related to the two-point function S_2:

$$S_2(x,y) \;=\; K(x,y) \;+\; \sum_{\substack{\omega_1 : x \to \bullet \\ \omega_2 : y \to \bullet}} \varrho_{11}(\omega_1, \omega_2) \tag{12.127}$$

As the last term in the RHS is positive, we have that for baby polymer-chain, BFS and ARW models:

$$K(x,y) \;\leq\; S_2(x,y) \,. \tag{12.128}$$

12.4.3 Monotonicity of $K(x,y)$ in h, and another upper bound on K

A bound sharper than (12.128) can be obtained from the properties discussed in Section 11.3. Indeed, it follows from (11.80) that $\partial K(x,y)/\partial h_z \leq 0$. On the other hand, $[K(x,y)]_{h=0} = [S_2(x,y)]_{h=0}$. Hence

$$K(x,y) \;\leq\; [K(x,y)]_{h=0} \;=\; [S_2(x,y)]_{h=0} \,. \tag{12.129}$$

for all nonnegative h.

12.4.4 The kernel $K(x,y)$ and the truncated two-point function [lower bound on $K(x,y)$]

The truncated two-point function is defined as

$$S_2^T(x_1,x_2) \;=\; S_2(x_1,x_2) \;-\; S_1(x_1)S_1(x_2) \,. \tag{12.130}$$

Introducing truncated kernels analogous to (12.8) — but in which some of the arguments can be dots "•" — we have:

$$S_2^T(x_1,x_2) = K(x_1x_2) + K^T(x_1\bullet|x_2\bullet) \tag{12.131}$$

with

$$K^T(x_1\bullet|x_2\bullet) = \sum_{\substack{\omega_1:x_1\to\bullet\\ \omega_2:x_2\to\bullet}} \varrho_{11}^T(\omega_1,\omega_2)\ , \tag{12.132}$$

where we denote, for $j_1, j_2 = 0, 1$:

$$\varrho_{j_1j_2}^T(\omega_1,\omega_2) = \varrho_{j_1j_2}(\omega_1,\omega_2) - \varrho_{j_1}(\omega_1)\varrho_{j_2}(\omega_2)\ . \tag{12.133}$$

Proceeding in the same way as for the truncated weights ϱ_{00}^T analyzed in Section 12.2, one obtains from Theorem 11.1 that for repulsive polymer chains

$$\varrho_{j_1j_2}^T(\omega_1,\omega_2) \le 0\ , \tag{12.134}$$

while for the BFS and ARW models (Theorems 11.2 and 11.5)

$$\sum_{\omega_2:x\to\bullet} \varrho_{j_11}^T(\omega_1,\omega_2) \le 0\ . \tag{12.135}$$

As a consequence we have the following analogue of Theorem 12.2.

Theorem 12.8 *Repulsive polymer-chain, BFS and ARW models satisfy*

$$K^T(x_1y|x_2\bullet) \le 0\ , \tag{12.136}$$

$$K^T(x_1\bullet|x_2\bullet) \le 0\ , \tag{12.137}$$

and hence, from (12.131)

$$S_2^T(x_1,x_2) \le K(x_1x_2)\ . \tag{12.138}$$

The "summed" version of (12.138) combined with (12.121) yields the so-called *weak GHS inequality* [223]:

Corollary 12.9 *(weak GHS inequality) For repulsive polymer-chain models, BFS and ARW models*

$$S_1(x) \ge \sum_y S_2^T(x,y)h_y\ . \tag{12.139}$$

The relation between (12.139) and the GHS inequality is provided by the formula[7]

$$\frac{\partial}{\partial h_z}\Big[\langle\varphi_x\rangle - \sum_y\langle\varphi_x;\varphi_y\rangle h_y\Big] = -\sum_y\langle\varphi_x;\varphi_y;\varphi_z\rangle h_y\ . \tag{12.140}$$

[7]We use the notation $\langle A;B\rangle \equiv \langle AB\rangle - \langle A\rangle\langle B\rangle$ and $\langle A;B;C\rangle \equiv \langle ABC\rangle - \langle A\rangle\langle BC\rangle - \langle B\rangle\langle AC\rangle - \langle C\rangle\langle AB\rangle + 2\langle A\rangle\langle B\rangle\langle C\rangle$.

This shows that (12.139) can be interpreted as an integrated version of the GHS inequality. Moreover, (12.139) implies that $M(h)/h$ is decreasing in h, i.e., that the magnetization curve lies above those of its secant lines which start at the origin. [By contrast, GHS implies that $M(h)$ is concave, i.e., that the magnetization curve lies above *all* of its secant lines]. This fact is sufficient to prove the continuity (in fact, local Lipschitz continuity) of $M(h)$ at $h \neq 0$, and hence the absence of phase transitions at $h \neq 0$ (see [223] for details).

For translation-invariant models (12.139) can be written in a more familiar form. Indeed, for all such models we extend (12.28) to the nonzero-field case by defining

$$\chi = \sum_y u_2(0,y) \tag{12.141}$$

where

$$u_2(x_1,x_2) = \begin{cases} S_2^T(x_1,x_2) & \text{for RW models} \\ \langle \varphi_{x_1}; \varphi_{x_2} \rangle & \text{for spin models} \end{cases} . \tag{12.142}$$

We remind the reader that for BFS models $S_2^T(x,y) = \langle \varphi_x; \varphi_y \rangle$, and hence

$$\chi_{\text{BFS-RW}} \equiv \sum_y S_2^T(x,y) = \chi_{\text{spin-model}} \equiv \sum_y \langle \varphi_x; \varphi_y \rangle . \tag{12.143}$$

The distinction between $S_2^T(x,y)$ and $\langle \varphi_x; \varphi_y \rangle$ is needed only for the ARW/Ising models (with $x = y$) because in this case

$$S_2^T(x,x) \underset{\neq}{>} \langle \sigma_x; \sigma_x \rangle = 1 - M^2 . \tag{12.144}$$

Hence,

$$\chi_{\text{ARW}} \equiv \sum_y S_2^T(x,y) \underset{\neq}{>} \chi_{\text{Ising}} \equiv \sum_y \langle \sigma_x; \sigma_x \rangle . \tag{12.145}$$

We notice that from (12.131) and (12.137), plus the fact that $K(x,x) = 1$, we have for the ARW model

$$S_2^T(x,x) \leq 1 , \tag{12.146}$$

and hence for the ARW model

$$\chi_{\text{Ising}} \underset{\neq}{<} \chi_{\text{ARW}} \leq \chi_{\text{Ising}} + M^2 . \tag{12.147}$$

With definition (12.141), (12.139) becomes, for translation-invariant models:

$$\chi_{\text{BFS/ARW}} \leq \frac{M}{h} , \tag{12.148}$$

which implies, from the leftmost inequality in (12.147)

$$\chi_{\text{Ising}} \leq \frac{M}{h} . \tag{12.149}$$

Henceforth we shall make no reference to χ_{ARW}; we therefore sometimes will denote χ_{Ising} simply by χ .

12.4.5 Non-trivial upper bound on $K(x,y)$

By combining (12.129) and (12.138) we see that the behavior of the kernel $K(x,y)$ as a function of $|x-y|$ is intermediate to that of the truncated two-point function and that of the untruncated two-point function at zero magnetic field. However, there is still a big gap between these bounds for temperatures below criticality, because in this range $S_2^T(x,y)$ vanishes when $|x-y| \to \infty$ while $S_2(x,y)$ does not. In this section we discuss inequalities complementary to that of Theorem 12.8 which give more information about the actual behavior of K.

Corollary 11.9 implies the following generalization of (12.15)

$$\varrho^T_{j_1j_2}(\omega_1,\omega_2) \geq \begin{cases} 0 & \text{if } \mathrm{supp}(\omega_1) \cap \mathrm{supp}(\omega_2) = \varnothing \\ -\varrho_{j_1}(\omega_1)\varrho_{j_2}(\omega_2) & \text{otherwise} \end{cases} \tag{12.150}$$

valid for the baby CIW, BFS and ARW models. This bound complements (12.134)–(12.135). From this and Proposition 11.11 one obtains, repeating the steps leading to Proposition 12.3:

Proposition 12.10
a) For the baby (generalized) CIW, BFS and ARW models

$$\begin{aligned} K^T(x_1\bullet|x_2\bullet) \;\geq\; & -\sum_z \Big[K(x_1z)h_z + \sum_{z_1} J_{zz_1}K(x_1z|z_1\bullet)\Big] \\ & \times\Big[K(x_2z)h_z + \sum_{z_2} J_{zz_2}K(x_2z|z_2\bullet)\Big] \end{aligned} \tag{12.151}$$

For the ARW model J_{zz_i} and h_z can be replaced by $\tanh(J_{zz_i})$ and $\tanh(h_z)$ respectively.
b) For the ARW model

$$K^T(x_1\bullet|x_2\bullet) \;\geq\; -\sum_z K(x_1z|z\bullet)K(x_2z|z\bullet) \tag{12.152}$$

Combining these expressions with (12.131) and (12.136) we arrive at the following analogue of Theorem 12.4:

Theorem 12.11

$$K(x_1,x_2) \;\leq\; S_2^T(x_1,x_2) + V(x_1\bullet|x_2\bullet) \tag{12.153}$$

with
a) For repulsive simple baby CIW, BFS and ARW models

$$V(x_1\bullet|x_2\bullet) \;=\; \sum_z K(x_1,z)K(x_2,z)\Big[h_z + \sum_{z_1} J_{zz_1}S_1(z_1)\Big]^2 . \tag{12.154}$$

b) For ARW models

$$V(x_1 \bullet | x_2 \bullet) = \sum_z K(x_1, z) K(x_2, z) [S_1(z)]^2 . \tag{12.155}$$

We remark that for ARW/Ising models (12.155) is a slightly better bound than (12.154), since

$$\begin{aligned} S_1(z) \equiv \langle \sigma_z \rangle &\leq \tanh h_z + \sum_{z_1 \neq z} (\tanh J_{zz_1}) \langle \sigma_{z_1} \rangle \\ &\leq h_z + \sum_{z_1} J_{zz_1} S_1(z_1) \end{aligned}$$

by Griffiths' third inequality [271] and the ghost-spin trick [270].

Results of this type were first proven in [223], and further discussion may be found there. We emphasize that the proof is virtually identical to that of the Aizenman-Fröhlich inequalities (12.18)-(12.25).

The following corollary is a consequence of the "summed" version of Theorem 12.11. Indeed, by summing (12.153) over x_2 and using (12.125) or (12.126) in the result (and (12.147) for ARW models), we obtain:

Corollary 12.12
a) For translation-invariant baby repulsive simple CIW and BFS models:

$$\frac{M}{h} \leq \chi + \left(\frac{M}{h}\right)^2 [h + |J| M]^2 . \tag{12.156}$$

b) For the translation-invariant Ising model

$$\frac{M}{\tanh h} \leq \chi_{\text{Ising}} + \left(\frac{M}{\tanh h}\right) 2 [M^2 \tanh h + M]^2 + M^2 . \tag{12.157}$$

Remark. *Using the random-current representation, one can prove* (12.157) *without the additive term* M^2.

Corollary 12.12 will lead to the critical-exponent inequalities $\delta \geq 3$ and $\gamma \geq 2\beta$, which are "mean-field bounds"; see Section 14.3 for details.

12.4.6 Once-improved upper bound on *K(x,y)*

The above inequalities are probably the best that can be obtained using only the random-walk expansion. However, by resorting to the more powerful random-current representation, Aizenman, Barsky and Fernández [10] were able to improve the upper bounds (12.157) and (for some one-component spin models) (12.156). Their result is the following:

Theorem 12.13 (Aizenman, Barsky and Fernández) [10]
a) For ARW models:

$$K(x,y) \;\le\; \langle\sigma_x;\sigma_y\rangle + \sum_{k,l} K(x,k)\tanh(J_{kl})\Big[(\langle\sigma_k\rangle^2 + \langle\sigma_l\rangle^2)\delta_{l,y} + \langle\sigma_k\rangle\langle\sigma_y;\sigma_k\sigma_l\rangle\Big]\,. \tag{12.158}$$

and

$$\langle\sigma_x\rangle \;\le\; \sum_z(\tanh h_z)\langle\sigma_x;\sigma_z\rangle + \langle\sigma_x\rangle^3 + \sum_{z,z'}(\tanh J_{zz'})\langle\sigma_x;\sigma_z\sigma_{z'}\rangle\langle\sigma_z\rangle^2\,. \tag{12.159}$$

b) For ferromagnetic 1-component spin models with single-spin measures in the Griffiths-Simon class [466] (in short "models in the GS class"; this class of models includes the Ising and 1-component φ^4 models, among others):

$$\begin{aligned}\langle\varphi_x\rangle \;\le\; & \sum_v h_v\langle\varphi_x;\varphi_v\rangle \;+\; h_x\langle\varphi_x\rangle^2 + \sum_y J_{xy}\langle\varphi_x\rangle^2\langle\varphi_y\rangle \\ & + \sum_{u,v}\Big[h_v\langle\varphi_v\rangle + \sum_y J_{vy}\langle\varphi_v\rangle\langle\varphi_y\rangle\Big]J_{uv}\langle\varphi_x;\varphi_u\varphi_v\rangle\,. \end{aligned} \tag{12.160}$$

Theorem 12.13 can be interpreted as a "once-improved" version of Theorem 12.11 in the same sense as the Aizenman-Graham inequality (12.153) is a "once-improvement" of the Aizenman-Fröhlich inequality (12.24). In fact inequality (12.160) is "almost" the Aizenman-Graham inequality with some sites identified with the site of a "ghost" spin. The only difference is that terms that were completely negligible in the situation of interest in [15] — and therefore were ignored in (12.32)–(12.33) — become now the dominant terms [10]. The fact that (12.158) is an improvement over (12.153)-(12.155) can be explicitly shown via the inequality

$$\begin{aligned}\langle\varphi_x;\varphi_k\varphi_l\rangle \;&\le\; \langle\varphi_k\rangle\langle\varphi_x;\varphi_l\rangle \;+\; (k \Longleftrightarrow l) \\ &\le\; \langle\varphi_k\rangle K(x,l) \;+\; (k \Longleftrightarrow l)\,. \end{aligned}$$

which follows from the GHS inequality and (12.138).

The "summed" versions of (12.158)-(12.160) are

Corollary 12.14 [10]
a) For ARW models:

$$\frac{M}{h} \;\le\; \chi \;+\; \left(\frac{M}{h}\right)^2\left[hM + h\beta\frac{\partial M}{\partial\beta}\right]\,. \tag{12.161}$$

b) For ferromagnetic models in the GS class:

$$\frac{M}{h} \;\le\; \chi \;+\; \left[|J|\left(\frac{M}{h}\right)^2 \;+\; \frac{M}{h}\right]\left[hM + h\beta\frac{\partial M}{\partial\beta}\right]\,. \tag{12.162}$$

These formulas are manifestly an improvement of Corollary 12.12, since

$$\beta\frac{\partial M}{\partial\beta} \;\leq\; |J|M\chi \;\leq\; |J|\frac{M^2}{h} \tag{12.163}$$

by the GHS inequality and the weak GHS inequality, respectively. This improvement is especially significant for temperatures below the critical temperature and for $h \simeq 0$, since $\beta\partial M/\partial\beta$ is finite as $h \downarrow 0$ while M^2/h diverges. Corollary 12.14 can thus be used directly on the coexistence curve ($T < T_c, h = 0^+$), while Corollary 12.12 must be used away from the coexistence curve and then combined with an "extrapolation principle" (see Section 14.3).

12.5 Truncated three-point function

The only general result available for the truncated three-point function is the GHS inequality [279, 491, 162]

$$\langle\varphi_x;\varphi_y;\varphi_z\rangle \;\leq\; 0\;, \tag{12.164}$$

valid for the 1-component spin models of the Ellis-Monroe-Newman class (which includes the Ising model as a limiting case). We note that the GHS inequality (12.164) is *not* true for general BFS models with single-spin measures of BFS class; this follows from a necessary and sufficient condition for the GHS inequality due to Ellis and Newman [163].

However, as for the case of S_4 for h=0, sharper (non-Gaussian) upper bounds are of interest. For the ARW models such improvement is possible because of the existence of the random-current representation. Resorting to the full power of such formalism, an improved bound was obtained in [12] through a procedure that included an application of the "dilution trick". A simplified version of this procedure yielded later the upper bound (12.115) on $\langle\sigma_0;\sigma_x;\sigma_y;\sigma_z\rangle$. One of the complications of the proof of the non-Gaussian upper bound on $\langle\sigma_0;\sigma_x;\sigma_y\rangle$ — as compared to that for the upper bound of $\langle\sigma_0;\sigma_x;\sigma_y;\sigma_z\rangle$ — is that it involves the kernel K whose dependence on the coupling constants is not known to be monotonic. A monotonic dependence on the couplings — which is crucial for the success of the proof — can only be restored upon summation via the upper bound on (12.126). For this reason, the upper bound on $\langle\sigma_0;\sigma_x;\sigma_y\rangle$ does not admit a "pointwise version" comparable to (12.115); there is only a "summed version" analogous to (12.118).

Let us denote

$$\frac{\partial\chi}{\partial h} \;=\; \sum_{x,y}\langle\sigma_0;\sigma_x;\sigma_y\rangle\;. \tag{12.165}$$

We then have [12]:

Theorem 12.15 *For translation-invariant Ising models*

$$\frac{\partial \chi}{\partial h} \leq -\frac{[1 - O(B_0 h/M)]_+}{48 B_0 (1 + 2|J| B_0)^2} (\tanh h) \chi^4 \tag{12.166}$$

where B_0 *denotes the "bubble diagram"* (12.49) *evaluated at* $h = 0$ *and* $[x]_+ = \max(x, 0)$.

On the other hand, it is even simpler to prove a *lower* bound on $\langle \sigma_0; \sigma_x; \sigma_y \rangle$ analogous to the "tree inequality" (12.26):

Theorem 12.16 [12] *For Ising models*

$$\begin{aligned} \langle \sigma_0; \sigma_x; \sigma_y \rangle \ \geq\ & -2 \Big\{ \sum_u K(y,u) (\tanh h_u) K(x,u) K(0,u) \\ & + \sum_{u,v} J_{uv} K(y,u) \langle \sigma_v \rangle \big[K(x,u) K(0,u) + K(x,v) K(0,v) \big] \Big\} . \end{aligned} \tag{12.167}$$

The RHS of (12.167) can be associated to to a tree diagram as in the RHS of (12.28), but with $0 \text{ —— } u \equiv K(0,u)$ [use (12.126)]. Summing over x and y and applying the leftmost inequality of (12.126) we obtain:

Corollary 12.17 *For translation-invariant Ising models*

$$-4|J| \, [1 + O(h/M)] \, [1 + M \tanh h]^3 \, \frac{M^4}{(\tanh h)^3} \leq \frac{\partial \chi}{\partial h} . \tag{12.168}$$

We note remarkable similarities between (12.166)/(12.168) and the bounds

$$-3\chi^4 \leq \overline{u}_4 \leq -\frac{\chi^4}{48 B_0 (1 + 2|J| B_0)^2} + \varepsilon \tag{12.169}$$

obtained respectively by summing (12.26) over three of the sites, and from (12.119). Both $\partial \chi / \partial h$ and $\overline{u}_4$ have "tree" lower bounds, and the respective upper bounds have basically the same form as the lower bound except for a factor B_0^{-3}. These similarities can be interpreted as follows, the random-walk expansion of S_3^T involves two terms: one with two walks and another with three walks ($k = 1$ and 2 respectively in (12.175) below). The former has a trivial h-dependence proportional to the first power of h while the trivial dependence of the latter is of order h^2. At small h it is not farfetched to expect that the term with two walks dominate, and hence the expansion for S_3^T becomes almost identical to that of S_4^T (with one lattice site transformed into a "ghost" spin).

Theorem 12.16 and Corollary 12.17 can be generalized to spins of the GS class [194]. To write these generalizations we need to introduce the kernel

$K^{GS}(x,y)$ for spin models in the GS class. This kernel is defined as for the Ising model, but using the representation of GS models as blocks of ferromagnetically coupled Ising spins [466, 5]. The resulting expression is not physically interesting because it includes references to the auxiliary Ising spins forming the blocks. Rather than entering into such details, we shall provide here only the most relevant (for our purposes) formula involving this kernel K^{GS} [cf. (12.122)]:

$$\sum_{y \in \mathbb{L}} K^{GS}(x,y) h_y \frac{a_y}{1 + S_1(y) h_y a_y} \;\leq\; S_1(x) \;\leq\; \sum_{y \in \mathbb{L}} K^{GS}(x,y) h_y a_y \,. \qquad (12.170)$$

Here the numbers $a_y \in (0,1]$ depend only on h_y and the single-spin measure; $a_y \to 1$ as $h_y \to 0$; and $a_y \equiv 1$ for φ^4 spins. As a result, we have in the translation-invariant case

$$\frac{M}{h(1-\varepsilon(h))} \;\leq\; \sum_y K^{GS}(x,y) \;\leq\; \frac{M}{h(1-\varepsilon(h))} \Big[1 + (1+\varepsilon(h)) M h\Big] \,, \qquad (12.171)$$

where $\varepsilon(h) \to 0$ as $h \to 0$ in general, and $\varepsilon(h) \equiv 0$ for φ^4 spins (and of course for Ising spins).

The generalization of Theorem 12.16 takes the form:

Theorem 12.18 [194] *For spin models in the GS class:*

$$\begin{aligned} \langle \varphi_0 ; \varphi_x ; \varphi_y \rangle \;\geq\; & -2 \Big\{ \sum_{u,v} K^{GS}(y,u) h_u K^{GS}(x,u) J_{uv} K^{GS}(0,v) \\ & + \sum_{u,v,p} J_{uv} K^{GS}(y,u) \langle \varphi_v \rangle [K^{GS}(x,u) J_{up} K^{GS}(0,p) \\ & \qquad + K^{GS}(x,v) J_{vp} K^{GS}(0,p)] \Big\} \,. \end{aligned} \qquad (12.172)$$

Using (12.171) one gets:

Corollary 12.19 [194] *For translation-invariant models in the GS class:*

$$-4\,|J|^2 \,[1 + O(h/M)] \left[\frac{1 + hM[1+\varepsilon(h)]}{1-\varepsilon(h)} \right]^3 \frac{M^4}{h^3} \;\leq\; \frac{\partial \chi}{\partial h} \qquad (12.173)$$

where $\varepsilon(h) \to 0$ as $h \to 0$. Here $\varepsilon(h)$ depends on the single-spin measure; it is identically zero for φ^4 and Ising spins.

Unfortunately, we do not have at present any analogous generalization of Theorem 12.15 for GS spins.

Inequality (12.166), combined with extrapolation principles derived from the GHS inequality [12], implies (together with the results of Sections 12.4.5 and 12.4.6) the following bounds on the critical exponents δ and $\widehat{\beta}$ of the Ising model [12]:

a) for $d > 4$, δ and $\widehat{\beta}$ take the mean-field values $\delta = 3$, $\widehat{\beta} = 1/2$;

b) for $d = 4$, δ and $\widehat{\beta}$ take the mean-field values with possible logarithmic corrections involving at most a logarithm to the first power for $\widehat{\beta}$ and to the $3/2$ power for δ.

See Section 14.3 for more details.

The lower bound in (12.166) has also some mild consequences on the triviality of the continuum limit taken through paths in the nonsymmetric region of the (β, h)-plane.

12.6 Truncated Green functions of higher order

The general expression for the truncated Green functions is

$$S_n^T(x_1,\ldots,x_n) = \sum_{i=1}^{n}(-1)^{i-1}(i-1)! \sum_{\substack{\{I_1,\ldots,I_i\} \\ \text{n.e.s.s. partition} \\ \text{of } (i,\ldots,n)}} \prod_{j=1}^{i} S_{|I_j|}(x_{I_j}) \,. \qquad (12.174)$$

Here "n.e.s.s. partition" means "partition by nonempty subsequences". It can be proven (see Section 12.7) that the "operation" of truncation "commutes" with the relation (10.10) in the sense that

$$S_n^T(x_1,\ldots,x_n) = \sum_{k=[(n+1)/2]}^{n} \sum_{\pi\in Q_{2k,n}} K^T(x_{\pi(1)}x_{\pi(2)}|\ldots|x_{\pi(2k-1)}x_{\pi(2k)}) \qquad (12.175)$$

where K^T corresponds to the truncation of the kernel K considered as a function over the sets of pairs:

$$K^T(x_1y_1|\ldots|x_ky_k) = \sum_{i=1}^{k}(-1)^{i-1}(i-1)! \sum_{\substack{\{I_1,\ldots,I_i\} \\ \text{n.e.s.s. partition} \\ \text{of } (i,\ldots,k)}} \prod_{j=1}^{i} K((xy)_{I_j}) \qquad (12.176)$$

(some of the "y" may be "•"). Moreover, "truncation" also "commutes" with the relation (10.8)

$$K^T(x_1y_1|\ldots|x_ky_k) = \sum_{\substack{\omega_1:x_1\to y_1 \\ \vdots \\ \omega_k:x_k\to y_k}} \varrho^T_{j_1\ldots j_k}(\omega_1,\ldots,\omega_k) \qquad (12.177)$$

with $j_i = j_i(y_i)$ equal to 1 if y_i is a dot and 0 otherwise. Here

$$\varrho^T_{j_1\ldots j_k}(\omega_1,\ldots,\omega_k) = \sum_{i=1}^{k}(-1)^{i-1}(i-1)! \sum_{\substack{\{I_1,\ldots,I_i\} \\ \text{n.e.s.s. partition} \\ \text{of } (i,\ldots,k)}} \prod_{t=1}^{i} \varrho_{j_{I_t}}(x_{I_t}) \,. \qquad (12.178)$$

The lowest-order nontrivial versions of (12.175)–(12.178) are contained in the formulas (12.6)–(12.10) for the zero-field case and (12.130)–(12.133) for nonzero field.

The above relations imply that the fundamental objects for the study of the truncated Green functions are the truncated weights (12.178). Unfortunately, very little is known for the functions involving truncated weights of order larger than two. The only results in this regard are those discussed in the previous section for the three-point truncated function, and the result of Shlosman [460] about the alternance of the signs of the truncated $2n$-point of the Ising model, which was derived via the random-current representation. Remarkably, an analogous result can be proven for the SAW model using the random-walk representation. Let us discuss this proof, which is based on well known results for truncated functions in the general sense discussed in Section 12.7.

It is not hard to prove, e.g. from (12.199), that if the walks "cluster" into two families of walks $\omega_{I_1}, \omega_{I_2}$ such that

$$\varrho_{j_{I_1} j_{I_2}}(\omega_{I_1}, \omega_{I_2}) \;=\; \varrho_{j_{I_1}}(\omega_{I_1}) \varrho_{j_{I_2}}(\omega_{I_2}) \tag{12.179}$$

then

$$\varrho^T_{j_{I_1} j_{I_2}}(\omega_{I_1}, \omega_{I_2}) \;=\; 0 \; . \tag{12.180}$$

In particular, this implies the following property for the weights of the baby CIW model.Let us define for any collection of walks $\omega_1, \ldots, \omega_n$ the graph $G = G(\omega_1, \ldots, \omega_n)$ with vertices $V(G) = \{1, \ldots, n\}$ and edges $E(g) = \{\{i, j\} | \operatorname{supp}(\omega_i) \cap \operatorname{supp}(\omega_j) \neq \varnothing\}$. We have:

Proposition 12.20 *For baby CIW models*

$$\varrho^T_{j_1 \ldots j_n}(\omega_1, \ldots, \omega_n) \;=\; 0 \tag{12.181}$$

whenever $G(\omega_1, \ldots, \omega_n)$ has two disconnected parts.

For the baby SAW we can in fact give an exact expression for the truncated weights. Indeed, for this model

$$\varrho_{j_1 \ldots j_k}(\omega_1, \ldots, \omega_k) \;=\; \left[\prod_{i=1}^{k} \varrho_{j_i}(\omega_i)\right] \left[\prod_{1 \le i < j \le k} f(\omega_i, \omega_j)\right] \tag{12.182}$$

with

$$f(\omega_i, \omega_j) = \begin{cases} 1 & \text{if } \operatorname{supp}(\omega_i) \cap \operatorname{supp}(\omega_j) \neq \varnothing \\ 0 & \text{otherwise} \end{cases} \tag{12.183}$$

(12.182) has precisely the form of the Boltzmann factor of a gas of walks with fugacities $\varrho_{j_i}(\omega_i)$ and pair interactions $\exp[-\beta\varphi(\omega_i, \omega_j)] = f(\omega_i, \omega_j)$. Hence the truncated weights are the Ursell functions of the corresponding Mayer expansion [447]. Furthermore, (12.183) tells us that this is a gas of walks with exclusion, and the Ursell functions for such system have been explicitly calculated by Gallavotti et al [234]. The result is:

Proposition 12.21 *For the baby SAW*

$$\varrho^T_{j_1\dots j_k}(\omega_1,\dots,\omega_k) = \left[\prod_{i=1}^{k} \varrho_{j_i}(\omega_i)\right] a_k(\omega_1,\dots,\omega_k) \tag{12.184}$$

with

$$a_k(\omega_1,\dots,\omega_k) \;=\; \sum_{\substack{C \text{ subgraph of } G(\omega_1,\dots,\omega_k) \\ C \text{ visiting all } \{1,\dots,k\} \\ C \text{ connected}}} (-1)^{\#\text{ of edges of } C} \;. \tag{12.185}$$

Moreover, it can be proven — directly from (12.185) or by even simpler arguments [447, page 94] — that

$$(-1)^{k-1} a_k \;\geq\; 0\;. \tag{12.186}$$

(12.175), (12.177) and (12.184) imply the following result:

Theorem 12.22 *In the zero-field case the truncated Green functions of the SAW model alternate in sign:*

$$(-1)^n S^T_{2n} \;\geq\; 0\;. \tag{12.187}$$

The analogous result for the Ising model and models in the GS class has been recently proven by Shlosman [460], using the random-current representation and more sophisticated combinatorics.

Another consequence of (12.186) is that it implies that the truncated Green functions for the baby SAW are polynomials in h with coefficients alternating in sign.

12.7 Relationship between the truncated Green functions, the truncated kernels and the truncated weights

This section is somewhat technical, and can be skipped without loss of continuity. It is strongly motivated by a set of lectures delivered by J. Slawny at Virginia Polytechnic Institute (1980-81).

12.7.1 Algebraic definition of truncation

The abstract setting to define the notion of truncation is the following [447, Chapter 4]. Let X be a set; consider

$$A(X) \;=\; \{F = (F_n)_{n\geq 0} \mid F_n\colon X^n \to \mathbf{C}\,,\; F_0 \in \mathbb{C}\}\;. \tag{12.188}$$

(Here $\mathbb{C}$ can be replaced by any field.) $A(X)$ is a vector space with the natural operations and an algebra with the convolution

$$(F * G)_n(x_1, \ldots, x_n) = \sum_{\substack{(I_1, I_2) \text{ s.s. part.} \\ \text{of } (1, \ldots, n)}} F_{|I_1|}(x_{I_1}) G_{|I_2|}(x_{I_2}) \tag{12.189}$$

where "s.s. part." stands for "partition by subsequences". The unit element of this convolution is the family of functions $1 = (\delta_n)_{n \geq 0}$. We will also use power notation: $F^{*n} = F * \ldots * F$ (n times); with the convention $F^{*0} \equiv 1$.

A collection $F^{(1)}, F^{(2)}, \ldots$ of elements of $A(X)$ is said *summable* if for each n the set of i's for which $F_n^{(i)}$ is not the zero function is finite. In such case it makes sense to define $\sum_{i \geq 1} F^{(i)} \in A(X)$ in the natural way:

$$\Big[\sum_{i \geq 1} F^{(i)}\Big]_n (x_1, \ldots, x_n) = \sum_{i \geq 1} F_n^{(i)}(x_1, \ldots, x_n) . \tag{12.190}$$

In particular, it is of interest to define power series in the sense (12.190). In this regard it is important to notice that

$$(F^{*n})_{n \geq 0} \text{ is summable } \iff F_0 = 0 \tag{12.191}$$

Therefore, for each element of

$$A_0(X) = \{F \in A(X) \colon F_0 = 0\} \tag{12.192}$$

we can define a power series (formal power series in the algebra $A(X)$). The two examples that concern us here are:

$$\exp F \equiv \sum_{k \geq 0} \frac{F^{*k}}{k!} \tag{12.193}$$

and

$$\log(1 + F) \equiv \sum_{k \geq 0} \frac{(-1)^{k-1}}{k - 1} F^{*k} . \tag{12.194}$$

If $F_0 = 1$, then $\log F \equiv \log[1 + (F - 1)]$ makes sense and we denote

$$F^T = \log F \tag{12.195}$$

or equivalently

$$F = \exp F^T \tag{12.196}$$

We say that F^T is the *truncation* of F, or that the functions F_n^T are the truncated functions corresponding to F_n. (12.194) and (12.195) yield the following explicit relation

$$F_n^T(x_1, \ldots, x_n) = \sum_{i=1}^{n} (-1)^{i-1} (i - 1)! \sum_{\substack{\{I_1, \ldots, I_i\} \\ \text{n.e.s.s. part.} \\ \text{of } (i, \ldots, n)}} \prod_{j=1}^{i} F_{|I_j|}(x_{I_j}) \tag{12.197}$$

where "n.e." means that the subsequences I_j must be nonempty. On the other hand, from (12.193) and (12.196) we obtain a complementary expression:

$$F_n(x_1,\ldots,x_n) = \sum_{i=1}^{n} \sum_{\substack{\{I_1,\ldots,I_i\} \\ \text{n.e.s.s. part.} \\ \text{of } (i,\ldots,n)}} \prod_{j=1}^{i} F^T_{|I_j|}(x_{I_j}) \,. \tag{12.198}$$

It is useful to keep in mind that as (12.197) and (12.198) are recursive relations in n, either of them determine the truncated functions *uniquely*. For instance, (12.198) can be explicitly written as the following recursive definition for F^T:

$$F^T_n(x_1,\ldots,x_n) = F(x_1,\ldots,x_n) - \sum_{i=2}^{n} \sum_{\substack{\{I_1,\ldots,I_i\} \\ \text{n.e.s.s. part.} \\ \text{of } (i,\ldots,n)}} \prod_{j=1}^{i} F^T_{|I_j|}(x_{I_j}) \,. \tag{12.199}$$

12.7.2 Relation between truncation in different algebras

The definitions of truncated Green functions, kernels and weights given in Chapter 12 [formulas (12.174)–(12.176) and (12.178)] correspond to truncations in algebras $A(X)$ for different sets X. The following lemma formalizes the concept of a relation "commuting with the operation of truncation" invoked in that chapter.

Lemma 12.23 *Let* $\Gamma: A(X_1) \to A(X_2)$ *be a map such that*

L1) Γ *is linear,*

L2) $\Gamma(F * G) = \Gamma(F) * \Gamma_2(G)$

L3) $F_0 = 0 \implies (\Gamma F)_0 = 0$

L4) $(\Gamma G)_n$ *involves only a finite set of* F_i*, i.e. for each* n *there exists* i_0 *such that*

$$F_i = 0 \text{ for } i \le i_0 \implies (\Gamma F)_n = 0 \tag{12.200}$$

Then, for any power series on $A(X)$*, and any* $F \in A_0(X)$*:*

$$\Gamma\Big(\sum_{k\ge 0} a_k F^{*k}\Big) = \sum_{k\ge 0} a_k (\Gamma F)^{*k} \,. \tag{12.201}$$

Proof. The hypotheses make the proof trivial because they guarantee the legitimacy of the obvious steps. ■

As an immediate consequence we have that if Γ is as in the lemma

$$(\Gamma F)^T = \Gamma(F^T) \tag{12.202}$$

Some remarks are in order. First, the hypotheses of the lemma and the notion of summability can be stated respectively as continuity of Γ and convergence of the series with respect to the topology generated by the sets $O_M = \{F \mid F_n = 0 \ \forall n \leq M\}$. Second, the lemma provides an operational way to check when the relations between functions is such that it survives truncation — like in (12.175) and (12.177). In fact, most of such relations can be stated as a suitable map Γ which is trivially seen to satisfy *L1*, *L3* and *L4*. Hence this lemma reduces a seemingly complicated proof [cf. the sheer definition (12.197)] to the verification of the relatively simpler statement *L2*. In a sense this lemma "zeroes in" the heart of the combinatorial difficulties for such a proof.

12.7.3 Relations between the truncation for Green functions, kernels and weights

12.7.3.1 Case $h = 0$ To present the basic ideas more simply, let us start with the case of no magnetic field. If

$$X_1 = \Omega \tag{12.203}$$

then, every family of weights defines a family

$$\varrho = (\varrho_n)_{n\geq 0} \in A(X_1) \quad , \varrho_0 = 1 \tag{12.204}$$

with

$$\varrho_n \equiv \varrho^{[n]} \tag{12.205}$$

On the other hand, if

$$X_2 = L^2 \tag{12.206}$$

then the kernels defined in (9.3) determine a family of functions

$$K = (K_k)_{k\geq 0} \in A(X_2) \quad , K_0 = 1 \tag{12.207}$$

with

$$K_k((x_1, y_1), \ldots, (x_k, y_k)) = K_k(x_1 y_1)| \ldots |x_k y_k) \; . \tag{12.208}$$

Finally, if

$$X_3 = L \tag{12.209}$$

then, the Green functions (9.7) define a family

$$S = (S_n)_{n\geq 0} \in A(X_3) \quad , S_0 = 1 \tag{12.210}$$

The respective truncated functions coincide with those defined via the algebraic formalism discussed above. The relations (9.3) and (9.7) can be expressed in the following form

$$\begin{array}{ccccc} A(X_1) & \xrightarrow{\Gamma_1} & A(X_2) & \xrightarrow{\Gamma_2} & A(X_3) \\ \varrho & \longmapsto & K & \longmapsto & S \end{array} \tag{12.211}$$

with

$$[\Gamma_1(\varrho)]_k\Big((x_1,y_1),\dots,(x_k,y_k)\Big) = \sum_{\substack{\omega_1: x_1 \to y_1 \\ \vdots \\ \omega_k: x_k \to y_k}} \varrho_k(\omega_1,\dots,\omega_k) , \tag{12.212}$$

$$[\Gamma_2(K)]_{2n}(x_1,\dots,x_{2n}) = \sum_{\pi\in Q_{2n}} K_n\Big((x_{\pi(1)},x_{\pi(2)}),\dots,(x_{\pi(2k-1)},x_{\pi(k)})\Big) , \tag{12.213}$$

$$[\Gamma_2(K)]_{2n+1} = 0 ; \tag{12.214}$$

and the identities (12.175), (12.177) are a consequence of:

Theorem 12.24

a) $[\Gamma_1\varrho]^T = \Gamma_1(\varrho^T)$

b) $[\Gamma_2 K]^T = \Gamma_2(K^T)$

Proof. It is enough to check that the maps Γ_1 , Γ_2 satisfy the hypothesis of Lemma 12.23. It is immediate to check L1, L3, L4. Let us first prove that Γ_1 verifies L2. If ϱ and ξ are elements of $A(X_1)$:

$$\begin{aligned}
&[\Gamma_1(\varrho*\xi)]_k\Big((x_1,y_1),\dots,(x_k,y_k)\Big) \\
&= \sum_{\substack{\omega_1: x_1 \to y_1 \\ \vdots \\ \omega_k: x_k \to y_k}} (\varrho*\xi)_k(\omega_1,\dots,\omega_k) \\
&= \sum_{\substack{\omega_1: x_1 \to y_1 \\ \vdots \\ \omega_k: x_k \to y_k}} \sum_{\substack{(I_1,I_2)\ \text{s.s. part.} \\ \text{of } (1,\dots,n)}} \varrho_{|I_1|}(\omega_{I_1})\xi_{|I_2|}(\omega_{I_2}) \\
&= \sum_{\substack{(I_1,I_2)\ \text{s.s. part.} \\ \text{of } (1,\dots,n)}} \Big[\sum_{(\omega:\to y)_{I_1}} \varrho_{|I_1|}(\omega_{I_1})\Big]\Big[\sum_{(\omega:\to y)_{I_2}} \xi_{|I_2|}(\omega_{I_2})\Big] \\
&= [(\Gamma_1\varrho)*(\Gamma_1\xi)]_k((x_1,y_1),\dots,(x_k,y_k)) .
\end{aligned} \tag{12.215}$$

This completes the proof of part a). With regard to Γ_2 we must compare

$$[\Gamma_2(F*G)]_{2n}(x_1,\ldots,x_{2n}) = \sum_{\pi\in Q_{2n}} (F*G)_n((x_{\pi(1)},x_{\pi(2)}),\ldots,(x_{\pi(2n-1)},x_{\pi(n)}))$$

$$= \sum_{\pi\in Q_{2n}} \sum_{\substack{(I_1,I_2)\text{ s.s. part.}\\ \text{of }(1,\ldots,n)}} F_{|I_1|}\Big(\big(x_{\pi(2i-1)},x_{\pi(2i)}\big)_{i\in I_1}\Big) G_{|I_2|}\Big(\big(x_{\pi(2i-1)},x_{\pi(2i)}\big)_{i\in I_2}\Big) \tag{12.216}$$

with

$$[(\Gamma_2F)*(\Gamma_2G)]_{2n}(x_1,\ldots,x_{2n}) = \sum_{\substack{(J_1,J_2)\text{ s.s. part.}\\ \text{of }(1,\ldots,2n)}} (\Gamma_2F)_{|J_1|}(x_{J_1})(\Gamma_2G)_{|J_2|}(x_{J_2})$$

$$= \sum_{\substack{(J_1,J_2)\text{ s.s. part.}\\ \text{of }(1,\ldots,2n)}} \Bigg[\sum_{\substack{\pi_1\in Q^{(J_1)}_{|J_1|}\\ |J_1|\text{ even}}} F_{|J_1|}\Big(\big(x_{\pi_1(j_{2i-1})},x_{\pi_1(j_{2i})}\big)_{1\le i\le|J_1|/2}\Big)\Bigg]$$

$$\times\Bigg[\sum_{\substack{\pi_2\in Q^{(J_2)}_{|J_2|}\\ |J_2|\text{ even}}} G_{|J_2|}\Big(\big(x_{\pi_2(j'_{2i-1})},x_{\pi_2(j'_{2i})}\big)_{1\le i\le|J_2|/2}\Big)\Bigg] \tag{12.217}$$

where in the last sums we denoted

$$J_1 = (j_1,\ldots,j_{|J_1|})\ ,\ J_2 = (j'_1,\ldots,j'_{|J_2|}) \tag{12.218}$$

and $Q^{(J_i)}_{|J_i|}$ is defined analogously to $Q_{|J_i|}$ but involving permutations of J_i. The restriction J_i even is due to (12.214). The reason why (12.216) and (12.217) are equal is more easily explained with words. If we read the sums starting outside in, we see that in (12.216) we first permute the $2n$ sites x_i, we pair sites that are contiguous in the permuted order and then we group the pairs into two sets: the pairs labelled by I_1 are assigned as arguments for F and those labelled by I_2 as arguments of G. On the other hand, in (12.217) we first group the sites into two sets labelled respectively by J_1 and J_2, we pair the sites inside each set, and then we permute the pairs in each set independently. After some thinking one can convince oneself that each of the terms presented in (12.216) is in (12.217) and vice versa. To finish the proof of b) we notice that if n is odd

$$[\Gamma_2(F*G)]_n = 0 = [(\Gamma_2F)*(\Gamma G)]_n \tag{12.219}$$

where the leftmost inequality is by (12.214), and the rightmost inequality is due to the fact that if n is odd at least one of the sets J_1 , J_2 in which $1,\ldots,n$ is partitioned has an odd number of elements. Hence, again by (12.214) all the terms in the rightmost expression are zero. ∎

12.7.3.2 Case $h \neq$ o The arguments are very similar to the previous ones. The Green functions again fit into the setting (12.210), but minor changes are needed for ϱ and K. Now we must consider

$$X_1 = \{0,1\} * \Omega \tag{12.220}$$

and

$$\varrho_n((j_1,\omega_1),\ldots,(j_n,\omega_n) = \varrho_{j_1\ldots j_n}(\omega_1,\ldots,\omega_k) \tag{12.221}$$

On the other hand K is as in (12.207)-(12.208) but with

$$X_2 = [L \cup \{\bullet\}]^2 \tag{12.222}$$

the map Γ_1 takes the form

$$[\Gamma_1(\varrho)]_k((x_1,y_1),\ldots,(x_k,y_k)) = \sum_{\substack{\omega_1 : x_1 \to y_1 \\ \vdots \\ \omega_k : x_k \to y_k}} \varrho^T_{j_1\ldots j_k}(\omega_1,\ldots,\omega_k) \tag{12.223}$$

with

$$j_1 = \begin{cases} 1 & \text{if } y_i = \bullet \\ 0 & \text{otherwise} \end{cases} \tag{12.224}$$

while

$$[\Gamma_2(K)]_n(x_1,\ldots,x_n) = \sum_{k=[(n+1)/2]}^{n} \sum_{\pi \in Q_{2k,n}} K_k(x_{\pi(1)}x_{\pi(2)}|\ldots|x_{\pi(2k-1)}x_{\pi(2k)}) \tag{12.225}$$

With these definitions Theorem 12.24 also holds. The proof of part a) for the present Γ_1 is identical to the one for the case $h=0$. The only delicate point in the proof of part b) is the comparison of $\Gamma_2(F*G)$ with $\Gamma_2(F)*\Gamma_2(G)$. We now have, in place of (12.216)–(12.217):

$$\begin{aligned}[][\Gamma_2(F*G)]_{2n}(x_1,\ldots,x_n) \\ = \sum_{k=[(n+1)/2]}^{n} \sum_{\pi\in Q_{2n}} \sum_{(I_1,I_2)\text{ s.s. part.of }(1,\ldots,k)} & F_{|I_1|}\Big(\big(x_{\pi(2i-1)},x_{\pi(2i)}\big)_{i\in I_1}\Big) \\ & \times G_{|I_2|}\Big(\big(x_{\pi(2i-1)},x_{\pi(2i)}\big)_{i\in I_2}\Big)\end{aligned} \tag{12.226}$$

and

$$\begin{aligned}[][(\Gamma_2F)*(\Gamma_2G)]_n(x_1,\ldots,x_n) \\ = \sum_{\substack{(J_1,J_2)\text{ s.s. part.} \\ \text{of }(1,\ldots,n)}} \Bigg[\sum_{k_1=[(|J_1|+1)/2]}^{|J_1|} \sum_{\pi_1\in Q^{(J_1)}_{2k_1,|J_1|}} F_{k_1}\Big(\big(x_{\pi_1(j_{2i-1})},x_{\pi_1(j_{2i})}\big)_{1\le i\le k_1}\Big)\Bigg] \\ \times\Bigg[\sum_{k_2=[(|J_2|+1)/2]}^{|J_2|} \sum_{\pi_2\in Q^{(J_2)}_{2k_2,|J_2|}} G_{k_2}\Big(\big(x_{\pi_2(j_{2i-1})},x_{\pi_2(j_{2i})}\big)_{1\le i\le k_2}\Big)\Bigg].\end{aligned} \tag{12.227}$$

The sets $Q^{(J_i)}_{2k,|J_i|}$ have a definition analogous to $Q_{2k,|J_i|}$ but for maps π with image in $J_i \cup \{\infty\}$. Reading the sums from left to right we see that in (12.226) one first permutes the sites and adds bullets "•" so to form $2k$ pairs; and then the resulting pairs are split into two groups labelled by I_1, I_2. In (12.227) these operations are performed in reverse order, namely we first split the sites into two groups labelled J_1, J_2, then to each group we add dots independently — so as to complete $2k_i$ pairs for the sites with subscripts in J_i — and finally we permute the pairs. The end product is the same as in (12.226).

Part III

Consequences for critical phenomena and quantum field theory

13. Background material

13.1 Models to be considered

In this section we summarize the basic definitions and notations to be used in the sequel. Most of these have already been introduced in earlier chapters, but we repeat them here in order to make Part III as self-contained as possible. We hope that Part III will be accessible to readers interested only in the physical consequences and not in the technical aspects of the random-walk representation.

In this monograph we have been considering two types of models: spin models (such as the Ising and φ^4 models), and random-walk models. The latter include models employed in polymer physics (such as the self-avoiding walk and the Edwards model), as well as "artificial" random-walk models which have been devised to yield random-walk representations of spin models (i.e. the BFS and ARW random-walk models defined in Chapters 9 and 10. Our aim has been to study all these random-walk models by unified methods, and so to derive results of physical interest (such as correlation inequalities) simultaneously for the polymer models and (via the relevant identities) for the spin models.

Let us be more precise about this point. The basic objects of spin models are the correlation functions $\langle\varphi_{x_1}\cdots\varphi_{x_n}\rangle$ and the truncated correlation functions $\langle\varphi_{x_1};\ldots;\varphi_{x_n}\rangle$. On the other hand, the basic objects of random-walk models are the Green functions $S_n(x_1,\ldots,x_n)$ [defined in Sections 9.1 and 10.1] and the truncated Green functions $S_n^T(x_1,\ldots,x_n)$ [defined for general n in Section 12.6]. A random-walk model constitutes a representation of a spin model whenever there is an identity

$$\langle\varphi_{x_1}\cdots\varphi_{x_n}\rangle \;=\; S_n(x_1,\ldots,x_n) \tag{13.1}$$

between the correlation functions of the spin models and the Green functions of the random-walk model. [Of course, (13.1) is equivalent to the corresponding identity

$$\langle\varphi_{x_1};\ldots;\varphi_{x_n}\rangle \;=\; S_n^T(x_1,\ldots,x_n) \tag{13.2}$$

for the truncated correlation and Green functions.] In particular, for rather general spin models such identities hold between the correlation functions of the spin model and the Green functions of the corresponding BFS random-walk model. Similarly, for Ising models such identities hold between the correlation functions of the Ising system and the Green functions of the corresponding

ARW model *provided* that the sites $x_1, \ldots, x_n$ are all distinct. This restriction to noncoincident sites is necessary except in two special cases: when $n = 1$, or when $n = 2$ and $h = 0$.

These identities were exploited in Chapters 11 and 12 to derive correlation inequalities for suitable classes of spin models. Having done so, we now suppress all further reference to the "artificial" BFS and ARW random-walk models, and discuss only the corresponding spin models. However, we do wish to discuss also the random-walk models arising in polymer physics. To be able to state compact formulas valid both for random-walk (RW) models and spin models, we adopt the notation

$$u_n(x_1, \ldots, x_n) = \begin{cases} S_n^T(x_1, \ldots, x_n) & \text{for RW models} \\ \langle \varphi_{x_1}; \ldots; \varphi_{x_n} \rangle & \text{for spin models} \end{cases} \tag{13.3}$$

Now let us review briefly the classes of models to which our analysis applies:

a) *Spin systems with ferromagnetic two-body interactions.* We consider one-component spin systems defined by a Hamiltonian

$$H \;=\; -\tfrac{1}{2} \sum_{x,y} J_{xy} \varphi_x \varphi_y \;-\; \sum_x h_x \varphi_x \;, \tag{13.4}$$

and an even single-spin measure $dP(\varphi)$. The two-body interaction $J_{xy} = J_{yx}$ is assumed to be ferromagnetic ($J_{xy} \geq 0$ for all x, y) and translation-invariant ($J_{xy} = J_{x+a,y+a}$ for all x, y, a). The magnetic field h is assumed to be nonnegative (this affects only the sign of various inequalities). We distinguish various classes of spin models according to the properties of the single-spin measure; for the reader's convenience we summarize the definitions of the classes in Table 13.1. The validity of different correlation inequalities depends on the class of $dP(\varphi)$.

We remind the reader that infinite-volume spin models must be defined as limits of finite-volume spin models with suitable boundary conditions. Our correlation inequalities are proven initially for finite-volume systems and are then carried over to the infinite-volume limit; see the remark at the beginning of Chapter 12. In the present chapter we are always considering an infinite-volume translation-invariant system.

b) *Polymer models.* The most general class of polymer models for which we have results is that of repulsive polymer-chain models (Section 9.2.1). Among these, the most important are the simple contact-interacting walk (CIW) models, in which the walks interact only if they intersect, and the interaction depends on the total number of visits at each lattice site. In particular, most of our results apply to repulsive CIW models. An important example of a repulsive CIW model is the self-avoiding walk (SAW). A more general example, which includes the SAW as a limiting case, is the Edwards model [160, 28, 97, 75] (see Section 9.2.1 above, Example d). Many of the results can be generalized to contact-interacting walk models in which the walks carry "color" and the interaction is color-dependent. In particular, we consider the "intersection properties of ordinary random walks" (IPORW) model in which intersections between different walks are forbidden but each walk may self-intersect freely. All of our results for

Table 13.1. Classes of spin models

Model	Single-Spin Measure	Remarks and References
Ising	$\delta(\varphi^2 - 1)$	
φ^4	$\text{const} \times e^{-\frac{\lambda}{4}\varphi^4 - \frac{\tau}{2}\varphi^2}$ $(\lambda \geq 0)$	
GS class	Sums of ferromagnetically coupled Ising spins, and limits thereof	Griffiths-Simon class: includes spin-l, uniformly distributed bounded spins, φ^4-spins ... [272], [466], [5], [15]
EMN class	$e^{-V(\varphi)}$ with V even and C^1, with V' convex on $[0,\infty)$; and limits of such measures	Ellis-Monroe-Newman class: includes Ising, φ^4, uniformly distributed bounded spins ... [162], [163]
BFS class	$e^{-f(\varphi^2)}$ with f convex; and limits of such measures	Brydges-Fröhlich-Spencer class: strictly includes EMN class[92], [97, Sections 5 and 6]

Examples

1) A BFS measure which is not EMN: $e^{-V(\varphi)}$ with $V(\varphi) = \varphi^8 - \varepsilon\varphi^6 + \varphi^4$ and $2\sqrt{14}/5 < \varepsilon \leq 2\sqrt{6}/3$.

2) A GS measure which is not BFS (hence not EMN): the equal-weight spin-1 measure $(1/3)[\delta(\varphi + 1) + \delta(\varphi) + \delta(\varphi - 1)]$.

3) An EMN measure which is not GS: $e^{-V(\varphi)}$ with $V(\varphi) = \varphi^6 + a\varphi^2$ and $a > 0$ large. (For large a, this measure fails to have the Lee-Yang property [414, 413]; but all GS measures have the Lee-Yang property [272, 466].)

polymer models refer to the "zero-magnetic-field case", i.e. to a system with a fixed finite number of polymers. The results for nonzero field — i.e. fixed density of polymers — only apply to the unphysical "baby" polymer models (Section 10.2.2).

13.2 Critical exponents

From now on we consider only translation-invariant models. The following quantities are of physical interest[1]:

The (magnetic) susceptibility

$$\chi = \sum_x u_2(0,x). \tag{13.5}$$

The magnetization

$$M = u_1(0). \tag{13.6}$$

The bubble diagram (at zero magnetic field)

$$B_0 = \sum_x S_2(0,x)^2_{h=0}. \tag{13.7}$$

The truncated bubble diagram

$$B^T = \sum_x u_2(0,x)^2. \tag{13.8}$$

The n-th order cumulant at zero momentum

$$\overline{u}_n = \sum_{x_2,\ldots,x_n} u_n(0,x_2,\ldots,x_n). \tag{13.9}$$

The correlation length. The (exponential or "true") correlation length ξ is defined so that $u_2(0,x) \sim e^{-|x|/\xi}$, as $|x| \to \infty$, where $|x|$ is, say, the Euclidean norm of x. More rigorously:

$$\xi = \limsup_{n\to\infty} \frac{-|x|}{\log u_2(0,x)}. \tag{13.10}$$

Other definitions of correlation length are also useful. For each $\phi > 0$ the correlation length of order ϕ is

$$\xi_\phi = \left\{ \frac{\sum_x |x|^\phi u_2(0,x)}{\chi} \right\}^{1/\phi}. \tag{13.11}$$

[1]The definitions given here (and employed in the remainder of this book) are the simplest from the statistical-mechanical point of view; they differ (by factors of temperature, which are inessential in the critical region) from the standard thermodynamic definitions expounded in Chapter 1.

By Hölder's inequality ξ_ϕ increases with ϕ , and for reflection-positive spin models[2] [218, 481]

$$c(d,\phi)\xi_\phi \ \leq \ \xi \tag{13.12}$$

for all $\phi > 0$.

The renormalized coupling constants for the lattice model:

$$g \ = \ \frac{-\overline{u}_4}{\chi^2 \xi^d} \tag{13.13}$$

$$g_\phi \ = \ \frac{-\overline{u}_4}{\chi^2 \xi_\phi^d} \quad . \tag{13.14}$$

These constants are a measure of the non-Gaussianness of the model (in zero magnetic field). It can be proven (modulo some technicalities) [481] that in the continuum limit g_2 becomes the renormalized coupling constant of the continuum quantum field theory.

The derivatives with respect to the "inverse temperature". Usually the coupling constants are parametrized in the form

$$J_{ij} \ = \ \beta \widetilde{J}_{ij} \ , \tag{13.15}$$

in which we fix the geometric structure of the interaction and vary only its "strength" β.[3] We therefore have the formula

$$\beta \frac{\partial}{\partial \beta} \ = \ \tfrac{1}{2} \sum_{x,y} J_{xy} \frac{\partial}{\partial J_{xy}} \ . \tag{13.16}$$

Let us make two remarks about the notation. First, we shall avoid (wherever possible) references to the coupling constants $\widetilde{J}_{ij}$, in order to use without changes the formulas developed in previous chapters in terms of the full couplings J_{ij}. We shall often use the quantity $|J| \equiv \sum_j J_{ij}$, which plays the same role as β. Second, we must face the unfortunate fact that the same letter β is used for one of the critical exponents. To prevent confusion without departing too much from the standard usage, we shall use the symbol $\widehat{\beta}$ for this critical exponent.

The specific heat (at zero magnetic field)

$$C_H \ = \ \tfrac{1}{4} \sum_{x,y,z} J_{0x} J_{yz} \frac{\partial u_2(0,x)}{\partial J_{yz}} \ . \tag{13.17}$$

In spin systems this coincides (at least formally) with the derivative of the energy per site with respect to the inverse temperature β.

Let us remark again that all the work done in previous chapters concerns systems in finite volume. The correlation inequalities not involving derivatives

[2]Here and in the sequel "reflection-positive models" stands for "infinite-volume translation-invariant Gibbs states which are reflection-positive for reflections in planes without sites, parallel to the coordinate planes". See Section 13.4.

[3]Note that we do *not* introduce an analogous parametrization for the magnetic field; i.e. our "h" corresponds to "βh" of the physics literature.

carry over immediately to the infinite-volume limit, but for differential inequalities some extra work is needed to prove that the "fluctuation-dissipation relations" (or "sum rules") implied formally by (13.16) remain valid in the thermodynamic limit. We shall ignore the details of such a proof (see for instance the appendix of [479]), and we shall freely use (13.16) for observables in the infinite system.

We make the assumption that there exists a critical point $(\beta = \beta_c, h = 0)$ such that $\chi(\beta_c, h = 0) = \infty$, and $\chi(\beta, h = 0) < \infty$ for $\beta < \beta_c$. For $\beta > \beta_c$ and $h = 0$ we consider the "+" state, i.e. define all the observables $A(\beta, h = 0)$ as the limit as $h \downarrow 0$ of $A(\beta, h)$. In principle, there might exist a second critical point at a certain $\beta_c' > \beta_c$ where $\beta_c' = \inf\{\beta\colon\ M(\beta, h) > 0\}$. In this case the range $\beta_c < \beta < \beta_c'$ would be occupied by an intermediate regime with infinite susceptibility and zero magnetization. The existence of such a regime has, however, been ruled out for spin systems with single-spin measures in the GS class [6, 9, 10]. We shall implicitly assume for our models that no such intermediate regime exists (otherwise the primed critical exponents defined below should be defined for $\beta > \beta_c'$, and there should be exponents δ, α_c and b_c for each isotherm in the range $\beta_c < \beta < \beta_c'$).

To define the critical exponents [487, 479], let us write $f(x) \sim x^\lambda$ as an abbreviation of $\lambda = \lim_{x \downarrow 0} \frac{\log f(x)}{\log x}$. Let us also define $t \equiv \beta_c - \beta$. We then have:

- For $t > 0$ $(T > T_c)$ and $h = 0$:

$$\begin{array}{rcl} \chi & \sim & t^{-\gamma} \\ \xi & \sim & t^{-\nu} \\ \xi_\phi & \sim & t^{-\nu_\phi} \\ C_H & \sim & t^{-\alpha} \\ \overline{u}_{2m} & \sim & t^{-\gamma - 2\sum_{k=2}^{m} \Delta_{2k}} \\ B_0 & \sim & t^{-b} \\ g & \sim & t^{\varrho} \equiv t^{d\nu - 2\Delta_4 + \gamma} \\ g_\phi & \sim & t^{\varrho_\phi} \equiv t^{d\nu_\phi - 2\Delta_4 + \gamma} \end{array} \tag{13.18}$$

- For $t < 0$ $(T < T_c)$ and $h = 0^+$:

$$\begin{array}{rcl} \chi & \sim & |t|^{-\gamma'} \\ \xi & \sim & |t|^{-\nu'} \\ \xi_\phi & \sim & |t|^{-\nu_\phi'} \\ C_H & \sim & |t|^{-\alpha'} \\ \overline{u}_n & \sim & |t|^{-\gamma' - \sum_{k=3}^{n} \Delta_k'} \\ M & \sim & |t|^{\widehat{\beta}} \\ B^T & \sim & t^{-b'} \end{array} \tag{13.19}$$

- For $t = 0$ ($T = T_c$) and $h > 0$:

$$\begin{aligned} M &\sim h^{1/\delta} \\ C_H &\sim h^{-\alpha_c/\delta} \\ B^T &\sim h^{-b_c/\delta} \end{aligned} \tag{13.20}$$

The exponents b, b' and b_c are not part of the standard zoo of critical exponents, but we find it convenient to introduce them here in view of the central role played in our analysis by "bubble diagrams".

For dimensions where the mean-field critical behavior is exact — i.e. $d > d_c$ — the specific heat is no longer divergent at β_c but only exhibits a jump discontinuity which is not described by the exponent α defined above (which becomes identically zero no matter the characteristics of the jump). For this reason, it is customary to define in addition an exponent α_{sing} which characterizes the behavior of the *singular part* of the specific heat. This exponent coincides with α for $d \leq d_c$ but it is expected to be strictly negative — and d-dependent — for $d > d_c$.

Finally, we note that the hyperscaling relation

$$\varrho \equiv d\nu - 2\Delta_4 + \gamma = 0 \tag{13.21}$$

(if valid without multiplicative logarithmic corrections) means that the renormalized coupling constant g tends to a nonzero value as $t \to 0$, so that the scaling (continuum) limit is non-Gaussian. Conversely, the failure of hyperscaling

$$\varrho \equiv d\nu - 2\Delta_4 + \gamma > 0 \tag{13.22}$$

means that g tends to zero as $t \to 0$, so that the scaling limit is Gaussian.

For d equal to the upper critical dimension d_c ($d_c = 4$ for short-range models), it is believed that the critical exponents take their mean-field values ($\gamma = \gamma' = 1$, $\nu = \nu' = 1/2$, etc.) with *multiplicative logarithmic corrections.* In this case, therefore, we define the logarithmic exponents $\overline{\gamma}$, $\overline{\nu}$, $\overline{b}$, $\overline{\varrho}$, $\overline{\overline{\beta}}$ and $\overline{\delta}$ by

- For $t > 0$ ($T > T_c$) and $h = 0$:

$$\begin{aligned} \chi &\sim t^{-1}|\log t|^{\overline{\gamma}} \\ \xi &\sim t^{-1/2}|\log t|^{\overline{\nu}} \\ \xi_\phi &\sim t^{-1/2}|\log t|^{\overline{\nu}_\phi} \\ C_H &\sim |\log t|^{\overline{\alpha}} \\ B_0 &\sim |\log t|^{\overline{b}} \\ g &\sim |\log t|^{-\overline{\varrho}} \\ g_\phi &\sim |\log t|^{-\overline{\varrho}_\phi} . \end{aligned} \tag{13.23}$$

- For $t < 0$ ($T < T_c$) and $h = 0^+$:

$$M \sim |t|^{1/2}|\log|t||^{\overline{\overline{\beta}}} . \tag{13.24}$$

- For $t = 0$ ($T = T_c$) and $h > 0$:

$$M \sim h^{1/3}|\log h|^{\overline{\delta}} . \tag{13.25}$$

Note that $\overline{\varrho} > 0$ means that hyperscaling is violated by logarithms, and hence that the continuum limit is trivial. Field-theoretic renormalization-group calculations [77] yield the predictions

$$\begin{aligned}
\overline{\gamma} &= \frac{n+2}{n+8} \\
\overline{\nu} &= \overline{\nu}_\phi = \frac{n+2}{2(n+8)} \\
\overline{\alpha} &= \frac{4-n}{n+8} \\
\overline{b} &= 1 \\
\overline{\varrho} &= \overline{\varrho}_\phi = 1 \\
\overline{\widehat{\beta}} &= \frac{3}{n+8} \\
\overline{\delta} &= \frac{1}{3}
\end{aligned} \tag{13.26}$$

for the n-vector model in $d = 4$ ($n = 0$ is the SAW, and $n = 1$ is the Ising model). For the $d = 4$ IPORW model, we have trivially $\overline{\gamma} = \overline{\nu} = 0$ and $\overline{\alpha} = \overline{b} = 1$; and it is a rigorous (but highly nontrivial) theorem [351, 185, 7] that $\overline{\varrho} = 1$ (see discussion below).

The above exponents correspond to the usual definitions adopted for the Ising systems. However, for the nearest-neighbor translation-invariant isotropic SAW model some of the exponents admit an alternative definition in terms of the magnitudes relevant to polymer physics. Indeed, if we denote

c_N = number of N-step nearest-neighbor SAWs starting at the origin and ending anywhere;

$c_N(x)$ = number of N-step nearest-neighbor SAWs starting at the origin and ending at x;

$\langle \omega(N)^2 \rangle$ = mean-square end-to-end distance $= \sum_x |x|^2 c_N(x)/c_N$;

c_{N_1,N_2} = number of pairs of walks (ω_1, ω_2) such that ω_1 starts at the origin, ω_2 starts anywhere, and they visit at least one common site;

then the critical exponents can be defined in the form:

$$\begin{aligned}
c_N &\sim \mu^N N^{\gamma-1} \\
c_N(x) &\sim \mu^N N^{\alpha_{\text{sing}}-2} \ (x \neq 0) \\
\langle \omega(N)^2 \rangle &\sim N^{2\nu} \\
c_{N_1,N_2} &\sim \mu^{N_1+N_2} N^{2\Delta_4+\gamma-2} g(N_1+N_2) \, .
\end{aligned} \tag{13.27}$$

Here $\mu = \beta_c^{-1}$ is the connective constant of the lattice, g is some scaling function, and the limits are taken for N, N_1 and N_2 tending to infinity. The equivalence of definitions (13.27) and (13.18) for the SAW is a consequence of the remark by deGennes that the CIW models are the $n \to 0$ limit of n-component spin models [see discussion around formulas (9.40)–(9.42)]. Indeed, this remark implies the identities

$$\chi = \sum_{N=0}^{\infty} \beta^N c_N , \tag{13.28}$$

$$S_2^T(0,x) = \sum_{N=0}^{\infty} \beta^N c_N(x) \tag{13.29}$$

and

$$\overline{u}_4 = 3 \sum_{N_1,N_2=0}^{\infty} \beta^{N_1+N_2} c_{N_1,N_2} . \tag{13.30}$$

The renormalized coupling constant g has a particularly clear physical interpretation for the SAW. Indeed, from (13.13) and (13.28))–(13.30),

$$g = \frac{3}{\xi^d} \sum_y \text{Prob}_y[\text{supp}(\omega_1) \cap \text{supp}(\omega_2) \neq \varnothing] , \tag{13.31}$$

where

$$\text{Prob}_y[f(\omega_1,\omega_2)] = \frac{1}{\chi^2} \sum_{\substack{\omega_1:0 \to \bullet \\ \omega_2:y \to \bullet}} \beta^{|\omega_1|+|\omega_2|} f(\omega_1,\omega_2) \tag{13.32}$$

and the sum is taken over all SAWs ω_1 [resp. ω_2] starting at 0 [resp. at y] and ending anywhere. Thus, roughly speaking, the renormalized coupling constant g measures the probability that two independent SAWs, starting at a distance of order ξ apart, will intersect.

13.3 Summary of correlation inequalities

Finally, let us list for the reader's convenience all the correlation inequalities that will be used in the present discussion. For spin systems we use the notation $\varphi^A = \prod_{x \in A} \varphi_x$. For simplicity, we use the phrases "BFS models" and "GS models" as a shorthand for "spin model of BFS class" and "spin model of GS class", respectively.

a) Griffiths' first and second inequalities:

$$\langle \varphi^A \rangle \geq 0 \tag{13.33}$$

and

$$\langle \varphi^A ; \varphi^B \rangle \equiv \langle \varphi^A \varphi^B \rangle - \langle \varphi^A \rangle \langle \varphi^B \rangle \geq 0 . \tag{13.34}$$

These inequalities are valid for all ferromagnetic spin models [491] with even single-spin measure, in particular for those with single-spin measures of the

BFS or GS class. An important consequence of (13.34) is that each correlation $\langle\varphi^A\rangle$ is an increasing function of each of the coupling constants J_{ij} — hence of β —and of the magnetic fields h_x.

b) Ginibre-Newman-Aizenman [410, 5] inequality:

$$u_4(x_1,x_2,x_3,x_4) \;\geq\; -2\,\min[S_2(x_1,x_2)S_2(x_3,x_4),\; S_2(x_1,x_3)S_2(x_2,x_4),\; S_2(x_1,x_4)S_2(x_2,x_3)]\,. \tag{13.35}$$

This inequality is valid for all ferromagnetic spin models with even single-spin measure (it is a consequence of the Ginibre inequality $\langle q_1q_2t_3t_4\rangle \geq 0$ and permutations [491]).

c) Lebowitz inequalities:
In zero magnetic field [362],

$$u_4(x_1,x_2,x_3,x_4) \;\leq\; 0\,. \qquad (12.14)/(13.36)$$

This is the simplest of the Gaussian upper bounds proved in Section 12.1 for repulsive polymer chains, BFS and GS models. Two immediate consequences of (13.36) are the inequalities

$$\overline{u}_4 \;\leq\; 0 \tag{13.37}$$

and

$$\beta\frac{\partial\chi}{\partial\beta} \;\leq\; |J|\chi^2\,. \tag{13.38}$$

For spin systems one also has the following generalization for nonzero magnetic field [491, 162]:

$$\begin{aligned}\langle\varphi_{x_1}\varphi_{x_2}\varphi_{x_3}\varphi_{x_4}\rangle - \langle\varphi_{x_1}\varphi_{x_2}\rangle\langle\varphi_{x_3}\varphi_{x_4}\rangle - \langle\varphi_{x_1}\varphi_{x_3}\rangle\langle\varphi_{x_2}\varphi_{x_4}\rangle - \langle\varphi_{x_1}\varphi_{x_4}\rangle\langle\varphi_{x_2}\varphi_{x_3}\rangle \\ \leq\; -2\langle\varphi_{x_1}\rangle\langle\varphi_{x_2}\rangle\langle\varphi_{x_3}\rangle\langle\varphi_{x_4}\rangle\,.\end{aligned} \tag{13.39}$$

This inequality is valid, for instance, for ferromagnetic spin systems with single-spin measures in the EMN or GS class [162, 163].

d) Aizenman-Fröhlich (AF) inequalities: For the repulsive simple CIW, BFS and GS models [213, 15],

$$\begin{aligned}u_4(x_1,x_2,x_3,x_4) \;\geq\; -\sum_z S_2(x_1,z)S_2(x_3,z)\Big[\delta_{z,x_2} + \sum_{z_1} J_{zz_1}S_2(z_1,x_2)\Big] \\ \times\Big[\delta_{z,x_4} + \sum_{z_2} J_{zz_2}S_2(z_2,x_4)\Big] \;-\; \{2 \text{ permutations}\}\,.\end{aligned} \qquad (12.24)/(13.40)$$

For Ising models one has the slight improvement [5]

$$u_4(x_1,x_2,x_3,x_4) \;\geq\; -2\sum_z\langle\sigma_{x_1}\sigma_z\rangle\langle\sigma_{x_3}\sigma_z\rangle\langle\sigma_z\sigma_{x_2}\rangle\langle\sigma_z\sigma_{x_4}\rangle\,. \tag{13.41}$$

The "summed" versions of (13.40) and (13.41) are, respectively,

$$\overline{u}_4 \geq -3|J|^2\chi^4 \left(1 + \frac{1}{|J|\chi}\right)^2 \qquad (12.30)/(13.42)$$

and

$$\overline{u}_4 \geq -2\chi^4 . \qquad (13.43)$$

e) Aizenman-Graham (AG) inequality ("once-improved AF inequality") [15]: For ferromagnetic spin models with single-spin measures in the GS class,

$$u_4(x_1,x_2,x_3,x_4) \geq -\sum_{z,z_1,z_2} S_2(x_1,z)J_{zz_1}S_2(z_1,x_2)J_{z_1z_2}\frac{\partial}{\partial J_{z_1z_2}}S_2(x_3,x_4)$$

$$- \varepsilon(x_1,x_2,x_3,x_4) \qquad (12.33)/(13.44)$$

where the error term $\varepsilon(x_1,x_2,x_3,x_4)$ becomes negligible in the critical region (see the discussion in Section 12.2). An analogous inequality holds for self-avoiding walks [28]. For Ising models one has the following slightly sharper version:

$$u_4(x_1,x_2,x_3,x_4) \geq -\sum_{z,z_1} S_2(x_1,z)S_2(x_2,z)\tanh(J_{zz_1})\frac{\partial}{\partial J_{zz_1}}S_2(x_3,x_4)$$

$$- \varepsilon(x_1,x_2,x_3,x_4) . \qquad (12.32)/(13.45)$$

These inequalities differ from (13.40) and (13.41) in that one of the products S_2S_2 has been replaced by $\frac{\partial S_2}{\partial J}$, which, as discussed in Section 12.2.3, is an improvement. We shall see that the improvement is especially relevant at (or below) the upper critical dimension. Summing (13.44) over three of the sites, we obtain

$$\overline{u}_4 \geq -|J|\chi^2\beta\frac{\partial\chi}{\partial\beta} - \varepsilon . \qquad (12.42)/(13.46)$$

Another consequence of (13.44) is (see Section 12.2.3)

$$\beta\frac{\partial\chi}{\partial\beta} \geq \frac{|J|\chi^2}{1+|J|^2B_0}[1 - \varepsilon'] , \qquad (12.50)/(13.47)$$

which complements (13.38).

f) Twice-improved AF inequality: For simple CIW models we can "doubly" improve the inequality (13.40), i.e. replace both factors S_2S_2 by factors $\partial S_2/\partial J$. We have (see (12.53))

$$u_4(x_1,x_2,x_3,x_4) \geq -\sum_{z}\Big[S_2(x_1,z)\delta_{z,x_2} + (1/2)\sum_{z_1} J_{zz_1}\frac{\partial}{\partial J_{zz_1}}S_2(x_1,x_2)\Big]$$

$$\times\Big[S_2(x_3,z)\delta_{z,x_4} + (1/2)\sum_{z_2} J_{zz_2}\frac{\partial}{\partial J_{zz_2}}S_2(x_3,x_4)\Big]$$

$$- \{2 \text{ permutations}\} . \qquad (13.48)$$

The corresponding "summed" inequality is

$$\overline{u}_4 \ \geq \ -3(\beta\frac{\partial\chi}{\partial\beta} \ + \ \chi)^2 \ . \qquad (12.56)/(13.49)$$

g) Non-Gaussian upper bound on $\overline{u}_4$ [194]: For translation- invariant Ising models, the use of the random-current representation and the "dilution trick" (Section 12.3.1) yields a strictly negative upper bound on $\overline{u}_4$:

$$\overline{u}_4 \ \leq \ - \ \frac{1}{48|J|^2 B_0}\left(\beta\frac{\partial\chi}{\partial\beta}\right)^2 [1-\varepsilon] \ . \qquad (12.118)/(13.50)$$

[This sharpens the Lebowitz inequality (13.37).] A notoriously less transparent "pointwise" version of this inequality was given in Section 12.3 (formula (12.115)). If inequality (13.50) is combined with (13.47) we obtain:

$$\overline{u}_4 \ \leq \ - \ \frac{\chi^4}{48B_0(1+|J|B_0)^2} \ [1-\varepsilon] \ . \qquad (12.119)/(13.51)$$

An inequality analogous to (13.50) can be proven also for the IPORW model (see Section 12.3.2).

h) The Ginibre inequality for the three-point function:

$$u_3(x,y,z) \ \geq \ -2 \ \min[\langle\varphi_x\rangle\,\langle\varphi_y;\varphi_z\rangle,\ \langle\varphi_y\rangle\,\langle\varphi_x;\varphi_z\rangle,\ \langle\varphi_z\rangle\,\langle\varphi_x;\varphi_y\rangle] \ . \qquad (13.52)$$

This inequality is valid for one-component ferromagnetic spin models with even single-spin measure (it is a consequence of the Ginibre inequality $\langle t_x q_y q_z\rangle \geq 0$ and permutations [491]).

i) The Griffiths-Hurst-Sherman (GHS) inequality [279]:

For one-component spin models with single-spin measures in the EMN or GS class,

$$u_3(x,y,z) \ \leq \ 0 \ . \qquad (12.164)/(13.53)$$

One immediate consequence is that

$$\overline{u}_3 \ \equiv \ \frac{\partial\chi}{\partial h} \ \leq \ 0 \ . \qquad (13.54)$$

Another consequence [409, 479] is that

$$\frac{\partial M}{\partial\beta} \ \leq \ |\tilde{J}|M\chi \ . \qquad (13.55)$$

j) The weak GHS inequality: For repulsive polymer-chain, BFS and GS models,

$$S_1(x) \;\geq\; \sum_y S_2^T(x,y)h_y \qquad (12.139)/(13.56)$$

and hence

$$M \;\geq\; h\chi\,. \qquad (13.57)$$

This can be interpreted as an integrated version of the GHS inequality (see Section 12.4.4).

k) Fröhlich-Sokal (FS) inequality [223]: For baby repulsive simple CIW models and BFS models,

$$\frac{M}{h} \;\leq\; \chi \;+\; \left(\frac{M}{h}\right)^2 [h+|J|M]^2\,. \qquad (12.156)/(13.58)$$

The "pointwise" inequality corresponding to (13.58) includes a rather mysterious kernel K; see Section 12.4.

l) Aizenman-Barsky-Fernández (ABF) inequality ("once-improved FS inequality") [10]: For models with single-spin measures in the GS class,

$$\frac{M}{h} \;\leq\; \chi \;+\; \left(\frac{M}{h}\right)^2 \Big(h+|J|M\Big)\Big(h+\frac{h}{M}\beta\frac{\partial M}{\partial\beta}\Big) \qquad (12.162)/(13.59)$$

while for the Ising model the following version is also valid:

$$\frac{M}{h} \;\leq\; \chi \;+\; \left(\frac{M}{h}\right)^2 \Big(hM+h\beta\frac{\partial M}{\partial\beta}\Big)\,. \qquad (12.161)/(13.60)$$

Substituting (13.55) into the ABF inequality, we obtain the "weak ABF inequality"

$$\frac{M}{h} \leq \chi + \frac{M^2}{h}(h+|J|M)(1+|J|\chi) \qquad (13.61)$$

for GS models, and

$$\frac{M}{h} \leq \chi + \frac{M^3}{h}(1+|J|\chi) \qquad (13.62)$$

for the Ising models. [The FS inequality is a further weakening of (13.61) obtained by substituting the weak GHS inequality $\chi \leq M/h$.]

m) Tree bound on $\bar{u}_3$ ($= \partial\chi/\partial h$): For the Ising model we have [12]

$$\bar{u}_3 \;\geq\; -4|J|\,[1+O(h/M)]\,[1+M\tanh h]^3\,\frac{M^4}{(\tanh h)^3}\,, \qquad (12.168)/(13.63)$$

which also has a "pointwise" version (formula (12.167)). For models in GS class there is an analogous inequality [194]:

$$\bar{u}_3 \;\geq\; -4\,|J|^2\,[1+O(h/M)]\left[\frac{1+Mh(1+\varepsilon(h))}{1-\varepsilon(h)}\right]^3\frac{M^4}{h^3}\,, \qquad (12.173)/(13.64)$$

where $\varepsilon(h) \to 0$ as $h \to 0$. Here $\varepsilon(h)$ depends on the single-spin measure; it is identically zero for φ^4 and Ising spins.

n) Non-Gaussian upper bound on $\bar{u}_3$: For the Ising model we have the Aizenman-Fernández (AFe) inequality [12]

$$\bar{u}_3 \leq -\frac{[1 - B_0(\tanh h)/M)]}{96B_0(1 + 2|J|B_0)^2} (\tanh h)\, \chi^4 . \qquad (12.166)/(13.65)$$

o) Schrader–Messager–Miracle-Solé inequality [453, 394, 302, 495]: These inequalities, which are valid for *nearest-neighbor* translation-invariant ferromagnetic spin models (with arbitrary single-spin measure), state that

$$\langle \varphi_0;\, \varphi_y \rangle \geq \langle \varphi_0;\, \varphi_x \rangle \qquad (13.66)$$

whenever $|y|_1 \equiv \sum_{i=1}^{d} |y_i| \leq |x|_\infty \equiv \max_{1 \leq i \leq d} |x_i|$. In particular, this holds whenever $y \leq d^{-1}|x|$ (Euclidean distance).

For easy reference we summarize the main conclusions of this discussion on inequalities in Tables 13.2, 13.3 and 13.4.

13.4 Reflection positivity, spectral representations and infrared bounds

A fundamental principle of quantum mechanics is the positivity of the inner product in the Hilbert space of states; physically this expresses the positivity of probabilities. When translated to a condition on the imaginary-time Green's functions (Schwinger functions), this "physical positivity" condition is known as *reflection positivity* (see Section 1.5).

Reflection positivity was first introduced, therefore, in axiomatic quantum field theory, as a necessary and sufficient condition on the (Euclidean-space) Schwinger functions to allow an analytic continuation back to a (Minkowski-space) quantum field theory. It was soon realized, however, that many interesting statistical-mechanical systems also possess reflection positivity; and while reflection positivity does not play the *fundamental* role in statistical mechanics that it does in quantum theory, it is nevertheless an extremely useful *technical* tool. In this section we explain what reflection positivity is, and summarize some of its principal consequences.

Let θ be a reflection in $\mathbb{R}^d$ that maps the lattice $\mathbb{Z}^d$ into itself; it will be a reflection in some hyperplane $\{x\colon\ \mathbf{n} \cdot x = c\}$ for suitable $\mathbf{n}$ and c. Then we denote by Θ be corresponding map on lattice field configurations,

$$(\Theta\varphi)_x \equiv \varphi_{\theta x} , \qquad (13.67)$$

and (by abuse of language) on observables,

$$(\Theta A)(\varphi) \equiv A(\Theta\varphi) . \qquad (13.68)$$

Table 13.2. Systems satisfying each inequality

Inequality **Name or abbreviation: (Eq. #)**	**Systems for which it holds**
Griffiths II: (13.34)	Ferromagnetic spin models with even single-spin measure
Ginibre-Newman-Aizenman: (13.35)	Ferromagnetic spin models with even single-spin measure
Lebowitz (zero magnetic field): (13.36)	Repulsive polymer-chain models; spin models in GS and BFS classes
Lebowitz (nonzero field): (13.39)	Spin models in GS and EMN classes
AF: (13.40), (13.41)	Repulsive simple CIW models; spin models in GS and BFS classes
AG: (13.44), (13.45)	Spin models in GS class; SAW
Twice improved AF: (13.48)	Simple CIW models
Non-Gaussian upper bound on $\overline{u}_4$: (13.50)	Ising, IPORW models
GHS: (13.53)	Spin models in GS and EMN classes
Weak GHS: (13.56)	Repulsive polymer-chain models; spin models in GS and BFS classes
FS: (13.58)	Baby repulsive CIW models; spin models in GS and BFS classes
ABF: (13.59), (13.60)	Spin models in GS class
AFe: (13.65)	Ising model
Schrader–Messager–Miracle-Solé: (13.66)	Nearest-neighbor translation-invariant ferromagnetic spin models

Table 13.3. Summed versions of the correlation inequalities for translation-invariant models in zero magnetic field

$$\overline{u}_4 \geq \begin{cases} -3|J|^2\chi^4\left(1+\dfrac{1}{|J|\chi}\right)^2 & \text{BFS models} \\[2ex] -|J|\chi^2\beta\dfrac{\partial\chi}{\partial\beta} - \varepsilon \geq -\dfrac{|J|^2\chi^4}{1+|J|^2B_0}[1-\varepsilon''] & \text{GS models} \\[2ex] -3\left(\beta\dfrac{\partial\chi}{\partial\beta}+\chi\right)^2 & \text{repulsive simple CIW models} \end{cases}$$

$$\overline{u}_4 \leq \begin{cases} 0 & \text{repulsive polymer-chain, BFS and GS models} \\[2ex] -\dfrac{1}{48|J|^2B_0}\left(\beta\dfrac{\partial\chi}{\partial\beta}\right)^2[1-\tilde{\varepsilon}] \leq -\dfrac{\chi^4}{48B_0(1+2|J|B_0)^2}[1-\tilde{\varepsilon}'] & \text{Ising models} \end{cases}$$

$$\beta\frac{\partial\chi}{\partial\beta} \leq |J|\chi^2 \qquad \text{repulsive polymer chains, BFS and GS models}$$

$$\beta\frac{\partial\chi}{\partial\beta} \geq \frac{|J|\chi^2}{1+|J|^2B_0}[1 - \varepsilon'] \qquad \text{repulsive simple CIW, GS models}$$

Table 13.4. Summed versions of the correlation inequalities for translation-invariant models in non-zero magnetic field

$$\chi \le \frac{M}{h} \le \begin{cases} \chi + \left(\frac{M}{h}\right)^2 [h + \lvert J\rvert M]^2 \\ \chi + \left(\frac{M}{h}\right)^2 \left[h + \lvert J\rvert M\right]\left[h + \frac{h}{M}\beta\frac{\partial M}{\partial\beta}\right] \\ \chi + \left(\frac{M}{h}\right)^2 \left[hM + h\beta\frac{\partial M}{\partial\beta}\right] \end{cases}$$	simple CIW, BFS and GS models GS models Ising models
$$\frac{\partial\chi}{\partial h} \ge \begin{cases} 0 \\ -\frac{[1 - O(B_0 h/M)]}{96B_0(1 + 2\lvert J\rvert B_0)^2}(\tanh h)\chi^4 \end{cases}$$	EMN and GS models Ising models
$$\frac{\partial\chi}{\partial h} \le -32\lvert J\rvert[1 + O(h/M)]\frac{M^4}{(\tanh h)^3}$$	Ising models

Let $\mathbb{Z}^d_{\geq}$ denote the half-space $\mathbb{Z}^d \cap \{x\colon\ \mathbf{n}\cdot x \geq c\}$, and let $\mathcal{F}_{\geq}$ denote the space of observables depending only on the fields in $\mathbb{Z}^d_{\geq}$. Then a probability measure μ on $\mathbb{R}^{\mathbb{Z}^d}$ is said to be *reflection-positive* (RP) with respect to θ if

$$\langle \overline{\Theta A}\, A\rangle_\mu \ \geq\ 0 \qquad \text{for all } A \in \mathcal{F}_{\geq}\,, \tag{13.69}$$

where $^-$ denotes complex conjugation.

In particular, we are interested in the case that μ is the Gibbs measure of a translation-invariant spin system with quadratic Hamiltonian

$$H \ =\ -\frac{1}{2}\sum_{x,y} J(x-y)\,\varphi_x\varphi_y\,. \tag{13.70}$$

The pair coupling $J = \{J(x)\}_{x\in\mathbb{Z}^d}$ is said to be *reflection-positive* (with respect to θ) if

$$\sum_{x,y\in\mathbb{Z}^d_{>}} \overline{f(x)}\, J(\theta x - y)\, f(y) \ \geq\ 0 \tag{13.71}$$

for all complex-valued functions f on $\mathbb{Z}^d$ whose support is finite and lies in $\mathbb{Z}^d_{>} \equiv \mathbb{Z}^d \cap \{x\colon\ \mathbf{n}\cdot x > c\}$. Then a fundamental theorem [218] states that (under mild technical conditions concerning the infinite-volume limit) if J is a reflection-positive coupling, then any translation-invariant Gibbs state μ for the Hamiltonian H is a reflection-positive measure.

Let us now make some general remarks about reflection-positive couplings, and give some examples:

1) If J is reflection-positive, then so is λJ for any $\lambda \geq 0$. Similarly, if J and J' are reflection-positive, then so is $\lambda J + (1-\lambda)J'$ for any $0 \leq \lambda \leq 1$. Therefore, the reflection-positive couplings form a *convex cone*, which we denote $\mathcal{R}$.

2) A slightly less trivial result [218] is that the pointwise product of two reflection-positive couplings is again reflection-positive: $(JJ')(x) \equiv J(x)J'(x)$. So $\mathcal{R}$ is in fact a *multiplicative convex cone*.

3) The two types of reflection most commonly considered are reflection in a *lattice hyperplane* (e.g. $x_1 = 0$), and reflection in a *bisector hyperplane* (e.g. $x_1 = -\frac{1}{2}$). In these two cases there exist spectral representations completely characterizing the class $\mathcal{R}$ of reflection-positive couplings [218]. Let us consider first the case of dimension $d = 1$. Then a coupling J is RP with respect to a *lattice* hyperplane if and only if

$$J(x) \ =\ a\delta_{x,0} + b\delta_{|x|,1} + c\delta_{|x|,2} + \int_{-1}^{1} \lambda^{|x|-2}\, d\varrho(\lambda) \tag{13.72}$$

where a, b are real, $c \geq 0$ and $d\varrho$ is a positive measure. A coupling J is RP with respect to a *bisector* hyperplane if and only if

$$J(x) \ =\ a\delta_{x,0} + b\delta_{|x|,1} + \int_{-1}^{1} \lambda^{|x|-1}\, d\varrho(\lambda) \tag{13.73}$$

where a is real, $b \geq 0$ and $d\varrho$ is a positive measure.[4] These representations generalize straightforwardly to dimension $d > 1$: the $d-1$ transverse dimensions simply "go for the ride". That is, J is RP with respect to reflection in the hyperplane $x_1 = 0$ if and only if

$$J(x_1, \mathbf{x}) = \int_{[-\pi,\pi]^{d-1}} d\mathbf{p}\, e^{i\mathbf{p}\cdot\mathbf{x}} \left[a_{\mathbf{p}}\delta_{x_1,0} + b_{\mathbf{p}}\delta_{|x_1|,1} + c_{\mathbf{p}}\delta_{|x_1|,2} + \int_{-1}^{1} \lambda^{|x_1|-2}\, d\varrho_{\mathbf{p}}(\lambda) \right] \tag{13.74}$$

where $a_{\mathbf{p}}, b_{\mathbf{p}}$ are real, $c_{\mathbf{p}} \geq 0$ and $d\varrho_{\mathbf{p}}$ is a positive measure for each $\mathbf{p} \in [-\pi, \pi]^{d-1}$. Similarly, J is RP with respect to reflection in the hyperplane $x_1 = -1/2$ if and only if

$$J(x_1, \mathbf{x}) = \int_{[-\pi,\pi]^{d-1}} d\mathbf{p}\, e^{i\mathbf{p}\cdot\mathbf{x}} \left[a_{\mathbf{p}}\delta_{x,0} + b_{\mathbf{p}}\delta_{|x_1|,1} + \int_{-1}^{1} \lambda^{|x_1|-1}\, d\varrho_{\mathbf{p}}(\lambda) \right] \tag{13.75}$$

where $a_{\mathbf{p}}$ is real, $c_{\mathbf{p}} \geq 0$ and $d\varrho_{\mathbf{p}}$ is a positive measure for each $\mathbf{p}$. A more complicated — and to our knowledge, still unsolved — problem is to characterize in similar terms the interactions that are simultaneously RP with respect to lattice (or bisector) hyperplanes in *all* directions x_i $(1 \leq i \leq d)$.

Examples.

(a) A nearest-neighbor *ferromagnetic* coupling

$$J(x-y) = \begin{cases} J & \text{if } |x-y| = 1 \\ 0 & \text{otherwise} \end{cases} \tag{13.76}$$

($J \geq 0$) is reflection-positive with respect to both lattice hyperplanes and bisector hyperplanes. A nearest-neighbor *antiferromagnetic* coupling ($J < 0$) is reflection-positive with respect to lattice hyperplanes but *not* bisector hyperplanes.

(b) Let Δ be the Laplacian on $\mathbb{R}^d$, and for each $a \geq 0$ let $G_a(x, y)$ be the integral kernel of the operator $(-\Delta + a)^{-1}$. Then, for any positive measure $d\varrho$ on $\mathbb{R}_+$, the coupling defined by

$$J(x-y) = \int d\varrho(a)\, G_a(x-y) \tag{13.77}$$

restricted to $x, y \in \mathbb{Z}^d$ is reflection-positive. An explicit example covered by this remark is

$$J(x-y) = \begin{cases} |x-y|^{-\kappa} & \text{for } x \neq y \\ 0 & \text{for } x = y \end{cases} \tag{13.78}$$

with $\kappa \geq d-2$.

[4] If we interpret $0^0 = 1$, then the terms $c\delta_{|x|,2}$ in (13.72) and $c\delta_{|x|,1}$ in (13.73) are just the contributions from a $\delta(\lambda)$ piece in $d\varrho(\lambda)$ and could therefore be omitted.

(c) Let J be a coupling that is reflection-positive with respect to bisectors on $\mathbb{Z}^1$. Then the coupling $\check{J}$ on $\mathbb{Z}^d$ defined by

$$\check{J}(x) \equiv J(|x|_1) \tag{13.79}$$

with $|x|_1 = \sum_{i=1}^{d} |x_i|$ is also reflection-positive [13].

We now discuss two important consequences of reflection positivity: *spectral representations* and *infrared bounds*.

Spectral representation [258, 449].

Let μ be a translation-invariant measure that is reflection-positive with respect to reflection in the lattice hyperplane $x_1 = 0$. Then it is not hard to show that the truncated two-point function

$$G(x) \equiv \langle \varphi_0 ; \varphi_x \rangle \tag{13.80}$$

has the spectral representation [258, 449]

$$G(x_1, \mathbf{x}) = \int_{[-\pi,\pi]^{d-1}} d\mathbf{p} \int_{-1}^{1} d\varrho_{\mathbf{p}}(\lambda)\, \lambda^{|x_1|} e^{i\mathbf{p}\cdot\mathbf{x}} \tag{13.81}$$

where $\mathbf{x} = (x_2, \ldots, x_d)$ and $d\varrho_{\mathbf{p}}(\lambda)$ is, for each $\mathbf{p}$, a *positive* measure on $[-1, 1]$. If, in addition, μ is RP with respect to reflection in the bisector hyperplane $x_1 = -1/2$, then each measure $d\varrho_{\mathbf{p}}(\lambda)$ is supported on $[0, 1]$.

If the exponential correlation length ξ [defined in (13.10)] is finite, then it is not hard to see that each measure $d\varrho_{\mathbf{p}}(\lambda)$ is supported on the interval $[-e^{-1/\xi}, e^{-1/\xi}]$.

Remark. The spectral representation (13.81) for the two-point function G is tantalizingly similar, but *not* identical, to the spectral representations (13.74) and (13.75) for the pair coupling J. This is because reflection positivity for a *measure* (hence also for that measure's moments) involves observables living in $\mathbb{Z}^d_{\geq}$, while reflection positivity for a *pair coupling* involves only observables living in $\mathbb{Z}^d_{>} \subsetneq \mathbb{Z}^d_{\geq}$.

Infrared bound [222, 218, 481]. Let

$$G(x) \equiv \langle \varphi_0 ; \varphi_x \rangle \tag{13.82}$$

be the truncated two-point function, and let

$$\widetilde{G}(p) \equiv \sum_x e^{-ip\cdot x}\, G(x) \tag{13.83}$$

be its Fourier transform. Similarly, let

$$\widetilde{J}(p) \equiv \sum_x e^{-ip\cdot x} J_{0x} \tag{13.84}$$

be the Fourier transform of the pair interaction. The infrared bound [222, 218] states that if J is a reflection-positive coupling, then

$$0 \le \widetilde{G}(p) \le [\widetilde{J}(0) - \widetilde{J}(p)]^{-1} . \tag{13.85}$$

In particular, for a nearest-neighbor interaction of strength J, we have

$$\begin{aligned} 0 \le \widetilde{G}(p) &\le \left[2J\sum_{i=1}^{d}(1-\cos p_i)\right]^{-1} \\ &\le \frac{\text{const}}{J|p|^2} \end{aligned} \tag{13.86}$$

(since $|p_i| \le \pi$).

There is also an x-space version of the infrared bound [481, Appendix A], which is a consequence of the p-space bound combined with the Schrader–Messager–Miracle-Solé correlation inequality. For nearest-neighbor ferromagnets, one gets

$$0 \le G(x) \le \frac{\text{const}}{J(|x|+1)^{d-2}} \tag{13.87}$$

for $d > 2$; the constant is universal.

Some further consequences. By combining the spectral representation (13.81) with the infrared bound (13.86), it is not difficult to prove the following bounds [481, section 2.2] relating χ to the correlation length ξ (or ξ_ϕ):

$$J\chi \le \text{const} \times \xi^2 \tag{13.88}$$

$$J\chi \le \text{const} \times \xi_\phi^2 \tag{13.89}$$

These are *universal* bounds, which hold for nearest-neighbor ferromagnets with arbitrary single-spin measure.

14. Inequalities for critical exponents

The principal goal of the theory of critical phenomena is to make quantitative predictions for universal features of critical behavior — critical exponents, universal ratios of critical amplitudes, equations of state, and so forth — as discussed in Section 1.1. (Non-universal features, such as critical temperatures, are of lesser interest.) The present status of the theory of critical phenomena is roughly the following:

- *Non-rigorous* renormalization-group calculations predict mean-field critical behavior for systems above their upper critical dimension d_c (e.g. $d_c = 4$ for short-range Ising-type models). For systems below their upper critical dimension (e.g. $d = 3$), RG methods predict exact scaling laws relating critical exponents, and give reasonably accurate numerical predictions of individual critical exponents (and other universal quantities).[1]

- *Rigorous* mathematical analysis has given a proof of (some aspects of) mean-field critical behavior for (certain) systems above their upper critical dimension (e.g. short-range Ising models for $d > 4$). For systems below their upper critical dimension, much less is known. Often one half of a scaling law can be proven as a rigorous *inequality*. Likewise, rigorous *upper or lower bounds* on individual critical exponents can in many cases be proven.

The goal of this chapter is to give an exposition of a significant part (but far from all) of the currently-available rigorous theory of critical exponents for spin and polymer systems. More precisely, we emphasize those results that are consequences of the correlation inequalities summarized in Section 13.3, and especially those that follow from the random-walk and random-current formalisms. Our main purpose is to explain the basic ingredients behind the proof of each result; thus, we shall often omit the detailed arguments, referring the reader to the original publication, except when no readily available reference exists. We emphasize that this chapter is *not* a comprehensive review of critical-exponent inequalities; for complementary material, see [199, 269, 415, 409, 464, 479, 480, 487].

---

[1]The situation in dimension $d = 2$ is special: the combination of exact solutions, Coulomb gas arguments and conformal invariance has allowed the presumably *exact* (through at present non-rigorous) calculation of numerous critical exponents.

The strongest available results concern the approach to critical point from the high-temperature side at zero magnetic field (the "symmetric regime"); we discuss these results in Section 14.1. In the subsequent three sections, we discuss all other approaches to the critical point (the "non-symmetric regime"). This regime includes, as a special case, the traditional paths to the critical point along the critical isotherm and along the coexistence curve. However, we believe that the traditional focus on these particular paths is too narrow, and one of our goals will be to establish bounds which are valid in a full neighborhood of the critical point in the (β, h)-plane. Examples of such bounds are the "extrapolation principles"of Section 14.2, and the bounds valid outside certain "horns" obtained in Section 14.5. These bounds will be of particular importance in our discussion of the scaling limit (Section 15.3).

It is important to distinguish between two types of bounds — called *universal* and *non-universal*[2] — which correspond to different ways of approaching the critical point:

1) *Universal bounds.* A universal bound is one that is valid for some large class of single-spin measures (e.g. GS class, BFS class, etc.), with constants that are universal numbers (independent of everything except possibly the dimensionality of space).[3] For example, the infrared bound

$$0 \le G(x;J) \le \frac{\text{const}}{J(|x|+1)^{d-2}} \qquad (13.87)/(14.1)$$

holds for all nearest-neighbor ferromagnetic spin models in dimension $d > 2$, irrespective of the single-spin measure. Note, in particular, that since the GS, BFS, ... classes are invariant under the field-strength rescaling $\varphi \to \alpha\varphi$, it follows that a universal bound must be dimensionally correct under such rescalings.Universal bounds are appropriate to the study of quantum field theory (Chapter 15), in which we must allow an arbitrary manner of approach to the critical surface (e.g. arbitrary variation of the bare φ^4 coupling constant λ).

2) *Non-universal bound.* These are bounds in which the constants are not universal, but depend on various parameters of the model. For example, consider a statistical-mechanical model in which one fixes the single-spin measure $dP(\varphi)$ and increases the nearest-neighbor coupling towards its critical value J_c. Now, for $J \le J_c$ we have

$$G(x;J) \le G(x,J_c) \le \frac{C}{(|x|+1)^{d-2+\eta}} \qquad (14.2)$$

[2]Our use here of the term "universal" is related to, but not identical to, its meaning in renormalization-group theory.

[3]That is, in a universal bound, "const" stands for some finite number that we could in principle calculate if we were less lazy. *Example:* In the infrared bound $0 \le \tilde{G}(p) \le \left[2J\sum_{i=1}^{d}(1-\cos p_i)\right]^{-1} \le \text{const}/J|p|^2$ for $|p_i| \le \pi$, "const" stands for $\pi^2/4$ (or any larger number).

by Griffiths' second inequality and (one) definition of the critical exponent η. However, the constant C depends on the single-spin measure. Thus, non-universal bounds are appropriate only for the study of certain restricted ways of approaching the critical surface (e.g. increasing J with fixed single-spin measure). These are, however, precisely the limits of interest in the statistical-mechanical theory of critical phenomena. [Note that since $\eta \geq 0$, (14.2) is in general stronger than (14.1), within its domain of validity.]

As another example, consider the summed Aizenman-Fröhlich inequalities in "intermediate bound" and "intermediate site" forms:

$$\overline{u}_4 \;\geq\; -3|J|^2\chi^4\left(1+\frac{1}{|J|\chi}\right)^2 \qquad (13.42)/(14.3)$$

$$\overline{u}_4 \;\geq\; -2\chi^4 \qquad (13.43)/(14.4)$$

The first form is valid for all single-spin measures in GS class, and is hence invariant under the rescaling $\varphi \to \alpha\varphi$ (which takes $\chi \to \alpha^2\chi$, $\overline{u}_4 \to \alpha^4\overline{u}_4$, $J \to \alpha^{-2}J$). In particular, it can be re-expressed in terms of the dimensionless combinations[4] $|J|\chi$ and $|J|^2\overline{u}_4$:

$$|J|^2\,\overline{u}_4 \;\geq\; -3(|J|\chi)^4\left(1+\frac{1}{|J|\chi}\right)^2 . \qquad (14.5)$$

The second form, by contrast, is valid only for *Ising* models (with their usual normalization $\sigma = 1$). In particular, this form is *not* invariant under field-strength rescalings. [However, (14.4) is somewhat stronger than (14.3) when it is valid.]

14.1 Symmetric regime

We start with the inequalities regarding exponents defined in the symmetric regime $h = 0$, $\beta < \beta_c$. The results of this section are summarized in Table 14.1.

14.1.1 Upper bound on the bubble diagram (critical exponent b)

These bounds are a preliminary result needed for practically all the subsequent inequalities. We shall present them in two different versions:

(a) *Universal form.* Let $\widetilde{G}(p) \equiv \sum_x e^{ipx}u_2(0,x)$, the two-point function in momentum space; then, by the Plancherel identity,

$$B_0 \;\equiv\; \sum_x u_2(0,x)^2 \;=\; (2\pi)^{-d}\int_{[-\pi,\pi]^d} \widetilde{G}(p)^2 dp\,. \qquad (14.6)$$

[4]For further discussion of dimensional analysis in quantum field theory, see [481, Section 2.1].

Table 14.1. Inequalities for critical exponents. Symmetric regime

Inequality	When saturated	References	Hypothesis	Models
$b \le \max\left(0, (2 - \frac{d}{2})\gamma\right)$ $\bar{b} \le 1$	Scaling with $\eta = 0$	[478], [481, App. A]	Griffiths II + ref. positivity	ref.-positive even ferromagnet
$b \le \max\left(0, (2 - \frac{d}{2-\eta})\gamma\right)$ $\bar{b} \le 1$	Scaling	[481, App. B]	Griffiths II	even ferromagnet
$b \ge \max\left(0, 2\gamma - d\nu\right)$	Scaling	[477]	Lebowitz (for $b \ge 0$); ref. positivity (for $b \ge 2\gamma - d\nu$)	ref.-positive BFS class, GS class
$\alpha \le b$ $\bar{\alpha} \le \bar{b}$	Gaussian	[478]	Lebowitz	BFS class, GS class
$\gamma \ge 1$ $\bar{\gamma} \ge 0$	Mean-field (Gaussian)	[256], [30]	Lebowitz	BFS class, GS class, repulsive polymer-chain
$\gamma \le b + 1$ $\bar{\gamma} \le \bar{b}$	Spherical model	[15], [12, App. A]	AG + Griffiths II	GS class
$\Delta_4 \le \frac{3}{2}\gamma$	Mean-field	[5], [213]	AF + Lebowitz	BFS class, GS class
$\Delta_4 \le \gamma + \frac{1}{2}$	Mean-field	[15]	AG	GS class, repulsive simple CIW
$\Delta_4 \le \frac{\gamma}{2} + 1$	Mean-field	[483]	Twice-improved AF	Repulsive simple CIW

(Continued on next page)

Table 14.1. Inequalities for critical exponents. Symmetric regime (Continued)

Inequality	When saturated	References	Hypothesis	Models
$\Delta_4 \geq 2 - \frac{b+\gamma}{2}$	Mean-field with $b = 0$		Ineq. (12.114)	SAW
$\Delta_4 \geq \frac{\gamma}{2} + 1 - b$	Mean-field with $b = 0$	[194]	Non-Gaussian upper bound on $\overline{u}_4$	Ising, IPORW
$d\nu_\phi - 2\Delta_4 + \gamma \geq 0$ ($\phi \geq d/2$)	Hyperscaling	[256], [30], [370], [5]	Ginibre-Newman-Aizenman; or Griffiths II	BFS class, GS class
$d\nu_\phi - 2\Delta_4 + \gamma \geq (d-4)\nu_\phi$	Mean-field	[5], [213], [481, App. B]	AF + ref. positivity	BFS class, GS class
$d\nu_\phi - 2\Delta_4 + \gamma \geq (d-4)\nu_\phi + (\gamma - 1)$	Mean-field	[15]	AG + ref. positivity	GS class
$d\nu_\phi - 2\Delta_4 + \gamma \geq d\nu_\phi - 2$	Mean-field	[483]	Twice-improved AF	Repulsive simple CIW

Now, for a large class of reflection-positive models (which includes nearest-neighbor ferromagnetic models), one has the infrared bound

$$0 \leq \widetilde{G}(p) \leq \frac{\text{const}}{Jp^2} . \qquad (13.86)/(14.7)$$

Moreover, the Griffiths II inequality $u_2(0, x) \geq 0$ implies that

$$\widetilde{G}(p) \leq \widetilde{G}(0) \equiv \chi . \qquad (14.8)$$

It follows easily (see e.g. [478] or [481, Appendix B]) that

$$B_0 \leq \text{const} \times J^{-2} \times \begin{cases} 1 & d > 4 \\ \log(J\chi) & d = 4 \\ (J\chi)^{2-d/2} & 2 < d < 4 \end{cases} \qquad (14.9)$$

In terms of critical exponents (14.9) implies that

$$b \leq \max\Bigl(0, \bigl(2 - \frac{d}{2}\bigr)\gamma\Bigr) \qquad \text{for } d > 2 \qquad (14.10)$$

$$\bar{b} \leq 1 \qquad \text{for } d = 4 \text{ (provided that } \gamma < \infty) \qquad (14.11)$$

(We remark that this method also yields bounds on b and $\bar{b}$ for certain long-range models in which an infrared bound stronger than (14.7) can be proven [12, 13].)

(b) *Non-universal form.* More generally, let us define an exponent η as the supremum of the values for which

$$\langle \varphi_0 \varphi_x \rangle_{\beta_c} \;\leq\; \frac{C}{(|x|+1)^{d-2+\eta}} \tag{14.12}$$

with some $C = C(\eta) < \infty$. Then, for those systems for which the Griffiths' second inequality is valid one has

$$\langle \varphi_0 \varphi_x \rangle_{\beta} \;\leq\; \langle \varphi_0 \varphi_x \rangle_{\beta_c} \;\leq\; \frac{C}{(|x|+1)^{d-2+\eta}} \tag{14.13}$$

for all $\beta \leq \beta_c$, from which it can be shown (see below) that

$$B_0 \;\leq\; \text{const} \times \begin{cases} 1 & d > 4-2\eta \\ \log \chi & d = 4-2\eta \\ \chi^{2-[d/(2-\eta)]} & 2-\eta < d < 4-2\eta \end{cases} \tag{14.14}$$

This implies the critical-exponent inequality

$$b \;\leq\; \max\Big(0, \big(2 - \frac{d}{2-\eta}\big)\gamma\Big) \qquad \text{for } d > 2-\eta \tag{14.15}$$

$$\bar{b} \;\leq\; 1 \qquad \text{for } d = 4-2\eta \tag{14.16}$$

In fact, (14.15)/(14.16) are *equalities* if the two-point function has the usual scaling behavior.

For systems obeying the infrared bound (14.7), we have

$$\eta \geq 0 \tag{14.17}$$

and hence (14.14)–(14.15) are stronger than (14.9)–(14.10). On the other hand (14.14)–(14.15) remain valid even if $\eta \geq 0$ is not true or not proven. We remark that the bounds (14.14) were previously proven [481, Appendix B] under an additional hypothesis which corresponds to requiring that $\langle \varphi_0; \varphi_x \rangle_{\beta_c}$ be roughly symmetric decreasing. In fact this requirement is not needed. Let us show this by sketching the proof of (14.14). We have

$$\begin{aligned} B_0 = \sum_x \langle \varphi_0 \varphi_x \rangle^2 \;&=\; \sum_{|x|\leq R} \langle \varphi_0 \varphi_x \rangle^2 + \sum_{|x|>R} \langle \varphi_0 \varphi_x \rangle^2 \\ &\leq\; \sum_{|x|\leq R} \left(\frac{C}{(|x|+1)^{d-2+\eta}} \right)^2 + \frac{C}{R^{d-2+\eta}} \sum_{|x|>R} \langle \varphi_0 \varphi_x \rangle \end{aligned}$$

since $d-2+\eta \geq 0$. Hence

$$B_0 \;\leq\; \text{const}\, C^2 \left\{ \begin{array}{ll} 1 & d > 4-2\eta \\ \log R & d = 4-2\eta \\ R^{4-2\eta-d} & 2-\eta < d < 4-2\eta \end{array} \right\} + \frac{C}{R^{d-2+\eta}}\,\chi \,. \tag{14.18}$$

We can now optimize over R: if $d \leq 4-2\eta$, take $R = \chi^{1/(2-\eta)}$ (this makes sense because $2-\eta \geq d/2 > 0$); if $d > 4-2\eta$, take $R \to \infty$. This yields (14.14).

14.1.2 Lower bound on the bubble diagram (critical exponent *b*)

Here we prove lower bounds on the bubble diagram [477] that are complementary to (14.9) and (14.14).

(a) *Bound useful for* $d > 4$. In [482] it is proven that

$$\chi \;\le\; \frac{\langle \varphi_0^2 \rangle_0}{1 - |J| \langle \varphi_0^2 \rangle_0} \;\le\; \frac{G(0)}{1 - |J| G(0)} \tag{14.19}$$

(here $\langle \,\cdot\, \rangle_0$ denotes the a priori single-spin measure) whenever the denominators are positive, for any ferromagnetic model satisfying the Lebowitz inequality. (This is a "mean-field bound".) Hence

$$\begin{aligned} G(0) \;&\ge\; |J|^{-1} \frac{|J|\chi}{|J|\chi + 1} \\ &\to\; |J_c|^{-1} \qquad \text{as } J \uparrow J_c \,, \end{aligned} \tag{14.20}$$

which immediately yields

$$\begin{aligned} B_0 \;\le\; G(0)^2 \;&\ge\; |J|^{-2} \left(1 - \frac{\text{const}}{|J|\chi} \right) \\ &\to\; |J_c|^{-2} \qquad \text{as } J \uparrow J_c \,. \end{aligned} \tag{14.21}$$

This proves that (14.9) is sharp for $d > 4$.

(b) *Bound useful for* $d < 4$. Let us assume that the model is reflection-positive with respect to both lattice and bisector hyperplanes. Then the spectral representation (13.81), translated to p-space, reads

$$\tilde{G}(p_1, \mathbf{p}) \;=\; \int\limits_{\cosh m - 1}^{\infty} \frac{d\varrho_{\mathbf{p}}(a)}{1 - \cos p_1 + a} \,, \tag{14.22}$$

where $m = \xi^{-1}$ is the mass gap. From this it easily follows that

$$\tilde{G}(p_1, \mathbf{p}) \;\ge\; \frac{\cosh m - 1}{\cosh m - \cos p_1} \tilde{G}(0, \mathbf{p}) \,. \tag{14.23}$$

Using this successively in each coordinate direction, we get

$$\begin{aligned} \tilde{G}(p) \;&\ge\; \left(\prod_{i=1}^{d} \frac{\cosh m - 1}{\cosh m - \cos p_i} \right) \tilde{G}(0) \\ &\ge\; \chi \prod_{i=1}^{d} \frac{m^2}{m^2 + p_i^2} \,. \end{aligned} \tag{14.24}$$

Squaring this and inserting it into the Plancherel identity (14.6), we obtain

$$B_0 \;\ge\; \text{const} \times \chi^2 \xi^{-d} \,. \tag{14.25}$$

This bound is most useful for $d < 4$. Indeed, if the usual scaling law $\chi \sim \xi^{2-\eta}$ holds, then both this bound and (14.14) are sharp for $2 - \eta < d < 4 - 2\eta$.

14.1.3 Upper bound on the specific heat (critical exponent α)

For systems satisfying the Lebowitz inequality (13.36) one readily obtains [478]

$$\begin{aligned} C_H &\le \tfrac{1}{2}\textstyle\sum_{x,y,z} J_{0x}J_{yz}\langle\varphi_0\varphi_z\rangle\langle\varphi_x\varphi_y\rangle \\ &\le \tfrac{1}{2}|J|^2 B_0 \,. \end{aligned} \tag{14.26}$$

In the last step we have used the Plancherel identity, or alternatively the Schwarz inequality in the space of summable functions of x, y, z with weight $J_{0x}J_{yz}$. This yields the inequality

$$\alpha \le b \tag{14.27}$$

and, if $\alpha = b = 0$ (as is expected in $d = d_c$), then

$$\overline{\alpha} \le \overline{b}\,. \tag{14.28}$$

If these inequalities are combined with (14.10), we conclude [478] that for systems satisfying Griffiths II, Lebowitz and the infrared bound (14.7),

$$\alpha \le \max\Bigl(0, \bigl(2 - \frac{d}{2}\bigr)\gamma\Bigr) \tag{14.29}$$

and, for $d = d_c$,

$$\overline{\alpha} \le 1\,. \tag{14.30}$$

More generally, if the long-distance bound (14.12) is assumed,

$$\alpha \le \max\Bigl(0, \bigl(2 - \frac{d}{2-\eta}\bigr)\gamma\Bigr)\,. \tag{14.31}$$

14.1.4 Bounds on the susceptibility (critical exponent γ)

14.1.4.1 Lower bound For systems satisfying the Lebowitz inequality, the differential inequality (13.38) can easily be integrated to yield [256, 30, 258]

$$\chi \ge \bigl(|J_c| - |J|\bigr)^{-1}\,, \tag{14.32}$$

which implies the critical exponent inequality

$$\gamma \ge 1 \tag{14.33}$$

and in the logarithmic case

$$\overline{\gamma} \ge 0\,. \tag{14.34}$$

This is a "mean-field (Gaussian) lower bound" in the sense that it is an equality for the Gaussian model — which is not surprising, since in this case the Lebowitz inequality is in fact an equality.

14.1.4.2 Upper bounds For systems satisfying the AG inequality, we can combine the differential inequality (13.47) with the bubble bound (14.14) to obtain

$$\frac{\partial\chi}{\partial\beta} \geq \text{const} \times \begin{cases} \chi^2 & d > 4-2\eta \\ \chi^2/\log\chi & d = 4-2\eta \\ \chi^{d/(2-\eta)} & 2-\eta < d < 4-2\eta \end{cases} \tag{14.35}$$

This differential inequality can then be integrated to yield

$$\chi \leq \text{const} \times \begin{cases} t^{-1} & d > 4-2\eta \\ t^{-1}(|\log t|+1) & d = 4-2\eta \\ t^{-(2-\eta)/(d-2+\eta)} & 2-\eta < d < 4-2\eta \end{cases}, \tag{14.36}$$

which implies

$$\gamma \leq \max\left(1, \frac{2-\eta}{d-(2-\eta)}\right) \qquad \text{for } d > 2-\eta \tag{14.37}$$

$$\bar{\gamma} \leq 1 \qquad \text{for } d = 4-2\eta. \tag{14.38}$$

For nearest-neighbor reflection-positive systems, the bound (14.17) implies that (14.36)–(14.38) hold with η replaced by 0.

Alternatively, if we *assume* that the bubble diagram satisfies a bound of the form

$$B_0 \leq \text{const} \times t^{-b} \tag{14.39}$$

with $b \geq 0$, then an integration of the AG inequality (13.47) yields [15], [12, Appendix A], [484]

$$\chi \leq \text{const} \times t^{-b-1}, \tag{14.40}$$

so that

$$\gamma \leq b+1\,. \tag{14.41}$$

In the logarithmic case

$$B_0 \leq \text{const} \times |\log t|^{\bar{b}} \tag{14.42}$$

(with $\bar{b} \geq 0$), we obtain

$$\chi \leq \text{const} \times t^{-1}|\log t|^{\bar{b}} \tag{14.43}$$

and hence

$$\bar{\gamma} \leq \bar{b}\,. \tag{14.44}$$

We note [484] that (14.36)–(14.44) are "spherical-model upper bounds" in the sense that they are equalities for the spherical model[5]. This can be explained by the fact that the lower bound was obtained by keeping only the

[5]The original spherical model was defined by Berlin and Kac [62]. Stanley [486] showed that it was equivalent (in at least some aspects) to the $N \to \infty$ limit of the N-vector model. In our opinion, the original spherical model is of mainly historical interest; we use here the term "spherical model" simply as a shorthand for "$N \to \infty$ limit of the N-vector model". The critical exponents for the N-vector model with large N can be found in [77] (cf. also (13.26) for the case $d = 4$).

bubble diagrams, and as $N \to \infty$ the geometric series of bubbles dominates the diagrammatic expansion of the N-vector model [110, 283].

Another interesting remark is that the combination of (13.38), (13.47) and (14.9) [or (14.14)] shows that for $d > 4$ (or $d > 4 - 2\eta$)

$$\text{const} \times t^{-2} \leq \frac{\partial \chi}{\partial \beta} \leq \text{const} \times t^{-2} \tag{14.45}$$

and hence

$$\text{const} \times t^{-1} \leq \chi \leq \text{const} \times t^{-1}, \tag{14.46}$$

which proves that for such dimensions the critical exponent γ really exists. On the other hand, the situation for $d < 4$ is less complete; the key missing element is a set of non-mean-field lower bounds on γ for $d < 4$.

Finally, by inserting (14.36) back into (14.14), we obtain an upper bound on the bubble diagram

$$B_0 \leq \text{const} \times \begin{cases} 1 & d > 4 - 2\eta \\ |\log t| & d = 4 - 2\eta \\ t^{-(4-2\eta-d)/(d-2+\eta)} & 2 - \eta < d < 4 - 2\eta \end{cases} \tag{14.47}$$

and hence

$$b \leq \max\left(0, \frac{4 - 2\eta - d}{d - 2 + \eta}\right) \qquad \text{for } d > 2 - \eta \tag{14.48}$$

$$\bar{b} \leq 1 \qquad \text{for } d = 4 - 2\eta\,. \tag{14.49}$$

14.1.5 Upper bounds on the renormalized coupling constant (critical exponent Δ_4)

In this subsection and the next, we study bounds on the connected four-point function at zero momentum, $\bar{u}_4$, or equivalently on the renormalized coupling constant $g \equiv -\bar{u}_4/\chi^2\xi^d$. We consider systems for which Lebowitz inequality (13.36) is true for them $g \geq 0$.

Bound for low dimensions By manipulation of the Ginibre-Newman-Aizenman inequality (13.35) [5] (or Griffiths' second inequality), it can be shown [5, 370] that

$$-\bar{u}_4 \leq \text{const} \times \chi^2 \xi_\phi^d \tag{14.50}$$

for any $\phi \geq d/2$. (For $\phi > d$ this result was previously obtained by Schrader [450], extending an earlier result by Glimm and Jaffe [257].) Therefore

$$g_\phi \leq \text{const} \qquad (\phi \geq d/2), \tag{14.51}$$

which implies

$$\varrho_\phi \equiv d\nu_\phi - 2\Delta_4 + \gamma \geq 0 \qquad (\phi \geq d/2) \tag{14.52}$$

and in the logarithmic case

$$\overline{\varrho}_\phi \geq 0 \qquad\qquad (\phi \geq d/2). \tag{14.53}$$

We refer to (14.51)–(14.53) as the Glimm-Jaffe upper bound on the renormalized coupling constant g_ϕ. This bound is good for $d < 4$ — where it is generally believed to be an equality (hyperscaling) — but very bad for $d > 4$.

Bound for high dimensions The Aizenman-Fröhlich inequalities (13.40) and (13.41) imply that in the vicinity of the critical point

$$-\overline{u}_4 \leq \text{const} \times \chi^4 , \tag{14.54}$$

hence

$$\Delta_4 \leq \frac{3}{2}\gamma , \tag{14.55}$$

which is a "mean-field bound". (We shall see below that this bound is sharp in $d > 4$ at least for the Ising and IPORW models.) In terms of the renormalized coupling constant, (14.54) implies that

$$g_\phi \leq \text{const} \times \frac{\chi^2}{\xi_\phi^d} . \tag{14.56}$$

If, in addition, the system is reflection-positive, it follows from (14.56) and (13.89) that

$$g_\phi \leq \frac{\text{const}}{\xi_\phi^{d-4}} , \tag{14.57}$$

which implies that

$$\varrho_\phi \geq (d-4)\nu_\phi . \tag{14.58}$$

This bound is complementary to (14.52) in the sense that it is a good bound for $d > 4$, but horrible for $d < 4$. We remark that if we use the non-universal bound $\gamma \leq (2-\eta)\nu$ [199] instead of the universal bound $\gamma \leq 2\nu$ implicit in (13.89) above, we get a slightly improved result in the case $\eta > 0$. But $\eta > 0$ is believed to occur only for $d < 4$, and in this case even the "improved" bound is bad.

Improvement using the AG inequality ("once-improved AF inequality"). For systems satisfying the Aizenman-Graham inequality (13.46), the bound (14.54) can be replaced by

$$-\overline{u}_4 \leq \text{const} \times \chi^2 \beta \frac{\partial \chi}{\partial \beta} , \tag{14.59}$$

which implies

$$\Delta_4 \leq \gamma + \frac{1}{2} . \tag{14.60}$$

Inequalities (14.59)–(14.60) are an improvement over (14.54)–(14.55) because, by (13.38), $\beta \partial\chi/\partial\beta \leq \text{const}\, \chi^2$, or equivalently $\gamma \geq 1$. For reflection-positive systems we obtain the following improvements of (14.56) and (14.57):

$$g_\phi \le \frac{\text{const}}{\xi_\phi^d}\,\beta\frac{\partial\chi}{\partial\beta} \le \frac{\text{const}}{\xi_\phi^{d-4}\chi^2}\,\beta\frac{\partial\chi}{\partial\beta} \tag{14.61}$$

which imply

$$\varrho_\phi \ge (d-4)\nu_\phi + (\gamma-1)\,. \tag{14.62}$$

This is equivalent to the bound (14.58) for $d > 4$; it is an improvement, but still disastrously bad, for $d < 4$. The main relevance of (14.61) is in dimension $d = 4$, where it shows that *if* χ *has a logarithm* — as permitted by (14.34)/(14.44), and as believed to occur, cf. (13.26) — so that $\chi^{-2}\partial\chi/\partial\beta \to 0$ as $\beta \uparrow \beta_c$, then $g \to 0$ as $\beta \uparrow \beta_c$, i.e. *the continuum limit is trivial.* In fact, (14.61) yields the bound

$$\overline{\varrho} \ge \overline{\varrho}_\phi \ge \overline{\gamma} \tag{14.63}$$

on the logarithmic exponent controlling the triviality of the continuum limit.

Bound for systems satisfying the twice-improved AF inequality. For such systems (13.49) implies

$$-\overline{u}_4 \le \text{const}\left(\beta\frac{\partial\chi}{\partial\beta}\right)^2 \tag{14.64}$$

in the vicinity of the critical point, hence

$$\Delta_4 \le \frac{\gamma}{2} + 1\,. \tag{14.65}$$

Moreover,

$$g_\phi \le \frac{\text{const}}{\xi_\phi^d\chi^2}\left(\beta\frac{\partial\chi}{\partial\beta}\right)^2 \tag{14.66}$$

implies

$$\varrho_\phi \ge d\nu_\phi - 2. \tag{14.67}$$

Therefore

$$d\nu_\phi > 2 \implies g_\phi \to 0 \text{ as } \xi_\phi \to \infty\,. \tag{14.68}$$

This result makes rigorous one half of a very beautiful heuristic argument about self-avoiding walks due to des Cloizeaux [125]. Recall from (13.31) that g measures, roughly speaking, the probability of intersection of two independent SAWs that start a distance of order ξ apart. Now, by the alternative definition of ν in (13.27), we can interpret a near-critical SAW as an object with "fractal dimension" $1/\nu$. Two independent such objects will "generically" intersect if and only if the sum of their fractal dimensions is at least as large as the dimension of the ambient space. So we expect g to be nonzero at the critical point if and only if

$$\frac{1}{\nu} + \frac{1}{\nu} \ge d\,, \quad \text{i.e. } d\nu \le 2\,. \tag{14.69}$$

Thus, (14.68) makes rigorous the "only if" half of this intuitive argument.[6]

[6]We remark that the correlation inequality (14.64), which holds for repulsive simple CIW models, presumably does *not* hold for N-component spin models with N greater than some

14.1.6 Non-Gaussian lower bound on the renormalized coupling constant (critical exponent Δ_4)

For the Ising and IPORW models, we can use the non-Gaussian bound (13.50) to obtain a lower bound

$$-\overline{u}_4 \;\geq\; \frac{\text{const}}{B_0}\left(\beta\frac{\partial\chi}{\partial\beta}\right)^2 , \tag{14.70}$$

which implies

$$\Delta_4 \;\geq\; \frac{\gamma}{2}+1-b\,. \tag{14.71}$$

With the bounds (14.10) or (14.15), this yields

$$\Delta_4 \;\geq\; \frac{\gamma}{2}+1-\max\Bigl(0,\bigl(2-\frac{d}{2-\eta}\bigr)\gamma\Bigr)\,. \tag{14.72}$$

For the Ising model these bounds are useful only for $d>4$: in that case we have $b=0$ and $\gamma=1$, so that (14.71) together with the upper bound (14.55) [or (14.60)] implies

$$\Delta_4 \;=\; \frac{3}{2}\,, \tag{14.73}$$

which is the mean-field value. For the Ising model in $d<4$, the bounds (14.70))–(14.72) are terrible (as is the bound $\Delta_4 \geq \gamma$ proven by Sokal [479]). On the other hand, for the IPORW model (14.70)–(14.72) are essentially sharp in all dimensions: indeed, (14.71) together with the exact results $\gamma=1, \nu=\nu_\phi=1/2$ and $b=\max(0,2-(d/2))$ and the upper bounds (14.52)/(14.55) implies

$$\Delta_4 \;=\; \min\left(\frac{3}{2},\frac{1}{2}+\frac{d}{4}\right) \tag{14.74}$$

and hence

$$\varrho_\phi \;=\; \max\left(\frac{d-4}{2},0\right)\,. \tag{14.75}$$

In other words, hyperscaling holds for the IPORW model in $d<4$, and is violated at exactly the mean-field rate for the IPORW model in $d\geq 4$. For the

---

critical N_c, and it is almost certain that $N_c<4$. Indeed, for $d=4-\varepsilon$ the renormalization group predicts [77]

$$\nu \;=\; \frac{1}{2}+\frac{(N+2)}{4(N+8)}\varepsilon+\frac{(N+2)(N+3)(N+20)}{8(N+8)^3}\varepsilon^2+O(\varepsilon^3)\,,$$

hence

$$d\nu \;=\; 2+\frac{N-4}{2(N+8)}\varepsilon+\frac{(N+2)^2(N+28)}{4(N+8)^3}\varepsilon^2+O(\varepsilon^3)\,,$$

which is >2 for sufficiently small $\varepsilon>0$ if $N\geq 4$. On the other hand, hyperscaling ($g\to$ const >0) is expected to hold for all N when $d<4$. It follows that the twice-improved AF inequality cannot be expected to hold for $N\geq 4$.

IPORW model in $d = 4$, hyperscaling is violated by a logarithm, i.e. $\overline{\varrho} = 1$. One half of this inequality, namely $\overline{\varrho} \le 1$, follows from (14.70). The other half can be proven by a beautiful renormalization-group argument, as discussed in Chapter 5.

Sharp results (i.e. hyperscaling) for the Ising model or SAW in $d < 4$ could be obtained if either the following two conjectured improvements of (14.70) could be proven:

$$-\overline{u}_4 \ \ge \ \frac{\text{const}}{B_0}\chi^4 \qquad (?) \tag{14.76}$$

$$-\overline{u}_4 \ \ge \ \frac{\text{const}}{C_H}\left(\beta\frac{\partial\chi}{\partial\beta}\right)^2 \qquad (?)\,. \tag{14.77}$$

(14.76) is an improvement over (14.70), since $\beta\partial\chi/\partial\beta \le \text{const} \times \chi^2$ by (13.38). It would imply

$$g \ \ge \ \text{const} \times \frac{\chi^2}{B_0\xi^d} \ \ge \ \text{const} \times \left[\frac{\chi^{1/(2-\eta)}}{\xi}\right]^d \tag{14.78}$$

for $d < 4 - 2\eta$ by the last line in (14.14). If we believe the "correlation-length scaling law" $\gamma = (2-\eta)\nu$ we obtain

$$g \ \ge \ \text{const} \ > \ 0 \quad \Longrightarrow \quad \textit{hyperscaling!} \tag{14.79}$$

for $d < 4 - 2\eta$. Similarly, (14.77) is an improvement of (14.70), since $C_H \le \text{const} \times B_0$ by (14.26). It would imply

$$\Delta_4 \ \ge \ \frac{\gamma - \alpha}{2} + 1 \qquad (?)\,. \tag{14.80}$$

The equality would correspond to a "thermodynamic scaling law" which is believed to be true in all dimensions.

Unfortunately, we do not have the slightest idea how to prove either (14.76) or (14.77) for the Ising model or the SAW. We can, however, prove them for the IPORW model, for which (14.70), (14.76) and (14.77) are equivalent ($\gamma = 1$ and $\alpha = b$ in all dimensions).

14.2 Extrapolation principles

In this section we begin our study of the non-symmetric regime. The first topic (Section 14.2.1) is a review of the Landau-Ginzburg theory (see e.g. [487, Chapter 10]), which is a version of the mean-field approximation. From a logical point of view, it may seem somewhat incongruous to consider a heuristic approximation method in the middle of a chapter devoted to rigorous inequalities. However, this study will soon pay off: in Section 14.4 we shall prove that, for Ising models in dimension $d > 4$, the Landau-Ginzburg prediction for the magnetization, $M_{LG}(t, h)$, provides a rigorous two-sided bound on the true magnetization

$M(t, h)$ in a full neighborhood of the critical point, and that similar (but slightly weaker) bounds hold also for the susceptibility. In other words, we shall prove that the Landau-Ginzburg theory predicts correctly (up to constant factors) the *exact critical behavior* of Ising models (and partial results for models in GS class) in dimension $d > 4$.

A key tool in this analysis will be a pair of *extrapolation principles*,derived in Sections 14.2.2 and 14.2.3, which allow lower or upper bounds on the magnetization (or in some cases the susceptibility) to be transferred from one region of the (t, h)-plane to another. These extrapolation principles are obtained by interpreting suitable correlation inequalities (here GHS or ABF) as *partial differential inequalities* which may then be integrated. In Section 14.6 we systematize this idea, which we suspect may have additional applications in the future.

The results of our study may be summarized as follows: Landau-Ginzburg theory satisfies (and can essentially be defined by) the two *identities*

$$\frac{\partial M_{LG}}{\partial \beta} = M_{LG}\chi_{LG} \tag{14.81}$$

and

$$\begin{aligned} M_{LG} &= h\chi_{LG} + 2M_{LG}^2 \frac{\partial M_{LG}}{\partial \beta} & (14.82) \\ &= h\chi_{LG} + 2M_{LG}^3 \chi_{LG} & (14.83) \end{aligned}$$

In the exact theory these identities are replaced by the *inequalities*

$$0 \le \beta \frac{\partial M}{\partial \beta} \le |J|\, M\chi \qquad \text{(Griffiths II, GHS)} \tag{14.84}$$

and

$$\begin{aligned} h\chi \le M &\le h\chi + |J|\, M^2 \beta \frac{\partial M}{\partial \beta} \left(1 + \frac{h}{|J|\, M}\right) \left(1 + \frac{M}{\beta \dfrac{\partial M}{\partial \beta}}\right) \\ &\qquad \text{(weak GHS, ABF)} & (14.85) \\ &\le h\chi + |J|^2 M^3 \chi \left(1 + \frac{h}{|J|\, M}\right) \left(1 + \frac{1}{|J|\chi}\right) \\ &\qquad \text{(weak ABF)} & (14.86) \end{aligned}$$

which allow some (but not all) of Landau-Ginzburg behavior to be derived for the exact theory.

14.2.1 Review of Landau-Ginzburg theory

The starting point of Landau-Ginzburg theory is the assumption that the Helmholtz free energy $F(t, M)$ — which is the Legendre transform of the Gibbs

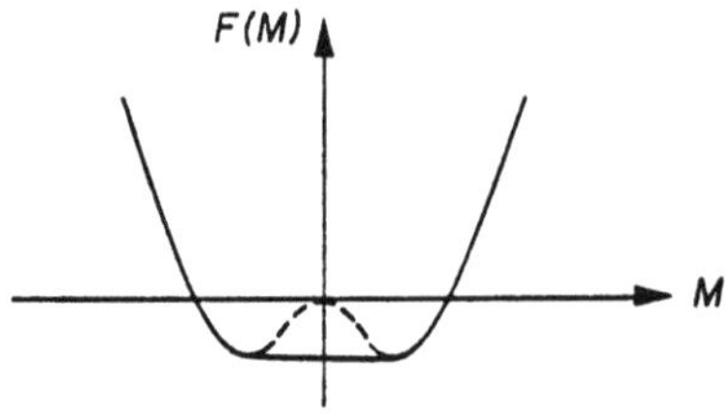

Fig. 14.1. The function (14.87) (broken line) and its convex hull (full line), the free energy (14.88)

free energy $\Phi(t, h)$ — is a smooth function of t and M near the critical point $(t = 0, M = 0)$.[7] It then follows from general arguments[8] that, after a smooth change of variables, F has the form

$$F(M) = \tfrac{1}{4}M^4 + \tfrac{1}{2}tM^2 . \tag{14.87}$$

However, it must be noticed that this functional form contradicts the general principles of statistical mechanics, which require that F be convex in M (and concave in t). Therefore, we make the minimal modification of (14.87) that has these properties, by defining F to be the convex hull of (14.87), i.e.

$$F(t, M) = \begin{cases} \tfrac{1}{4}M^4 + \tfrac{1}{2}tM^2 & \text{for } t \geq 0 \text{ or } |M| \geq \sqrt{-t} \\ -\tfrac{1}{4}t^2 & \text{for } t < 0 \text{ and } |M| < \sqrt{-t} \end{cases} \tag{14.88}$$

(see Figure 14.1).

By standard thermodynamics, the magnetic field is the derivative of F with respect to M:

$$h = \frac{\partial F}{\partial M} . \tag{14.89}$$

The Landau-Ginzburg magnetization, $M_{LG}(t, h)$, is therefore the solution of the *Landau-Ginzburg equation*,

$$M_{LG}^3 + t\, M_{LG} = h \tag{14.90}$$

that lies in the region $M^2 \geq -t$. (For $h = 0$ and $t < 0$, M can be any number between $-\sqrt{-t}$ and $\sqrt{-t}$. The interior points correspond to mixed phases.) Introducing (for $h > 0$) the *scaling variables*

$$x = th^{-2/3} \tag{14.91}$$

$$y = Mh^{-1/3}, \tag{14.92}$$

[7] This assumption turns out to be *false* in dimension $d \leq 4$, which is why the Landau-Ginzburg theory fails in that case.

[8] Let the qualitative behavior be as in the Ising model, i.e. a second-order transition at $(t = 0, h = 0)$ and a line of first-order transitions at $(t < 0, h = 0)$. Then it follows from catastrophe theory [29] that F can be reduced by a smooth change of variables to the *normal form* (14.87).

we can write the Landau-Ginzburg magnetization in the *scaling form*

$$M_{LG}(t,h) = h^{1/3} f_{LG}(t h^{-2/3}) , \tag{14.93}$$

where $y = f_{LG}(x)$ is defined implicitly by the equation

$$y^3 + xy = 1 \tag{14.94}$$

(with $y^2 \geq -x$). It is not hard to show that f_{LG} is an analytic, strictly decreasing function of $x \in (-\infty, \infty)$, with the following asymptotic behavior:

$$f_{LG}(x) \;=\; \begin{cases} 1 - \frac{1}{3}x + O\left(x^2\right) & x \simeq 0 \\ x^{-1}\left[1 + O\left(x^{-3}\right)\right] & x \to +\infty \\ (-x)^{1/2} + \frac{1}{2}(-x)^{-1} + O\left((-x)^{-5/2}\right) & x \to -\infty \end{cases} \tag{14.95}$$

$$f'_{LG}(x) \;=\; \begin{cases} -\frac{1}{3} + O(x) & x \simeq 0 \\ -x^{-2}\left[1 + O\left(x^{-3}\right)\right] & x \to +\infty \\ -\frac{1}{2}(-x)^{-1/2} + \frac{1}{2}(-x)^{-2} + O\left((-x)^{-7/2}\right) & x \to -\infty \end{cases} \tag{14.96}$$

It follows that the Landau-Ginzburg magnetization $M_{LG}(t,h)$ behaves as

$$M_{LG}(t,h) \sim \begin{cases} h^{1/3}\left[1 + O(x)\right] & x \simeq 0 \\ h\,t^{-1}\left[1 + O\left(x^{-3}\right)\right] & x \to +\infty \\ (-t)^{1/2}\left[1 + O\left((-x)^{-3/2}\right)\right] & x \to -\infty \end{cases} \tag{14.97}$$

where $x \equiv t h^{-2/3}$, and that the Landau-Ginzburg susceptibility $\chi_{LG}(t,h) \equiv \partial M_{LG}(t,h)/\partial h$ behaves as

$$\chi_{LG}(t,h) \sim \begin{cases} \frac{1}{3}h^{-2/3}\left[1 + O(x)\right] & x \simeq 0 \\ t^{-1}\left[1 + O\left(x^{-3}\right)\right] & x \to +\infty \\ \frac{1}{2}(-t)^{-1}\left[1 + O\left((-x)^{-3/2}\right)\right] & x \to -\infty \end{cases} \tag{14.98}$$

The first line ($x \simeq 0$) is applicable close to the critical isotherm; the second line ($x \to +\infty$) corresponds to the high-temperature/small-field region; and the third line ($x \to -\infty$) corresponds to the low-temperature/small-field region. In particular, along the three "standard paths" we have

$$M_{LG} \sim \begin{cases} h^{1/3} & h \downarrow 0, t = 0 \\ (-t)^{1/2} & t \uparrow 0, h = 0^+ \end{cases} \tag{14.99}$$

and

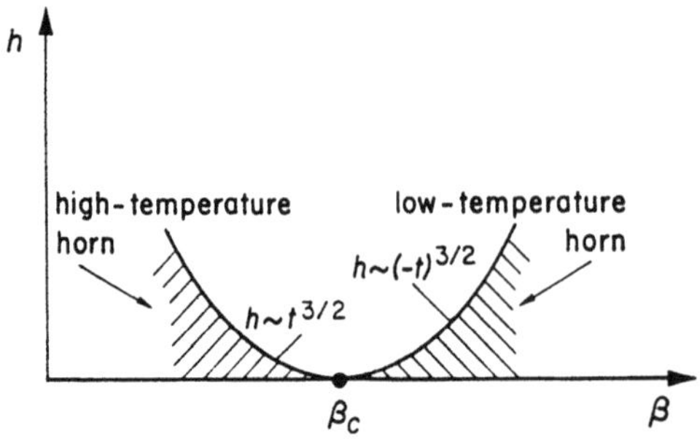

Fig. 14.2. High-temperature horn and low-temperature horn in the (β, h)-plane. Here $t \sim \beta_c - \beta$

$$\chi_{LG} \sim \begin{cases} t^{-1} & t \downarrow 0, h = 0 \\ h^{-2/3} & h \downarrow 0, t = 0 \\ (-t)^{-1} & t \uparrow 0, h = 0 \end{cases} \tag{14.100}$$

from which we derive the Landau-Ginzburg (or mean-field) predictions for critical exponents,

$$\delta = 3, \quad \widehat{\beta} = \frac{1}{2}, \quad \gamma = \gamma' = 1. \tag{14.101}$$

In fact, the behaviors $M_{LG} \sim h^{1/3}$ and $\chi_{LG} \sim h^{-2/3}$ hold not only on the critical isotherm, but in any region of the form

$$-C_1 \leq x \equiv th^{-2/3} \leq C_2 \tag{14.102}$$

$(C_1, C_2 < \infty)$. The set (14.102) covers the entire (t, h)-plane with the exception of two "horn-shaped" regions, one at high temperature and one at low temperature (see Figure 14.2). These horns will play an important role in the sequel. In particular, along any ray $t = ah$ (a finite) we have $x \to 0$ as $h \to 0$, so that $M_{LG} \sim h^{1/3}$ and $\chi_{LG} \sim h^{-2/3}$ hold with a critical amplitude that is independent of a.

Let us also mention some important combinations of $M_{LG}(t, h)$ and $\chi_{LG}(t, h)$ that depend only on the scaling variable $x = th^{-2/3}$:

$$\frac{M_{LG}(t,h)}{h^{1/3}} = f_{LG}(x) \tag{14.103}$$

$$\frac{h\chi_{LG}(t,h)}{M_{LG}(t,h)} = g(x) \equiv \frac{1}{3}\left[1 - 2x\frac{f'_{LG}(x)}{f_{LG}(x)}\right] \tag{14.104}$$

$$M_{LG}(t,h)^2 \chi_{LG}(t,h) = k(x) \equiv f_{LG}(x)^3 g(x) \tag{14.105}$$

The behavior of the function f_{LG} has already been discussed. The function g is strictly increasing and satisfies $0 \leq g(x) \leq 1$ for all x; its asymptotic behavior is

$$g(x) = \begin{cases} \frac{1}{3} + O(x) & x \simeq 0 \\ 1 + O(x^{-3}) & x \to +\infty \\ \frac{1}{2}(-x)^{-3/2} + O\left((-x)^{-3}\right) & x \to -\infty \end{cases} \tag{14.106}$$

In particular, $g(x)$ is bounded above everywhere by 1, and is bounded below by a nonzero constant in any region $x \geq -C_1$. This means that χ_{LG} is always less than M_{LG}/h, and it is of the same order of magnitude as M_{LG}/h throughout the (t, h)-plane *except* in a "low-temperature horn". Similarly, the function k is strictly decreasing and satisfies $0 \leq k(x) \leq \frac{1}{2}$ for all x; its asymptotic behavior is

$$k(x) = \begin{cases} \frac{1}{3} + O(x) & x \simeq 0 \\ x^{-3} + O\left(x^{-6}\right) & x \to +\infty \\ \frac{1}{2} + O\left((-x)^{-3/2}\right) & x \to -\infty \end{cases} \tag{14.107}$$

In particular, $k(x)$ is bounded above everywhere by $\frac{1}{2}$, and is bounded below by a nonzero constant in any region $x \leq C_2$. This means that $M_{LG}^2\chi_{LG}$ is everywhere bounded above (by $\frac{1}{2}$), and it is bounded below throughout the (t, h)-plane *except* in a "high-temperature horn". These combinations of M, χ and h will play a crucial role in our study of the non-symmetric regime.

Next, we observe that $M_{LG}(t, h)$ satisfies the pair of partial differential equations

$$\frac{\partial M_{LG}}{\partial t} = -M_{LG}\frac{\partial M_{LG}}{\partial h} \tag{14.108}$$

$$M_{LG} = h\frac{\partial M_{LG}}{\partial h} - 2M_{LG}^2\frac{\partial M_{LG}}{\partial t} \tag{14.109}$$

[Using the definition $M_{LG} = h^{1/3}f(th^{-2/3})$, these PDEs are equivalent to the ODEs

$$f'_{LG} = -\frac{f_{LG}}{x + 3f_{LG}^2} \tag{14.110}$$

$$f'_{LG} = \frac{f_{LG}^2}{2xf_{LG} - 3}, \tag{14.111}$$

respectively. But the latter are easily derived by differentiating the defining relation $f_{LG}^3 + xf_{LG} = 1$.] Conversely, this pair of PDEs *completely characterizes* $M_{LG}(t, h)$, except for a single constant of integration which corresponds to setting the critical temperature to be $t = 0$. To see this, first solve (14.108)/(14.109) for $\partial M_{LG}/\partial h$ and $\partial M_{LG}/\partial t$,

$$\frac{\partial M_{LG}}{\partial h} = \frac{M_{LG}}{h + 2M_{LG}^3} \tag{14.112}$$

$$\frac{\partial M_{LG}}{\partial t} = \frac{M_{LG}^2}{h + 2M_{LG}^3} \tag{14.113}$$

and then rewrite these as

$$\frac{\partial}{\partial h}\left(M_{LG}^2 - \frac{h}{M_{LG}}\right) = 0 \tag{14.114}$$

$$\frac{\partial}{\partial t}\left(M_{LG}^2 - \frac{h}{M_{LG}}\right) = -1\,. \tag{14.115}$$

The general solution is

$$M_{LG}^2 - \frac{h}{M_{LG}} = t_c - t \tag{14.116}$$

where t_c is an arbitrary constant of integration (the "critical temperature"). Taking $t_c = 0$, we recover the Landau-Ginzburg equation $M_{LG}^3 + tM_{LG} = h$.

Numerous second-order PDEs can also be derived from (14.108)/(14.109). For example, by applying $\partial/\partial h$ to (14.112) and doing some algebra, one obtains

$$\frac{\partial^2 M_{LG}}{\partial h^2} = -6(h + 2M_{LG}^3)\left(\frac{\partial M_{LG}}{\partial h}\right)^4 \tag{14.117}$$

$$\frac{\partial^2 M_{LG}}{\partial h^2} = -6M_{LG}\left(\frac{\partial M_{LG}}{\partial h}\right)^3 \tag{14.118}$$

$$\frac{\partial^2 M_{LG}}{\partial h^2} = -6\frac{M_{LG}^4}{(h + 2M_{LG}^3)^3} \tag{14.119}$$

and many others. And one can go to higher order, e.g.

$$\frac{\partial^3 M_{LG}}{\partial h^3} = -6\left(\frac{\partial M_{LG}}{\partial h}\right)^4 \left[1 + 18M_{LG}^2\frac{\partial M_{LG}}{\partial h}\right] \tag{14.120}$$

et cetera ad nauseam.

It is amusing to note [409] that (14.108) is the zero-viscosity limit of Burgers' equation in fluid dynamics,

$$\frac{\partial u}{\partial t} = -u\frac{\partial u}{\partial x}, \tag{14.121}$$

if one makes the identifications

magnetization M	$\longleftrightarrow$	fluid velocity u
temperature t	$\longleftrightarrow$	time t
magnetic field h	$\longleftrightarrow$	spatial position x.

We can then make the interpretations

phase transition	$\longleftrightarrow$	shock wave
critical point	$\longleftrightarrow$	shock formation
Maxwell's rule	$\longleftrightarrow$	jump construction

and so on.

More important for our purposes, however, is the observation that the *true* magnetization $M(t,h)$ satisfies the above PDEs as *inequalities* (except for small error terms in some cases). Indeed, corollary (13.55) of the GHS inequality is

$$\frac{\partial M}{\partial t} \geq - \text{ const} \times M \frac{\partial M}{\partial h} , \tag{14.122}$$

which corresponds to (14.108), while the ABF inequality (13.60) is

$$M \leq h \frac{\partial M}{\partial h} - (\approx \text{const}) \, M^2 \frac{\partial M}{\partial t} + M^3 , \tag{14.123}$$

which corresponds to (14.109). By combining these two inequalities, one gets the weak ABF inequality (13.61), which corresponds to (14.112). Similarly, for $d > 4$ and $\beta \leq \beta_c$ (where the "bubble graph" B_0 is finite), the AFe inequality (13.65) reads

$$\frac{\partial^2 M}{\partial h^2} \leq -(\approx \text{const}) \, h \left(\frac{\partial M}{\partial h} \right)^4 , \tag{14.124}$$

which corresponds to (14.117). The tree inequality (13.63) corresponds to (14.119); the Aizenman-Fröhlich inequality (13.42)/(13.43) corresponds to (14.120) evaluated at $h = 0$, $\beta \leq \beta_c$; and so on.

In view of these strong similarities between the *equalities* of the Landau-Ginzburg theory and the *inequalities* of the true theory, one might conjecture that Landau-Ginzburg theory could provide some kind of *one-sided bound* on the true magnetization, and that under certain conditions (e.g. finiteness of the bubble) it might even provide a *two-sided bound*. Much of the rest of this chapter is devoted to showing that these conjectures are correct. That is, we shall prove that in a full neighborhood of the critical point,

$$M(t,h) \geq \text{ const } \times M_{LG}(t,h) \tag{14.125}$$

for Ising and φ^4 models in any dimension d, while

$$M(t,h) \leq \text{ const } \times M_{LG}(t,h) \tag{14.126}$$

for Ising models in dimension $d > 4$.

Of course, the mean-field critical exponents (14.101) are *not* correct in dimension $d < 4$, so the Landau-Ginzburg theory must be taken with a large grain of salt. Two decades ago[9], Widom [508] and Domb and Hunter [137] proposed a phenomenological generalization of (14.93), namely

$$M(t,h) = h^{1/\delta} f\left(t h^{-1/\widehat{\beta}\delta}\right), \tag{14.127}$$

where $\widehat{\beta}$ and δ are unknown critical exponents and f is an unknown *scaling function* having the following properties:

[9] It was two decades ago when this was first written. It is now two decades and a half.

- $f(x) > 0$ for all real x
- f is analytic in a neighborhood of the real axis
- f is strictly decreasing
- f has the asymptotic behavior

$$f(x) \sim \begin{cases} (-x)^{\widehat{\beta}} & x \to -\infty \\ x^{-\widehat{\beta}(\delta-1)} & x \to +\infty \end{cases} \tag{14.128}$$

It is now believed that (14.127) describes correctly the asymptotic behavior of $M(t, h)$ in a small neighborhood of the critical point.[10] The Ansatz (14.127) implies the scaling relation among critical exponents,

$$\widehat{\beta}(\delta - 1) = \gamma = \gamma'. \tag{14.129}$$

It is believed that (14.129) holds in all dimensions, with $\gamma = \gamma' = 1$ in dimensions $d > 4$, and $\gamma = \gamma' > 1$ in dimensions $d < 4$.

Let us examine more closely the relation between the scaling Ansatz (14.127), the partial differential equations (14.108)/(14.109) and the partial differential inequalities (14.122)/(14.123). It is not hard to see that the scaling Ansatz satisfies the partial differential equation (14.108) if and only if

$$\widehat{\beta}(\delta - 1) = 1 \tag{14.130}$$

and

$$cf^{\delta} + xf = 1 \tag{14.131}$$

for some constant $c > 0$. We thus have a family of scaling solutions to (14.108), indexed by a parameter δ, which generalize the Landau-Ginzburg solution (and reduce to it in case $\delta = 3$). Similarly, the scaling Ansatz satisfies the partial differential equation (14.109) if and only if

$$\widehat{\beta} = \frac{1}{2} \tag{14.132}$$

and

$$f^{2\delta/(\delta-1)} + xf^{2/(\delta-1)} = \widetilde{c} \tag{14.133}$$

for some constant $\widetilde{c} > 0$. We thus have a family of scaling solutions to (14.109), indexed by a parameter δ, which generalize the Landau-Ginzburg solution (and reduce to it in case $\delta = 3$).

It is convenient to combine these two one-parameter families into a single two-parameter *generalized Landau-Ginzburg equation*

$$f^p + xf^q = 1\,, \tag{14.134}$$

---

[10] Except for possible multiplicative logarithmic corrections.

which defines the *generalized Landau-Ginzburg scaling function* $f_{p,q}(x)$. For $p > q$, the function $f_{p,q}$ is real-analytic and strictly decreasing, with the asymptotic behavior

$$f_{p,q}(x) = \begin{cases} 1 - \frac{1}{p}x + O\left(x^2\right) & x \simeq 0 \\ x^{-1/q}\left[1 + O\left(x^{-p}\right)\right] & x \to +\infty \\ (-x)^{1/(p-q)} + \frac{1}{p-q}(-x)^{-(p-1)/(p-q)} \\ \qquad +O\left((-x)^{-(2p-1)/(p-q)}\right) & x \to -\infty \end{cases} \tag{14.135}$$

$$f'_{p,q}(x) = \begin{cases} -\frac{1}{p} + O(x) & x \simeq 0 \\ -\frac{1}{q}x^{-(q+1)/q}\left[1 + O\left(x^{-p}\right)\right] & x \to +\infty \\ -\frac{1}{p-q}(-x)^{-(p-q-1)/(p-q)} + \frac{p-1}{(p-q)^2}(-x)^{-(2p-q-1)/(p-q)} \\ \qquad +O\left((-x)^{-(3p-q-1)/(p-q)}\right) & x \to -\infty \end{cases} \tag{14.136}$$

There is an obvious relation between $f_{p,q}$ and $f_{p',q'}$ whenever $p/q = p'/q'$, namely $f_{p,q} = [f_{p',q'}(x)]^{p'/p}$. The scaling Ansatz

$$M(t,h) = h^{1/\delta} f_{p,q}\left(th^{-1/(\widehat{\beta}\delta)}\right) \tag{14.137}$$

satisfies the partial differential equation (14.108) in case

$$p = \delta\,, \quad q = 1\,, \quad \widehat{\beta} = \frac{1}{\delta - 1}\,, \quad \delta \text{ arbitrary}\,,$$

while it satisfies (14.109) in case

$$p = \frac{2\delta}{\delta - 1}\,, \quad q = \frac{2}{\delta - 1}\,, \quad \widehat{\beta} = \frac{1}{2}\,, \quad \delta \text{ arbitrary}\,.$$

The intersection of these two one-parameter families is the ordinary Landau-Ginzburg theory

$$p = 3\,, \quad q = 1\,, \quad \widehat{\beta} = \frac{1}{2}\,, \quad \delta = 3\,,$$

and $f_{LG} = f_{3,1}$.

On the other hand, the scaling Ansatz (14.127) satisfies the partial differential *inequality* (14.27) for small h, t if and only if

$$\widehat{\beta}(\delta - 1) \geq 1 \tag{14.138}$$

[together with a condition on f in case $\widehat{\beta}(\delta - 1) = 1$]; and it satisfies the partial differential inequality (14.123) for small h, t if and only if

$$\widehat{\beta} \leq \frac{1}{2} \tag{14.139}$$

[together with a condition on f in case $\widehat{\beta} = 1/2$]. We have thus "morally proven" the critical-exponent inequalities $\widehat{\beta}(\delta - 1) \geq 1$ and $\widehat{\beta} \leq 1/2$; a proof which does not rely on the scaling Ansatz (14.127) will be given in Sections 14.3.4 and 14.4.1, respectively.

14.2.2 Consequences of the GHS inequality

In this section we prove an important *extrapolation principle* [409, 12] which allows lower or upper bounds on the magnetization to be transferred from one region of the (β, h)-plane to another. This principle is a consequence of the GHS inequality (together with Griffiths' second inequality). In the sequel we shall use this extrapolation principle to relate the critical behavior along different lines of approach to the critical point.

The starting point is the observation [409] that the inequality

$$\frac{\partial M}{\partial \beta} \leq |\widetilde{J}| M \frac{\partial M}{\partial h}, \tag{13.60)/(14.140}$$

which is an easy consequence of the GHS inequality [409, 479], can be rephrased as saying that the curves of constant magnetization

$$M\left(\beta, h(\beta)\right) = m \tag{14.141}$$

have a slope which is no more negative than $-|\widetilde{J}|m$. Indeed, differentiating (14.141) and using (14.140), we obtain

$$0 = \frac{\partial M}{\partial \beta} + \frac{\partial M}{\partial h}\frac{dh}{d\beta}\Big|_{M=\text{const}} \tag{14.142}$$

$$\leq \left(\frac{\partial M}{\partial h}\right)\left[|\widetilde{J}|M + \frac{dh}{d\beta}\Big|_{M=\text{const}}\right]. \tag{14.143}$$

Since $\partial M/\partial h \geq 0$ by Griffiths' second inequality, we have

$$\frac{dh}{d\beta}\Big|_{M=\text{const}} \geq -|\widetilde{J}|M. \tag{14.144}$$

(We ignore here all subtleties associated with the infinite-volume limit; see [12] for a detailed treatment.)

The following result is an almost immediate consequence of (14.144) together with Griffiths' second inequality:

Lemma 14.1 GHS extrapolation principles (Based on [12, Appendix].) *Consider some* $\beta_0, h_0 \geq 0$.

(a) *Assume that* $M(\beta_0, h_0) \geq m \geq 0$. *Then* $M(\beta, h) \geq m$ *for every* (β, h) *in the region*

$$\left\{(\beta, h): \quad h \geq \max\left[h_0, h_0 - |\widetilde{J}|m(\beta - \beta_0)\right]\right\}.$$

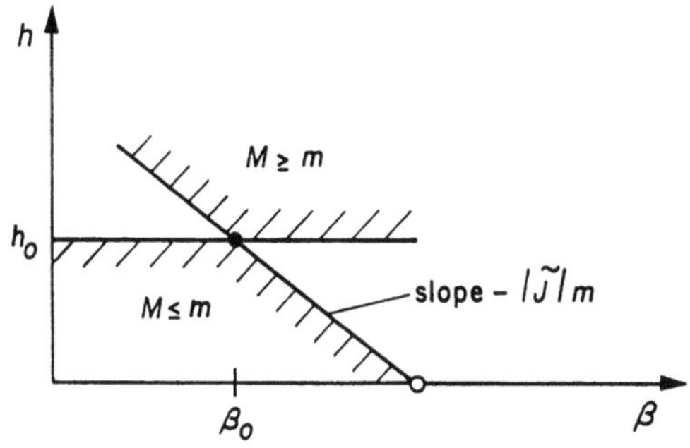

Fig. 14.3. Graphical summary of Lemma 14.1

(b) *Assume that $M(\beta_0, h_0) \le m$ with $m > 0$. Then $M(\beta, h) \le m$ for every (β, h) in the region*

$$\left\{(\beta, h): \quad h \le \min\left[h_0, h_0 - |\tilde{J}|m(\beta - \beta_0)\right]\right\}$$

except possibly at the "corner" $\beta = \beta_0 + [|\tilde{J}|m(h_0)]^{-1}h_0$, $h = 0$.

Lemma 14.1 is summarized graphically in Figure 14.3.

Proof. We find it convenient to integrate (14.140) *not* along the curves of constant M, but rather along the curves in the (β, h)-plane which *would* be curves of constant M if (14.140) were an *equality*, namely the straight lines of slope $-|\tilde{J}|m$. More generally, let us consider M along an arbitrary curve $h = h(\beta)$, and write $\mathcal{M}(\beta) \equiv M(\beta, h(\beta))$. We then have

$$\frac{d\mathcal{M}}{d\beta} = \frac{\partial M}{\partial \beta} + \frac{\partial M}{\partial h}\frac{dh}{d\beta} \tag{14.145}$$

$$\le \frac{\partial M}{\partial h}\left[|\tilde{J}|M + \frac{dh}{d\beta}\right], \tag{14.146}$$

where $\partial M/\partial h \ge 0$ by Griffiths' second inequality. In particular, if $h = h(\beta)$ is a straight line of slope $-|\tilde{J}|m$, we have

$$\frac{d\mathcal{M}}{d\beta} \le |\tilde{J}|\frac{\partial M}{\partial h}[\mathcal{M} - m]. \tag{14.147}$$

(a) Let $M(\beta_0, h_0) \ge m > 0$ and consider $\mathcal{M}$ along the straight line $\beta \le \beta_0$, $h = h_0 - |\tilde{J}|m(\beta - \beta_0)$. We claim that $\mathcal{M}(\beta) \ge m$ for all $\beta \le \beta_0$. Suppose the contrary, i.e. suppose that there exists $\beta_1 < \beta_0$ such that $\mathcal{M}(\beta_1) < m$. Then (assuming that $\mathcal{M}$ is once continuously differentiable[11]) there must exist $\beta \in$

[11] A careful argument would prove Lemma 14.1 first in finite volume (with periodic boundary conditions), where all thermodynamic quantities are smooth, and then pass to the infinite-volume limit in the final result.

(β_1, β_0) such that $\mathcal{M}(\beta) < m$ and $d\mathcal{M}/d\beta > 0$. But this contradicts (14.147). The claim is therefore proven; the rest follows from Griffiths' second inequalities $\partial M/\partial h \geq 0$ and $\partial M/\partial \beta \geq 0$.

(b) Assume first the *strict* inequality $M(\beta_0, h_0) < m$ with $m > 0$, and consider M along the straight line $\beta \geq \beta_0$, $h = h_0 - |\tilde{J}| m (\beta - \beta_0)$. We claim that $\mathcal{M}(\beta) < m$ and $d\mathcal{M}/d\beta \leq 0$ for all $\beta \geq \beta_0$. Indeed, suppose that the first of these claims is false, i.e. suppose that there exists $\beta_1 > \beta_0$ such that $\mathcal{M}(\beta_1) \geq m$. Then (assuming that $\mathcal{M}$ is once continuously differentiable there must exist $\beta \in (\beta_0, \beta_1)$ such that $\mathcal{M}(\beta) < m$ and $d\mathcal{M}/d\beta > 0$. But this contradicts (14.147). Hence we must have $\mathcal{M}(\beta) < m$ for all $\beta \geq \beta_0$; and $d\mathcal{M}/d\beta \leq 0$ then follows immediately from (14.147). The rest follows from Griffiths' second inequalities $\partial M/\partial h \geq 0$ and $\partial M/\partial \beta \geq 0$.

If we only have $M(\beta_0, h_0) = m$, then we use the above argument with m replaced by $m + \varepsilon$, and take the limit $\varepsilon \downarrow 0$. Since the (weak) GHS inequality implies that $\mathcal{M}$ is a continuous function of h except possibly at $h = 0$, we can pass to the limit in the inequality except possibly at the "corner." ∎

Remark. A careful reader will note that the proofs of parts (a) and (b) are different in an essential way; they are *not* merely related by a few changes of signs.

Using Lemma 14.1, lower bounds on the magnetization can be extrapolated "upwards and to the left" in the (β, h)-plane, while upper bounds on the magnetization can be extrapolated "downwards and to the right."

As a first example, suppose that the spontaneous magnetization $M_0(\beta) \equiv M(\beta, h = 0^+)$ is bounded below by some function $m(\beta)$, for β in some interval $[0, \beta_1]$. Without loss of generality we can assume that $m(\beta)$ is nonnegative and nondecreasing, and for simplicity let us assume also that $m(\beta)$ is continuous. Now let $m^*(\beta, h)$ be the unique solution to the partial differential *equation*

$$\frac{\partial m^*}{\partial \beta} = |\tilde{J}| m^* \frac{\partial m^*}{\partial h} \tag{14.148}$$

in the region

$$\left\{(\beta, h): \quad 0 < h \leq |\tilde{J}| m(\beta_1)(\beta_1 - \beta)\right\} \tag{14.149}$$

with initial condition $m^*(\beta, h = 0^+) = m(\beta)$. (The existence and uniqueness of this solution is easily demonstrated: the point is that the curves of constant m^* are straight lines of slope $-|\tilde{J}| m^*$, and every point in the region (14.149) can be connected to the initial surface $\{0 \leq \beta \leq \beta_1, h = 0\}$ by one and only one such line — see Figure 14.4.) Then we claim that $m^*(\beta, h)$ is a lower bound for $M(\beta, h)$ everywhere in the region (14.149):

Proposition 14.2 *Suppose that $M(\beta, h = 0^+) \geq m(\beta) \geq 0$ for β in some interval $[0, \beta_1]$, where the function $m(\beta)$ is nondecreasing and continuous. Let $m^*(\beta, h)$ be the unique solution to the partial differential equation (14.148) in*

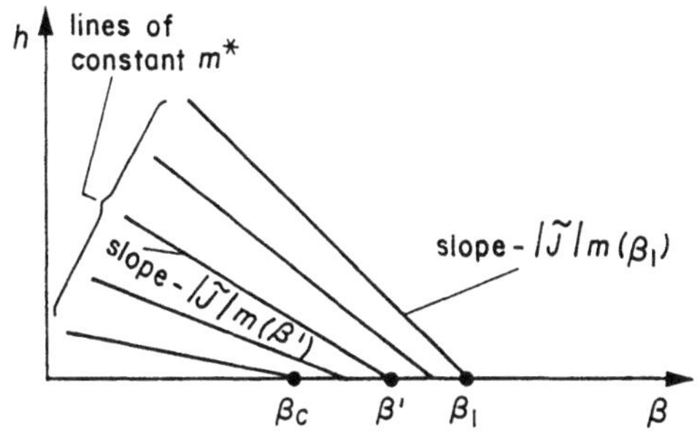

Fig. 14.4. Construction of the function $m^*(\beta, h)$ in the region (14.149). The line originating at $\beta' \in [0, \beta_1]$ has slope $-|\widetilde{J}|m(\beta')$, and along this line $m^*(\beta, h) = m(\beta')$

the region (14.149) with the initial condition $m^*(\beta, h = 0^+) = m(\beta)$. *Then, for all* (β, h) *in the region (14.149),*

$$M(\beta, h) \geq m^*(\beta, h). \tag{14.150}$$

Proof. For each (β, h) in the region (14.149), there exists a unique $\beta_0 \in [0, \beta_1]$ such that $h = |\widetilde{J}|m(\beta_0)(\beta_0 - \beta)$, and moreover $m^*(\beta, h) = m(\beta_0)$. The claim follows immediately from Lemma 14.1(a). ∎

Clearly Proposition 14.2 gives the *best possible* bound based only on the partial differential inequality (14.140), since $m^*(\beta, h)$ is itself a solution to this inequality — as an *equality*!

Typically we shall use Proposition 14.2 to extrapolate a lower bound on the magnetization from the coexistence curve $\{\beta > \beta_c, h = 0^+\}$ to the remainder of the (β, h)-plane. In this case we shall use functions $m(\beta)$ of the form

$$m(\beta) = \begin{cases} 0 & \text{if } \beta \leq \beta_c \\ c(\beta - \beta_c)^\lambda & \text{if } \beta_c \leq \beta \leq \beta_1 \end{cases} \tag{14.151}$$

with $c, \lambda > 0$ and some $\beta_1 > \beta_c$. It is a calculus exercise to compute the behavior of the corresponding $m^*(\beta, h)$:

$$m^*(\beta, h) = c \left(\frac{h}{c|\widetilde{J}|} \right)^{\lambda/(1+\lambda)} f_{1+\frac{1}{\lambda},1} \left(\left(\frac{c|\widetilde{J}|}{h} \right)^{\lambda/(1+\lambda)} t \right) \tag{14.152}$$

where $t \equiv \beta_c - \beta$ and $f_{1+\frac{1}{\lambda},1}(x)$ is the generalized Landau-Ginzburg scaling function (14.134) with $p = 1 + 1/\lambda$ and $q = 1$, defined by the equation

$$f^{1+\frac{1}{\lambda}} + xf = 1\,. \tag{14.153}$$

(In particular, if $\lambda = \frac{1}{2}$, then $f_{1+\frac{1}{\lambda},1}$ is the ordinary Landau-Ginzburg scaling function $f_{LG} = f_{3,1}$.) Using (14.135) and Proposition 14.2, we conclude that hence

$$
\begin{aligned}
M(\beta,h) &\geq m^*(\beta,h) \\
&= c\left(\frac{h}{c|\widetilde{J}|}\right)^{\lambda/(1+\lambda)} f_{1+\frac{1}{\lambda},1}\left(\left(\frac{c|\widetilde{J}|}{h}\right)^{\lambda/(1+\lambda)} t\right) \\
&= \begin{cases} c(c|\widetilde{J}|)^{-\lambda/(1+\lambda)}\, h^{\lambda/(1+\lambda)}\,[1+O(x)] & x \simeq 0 \\ |\widetilde{J}|^{-1}\, h\, t^{-1}\left[1+O\left(x^{-(1+\frac{1}{\lambda})}\right)\right] & x \to +\infty \\ c(-t)^{\lambda}\left[1+O\left(|x|^{-(1+\lambda)}\right)\right] & x \to -\infty \end{cases}
\end{aligned}
\tag{14.154}
$$

where $x \equiv \left(c|\widetilde{J}|/h\right)^{\lambda/(1+\lambda)} t$. The first line is useful close to the critical isotherm; the second line is useful in the high-temperature/small-field region; and the third line reproduces the initial data in the low-temperature/small-field region.It is worth emphasizing that neither c nor λ appears in the leading term on the second line — we thus get a *universal (mean-field) lower bound on the high-temperature zero-field susceptibility* [which is precisely the Glimm-Jaffe bound (14.32)].

Complementary results are obtained from part (b) of Lemma 14.1. Let us consider, for example, the extrapolation from the initial isotherm to the coexistence curve. The basic result is the following complement of Proposition 14.2, which is proven in exactly the same way but using Lemma 14.1(b).

Proposition 14.3 *Suppose that $M(\beta=\beta_c,h) \leq m(h)$ for h in some interval $[0,h_1]$, where the function $m(h)$ is nondecreasing and continuous. Let $m^*(\beta,h)$ be the unique solution to the partial differential equation (14.148) in the region*

$$
\left\{(\beta,h)\colon\quad 0 \leq h \leq h_1 \text{ and } 0 \leq \beta-\beta_c \leq \frac{h-h_1}{|\widetilde{J}|\, m(h_1)}\right\}
\tag{14.155}
$$

with the initial condition $m^*(\beta=\beta_c,h)=m(h)$. Then

$$
M(\beta,h) \leq m^*(\beta,h)
\tag{14.156}
$$

for all (β,h) in the region (14.155) except possibly at the "corner" $\beta=\beta_c+[|\widetilde{J}|m(h_1)]^{-1}h_1, h=0$.

The relevant application of this proposition is for a magnetization on the critical isotherm bounded above by a function of the form

$$
m(h) \;=\; ch^{1/\mu}
\tag{14.157}
$$

with $c>0$ and $\mu>1$. We then obtain

$$
m^*(\beta,h) = ch^{1/\mu} f_{\mu,1}\left(\frac{c|\widetilde{J}|t}{h^{1-\frac{1}{\mu}}}\right)
\tag{14.158}
$$

where $t = \beta_c - \beta$ as above and $f_{\mu,1}$ is the generalized Landau-Ginzburg scaling function satisfying

$$f^{\mu} + xf = 1\,. \tag{14.159}$$

Using (14.135) and Proposition 14.3, we conclude that for $\beta \geq \beta_c$ the magnetization satisfies

$$\begin{aligned} M(\beta,h) &\leq m^*(\beta,h) \\ &= ch^{1/\mu} f_{\mu,1}\left(\frac{c|\widetilde{J}|t}{h^{1-\frac{1}{\mu}}}\right) \\ &= \begin{cases} ch^{1/\mu}\left[1+O(x)\right] & x \lesssim 0 \\ \left(c^{\mu}|\widetilde{J}|\right)^{1/(\mu-1)} (-t)^{1/(\mu-1)} \left[1+O\left(|x|^{-\mu/(\mu-1)}\right)\right] & x \to -\infty \end{cases} \end{aligned} \tag{14.160}$$

where $x \equiv c|\widetilde{J}|t/h^{1-\frac{1}{\mu}}$. The first line reproduces the initial data close to the critical isotherm, and the second line is useful in the low-temperature/small-field region.

Another important consequence of Lemma 14.1 (and Griffiths' second inequality) follows from the fact that the bounds on M extrapolate along lines of slope proportional to the magnetization. Therefore, close to the critical point where the magnetization vanishes, such lines become almost horizontal. This implies that the regions

$$C^{+}_{c,\kappa} = \{(t,h)\colon t \geq -ch^{\kappa}\} \tag{14.161}$$

$$C^{-}_{c,\kappa} = \{(t,h)\colon t \leq ch^{\kappa}\} \tag{14.162}$$

— where $t = \beta_c - \beta$ as before and $c > 0$ — have the property that certain power-law bounds on the magnetization propagates throughout the whole region. The complement of a region $C^{+}_{c,\kappa}$ is a "low-temperature horn", while the complement of a region $C^{-}_{c,\kappa}$ is a "high-temperature horn". [In particular the regions $C^{\pm}_{c,2/3}$ are of relevance in the discussion of triviality (Sections 14.5 and 15.3).

Proposition 14.4 *Fix $\kappa > 0$ and let Γ be a continuous curve of the form $t = g(h)$, $0 \leq h \leq \tilde{\varepsilon}$, with $g(0) = 0$. Suppose that $\mathcal{M}(h) = M(g(h),h)$ is a continuous function satisfying*

$$\mathcal{M}(h) \leq C_+ h^{\mu_+}\left[1+o(1)\right] \tag{14.163}$$

as $h \downarrow 0$, with

$$\mu_+ > \max(1-\kappa, 0)\,. \tag{14.164}$$

(a) If Γ lies entirely in some region $C^{-}_{c',\kappa}$, then

$$M(t,h) \le C_+ h^{\mu_+} [1 + o(1)] \tag{14.165}$$

uniformly in $C^+_{c,\kappa}$.

(b) If Γ *lies entirely in some region* $C^+_{c',\kappa}$ *and* $\mathcal{M}(h)$ *satisfies*

$$\mathcal{M}(h) \ge C_- h^{\mu_-} [1 + o(1)] \tag{14.166}$$

as $h \downarrow 0$, *then*

$$M(t,h) \ge C_- h^{\mu_-} [1 + o(1)] \tag{14.167}$$

uniformly in $C^-_{c,\kappa}$.

Remarks. In words, part (a) of the proposition states that power-law upper bounds of the form $\mathcal{M} \le \text{const} \times h^{\mu_+}$ propagate from a curve Γ to a neighborhood of the critical point which *excludes* a low-temperature horn. This exclusion is to be expected, because on the low-temperature side of criticality ($t < 0$) the magnetization is strictly positive as $h \downarrow 0$; it is impossible to get an upper bound on M in terms of h alone. The restriction $\Gamma \in C^-_{c',\kappa}$ means that the initial curve Γ cannot be too close to the high-temperature $h = 0$ regime. This restriction is understandable, because as we shall see [cf. (14.285) below], close to this regime $M \lesssim ht^{-\gamma}$ and hence by taking $t = g(h)$ "flat" enough ($t \sim h^\kappa$ with κ very small), we get bounds of the type $M \le \text{const} \times h^{1-\varepsilon'}$. These bounds cannot possibly propagate to (for instance) the critical isotherm, where $M \sim h^{1/\delta}$ with $\delta \ge 3$.

Analogous arguments can be made for the restrictions and conclusions of part (b).

We point out, finally, that straight lines $t = ah$ lie asymptotically in $C^+_{c_1,\kappa_1} \cap C^-_{c_2,\kappa_2}$ for every $c_1, c_2 > 0$ and every $0 < \kappa_1, \kappa_2 < 1$.

Proof. (a) The curve Γ divides the (β, h) plane in two regions: left and right. The left region $[t \ge g(h)]$ is the "easy" part; the bound (14.163) propagates there just by Griffiths second inequality:

$$M(t,h) \;\le\; M(g(h),h) \;\le\; C_+ h^{\mu_+} [1 + o(1)] \qquad [t \ge g(h)]\,. \tag{14.168}$$

The proof for the points to the right of the curve is based on the fact that each such point can be joined to a point in Γ via a straight line of slope $-|J|\,\mathcal{M}$ (the extrapolation lines of Lemma 14.1):

Claim: For every point $P = (t_1, h_1) \in C^+_{c,\alpha}$, with $t_1 < g(h_1)$ and h_1 small enough, there exists a point $Q = (g(h_0), h_0) \in \Gamma$ such that the straight line through Q of slope $-|\tilde{J}|\mathcal{M}(h_0)$ passes through P (Figure 14.5). Moreover,

$$h_1 \;\le\; h_0 \;\le\; h_1[1 + o(1)]\,. \tag{14.169}$$

This claim can be proven for instance by showing that for fixed small enough h_1, the function

$$F(h) \equiv h - h_1 - |J|\,\mathcal{M}(h)\,[g(h) - t_1] \tag{14.170}$$

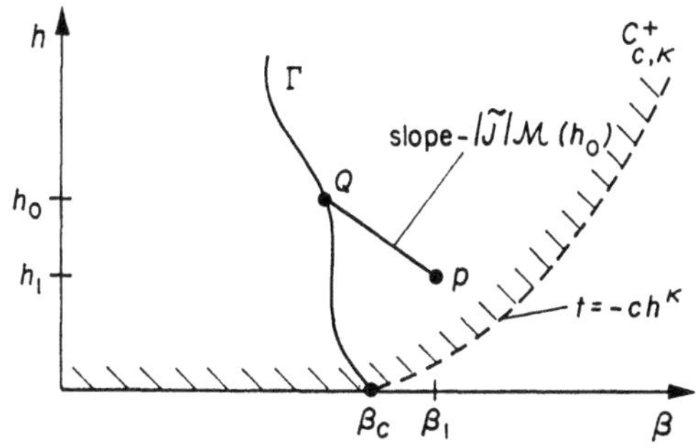

Fig. 14.5. Extrapolations used in Proposition 14.4(a). Every point $P \in C^+_{c,\kappa}$ can be connected to some point $Q(g(h_0), h_0)$ on the curve Γ by a line of slope $-|\widetilde{J}|\mathcal{M}(h_0)$. Note that $t = \beta_c - \beta$

has some h_0, very close to h_1, for which $F(h_0) = 0$. Indeed, on the one hand for $t_1 < g(h_1)$,

$$F(h_1) \;=\; -\;|J|\,\mathcal{M}(h_1)[g(h_1) - t_1] \;<\; 0\,. \tag{14.171}$$

and on the other hand

$$F(2h_1) \;=\; h_1\left[1 - \frac{|J|\,\mathcal{M}(2h_1)[g(2h_1) - t_1]}{h_1}\right] \tag{14.172}$$

$$\;=\; h_1[1 - \varepsilon]\,. \tag{14.173}$$

But by (14.163) and the hypotheses $g(h) \le c'h^\kappa$ and $t_1 \ge -ch^\kappa$:

$$\varepsilon \le \text{const} \times h_1^{\mu_+ + \kappa - 1} \underset{h_1 \to 0}{\longrightarrow} 0$$

if $\mu_+ > \max(1-\kappa, 0)$. Therefore $F(2h_1) > 0$ for h_1 small enough, and from (14.171) and the continuity of F we find that there is an h_0 as claimed, with

$$h_1 \;\le\; h_0 \;\le\; 2h_1\,. \tag{14.174}$$

However, from (14.174), (14.163) and the assumed bounds on $f(h)$ and t:

$$\begin{aligned} h_0 - h_1 \;&=\; |J|\,\mathcal{M}(h_0)\,[g(h_0) - t_1] \\ &\le\; [\text{const}\, h_1^{\mu_+ + \kappa - 1}]\, h_1 \\ &=\; o(1)\, h_1\,. \end{aligned} \tag{14.175}$$

This concludes the proof of the claim. Part (a) of the proposition now follows immediately from Lemma 14.1 which implies

$$M(t_1, h_1) \le \mathcal{M}(h_0)$$

and by (14.163) and (14.169)

$$M(t_1, h_1) \;\le\; C_+ h_1^{\mu_+}\,[1 + o(1)] \qquad [t_1 < g(h_1)]\,. \tag{14.176}$$

(b) The proof is analogous and is left as an exercise. ∎

14.2.3 Consequences of the ABF inequality

In this section we show how the ABF inequality

$$\frac{M}{h} \le \frac{\partial M}{\partial h} + M^2 \left[1 + \frac{|J|\, M}{h}\right] \left[1 + \frac{\beta}{M}\frac{\partial M}{\partial \beta}\right], \qquad (13.59)/(14.177)$$

(together with the weak GHS and Griffiths II inequalities) can be employed to prove an extrapolation principle which goes in the opposite direction to the one derived from the GHS inequality. [Analogous results can be obtained using the Ising-model ABF inequality (13.60).] Defining R to be the dimensionless combination

$$R = \frac{|J|\, M}{h}, \qquad (14.178)$$

we can rewrite the ABF inequality (14.177) as

$$\frac{\partial R}{\partial \beta} \ge -\frac{|\widetilde{J}|}{h(R^2 + R)} \frac{\partial R}{\partial h} \qquad (14.179)$$

(recall that $|J| = \beta|\widetilde{J}|$). Now consider R along some curve $h = h(\beta)$, and write $\mathcal{R}(\beta) \equiv R(\beta, h(\beta))$. We then have

$$\frac{d\mathcal{R}}{d\beta} = \frac{\partial R}{\partial \beta} + \frac{\partial R}{\partial h}\frac{dh}{d\beta} \qquad (14.180)$$

$$\ge -\frac{\partial R}{\partial h}\left[\frac{|\widetilde{J}|}{h(R^2 + R)} - \frac{dh}{d\beta}\right], \qquad (14.181)$$

where $-\partial R/\partial h = -|J|\, \partial(M/h)/\partial h \ge 0$ by the weak GHS inequality. In particular, fix $\beta_0, h_0 \ge 0$ and $0 < r < \infty$, and consider the curve

$$h = h^*(\beta) \equiv \left[h_0^2 + \frac{2|\widetilde{J}|}{r^2 + r}(\beta - \beta_0)\right]^{1/2}, \qquad (14.182)$$

which solves the equation $dh/d\beta = |\widetilde{J}|/h(r^2 + r)$ with initial condition $h(\beta_0) = h_0$. With this choice of $h(\beta)$, (14.180) becomes

$$\frac{d\mathcal{R}}{d\beta} \ge -\frac{\partial R}{\partial h}\frac{|\widetilde{J}|}{h}\left[\frac{1}{\mathcal{R}^2 + \mathcal{R}} - \frac{1}{r^2 + r}\right]. \qquad (14.183)$$

We can now prove an extrapolation principle which is analogous (but opposite) to Lemma 14.1:

Lemma 14.5 ABF extrapolation principles *Consider some $\beta_0, h_0 \ge 0$.*

(a) Assume that $R(\beta_0, h_0) \equiv \beta_0|\widetilde{J}|M(\beta_0, h_0)/h_0 \ge r > 0$. Then $R(\beta, h) \ge r$ for every (β, h) in the region

$$\left\{(\beta, h)\colon \quad \beta \ge \beta_0, h \le \left[h_0^2 + \frac{2|\widetilde{J}|}{r^2 + r}(\beta - \beta_0)\right]^{1/2}\right\}. \qquad (14.184)$$

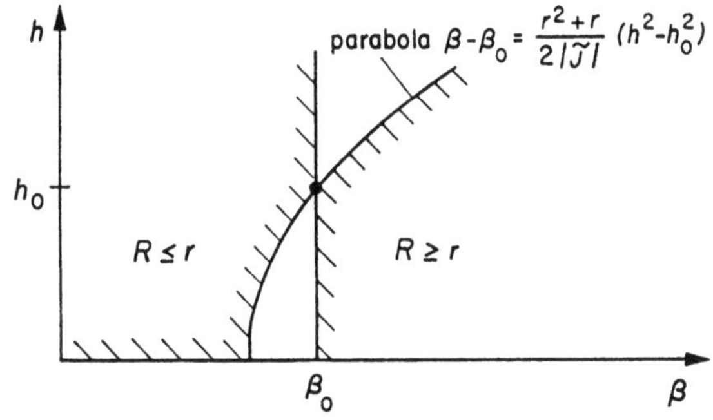

Fig. 14.6. Graphical summary of Lemma 14.5

(b) *Assume that* $R(\beta_0, h_0) \equiv \beta_0|\widetilde{J}|M(\beta_0,h_0)/h_0 \le r$ *with* $r > 0$. *Then* $R(\beta, h) \le r$ *for every* (β, h) *in the region*

$$\left\{ (\beta, h): \quad 0 \le \beta \le \beta_0 \text{ and } h \ge \left[h_0^2 + \frac{2|\widetilde{J}|}{r^2+r}(\beta-\beta_0) \right]_+^{1/2} \right\} \tag{14.185}$$

where $[x]_+ = \max(x, 0)$.

Lemma 14.5 is summarized graphically in Figure 14.6.

Proof.

(a) Let $R(\beta_0, h_0) \ge r > 0$ and consider $\mathcal{R}(\beta) \equiv R(\beta, h^*(\beta))$. We claim that $\mathcal{R}(\beta) \ge r$ for all $\beta \ge \beta_0$. Suppose the contrary, i.e. suppose that there exists $\beta_1 > \beta_0$ such that $\mathcal{R}(\beta_1) < r$. Then (assuming that $\mathcal{R}$ is once continuously differentiable[12]) there must exist $\beta \in (\beta_0, \beta_1)$ such that $\mathcal{R}(\beta) < r$ and $d\mathcal{R}/d\beta < 0$. But this contradicts (14.183).

We have thus shown that $R(\beta, h^*(\beta)) \ge r$ for all $\beta \ge \beta_0$. The weak GHS inequality $\partial\mathcal{R}/\partial h \le 0$ then implies that $R(\beta, h) \ge R(\beta, h^*(\beta)) \ge r$ whenever $\beta \ge \beta_0$ and $0 \le h \le h^*(\beta)$.

(b) Assume first that the *strict* inequality $R(\beta_0, h_0) < r$ with $r > 0$, and consider $\mathcal{R}(\beta) \equiv R(\beta, h^*(\beta))$. We claim that $\mathcal{R}(\beta) < r$ and $d\mathcal{R}/d\beta \ge 0$ for all β in the interval $\left[\beta_0 - (r^2+r)h_0^2/2|\widetilde{J}|\right]^{1/2} \le \beta \le \beta_0$. Indeed, suppose that the first of these claims is false, i.e. suppose that there exists β_1 in this interval such that $\mathcal{R}(\beta_1) \ge r$. Then (assuming that R is once continuously differentiable[12]) there must exist $\beta \in (\beta_1, \beta_0)$ such that $\mathcal{R}(\beta) < r$ and $d\mathcal{R}/d\beta < 0$. But this contradicts (14.183). Hence we must have $\mathcal{R}(\beta) < r$ for all β in the interval $\left[\beta_0 - (r^2+r)h_0^2/2|\widetilde{J}|\right]^{1/2} \le \beta \le \beta_0$; and $d\mathcal{R}/d\beta \ge 0$ then follows immediately from (14.183). The rest follows from the weak GHS inequality $\partial R/\partial h \le 0$ and Griffiths' second inequality $\partial R/\partial\beta \ge 0$.

[12] A careful argument would prove Lemma 14.5 first in finite volume (with periodic boundary conditions), where all thermodynamic quantities are smooth, and then pass to the infinite-volume limit in the final result.

If we only have $R(\beta_0, h_0) = r$, then we use the above argument with r replaced by $r + \varepsilon$, and take the limit $\varepsilon \downarrow 0$. Since the weak GHS inequality implies that R is a continuous and decreasing function of h for $h > 0$, and we are considering R to be *defined* at $h = 0$ by its limit as $h \downarrow 0$, we can pass to the limit in the inequality. ∎

Remark. If we start instead from the Ising-model form of the ABF inequality, (13.60), we can derive a completely analogous extrapolation principle: Define

$$R = \frac{\beta M}{h},$$

so that the ABF inequality reads

$$\frac{\partial R}{\partial \beta} \geq -\frac{\beta}{hR^2}\frac{\partial R}{\partial h}. \tag{14.186}$$

Considering R along a curve $h = h(\beta)$, and setting $\mathcal{R}(\beta) \equiv R(\beta, h(\beta))$, we have

$$\frac{d\mathcal{R}}{d\beta} \geq -\frac{\partial R}{\partial h}\left[\frac{\beta}{hR^2} - \frac{dh}{d\beta}\right]. \tag{14.187}$$

In particular, we can consider the curve

$$h = h^*(\beta) \equiv \left[h_0^2 + \frac{1}{r^2}\left(\beta^2 - \beta_0^2\right)\right]^{1/2}, \tag{14.188}$$

which solves the equation $dh/d\beta = \beta/hr^2$ with initial condition $h(\beta_0) = h_0$. With this choice of $h(\beta)$, (14.187) becomes

$$\frac{d\mathcal{R}}{d\beta} \geq -\frac{\partial R}{\partial h}\frac{\beta}{h}\left[\frac{1}{\mathcal{R}^2} - \frac{1}{r^2}\right]. \tag{14.189}$$

These formulae are completely analogous to (14.179)–(14.183) with the following changes:

1) $R^2 + R$ is replaced by R^2, a slight simplification. In the applications, R will be very large (or at least bounded away from zero) near the critical point, so R^2 behaves essentially the same as $R^2 + R$.

2) $|\widetilde{J}|$ is replaced by β. *This change destroys the dimensional correctness of the formulae*; they are valid *only* for the Ising model with its standard normalization $\sigma = \pm 1$.

One can therefore prove an extrapolation principle completely analogous to Lemma 14.5, in which $h_0^2 + 2|\widetilde{J}|(\beta - \beta_0)/(r^2 + r)$ is replaced by $h_0^2 + (\beta^2 - \beta_0^2)/r^2$ in defining the regions (14.184)–(14.185). The proof is identical to that of Lemma 14.5; the key fact is that r^2 is (like $r^2 + r$) an *increasing* function of r.

Using Lemma 14.5, lower bounds on the magnetization can be extrapolated "upwards and to the right" in the (β, h)-plane, while upper bounds can be extrapolated "downwards and to the left." Note that this direction of extrapolation is *opposite* to that given by the GHS extrapolation principle (Lemma 14.1).

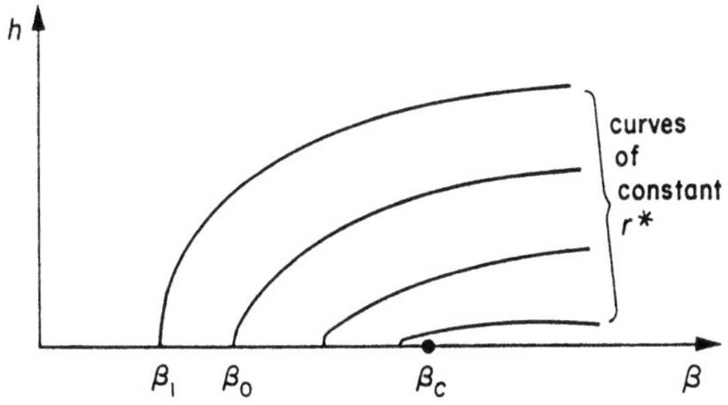

Fig. 14.7. Construction of the function $r^*(\beta, h)$ in the region (14.191). The curve originating at $\beta_0 \in [\beta_1, \beta_c)$ is the parabola $\beta - \beta_0 = (2|\widetilde{J}|)^{-1}[r(\beta_0)^2 + r(\beta_0)]h^2$, and along this curve $r^*(\beta, h) = r(\beta_0)$

As a first example, suppose that the zero-field susceptibility $\chi_0(\beta) = \chi(\beta, h = 0)$ is bounded below by some function $c(\beta)$, for β in some interval $[\beta_1, \beta_c)$ on the high-temperature side of criticality. Without loss of generality we can assume that $c(\beta)$ is nonnegative, nondecreasing, continuous, and tends to $+\infty$ as $\beta \to \beta_c$ (since $\chi_0(\beta)$ has these properties). For convenience let us work instead with the function $R_0(\beta) \equiv R(\beta, h = 0^+) = \beta|\widetilde{J}|\chi_0(\beta)$; then our assumption becomes $R_0(\beta) \geq r(\beta) \equiv \beta|\widetilde{J}|c(\beta)$ for $\beta \in [\beta_1, \beta_c)$.

Now let $r^*(\beta, h)$ be the unique solution to the partial differential *equation*

$$\frac{\partial r^*}{\partial \beta} = -\frac{|\widetilde{J}|}{h(r^{*2} + r^*)} \frac{\partial r^*}{\partial h} \tag{14.190}$$

in the region

$$\left\{(\beta, h): \quad \beta \geq \beta_1,\ 0 < h \leq \left[2|\widetilde{J}|[r(\beta_1)^2 + r(\beta_1)]^{-1} (\beta - \beta_1)\right]^{1/2}\right\} \tag{14.191}$$

with initial condition $r^*(\beta, h = 0^+) = r(\beta)$ for $\beta \in [\beta_1, \beta_c)$. (The existence and uniqueness of this solution is easily demonstrated: the point is that the curves of constant r^* are the parabolae $\beta - \beta_0 = (2|\widetilde{J}|)^{-1}[r(\beta_0)^2 + r(\beta_0)]h^2$, and every point in the region (14.191) can be connected to the initial surface $\{\beta_1 \leq \beta \leq \beta_c, h = 0\}$ by one and only one such curve — see Figure 14.7.) Then we claim that $r^*(\beta, h)$ is a lower bound for $R(\beta, h)$ everywhere in the region (14.191):

Proposition 14.6 *Suppose that* $\beta|\widetilde{J}|\chi_0(\beta) \equiv R(\beta, h = 0^+) \geq r(\beta) \geq 0$ *for* β *in some interval* $[\beta_1, \beta_c)$*, where the function* $r(\beta)$ *is nondecreasing, continuous, and tends to* $+\infty$ *as* $\beta \to \beta_c$*. Let* $r^*(\beta, h)$ *be the unique solution to the partial differential equation (14.190) in the region (14.191) with the initial condition* $r^*(\beta, h = 0^+) = r(\beta)$*. Then, for all* (β, h) *in the region (14.191),*

$$R(\beta, h) \geq r^*(\beta, h) \tag{14.192}$$

and hence

$$M(\beta,h) \equiv \frac{h}{\beta|\widetilde{J}|} R(\beta,h) \geq \frac{h}{\beta|\widetilde{J}|} r^*(\beta,h). \tag{14.193}$$

Proof. For each (β, h) in the region (14.191), there exists a unique $\beta_0 \in [\beta_1, \beta_c)$ such that $h = \left[2|\widetilde{J}|[r(\beta_0)^2 + r(\beta_0)]^{-1}(\beta-\beta_0)\right]^{1/2}$, and moreover $r^*(\beta,h) = r(\beta_0)$. The claim follows immediately from Lemma 14.5(a). ∎

Clearly Proposition 14.6 gives the *best possible* bound based only on the partial differential inequality (14.179), since $r^*(\beta,h)$ is itself a solution to this inequality — as an *equality*!

Typically we shall assume a power-law behavior for the lower bound $r(\beta)$, i.e.

$$r(\beta) = c\,(\beta_c - \beta)^{-\kappa} \tag{14.194}$$

with $c, \kappa > 0$ and some $\beta_1 < \beta_c$. It is now a calculus exercise to compute the behavior of the corresponding $r^*(\beta,h)$: it is the unique positive solution of the equation

$$\frac{h^2}{2|\widetilde{J}|c^{1/\kappa}}\,(r^*)^{2+1/\kappa}\left[1+\frac{1}{r^*}\right] + \frac{t}{c^{1/\kappa}}\,(r^*)^{1/\kappa} = 1 \tag{14.195}$$

where $t \equiv \beta_c - \beta$. In particular, as $(t,h) \to (0,0)$ the solution r^* tends to infinity. Therefore, the term $1/r^*$ becomes negligible in comparison with 1, and r^* satisfies asymptotically the equation

$$\frac{h^2}{2|\widetilde{J}|c^{1/\kappa}}\,(r^*)^{2+1/\kappa} + \frac{t}{c^{1/\kappa}}\,(r^*)^{1/\kappa} = 1\,. \tag{14.196}$$

Defining the dimensionless combinations

$$f = (2|\widetilde{J}|)^{-\kappa/(2\kappa+1)} C^{-1/(2\kappa+1)}\, h^{2\kappa/(2\kappa+1)}\, r^* \tag{14.197}$$

$$x = \left(\frac{2|\widetilde{J}|}{c^2}\right)^{1/(2\kappa+1)} t\, h^{-2/(2\kappa+1)} \tag{14.198}$$

we find that f satisfies asymptotically the generalized Landau-Ginzburg equation

$$f^{2+\frac{1}{\kappa}} + x f^{\frac{1}{\kappa}} = 1\,. \tag{14.199}$$

That is, $f = f(x)$ is precisely the generalized Landau-Ginzburg scaling function $f_{2+\frac{1}{\kappa},\frac{1}{\kappa}}$ defined in (14.134) ff. (In particular, if $\kappa = 1$, then $f_{2+\frac{1}{\kappa},\frac{1}{\kappa}}$ equals the ordinary Landau-Ginzburg scaling function $f_{LG} = f_{3,1}$.) Using (14.135) and Proposition 14.6, we obtain

$$M(\beta,h)$$

$$\geq \frac{h}{\beta|\widetilde{J}|}\, r^*(\beta,h)$$

$$= \frac{1}{\beta}\left(\frac{2^\kappa c}{|\widetilde{J}|^{\kappa+1}}\right)^{1/(2\kappa+1)} h^{1/(2\kappa+1)} f_{2+\frac{1}{\kappa},\frac{1}{\kappa}}\left(\left(\frac{2|\widetilde{J}|}{c^2}\right)^{1/(2\kappa+1)} t\, h^{-2/(2\kappa+1)}\right)$$

$$= \begin{cases} \frac{1}{\beta}\left(\frac{2^\kappa c}{|\widetilde{J}|^{\kappa+1}}\right)^{1/(2\kappa+1)} h^{1/(2\kappa+1)}\,[1-O(x)] & x \simeq 0 \\ \frac{c}{\beta|\widetilde{J}|}\, h t^{-\kappa}\left[1+O\left(x^{-(2\kappa+1)}\right)\right] & x \to +\infty \\ \frac{1}{\beta}\left(\frac{2}{|\widetilde{J}|}\right)^{1/2}(-t)^{1/2}\left[1+O\left(|x|^{-(1+\frac{1}{2\kappa})}\right)\right] & x \to -\infty \end{cases} \tag{14.200}$$

The first line is useful close to the critical isotherm; the second line reproduces the initial data in the high-temperature/small-field region; and the third line yields information on the spontaneous magnetization (by taking $h \downarrow 0$ at fixed $t < 0$) and more generally on the low-temperature/small-field region. It is worth emphasizing that neither c nor κ appears in the leading term on the third line — we thus get a *universal (mean-field) lower bound on the spontaneous magnetization* (in particular, $\widehat{\beta} \le \frac{1}{2}$).

Remark. If we use instead the Ising-model form of the extrapolation principle and define $r(\beta) = c(\beta_c^2 - \beta^2)^{-\kappa}$ [which is equivalent to (14.194) near the critical point], then with the definitions $\hat{t} = \beta_c^2 - \beta^2$, $x = (ch)^{-2/(2\kappa+1)}\hat{t}$ and $f = c^{-1}(ch)^{2\kappa/(2\kappa+1)} r^*$, we find that f satisfies *exactly* the generalized Landau-Ginzburg equation (14.199). So the Ising-model form of the extrapolation principle leads to slightly simpler algebra.

A second example of the consequences of Lemma 14.5 is the extrapolation from the critical isotherm down to the high-temperature zero-field region.

Proposition 14.7 *Suppose that $R(\beta = \beta_c, h) \le r(h)$ for h in some interval $(0, h_1]$ where the function $r(h)$ is nonincreasing, continuous and tends to $+\infty$ as $h \to 0$. Let $r^*(\beta, h)$ be the unique solution of the partial differential equation (14.190) in the region*

$$\left\{(\beta, h):\quad 0 < h \le h_1, \text{ and } \beta_c + \frac{r(h_1)^2 + r(h_1)}{2|\widetilde{J}|}(h^2 - h_1^2) \le \beta \le \beta_c\right\} \tag{14.201}$$

with the initial condition $r^*(\beta = \beta_c, h) = r(h)$. Then for all (β, h) in the region (14.201)

$$R(\beta, h) \le r^*(\beta, h) \tag{14.202}$$

This proposition is proven in the same way as Proposition 14.6 but using part (b) of Lemma 14.5.

In particular, if we assume a power-law behavior

$$r(h) = ch^{-\psi} \tag{14.203}$$

with $c > 0$ and $0 < \psi < 1$, we find that $r^*(\beta, h)$ is the unique positive solution of the equation

$$\frac{h^2}{c^{2/\psi}}(r^*)^{2/\psi} + \frac{2|\widetilde{J}|t}{c^{2/\psi}}\frac{(r^*)^{(2/\psi)-2}}{1+(r^*)^{-1}} = 1 \tag{14.204}$$

where $t \equiv \beta_c - \beta$. In particular, as $(t, h) \downarrow (0, 0)$ the solution r^* tends to infinity. Therefore, the term $(r^*)^{-1}$ becomes negligible in comparison with 1, and r^* satisfies asymptotically the equation

$$\frac{h^2}{c^{2/\psi}}(r^*)^{2/\psi} + \frac{2|\widetilde{J}|t}{c^{2/\psi}}(r^*)^{(2/\psi)-2} = 1\,. \tag{14.205}$$

Defining the dimensionless combinations

$$f = \frac{1}{c}h^{\psi}r^* \tag{14.206}$$

$$x = \frac{2|\widetilde{J}|}{c^2}th^{-(2-2\psi)} \tag{14.207}$$

we find that f satisfies asymptotically the generalized Landau-Ginzburg equation

$$f^{2/\psi} + xf^{(2/\psi)-2} = 1\,. \tag{14.208}$$

That is, $f = f(x)$ is precisely the generalized Landau-Ginzburg scaling function $f_{\frac{2}{\psi},\frac{2}{\psi}-2}$ defined in (14.134) ff. (In particular, if $\psi = \frac{2}{3}$, then $f_{\frac{2}{\psi},\frac{2}{\psi}-2}$ equals the ordinary Landau-Ginzburg scaling function $f_{LG} = f_{3,1}$.) Using (14.135) and Proposition 14.7, we conclude that for $\beta \le \beta_c$ ($t \ge 0$) the magnetization satisfies

$$\begin{aligned} & M(\beta,h) \\ \le\ & \frac{h}{\beta|\widetilde{J}|}r^*(\beta,h) \\ =\ & \frac{c}{\beta|\widetilde{J}|}h^{1-\psi}f_{\frac{2}{\psi},\frac{2}{\psi}-2}\left(\frac{2|\widetilde{J}|}{c^2}th^{-(2-2\psi)}\right) \\ =\ & \begin{cases} \frac{c}{\beta|\widetilde{J}|}h^{1-\psi}\,[1-O(x)] & x \simeq 0 \\ \frac{1}{\beta}\left(\frac{c^2}{2^{\psi}|\widetilde{J}|^{2-\psi}}\right)^{1/(2-2\psi)} t^{-\psi/(2-2\psi)}\,h\left[1+O(x^{-2/(2-2\psi)})\right] & x \to +\infty \end{cases} \end{aligned} \tag{14.209}$$

The first line reproduces the initial data close to the critical isotherm, and the second line is useful in the small-field/high-temperature region.

Remark. If we use instead the Ising-model form of the extrapolation principle, then with the definitions $\hat{t} = \beta_c^2 - \beta^2$, $x = c^{-2}h^{2\psi-2}\hat{t}$ and $f = c^{-1}h^{\psi}r^*$, we find that f satisfies *exactly* the generalized Landau-Ginzburg equation (14.208). So the Ising-model form of the extrapolation principle leads to slightly simpler algebra.

14.3 Non-symmetric regime: Standard approaches to the critical point

We now consider bounds for the region $h > 0$ or $\beta > \beta_c$. In this section we focus on the critical exponents for the standard approaches to the critical point, namely the critical isotherm ($\beta = \beta_c, h \downarrow 0$) and the two-phase coexistence curve ($h = 0^+, \beta \downarrow \beta_c$). In Section 14.4 we discuss the much deeper bounds that hold in a full neighborhood of the critical point in the (β, h)-plane.

To simplify the notation we shall abbreviate

$$M_0 = \lim_{h \downarrow 0} M(h) , \qquad \chi_0 = \lim_{h \downarrow 0} \chi(h) . \tag{14.210}$$

Table 14.2 summarizes the critical-exponent inequalities proven in this section and in Section 14.4.

Table 14.2. Inequalities for critical exponents. Non-symmetric regime

Inequality	When saturated	References	Hypothesis	Models
$b' \le \max\left(0, (2 - \frac{d}{2})\gamma'\right)$	Scaling with $\eta = 0$	[478], [481, App. A]	Griffiths II + ref. positivity	ref.-positive even ferromagnet
$b' \ge \max\left(0, 2\gamma' - d\nu'\right)$	Scaling	[477]	Lebowitz $(h = 0)$ + M continuous (for $b' \ge 0$); ref. positivity (for $b' \ge 2\gamma' - d\nu'$)	BFS class, GS class
$b_c \le \max\left(0, (2 - \frac{d}{2})(\delta - 1)\right)$	Scaling with $\eta = 0$	[478], [481, App. A]	Griffiths II + ref. positivity	ref.-positive even ferromagnet
$b_c \le \max\left(0, (2 - \frac{d}{2-\eta})(\delta - 1)\right)$	Scaling	[481, App. B]	Griffiths II + GHS	even ferromagnet
$\alpha' \le \max(b', \gamma' - 2\widehat{\beta})$		[478]	Lebowitz $(h \neq 0)$	EMN class, GS class
$\alpha_c \le \max(b_c, \delta - 3)$		[478]	Lebowitz $(h \neq 0)$	EMN class, GS class
$\gamma' \ge 1$	Mean-field	[479]	GHS + Griffiths II + M continuous	EMN class, GS class
$\gamma' \ge \widehat{\beta}(\delta - 1)$	Scaling	[269]	GHS + Griffiths II	EMN class, GS class

(Continued on next page)

Table 14.2. Inequalities for critical exponents. Non-symmetric regime (Continued)

Inequality	When saturated	References	Hypothesis	Models
$\widehat{\beta}(\delta - 1) \geq 1$	Mean-field	[409]	GHS + Griffiths II	EMN class, GS class
$\widehat{\beta} \leq \frac{1}{2}$	Mean-field, spherical model	[6], [8], [10]	ABF + Griffiths II + weak GHS	GS class
$\widehat{\beta} \geq \min\left(\frac{1}{2}, \frac{1-3b}{2}\right)$		[12]	AFe + AG + Griffiths II + weak GHS + restriction $b < 1/3$	large-d Ising
$\delta \geq 3$	Mean-field	[218]	FS + weak GHS	BFS class
		[10]	ABF + Griffiths II + weak GHS	GS class
$\delta \geq 2\gamma + 1$	Mean-field, spherical model		ABF + Griffiths II + weak GHS	GS class
$\delta \leq \min\left(3, \frac{3(1-b)}{1-3b}\right)$		[12]	AFe + AG + Griffiths II + weak GHS + restriction $b < 1/3$	large-d Ising
$\Delta_3' \leq d\nu_\phi' - \beta$ for $\phi \geq d$	Hyperscaling	[479]	GHS + Lebowitz (or Ginibre)	EMN class, GS class

(Continued on next page)

Table 14.2. Inequalities for critical exponents. Non-symmetric regime (Continued)

Inequality	When saturated	References	Hypothesis	Models
$d\nu' \geq \gamma' + 2\beta$	Hyperscaling	[479]	GHS + Griffiths II + new Lebowitz (or Ginibre) + ref. positivity	ref.-positive EMN class, GS class
$\gamma' \geq \dfrac{2(2-\eta)\beta}{d-2+\eta}$	Hyperscaling	[199]	Griffiths II + M continuous	even ferromagnets
$\nu' \geq \dfrac{2\beta}{d-2+\eta}$	Hyperscaling	[199], [479]	GHS + Griffiths II + new Lebowitz (or Ginibre) + M continuous + ref. positivity	ref.-positive EMN class, GS class

14.3.1 Upper bound on the truncated bubble diagram (critical exponents b' and b_c)

These preliminary results are the basis for many of the subsequent inequalities. As in the symmetric regime, we distinguish two versions:

(a) For systems satisfying Griffiths II and the infrared bound (14.7) for the Fourier transform of $\langle\varphi_0;\varphi_x\rangle$, we have [481, Appendix B]

$$B^T \;\leq\; \text{const} \times \begin{cases} 1 & d>4 \\ \log\chi & d=4 \\ \chi^{2-d/2} & 2<d<4 \end{cases} \tag{14.211}$$

This implies

$$b' \;\leq\; \max\Big(0, \Big(2-\frac{d}{2}\Big)\gamma'\Big) \tag{14.212}$$

and

$$b_c \;\leq\; \max\Big(0, \Big(2-\frac{d}{2}\Big)\Big(\delta-1\Big)\Big) \tag{14.213}$$

In (14.213) we used that $\chi \leq M/h$ by the (weak) GHS inequality.

(b) If Griffiths II and GHS inequalities are satisfied, we have that for $\beta \leq \beta_c$ (in particular on the critical isotherm)

$$u_2(0,x)_{\beta,h} \;\leq\; u_2(0,x)_{\beta,h=0} \;=\; S_2(0,x)_{\beta,h=0} \;\leq\; S_2(0,x)_{\beta_c,h=0} \tag{14.214}$$

where the first inequality is due to GHS and the last one to Griffiths II. Therefore, the bounds (14.14) are also valid for B^T:

$$B^T(\beta \le \beta_c) \;\le\; \text{const} \times \begin{cases} 1 & d > 4 - 2\eta \\ \log \chi & d = 4 - 2\eta \\ \chi^{2-[d/(2-\eta)]} & 2 - \eta < d < 4 - 2\eta \end{cases} \tag{14.215}$$

In particular for the critical isotherm this implies

$$b_c \;\le\; \max\Bigl(0, \bigl(2 - \frac{d}{2-\eta}\bigr)(\delta - 1)\Bigr) . \tag{14.216}$$

No analogous bound has been proven for b' because (14.215) is *not* known to be true on the low-temperature side of the critical point ($\beta > \beta_c$).

14.3.2 Lower bound on the truncated bubble diagram (critical exponents b' and b_c)

As in the symmetric regime, we prove two versions of the bound:

(a) *Bound useful for $d > 4$.* The mean-field bound (14.19) implies that $|J_c|\langle\varphi_0^2\rangle_0 \ge 1$, and hence

$$\langle\varphi_0^2\rangle \;\ge\; \langle\varphi_0^2\rangle_0 \;\ge\; |J_c|^{-1} \tag{14.217}$$

for all $J \ge 0$. Assuming that the magnetization is continuous at the critical point (this is proven in Section 14.4.2 for $d \ge 4$), we have

$$u_2(0,0) \;=\; \langle\varphi_0^2\rangle - M^2 \;\longrightarrow \ge |J_c|^{-1} \tag{14.218}$$

as the critical point is approached. It follows immediately that

$$B^T \;\ge\; u_2(0,0)^2 \;\longrightarrow \ge |J_c|^{-2} . \tag{14.219}$$

This proves that (14.211) is sharp for $d > 4$.

(b) *Bound useful for $d < 4$.* The reasoning (14.22)–(14.25) applies without change to the *truncated* two-point function in the non-symmetric regime. Hence

$$B^T \;\ge\; \text{const} \times \chi^2 \xi^{-d} . \tag{14.220}$$

If the usual scaling law $\chi \sim \xi^{2-\eta}$ holds, then both this bound and (14.215) are sharp for $2 - \eta < d < 4 - 2\eta$.

14.3.3 Upper bound on the specific heat (critical exponents α' and α_c)

For systems satisfying the Lebowitz inequality in nonzero field (13.39) one obtains the following generalization of (14.26) [478]

$$\begin{aligned} C_H &\leq (1/2)\sum_{x,y,z} J_{0x}J_{yz}\Big[\langle\varphi_0;\varphi_y\rangle\langle\varphi_x;\varphi_z\rangle + \langle\varphi_0;\varphi_y\rangle\langle\varphi_x\rangle\langle\varphi_z\rangle \\ &\qquad + \langle\varphi_x;\varphi_z\rangle\langle\varphi_0\rangle\langle\varphi_y\rangle\Big] \\ &= (1/2)|J|^2 B_T + |J|^2 M^2\chi , \end{aligned} \tag{14.221}$$

which implies

$$\alpha' \leq \max(b', \gamma' - 2\widehat{\beta}) \tag{14.222}$$

$$\alpha_c \leq \max(b_c, \delta - 3) \tag{14.223}$$

The combination of these results with the bounds (14.212) and (14.213) for the bubble exponents yields

$$\alpha' \leq \max\Big(0, \big(2 - \frac{d}{2}\big)\gamma', \gamma' - 2\widehat{\beta}\Big) \tag{14.224}$$

$$\alpha_c \leq \max\Big(0, \big(2 - \frac{d}{2}\big)(\delta - 1), \delta - 3\Big) , \tag{14.225}$$

or, from (14.216),

$$\alpha_c \leq \max\Big(0, \big(2 - \frac{d}{2-\eta}\big)(\delta - 1), \delta - 3\Big) . \tag{14.226}$$

14.3.4 Bounds on the susceptibility (critical exponent γ')

The only rigorous bound for γ' available at present is the "mean-field" bound

$$\gamma' \geq 1 \tag{14.227}$$

proven for systems satisfying Griffiths II and GHS inequalities and for which the magnetization is continuous at β_c. This is the analogue of the bound $\gamma \geq 1$ stated above. We do not have a proof that $\gamma' = 1$ for $d > 4$ as we do for γ , because no inequality of the form of the AG inequality (involving B^T instead of B_0) has been proven for the non-symmetric regime. We discuss two different proofs of (14.227).

(a) The first proof employs an argument [479] very similar to the proof that $\gamma \geq 1$, but using (13.55) instead of (13.38). If we integrate (13.55) between $\beta_c + \varepsilon$ and some fixed $\beta^* > \beta_c + \varepsilon$), we obtain

$$\log M_0(\beta^*) - \log M_0(\beta_c + \varepsilon) \leq |\widetilde{J}| \int_{\beta_c+\varepsilon}^{\beta^*} \chi_0(\beta')d\beta' . \tag{14.228}$$

Since $M_0(\beta^*) > 0$, we conclude that *if* $M_0 \to 0$ as $\beta \downarrow \beta_c$, then the RHS diverges as $\varepsilon \downarrow 0$. Under the additional assumption that γ' exists, this implies that $\gamma' \geq 1$.

(b) If we assume that the exponents $\gamma', \widehat{\beta}$ and δ exist, we can obtain (14.227) by proving two intermediate inequalities which show more clearly what is going on:

(i) Griffiths [269]

$$\gamma' \geq \widehat{\beta}(\delta - 1) . \tag{14.229}$$

(This inequality is an equality if thermodynamic scaling holds, as is believed to occur in all dimensions. In particular, it is an equality in mean-field theory.)

Proof. For $h \geq 0$, $\beta \geq \beta_c$:

$$M_0(\beta) + \chi_0(\beta)h \geq M(\beta, h) \geq M(\beta_c, h) \geq \text{const} \times h^{1/\delta} . \tag{14.230}$$

The first inequality follows from GHS, the second from Griffiths II and the third one by the assumed existence of δ . Therefore

$$M_0(\beta) \geq \text{const} \times h^{1/\delta} - \chi_0(\beta)h \tag{14.231}$$

for all $h \geq 0$. We can now optimize over h by taking $h = \text{const} \times \chi_0^{-\delta/(\delta-1)}$ to obtain

$$M_0(\beta) \geq \text{const} \times \chi_0(\beta)^{-1/(\delta-1)} \tag{14.232}$$

which implies $\widehat{\beta} \leq \gamma'/(\delta - 1)$. This can be written as (14.229) because the assumed divergence of χ implies that $\delta \geq 1$. ∎

(ii) Newman [409]

$$\widehat{\beta}(\delta - 1) \geq 1 . \tag{14.233}$$

(This is a "mean-field bound".)

Proof. This is an immediate consequence of the GHS extrapolation principle (Lemma 14.1 and Proposition 14.3). Indeed, setting $\mu = \delta$ in (14.157), we conclude from (14.160) that $M(\beta, h = 0^+) \leq \text{const} \times (\beta - \beta_c)^{1/(\delta-1)}$, hence that $\widehat{\beta} \geq 1/(\delta - 1)$. [Alternatively we could use the GHS extrapolation principle in the other direction: applying Proposition 14.2 and setting $\lambda = \widehat{\beta}$ in (14.151), we conclude from (14.154) that $M(\beta_c, h) \geq \text{const} \times h^{\widehat{\beta}/(1+\widehat{\beta})}$, hence that $\delta \geq (1 + \widehat{\beta})/\widehat{\beta}$.] ∎

In Section 14.4.2 we shall present an extended version of Newman's argument, which proves an upper bound on $M(\beta, h)$ in the entire region $\{\beta \geq \beta_c, h \geq 0\}$.

14.3.5 Other hyperscaling inequalities

Let us mention a few other critical-exponent inequalities that can be obtained by clever use of the Griffiths, GHS and related inequalities:

$$\Delta_3' \leq d\nu_\phi' - \beta \qquad \text{for } \phi \geq d \tag{14.234}$$
$$d\nu' \geq \gamma' + 2\beta \tag{14.235}$$
$$\gamma' \geq \frac{2(2-\eta)\beta}{d-2+\eta} \tag{14.236}$$
$$\nu' \geq \frac{2\beta}{d-2+\eta} \tag{14.237}$$

Here (14.234) and (14.235) are due to Sokal [479], (14.236) is due to Fisher [199], and (14.237) is an immediate corollary of (14.235) and (14.236). All of these inequalities become equality when hyperscaling holds, and (14.235) has a beautiful physical interpretation [509, 198, 510, 288]. In Section 14.4.1 we shall prove that (14.235)–(14.237) are in fact *strict* inequalities for $d > 4$, i.e. hyperscaling is violated. Note also that (14.234) is an absolute upper bound on the "dimensionless renormalized three-point coupling constant" $g_\phi^{(3)} \equiv -\bar{u}_3/M\chi\xi_\phi^d$, completely analogous to the Glimm-Jaffe bound (14.50)/(14.51) for the symmetric phase.

14.4 Non-symmetric regime: Bounds in a full neighborhood of the critical point

In this section we do not limit ourselves to any particular path of approach to the critical point; rather, we prove bounds that hold in a full neighborhood of the critical point in the (β, h) plane. (This is particularly natural for applications to quantum field theory, as discussed in Section 15.3.) The main results of this section are the "Landau-Ginzburg" bounds on the magnetization,

$$M(t,h) \geq \text{const} \times M_{LG}(t,h) \qquad [\text{GS class, all } d] \tag{14.238}$$

$$M(t,h) \leq \text{const} \times M_{LG}(t,h) \qquad [\text{Ising, } d > 4] \tag{14.239}$$

along with similar but weaker inequalities for the susceptibilities.

Let us remark on the difference between (14.239) and the "mean-field upper bounds on the magnetization" previously obtained in the literature [427, 474]. The usual bounds are of the form

$$M(\beta,h) \leq M_{MF}(\beta,h)\,. \tag{14.240}$$

This proves, for example that the mean-field prediction for the inverse critical temperature, $\beta_{c,MF}$, is a lower bound for the true critical inverse temperature β_c; but if $\beta_{c,MF} \neq \beta_c$ (as is almost always the case), the bound (14.240) gives *no* information about the critical region.

14.4.1 Lower bounds on the magnetization

The best lower bounds on the magnetization can be proven for models in GS class (which includes Ising and φ^4 models), using the ABF inequality (13.59).

For these models, we can prove (among other things) the universal Landau-Ginzburg lower bound (14.238) in a full neighborhood of the critical point. Somewhat weaker inequalities can be obtained for models of BFS class, using the FS inequality (13.58).

Consequences of the ABF inequality The main tool in our analysis is the extrapolation principle derived in Section 14.2.3 from the ABF inequality. Starting from a lower bound on the zero-field susceptibility on the high-temperature side of criticality,

$$\chi_0(\beta) \geq C\,(\beta_c - \beta)^{-\kappa}\,, \tag{14.241}$$

the extrapolation principle (Proposition 14.6) implies a lower bound on the magnetization $M(\beta, h)$ in a full neighborhood of the critical point in the (β, h)-plane:

$$M(\beta,h)$$

$$\geq \; C\left(\frac{2}{C^2\beta^2|\widetilde{J}|}\right)^{\kappa/(2\kappa+1)} h^{1/(2\kappa+1)} f_{2+\frac{1}{\kappa},\frac{1}{\kappa}}\left(\left(\frac{2}{C^2\beta^2|\widetilde{J}|}\right)^{1/(2\kappa+1)} th^{-2/(2\kappa+1)}\right) \tag{14.242}$$

$$\sim \begin{cases} h^{1/(2\kappa+1)} & \text{for } th^{-2/(2\kappa+1)} \text{ bounded} \\ ht^{-\kappa} & \text{for } th^{-2/(2\kappa+1)} \to +\infty \\ (-t)^{1/2} & \text{for } th^{-2/(2\kappa+1)} \to -\infty \end{cases} \tag{14.243}$$

[cf. (14.200)]; here $f_{2+\frac{1}{\kappa},\frac{1}{\kappa}}$ is the generalized Landau-Ginzburg scaling function defined in (14.199).

We can use (14.242)/(14.243) in two ways:

(a) *Universal bounds.* Firstly, (14.241) always holds *as a universal bound* with $\kappa = 1$ and $C = 1/|\widetilde{J}|$; this is the Glimm-Jaffe bound (14.32) on the high-temperature susceptibility. But for $\kappa = 1$, the generalized Landau-Ginzburg function $f_{2+\frac{1}{\kappa},\frac{1}{\kappa}}$ becomes the *ordinary* Landau-Ginzburg function $f_{3,1} \equiv f_{LG}$, and (14.242) becomes

$$M(\beta,h) \;\geq\; 2^{1/3}|J|^{-1/2}\, M_{LG}\left(\frac{2^{1/3}t}{\beta}, \frac{h}{|J|^{1/2}}\right)\,. \tag{14.244}$$

We have thus proven a *universal Landau-Ginzburg lower bound on the magnetization in a full neighborhood of the critical point.* Note that this expression is dimensionally correct[13]: it bounds the dimensionless magnetization $|J|^{1/2}M$ in terms of the dimensionless parameters

[13]Since in our development of Landau-Ginzburg theory we set the coefficients of $F(M)$ to be pure numbers [cf. (14.87)], all "Landau-Ginzburg" quantities — such as M_{LG} and its arguments — are *dimensionless*.

$$\tilde{t} = \frac{2^{1/3}t}{\beta} = 2^{1/3}\left(\frac{\beta_c}{\beta} - 1\right) \tag{14.245}$$

$$\tilde{h} = \frac{h}{|J|^{1/2}} \tag{14.246}$$

The bound (14.244) can be summarized rather sloppily (i.e. ignoring dimensionful constants) as

$$M(\beta, h) \geq \text{const} \times M_{LG}(t, h) \tag{14.247}$$

Interesting special cases of (14.242) can be obtained by evaluating it on the critical isotherm or on the coexistence curve, yielding

$$M(\beta_c, h) \geq \text{const} \times h^{1/3} \tag{14.248}$$

$$M(\beta, h = 0^+) \geq \text{const} \times (\beta - \beta_c)^{1/2} \tag{14.249}$$

(again we are being sloppy about dimensionful constants). These bounds can be summarized in the critical-exponent inequalities

$$\delta \geq 3 \tag{14.250}$$

$$\widehat{\beta} \leq \frac{1}{2}\,. \tag{14.251}$$

(14.250) and (14.251) are mean-field bounds which complement the Buckingham-Gunton [285, 286, 199, 415], Fisher [199] and Sokal [479] inequalities

$$\delta \geq \frac{d+2-\eta}{d-2+\eta} \tag{14.252}$$

$$\widehat{\beta} \leq \frac{d-2+\eta}{4-2\eta}\gamma' \tag{14.253}$$

$$\widehat{\beta} \leq \frac{1}{2}(d\nu' - \gamma') \tag{14.254}$$

in the same sense that the Aizenman-Fröhlich mean-field bound $\Delta_4 \leq \frac{3}{2}\gamma$ [cf. (14.55)] complements the Glimm-Jaffe inequality $\Delta_4 \leq \frac{1}{2}(\gamma + d\nu_\phi)$. In all four cases, the latter (hyperscaling) inequality is expected to be better for low dimension, while the former (mean-field) inequality is better for high dimension. In particular, if $d > 4 - 2\eta$ (e.g. $d > 4$ for reflection-positive models), then the Buckingham-Gunton, Fisher and Glimm-Jaffe inequalities are all *strict* inequalities — that is, hyperscaling is violated. Likewise, if $d > 4$ (and the model is reflection-positive), then the Sokal inequality is strict[14] — hyperscaling is again violated.

Let us also remark that the bound $\widehat{\beta} \leq 1/2$ is saturated in the spherical model (in *all* dimensions).

[14] Here we use the bound $\gamma' \leq 2\nu'$, which can be obtained for reflection-positive models using the spectral representation and the infrared bound: see (13.88).

A further consequence of (14.249) is that for spin models of GS class, there is a *single* critical temperature β_c which signals both the divergence of the susceptibility and the onset of spontaneous magnetization — that is, there does *not* exist an "intermediate phase" in which $\chi_0 = \infty$ but $M_0 = 0$.

Historical remark. Aizenman [9, 6] used the AG inequality together with an ingenious finite-size argument to derive $\widehat{\beta} \le 1/2$ (and many other interesting results) under an additional hypothesis of "regularity". With the ABF inequality the regularity hypothesis can be dispensed with, at least for getting $\widehat{\beta} \le 1/2$.

(b) *Non-universal bounds.* A second way of using (14.242)/(14.243) is to note that (14.241) holds by definition with $\kappa = \gamma$ (or $\kappa = \gamma - \varepsilon$ for any $\varepsilon > 0$ if there are multiplicative logarithmic corrections). Therefore, evaluating (14.242)/(14.243) on the critical isotherm, we have

$$M(\beta_c, h) \ge \text{const} \times h^{1/(2\gamma+1)} , \tag{14.255}$$

which proves the new critical-exponent inequality

$$\delta \ge 2\gamma + 1 . \tag{14.256}$$

This bound becomes equality in mean-field theory and also for the spherical model; it can be looked at as a "spherical-model upper bound" on γ, somewhat similar to the Aizenman-Graham bounds (14.37) *et seq.*

Consequences of the FS inequality (and weak GHS inequality) Similar but weaker lower bounds on M can be proven[15] for a slightly larger class of models — the BFS class — using the FS inequality (13.58). We sketch the analysis, leaving many of the details to the reader.

Our reasoning is based on the inequalities

$$M_0 \le M \tag{14.257}$$

$$h\chi \le M \tag{14.258}$$

$$M \le h\chi + \left(\frac{M^2}{h}\right)(h + |J|M)^2 , \tag{14.259}$$

which follow from the Griffiths II, weak GHS and FS inequalities, respectively. The method consists in using $\chi = \partial M/\partial h$ and considering (14.258) and (14.259) at fixed β as differential inequalities for the function $M(h)$.

[15] These results were obtained by Fröhlich and Sokal in 1984-85, and were supposed to appear in [223]. However, the writing of this paper was delayed by "unforeseen circumstances", and in the meantime stronger versions of these results were obtained (for a slightly smaller class of models) by Aizenman, Barsky and Fernández [12, 10]. We have incorporated the principal results of [223] into this book (Sections 10.2.3, 11.1.1, 11.1.2, 12.4 and the present paragraph), thereby relieving Fröhlich and Sokal of the responsibility of writing [223].

First we define $F(h) \equiv M(h)/h$ and note that the weak GHS inequality (14.258) implies that

$$\frac{dF}{dh} \le 0 . \tag{14.260}$$

It follows that

$$F_0 \equiv \lim_{h \downarrow 0} F(h) \tag{14.261}$$

exists (it may be $+\infty$). In fact:

a) If $M_0 > 0$, then $F_0 = +\infty$.

b) If $M_0 = 0$ and $\chi_0 \equiv \lim_{h\downarrow 0} \chi(h)$ exists (as it does, in particular, if the GHS inequality (13.54) holds), then $F_0 = \chi_0$ (which may be $+\infty$).

The FS inequality (14.259) can now be written as

$$\frac{dF}{dh} \ge -hF^2(1+|J|F)^2 , \tag{14.262}$$

valid for $h \ge 0$. Integrating this from F_0 to $F(h)$, we find

$$-\int_{F_0}^{F(h)} \frac{dF'}{F'^2(1+|J|F')^2} \le \int_0^h h'dh' \tag{14.263}$$

and hence

$$\psi(|J|F(h)) - \psi(|J|F_0) \le \frac{h^2}{2|J|} , \tag{14.264}$$

where

$$\psi(x) \equiv \int_x^\infty \frac{dx}{x^2(1+x)^2} \tag{14.265}$$

$$= \frac{1}{x} + \frac{1}{1+x} - 2\log\left(1+\frac{1}{x}\right) . \tag{14.266}$$

Since ψ is a decreasing function of its argument, it follows that

$$F(h) \ge \frac{1}{|J|}\psi^{-1}\left[\psi(|J|\,F_0) + \frac{h^2}{2|J|}\right] . \tag{14.267}$$

We content ourselves with two weakened versions of (14.267) which are sufficient for our purposes.

(i) Assume that $F_0 < \infty$ (i.e. $\beta < \beta_c$). The function ψ is convex and decreasing, hence so is ψ^{-1}. It then follows from (14.267) [or directly from (14.264)] that

$$F(h) \ge F_0 - \frac{F_0^2(1+|J|F_0)^2}{2}h^2 \tag{14.268}$$

and hence that

$$M(h) \ge F_0 h - \frac{F_0^2(1+|J|F_0)^2}{2}h^3 . \tag{14.269}$$

(This is a good bound only if $h \ll |J|^{-1}F_0^{-3/2}$.) On the other hand, the weak GHS inequality (14.260) implies that

$$M(h) \;\le\; F_0 h \;. \tag{14.270}$$

We see that (14.269) and (14.270) agree through second order in h. It follows that

$$(\overline{u}_3)_{h=0} \;\equiv\; \left.\frac{d^2M}{dh^2}\right|_{h=0} \;=\; 0 \tag{14.271}$$

and

$$0 \;\ge\; (\overline{u}_4)_{h=0} \;\equiv\; \left.\frac{d^3M}{dh^3}\right|_{h=0} \;\ge\; -3F_0^2(1+|J|F_0)^2 \;. \tag{14.272}$$

Since $F_0 = \chi_0$ in the case at hand (assuming that χ_0 exists), we see that we have recovered the summed versions of the Lebowitz and AF inequalities, (13.37) and (13.42), respectively.

(ii) Assume that $|J|\,F_0$ is large (in particular this includes the case $F_0 = \infty$ (i.e. $\beta \ge \beta_c$). Using simple bounds in (14.266) for x large, it is not hard to show that

$$M(h) \;=\; h\,F(h) \;\ge\; \left(F_0^{-3} + \frac{3\,|J|^2}{2}h^2\right)^{-1/3} h \;-\; O\!\left(\frac{h}{|J|}\right) \tag{14.273}$$

provided that $(|J|\,F_0)^{-1}$ and $|J|^{-1/2}\,h$ are sufficiently small. In particular, if $F_0 = \infty$ (i.e. $\beta \ge \beta_c$) we have

$$M(h) \;\ge\; (\frac{2}{3|J|^2})^{1/3} h^{1/3} \;-\; O(h) \;. \tag{14.274}$$

For $\beta = \beta_c$, (14.274) implies the critical-exponent inequality

$$\delta \;\ge\; 3 \;, \tag{14.275}$$

which proves the failure of the hyperscaling law $\delta = (d+2-\eta)/(d-2+\eta)$ for BFS models in dimension $d > 4 - 2\eta$. This together with the GHS inequality implies (after a little work) the bound

$$\gamma' \;\ge\; 2\widehat{\beta} \;. \tag{14.276}$$

But this is nothing new, because it follows already from $\gamma' \ge \widehat{\beta}(\delta - 1)$ [eq. (14.229)] and $\delta \ge 3$ [eq. (14.275)]. In any case, (14.276) combined with $\gamma' \le 2\nu'$ [see comments following (14.254)] proves the failure of the hyperscaling law $d\nu' > \gamma' + 2\widehat{\beta}$ for $d > 4$ for reflection-positive BFS models in dimension $d > 4$.

14.4.2 Upper bounds on the magnetization

For the nearest-neighbor Ising model, the AFe inequality (13.65) can be combined with the GHS extrapolation principle (Lemma 14.1) to derive an *upper* bound on the magnetization in a full neighborhood of the critical point in the (β, h)-plane. For $d > 4$, this bound is precisely of Landau-Ginzburg form. For $d \le 4$, this is of course weaker (since the Landau-Ginzburg exponents are *not* correct), but we do in any case obtain bounds on the exponents δ and $\widehat{\beta}$ that are complementary to (14.250) and (14.251). We expect that these bounds are true also for φ^4 and other GS models, but they have been proven so far only for the Ising model.

The argument proceeds in two steps: The first step is to obtain an upper bound on M valid in a full *high-temperature* neighborhood of the critical point,

$$\mathcal{N}_+ \equiv \{t \geq 0, h \geq 0,\ t, h \text{ small}\}\ . \tag{14.277}$$

The restriction to $t \geq 0$ arises from the B_0 factors in the AFe inequality, which render it useless at temperatures where there is spontaneous magnetization (i.e. $t < 0$).[16] The second step is to remove the restriction to $t \geq 0$, by applying the GHS extrapolation principle. The result is a bound in a full neighborhood of the critical point.

Step 1. The main tool is the AFe inequality (13.65),

$$\frac{\partial \chi}{\partial h} \;\leq\; -\frac{[1 - B_0(\tanh h)/M)]}{96 B_0 (1 + 2|J|B_0)^2}\,(\tanh h)\,\chi^4\ ,$$

which at fixed β is a first-order differential inequality for the function $\chi(h)$. Near the critical point, $\tanh h \approx h$ and $|J|$ is bounded, so we can simplify the inequality to

$$\frac{\partial \chi}{\partial h} \;\leq\; -\frac{c}{B_0^3}\Big[1 - B_0 h/M\Big] h \chi^4\ . \tag{14.278}$$

Let us fix $(t, h) \in \mathcal{N}_+$ and divide into two cases according to the size of $B_0(t)h/M(t,h)$:

Case 1. $B_0(t)h/M(t,h) > 1/2$.
Clearly in this case we have

$$M(t,h) \;<\; 2B_0(t)h \tag{14.279}$$

by definition.

Case 2. $B_0(t)h/M(t,h) \leq 1/2$.
By the weak GHS inequality, $h/M(t,h)$ is an increasing function of h (at fixed t), so we have

$$B_0(t)h'/M(t,h') \leq 1/2$$

for all h' in the interval $0 \leq h' \leq h$. Substituting this into (14.278) and integrating from $h'' = 0$ to $h'' = h'$, we obtain

$$\begin{aligned} \chi(t,h') \;&\leq\; \left(\chi_0(t)^{-3} + \frac{3c}{4B_0(t)^3}\, h'^2\right)^{-1/3} \\ &\leq\; \text{const} \times \min\Big[\chi_0(t),\, B_0(t) h'^{-2/3}\Big] \end{aligned} \tag{14.280}$$

[16] We may dream of an analogous inequality with B^T or B_0^T in place of B_0, but we have no idea how to prove such an inequality.

for $0 \le h' \le h$. Integrating once more, we obtain

$$M(t,h) \;\le\; \text{const} \times \min\left[\chi_0(t)h, B_0(t)h^{1/3}\right] . \qquad (14.281)$$

Now $B_0(t) \le \chi_0(t)$ for the Ising model (since $\langle\sigma_0\sigma_x\rangle \le 1$), and clearly $h \le \text{const} \times h^{1/3}$ for h small. Therefore, the bound in Case 1 is *stronger* than that in Case 2, so we can conclude that (14.281) holds *throughout the region* $\mathcal{N}_+$.

Step 2. The next step is to extrapolate the upper bound (14.281) "downwards and to the right" in the (β, h)-plane, using the GHS extrapolation principle. Let us consider separately the cases $d > 4$ and $d \ge 4$.

(a) $d > 4$. We know already that $B_0(t) \le \text{const}$ [bubble bound (14.9)] and $\chi_0(t) \le \text{const} \times t^{-1}$ [Aizenman-Graham bound (14.36)] uniformly for $t \ge 0$. Therefore, (14.281) implies that

$$\begin{aligned} M(t,h) \;&\le\; \text{const} \times \min\left[ht^{-1}, h^{1/3}\right] \\ &\le\; \text{const} \times M_{LG}(t,h) \end{aligned} \qquad (14.282)$$

uniformly for $(t,h) \in \mathcal{N}_+$. Now we use the GHS extrapolation principle (Proposition 14.3) to extrapolate this upper bound from the critical isotherm to a low-temperature neighborhood

$$\mathcal{N}_- \equiv \{t \le 0, h \ge 0,\ t,h \text{ small}\} . \qquad (14.283)$$

We have already performed the relevant computation [cf. (14.157) ff. with $\mu = 3$]; the result is that

$$M(t,h) \le \text{const} \times M_{LG}(t,h) \qquad (14.284)$$

also in $\mathcal{N}_-$. So we have proven the promised Landau-Ginzburg upper bound on the magnetization for Ising (but unfortunately not φ^4) models in dimension $d > 4$, in a full neighborhood of the critical point.

(b) $d < 4$. Let us assume that

$$\begin{aligned} B_0(t) \;&\le\; \text{const} \times t^{-b} \\ \chi_0(t) \;&\le\; \text{const} \times t^{-\gamma} \end{aligned}$$

for $t \ge 0$ small[17]; then (14.281) states that

$$M(t,h) \le \text{const} \times \min\left[ht^{-\gamma}, t^{-b}h^{1/3}\right] \qquad (14.285)$$

uniformly for $(t,h) \in \mathcal{N}_+$. Clearly, (14.285) is useless on and near the critical isotherm (if $b > 0$), but the bound $t^{-b}h^{1/3}$ can be improved in large parts

[17]Multiplicative logarithmic corrections, if present, can be handled by similar methods; see [12].

of $\mathcal{N}_+$ by using the GHS extrapolation principle. The same extrapolation will also produce a bound valid in the low-temperature neighborhood $\mathcal{N}_-$. So fix $(t_0, h_0) \in \mathcal{N}_+ \cup \mathcal{N}_-$ and define $m = M(t_0, h_0)$. Then, by Lemma 14.1(a), we have

$$\begin{aligned} m &\le \inf_{t \ge t_0} M(t, h_0 + (t - t_0)m|\tilde{J}|) \\ &\le \inf_{t \ge \max(t_0,0)} \text{const} \times t^{-b} \left[h_0 + (t - t_0)m|\tilde{J}|\right]^{1/3} \end{aligned} \tag{14.286}$$

[Note that $h \equiv h_0 + (t - t_0)m|\tilde{J}|$ will be small whenever h_0 and t are, since $m \le 1$ and $|\tilde{J}|$ is bounded; so (t, h) will lie in $\mathcal{N}_+$ provided that $t \ge \max(t_0, 0)$.] Simple calculus shows that *if* $b < 1/3$ — as we shall assume henceforth — then the minimum of (14.286) is attained at

$$t = \frac{3b}{1 - 3b} \left(\frac{h_0}{m|\tilde{J}|} - t_0 \right) , \tag{14.287}$$

provided that this point lies within the allowable region $t \ge \max(t_0, 0)$. For $t_0 \le 0$, this calculation is always satisfied; and for $t_0 > 0$, we can use the bound $m \le \text{const} \times t_0^{-b} h_0^{1/3}$ to conclude that $t \ge t_0 > 0$ whenever

$$t_0 \le \text{const} \times h_0^{2/(3-3b)} . \tag{14.288}$$

This defines the complement of a "high-temperature horn". Inserting (14.287) into (14.286), we obtain

$$m^{(3-3b)/(1-3b)} + c|\tilde{J}|t_0 m \le ch_0 \tag{14.289}$$

where $c = \left(|\tilde{J}|/3b\right)^{3b/(1-3b)}/(1 - 3b)$. Defining now the scaling variables

$$\begin{aligned} f &= (ch_0)^{-(1-3b)/(3-3b)} m \\ x &= c(ch_0)^{-2/(3-3b)} t_0 \end{aligned} \tag{14.290–14.291}$$

we obtain

$$f^{(3-3b)/(1-3b)} + xf \le 1 \tag{14.292}$$

— a generalized Landau-Ginzburg *inequality*! It is not hard to show[18] that all solutions of this inequality are bounded above by the largest solution of the corresponding equality, i.e.

$$f \le f_{\frac{3-3b}{1-3b},1}(x) , \tag{14.293}$$

where $f_{p,q}(x)$ is the generalized Landau-Ginzburg function (14.134).

[18] For $x \ge 0$, the left-hand side of (14.292) is an increasing function of f. For $x < 0$, the left-hand side is also an increasing function of f at least for $f \ge 0$, $f \ge f_{\frac{3-3b}{1-3b},1}$. Indeed, if we denote $\varepsilon = (3 - 3b)/(1 - 3b)$, we observe that $\varepsilon \ge 3$ because $0 \le b < 1/3$. Then the definition $f_{\varepsilon,1}^{\varepsilon} + xf_{\varepsilon,1} = 1$ implies $\partial(\text{LHS})/\partial f = f_{\varepsilon,1} = (\varepsilon/f_{\varepsilon,1}) - (\varepsilon - 1)x \ge 0$. As $\partial^2(\text{LHS})/\partial f^2 = \varepsilon(\varepsilon - 1)f^{\varepsilon-2} \ge 0$, we conclude that $\partial(\text{LHS})/\partial f \ge 0$ for $f \ge f_{\varepsilon,1}$.

We conclude that

$$\begin{aligned} M(t,h) &\leq (ch)^{(1-3b)/(3-3b)} f_{\frac{3-3b}{1-3b},1}\left(c(ch)^{-2/(3-3b)}t\right) \\ &= \begin{cases} (ch)^{(1-3b)/(3-3b)}\,[1+O(x)] & x \simeq 0 \\ c(-t)^{-(1-3b)/2}\left[1+O\left(x^{-(3-3b)/2}\right)\right] & x \to -\infty \end{cases} \end{aligned} \tag{14.294}$$

for (t,h) in the region $t \leq \text{const}\, h^{2/(3-3b)}$. For points outside this region we have to resort to the original bound (14.285). Indeed, each of the three bounds in (14.285)/(14.294) is sharper than the other two in some nonempty region of the (t,h)-plane.

(c) $d = 4$. In this case the bubble exhibits a logarithmic divergence:

$$B_0(t) \leq |\log t|^{\bar{b}} \tag{14.295}$$

with $\bar{b} \leq 1$. The same procedure as for the $d < 4$ case can be performed. For $b < 1/3, \bar{b} = 1$ the result is [12]

$$M(\beta = \beta_c, h \downarrow 0) \leq \text{const} \times h^{1/3} |\log h| \tag{14.296}$$

for $h \downarrow 0$, and

$$M_0(\beta) \leq \text{const} \times |t|^{1/2} |\log |t||^{3/2} \tag{14.297}$$

for $\beta \downarrow \beta_c$.

From the discussion in this section we conclude that if $b < 1/3$ we obtain the critical-exponent inequalities

$$\delta \leq 3 \min\left(1, \frac{1-b}{1-3b}\right) \tag{14.298}$$

$$\widehat{\beta} \geq \frac{1}{2} \min(1, 1-3b)\ . \tag{14.299}$$

In particular, if we invoke the bound (14.48) on the bubble exponent b, we obtain

$$\delta \leq 3 \min\left(1, \frac{2d-3(2-\eta)}{4d-7(2-\eta)}\right) \tag{14.300}$$

$$\widehat{\beta} \geq \frac{1}{2} \min\left(1, \frac{4d-7(2-\eta)}{d-(2-\eta)}\right) \tag{14.301}$$

valid for $d > \frac{7}{4}(2-\eta)$. For nearest-neighbor Ising models the infrared bound (14.17) implies that inequalities (14.300) and (14.301) hold with η replaced by 0. Combining this with (14.250) and (14.251), we conclude that for the nearest-neighbor Ising model in dimension $d \geq 4$, the critical exponents δ and $\widehat{\beta}$ exist and take their mean-field values

$$\delta = 3 \quad \text{and} \quad \widehat{\beta} = 1/2\ , \tag{14.302}$$

with possible logarithmic corrections for $d = 4$.

This results also imply the continuity of the magnetization at the critical temperature whenever $b < 1/3$, and in particular whenever $d > \frac{7}{4}(2-\eta)$. For a nearest-neighbor Ising model, which has $\eta \geq 0$, this corresponds to "$d > 3.5$", which unfortunately does not include any physical dimension strictly less than four. The trouble here is the appearance of B_0^3 in the denominator of the AFe inequality, where one would hope for a single power of B_0. In any case, these bounds can be applied to some long-range reflection-positive Ising models in dimensions $d = 1, 2, 3$ [13], and the bounds obtained for the critical exponents are consistent with predictions based on renormalization-group arguments [207].

14.5 Horns and all that

14.5.1 The fundamental quantities

In Section 14.2.1 we saw that in Landau-Ginzburg theory the combinations M^3/h and $h\chi/M$ (and their product $M^2\chi$) play a special role, in that they depend only on the scaling variable $x \equiv th^{-2/3}$. In this section we study the relations which hold between these quantities in the exact theory.

The "dimensionless" combinations of interest to us are $|J|^2 M^3/h$ and $h\chi/M$, together with their product $|J|^2 M^2\chi$. By the weak GHS and weak ABF inequalities, we have

$$\frac{h\chi}{M} \leq 1 \leq \frac{h\chi}{M} + |J|^2 M^2 \chi \left(1 + \frac{h}{|J|\, M}\right)\left(1 + \frac{1}{|J|\, \chi}\right) \tag{14.303}$$

$$\leq \frac{h\chi}{M} + |J|^2 M^2 \chi \left(1 + \frac{1}{|J|\, \chi}\right)^2 . \tag{14.304}$$

Therefore,

$$\left[1 + \frac{|J|^2 M^3}{h}(1+\varepsilon)^2\right]^{-1} \leq \frac{h\chi}{M} \leq 1 \tag{14.305}$$

where $\varepsilon \equiv 1/(|J|\, \chi)$. Note that ε is expected to be small everywhere in a neighborhood of the critical point (later we will prove this for Ising models). In any case, we have proven:

Proposition 14.8 *If $|J|^2 M^3/h$ and $1/(|J|\, \chi)$ are bounded above, then $h\chi/M$ is bounded below away from zero (with universal constants). The precise bound is (14.305).*

We remark that the lower bound in (14.305) is (up to constants) an *equality* in Landau-Ginzburg theory.

We can also get an interesting result by multiplying (14.305) by $|J|^2 M^3/h$:

$$\frac{\dfrac{|J|^2 M^3}{h}}{1+\dfrac{|J|^2 M^3}{h}(1+\varepsilon)^2} \;\le\; |J|^2 M^2\chi \;\le\; \frac{|J|^2 M^3}{h} \tag{14.306}$$

Thus, we have proven:

Proposition 14.9

(a) If $|J|^2 M^3/h$ is bounded above, then so is $|J|^2 M^2\chi$.

(b) If $|J|^2 M^3/h$ is bounded below and $1/(|J|\,\chi)$ is bounded above, then $|J|^2 M^2\chi$ is bounded below.

(c) If $|J|^2 M^2\chi$ is bounded below, then so is $|J|^2 M^3/h$.

(d) If $|J|^2 M^2\chi$ is sufficiently small *[namely, $< (1+\varepsilon)^{-2} < 1$], then $|J|^2 M^3/h$ is bounded above.*

In all cases the precise bound is (14.306).

In summary, we have the following implications (assuming that $\varepsilon \equiv 1/(|J|\,\chi)$ is bounded above):

1)

$$\frac{|J|^2 M^3}{h} \ge \text{const} \iff |J|^2 M^2\chi \ge \text{const}$$

We shall see in the next subsection that for *Ising* models these equivalent conditions hold everywhere in the (β, h)-plane *except* in a "high-temperature horn".

2)

$$\begin{array}{ccc} \dfrac{|J|^2 M^3}{h} \le \text{const} & \overset{\Leftarrow}{\Longrightarrow} & |J|^2 M^2\chi \le \text{const} \\ \Downarrow & & \\ \frac{h\chi}{M} \ge \text{const} & & \end{array}$$

where the little arrow denotes an implication which holds only if $|J|^2 M^2\chi$ is sufficiently small.

We shall see in the next subsection that all of these three (inequivalent) conditions hold everywhere in the (β, h)-plane *except* in a "low-temperature horn", *provided that the dimension is* > 4.

14.5.2 Horns and the susceptibility

The next step is to verify the hypotheses of Proposition 14.8 or 14.9 for some explicit region of the (β, h)-plane. But we have already done the hard work:

- In Section 14.4.1 we proved, for GS-class models in any dimension d, a Landau-Ginzburg *lower* bound on the magnetization, valid in a full neighborhood of the critical point.

- In Section 14.4.2 we proved, for Ising models in dimension $d > 4$, a Landau-Ginzburg *upper* bound on the magnetization, valid in a full neighborhood of the critical point.

It remains only to put the ingredients together.

The first order of business is to show that the "error terms" in the weak ABF inequality (14.303) are indeed negligible in a full neighborhood of the critical point. We argue as follows:

1) $h/M \to 0$ along any path of approach to the critical point. This is a consequence of the Landau-Ginzburg lower bound on M, (14.244), together with the properties (14.95) of the Landau-Ginzburg function.

2) *If* $b < \frac{1}{3}$ (e.g. $d > \frac{7}{2}$ for nearest-neighbor models), then the generalized Landau-Ginzburg upper bound on M, (14.284)/(14.294), implies that $M \to 0$ along any path of approach to the critical point.

3) The weak ABF inequality (14.303) can be written as

$$\chi \geq \frac{1 - |J|\, M^2 \left(1 + \dfrac{h}{|J|\, M}\right)}{\dfrac{h}{M} + |J|^2\, M^2 \left(1 + \dfrac{h}{|J|\, M}\right)}. \tag{14.307}$$

The preceding results $h/M \to 0$ and $M \to 0$ then imply that $\chi \to \infty$ along any path of approach to the critical point, *provided that* $b < \frac{1}{3}$. (We suspect that $\chi \to \infty$ can be proven without this assumption, but we have so far been unable to do so.)

Next we invoke the Landau-Ginzburg lower bound on the magnetization (14.244). The properties (14.95) of the Landau-Ginzburg scaling function imply that $|J|^2\, M^3/h$ is bounded below in any region of the form $x \equiv th^{-2/3} \leq C_2$, i.e. the complement of a "high-temperature horn". By Proposition 14.9(b), we conclude that $|J|^2\, M^2\chi$ is also bounded below in this region. (And of course $h\chi/M$ is bounded above, by 1.)

Finally, we invoke the Landau-Ginzburg upper bound on the magnetization, (14.284), which holds whenever the bubble B_0 is bounded as the critical point is approached from the high-temperature side (e.g. $d > 4$ for nearest-neighbor models). The properties (14.95) of the Landau-Ginzburg scaling function imply that $|J|^2\, M^3/h$ is bounded above in any region of the form $x \equiv th^{-2/3} \geq -C_1$, i.e. the complement of a "low-temperature horn." By Propositions 14.9(a) and 14.8, we conclude that $|J|^2\, M^2\chi$ is bounded above and $h\chi/M$ is bounded below in this region.

We can now put these facts together to deduce Landau-Ginzburg-type bounds on the susceptibility for dimension $d > 4$:

1) In any region of the form $x \equiv th^{-2/3} \leq C_2$, i.e. away from a "high-temperature horn," we have

$$\chi \geq \frac{\text{const}}{|J|^2 M^2} \geq \frac{\text{const}}{M_{LG}^2} \geq \text{const} \times \chi_{LG}$$

and

$$\chi \leq \frac{M}{h} \leq \text{const} \times \frac{M_{LG}}{h}.$$

If we also stay away from a "low-temperature horn", then $\chi_{LG} \sim M_{LG}/h$, and these bounds are sharp. However, inside the low-temperature horn we have $\chi_{LG} \ll M_{LG}/h$, and our bounds on χ are not sharp.

2) In any region of the form $x \equiv th^{-2/3} \geq -C_1$, i.e. away from a "low-temperature horn," we have two-sided bounds of the form

$$\chi \sim \frac{M}{h} \sim \frac{M_{LG}}{h} \sim \chi_{LG} \, .$$

In summary: For $d > 4$ we have proven that

- Away from a low-temperature horn,

$$\chi \sim \frac{M}{h} \sim \frac{M_{LG}}{h} \sim \chi_{LG} \, .$$

- Inside the low-temperature horn,

$$\chi_{LG} \lesssim \chi \lesssim \frac{M}{h} \sim \frac{M_{LG}}{h} \, .$$

[The true behavior is believed to be $\chi_{LG} \sim \chi \ll \frac{M}{h} \sim \frac{M_{LG}}{h}$.]

14.6 General considerations on extrapolation principles

As remarked above, Landau-Ginzburg theory can be defined by a system of first order partial differential equations (PDEs), while the exact theory only satisfies the corresponding partial differential inequalities (PDIs). In this sense, a comparison between exact and mean-field theory corresponds to a comparison between solutions of PDIs and solutions of the corresponding PDEs. In this section we present a general discussion of this latter point of view, showing how extrapolation principles, or the resulting comparison theorems, become a natural tool.

14.6.1 Extrapolation principles for a single ordinary differential equation

For pedagogical reasons, as well as for future use, let us discuss first the case of the simple ordinary differential equation (ODE)

$$f'(\tau) + b(\tau)f(\tau) = 0 . \tag{14.308}$$

This equation can be easily integrated (if b is locally integrable); its solutions are exponential. We are interested in the properties of smooth functions F constrained only by the ordinary differential *inequality* (ODI)

$$F'(\tau) + b(\tau)F(\tau) \leq 0 , \tag{14.309}$$

or by

$$F'(\tau) + b(\tau)F(\tau) \geq 0 . \tag{14.310}$$

We shall only consider the case in which $b(\tau)$ never changes sign. Let us start by (14.310), which we write as

$$F' \geq -bF . \tag{14.311}$$

We see that certain combinations of signs of F and F' are forbidden. For example, if $b \geq 0$ we can *not* have a function that is *negative* and *strictly decreasing* at the same time. By continuity this implies that such F can not pass from positive or zero to negative values as τ increases. That is, $F(\tau_0) \geq 0$ implies $F(\tau) \geq 0$ for all $\tau \geq \tau_0$. Analogously, $F(\tau_0) < 0$ implies $F(\tau) < 0$ for all $\tau \leq \tau_0$ and, moreover, by (14.311) F has to be increasing for $\tau \leq \tau_0$. We leave to the reader the argument for the case $b \leq 0$ [which, for instance, follows from the $b \geq 0$ case by a time reflection], and for the cases corresponding to (14.309) [which correspond to those for (14.310) applied to $-F$]. The results can be summarized as follows:

Lemma 14.10 Extrapolation principles *Consider a function $b(\tau)$ of constant sign and locally integrable.*
a) For a solution of $F' + bF \leq 0$:

a1) If $b \geq 0$

i) $F(\tau_0) \leq 0$ implies $F(\tau) \leq 0$ for all $\tau \geq \tau_0$.

ii) $F(\tau_0) > 0$ implies $F(\tau) > 0$ for all $\tau \leq \tau_0$; moreover F is decreasing for all $\tau \leq \tau_0$.

a2) If $b \leq 0$

i) $F(\tau_0) < 0$ implies $F(\tau) < 0$ for all $\tau \geq \tau_0$; moreover F is decreasing for all $\tau \geq \tau_0$.

ii) $F(\tau_0) \geq 0$ implies $F(\tau) \geq 0$ for all $\tau \leq \tau_0$.

b) For a solution of $F' + bF \geq 0$*:*

b1) If $b \geq 0$

- *i)* $F(\tau_0) \geq 0$ *implies* $F(\tau) \geq 0$ *for all* $\tau \geq \tau_0$.
- *ii)* $F(\tau_0) < 0$ *implies* $F(\tau) < 0$ *for all* $\tau \leq \tau_0$*; moreover* F *is increasing for all* $\tau \leq \tau_0$.

b2) If $b \leq 0$

- *i)* $F(\tau_0) > 0$ *implies* $F(\tau) > 0$ *for all* $\tau \geq \tau_0$*; moreover* F *is increasing for all* $\tau \geq \tau_0$.
- *ii)* $F(\tau_0) \leq 0$ *implies* $F(\tau) \leq 0$ *for all* $\tau \leq \tau_0$.

In all cases the result is that certain bounds "extrapolate" to $\tau \geq \tau_0$ or $\tau \leq \tau_0$. In fact, the above lemma can be summarized by the following extrapolation principles: If $F' + bF \leq 0$, then upper bounds (by 0) extrapolate to larger τ, and lower bounds (by 0) extrapolate to smaller τ (irrespective of the sign of b). If $F' + bF \geq 0$ the directions of extrapolation reverse. The sign of b determines in which direction the solution exhibits monotonic behavior.

Instead of comparing F with 0 we can compare it with a solution f^* of the ODE (14.308). Indeed, F satisfies (14.309) or (14.310) if so does $F - f^*$, hence we can apply Lemma 14.10 to the latter. The resulting comparison theorem is stated next. To abbreviate — and also to exhibit more clearly which hypotheses determine which behavior — we shall deal simultaneously with both cases: one corresponds to the upper line inside each pair of curly brackets, while the other corresponds to the lower line.

Theorem 14.11 Comparison theorem *Let* $b(\tau)$ *be a function with constant sign and locally integrable, and let* f^* *be the solution of* $f'(\tau) + b(\tau)f(\tau) = 0$ *with initial condition* $f^*(\tau_0) = f_0$*. Let* F *be a solution of*

$$F' + bF \left\{ \begin{matrix} \leq \\ \geq \end{matrix} \right\} 0 .$$

Then

i) $F(\tau_0) \left\{ \begin{matrix} \leq \\ \geq \end{matrix} \right\} f_0$ *implies* $F(\tau) \left\{ \begin{matrix} \leq \\ \geq \end{matrix} \right\} f^*(\tau)$ *for all* $\tau \geq \tau_0$*; moreover, if* $b \leq 0$ *and* $F(\tau_0) \left\{ \begin{matrix} < \\ > \end{matrix} \right\} f_0$, $F - f^*$ *is* $\left\{ \begin{matrix} \text{decreasing} \\ \text{increasing} \end{matrix} \right\}$ *for* $\tau \geq \tau_0$.

ii) $F(\tau_0)\left\{\begin{array}{c}\geq\\=\\\leq\end{array}\right\} f_0$ *implies* $F(\tau)\left\{\begin{array}{c}\geq\\=\\\leq\end{array}\right\} f^*(\tau)$ *for all* $\tau \leq \tau_0$*; moreover, if* $b \geq 0$ *and* $F(\tau_0)\left\{\begin{array}{c}>\\=\\<\end{array}\right\} f_0$, $F - f^*$ *is* $\left\{\begin{array}{c}\text{decreasing}\\\text{increasing}\end{array}\right\}$ *for* $\tau \leq \tau_0$.

Proof. Most of the statements follow from applying Lemma 14.10 to $F - f^*$. The only exceptions are the "equality" part for those signs of b for which there are monotonicity statements. These require an extra argument based on the continuous dependence of f^* on the initial condition. As an example, let us prove the lower line of Part ii) for $b \geq 0$: If $F' + bF \geq 0$ with $b \geq 0$,

$$F(\tau_0) \leq f_0 \quad \Longrightarrow F(\tau) \leq f^* \quad \forall \tau \leq \tau_0\,. \tag{14.312}$$

If $F(\tau_0) < f_0$ then the result follows from Lemma 14.10. If $F(\tau_0) = f_0$ then $F(\tau_0) < f_0 + \varepsilon$ for every $\varepsilon > 0$. Hence, by Lemma 14.10 $F(\tau) < f^*_\varepsilon(\tau)$ for every $\tau \leq \tau_0$ and every $\varepsilon > 0$, where f^*_ε is the solution of the ODE (14.308) with initial condition $f^*_\varepsilon(\tau_0) = f_0 + \varepsilon$. By letting $\varepsilon \to 0$ we obtain $F(\tau) \leq f^*(\tau)$. ∎

In particular, since $f^* = 0$ is the solution of (14.308) for $f_0 = 0$, this theorem implies Lemma 14.10 but with all strict inequalities replaced by non-strict ones.

As we shall see in the rest of this section, the ideas discussed above generalize to the types of PDIs in two variables of interest here — sub- and supra-conservation laws: there appear curves along which bounds extrapolate (and there is monotonic behavior in one direction), which yield comparison theorems. This generalizes the discussion of Sections 14.2.2 and 14.2.3.

14.6.2 Conservation laws. Characteristics

We start by an overview of the PDE approach to Landau-Ginzburg theory. As shown above, this theory can be reconstructed (except for the position of the critical point) from the system of PDEs

$$\frac{\partial M}{\partial \beta} - |\tilde{J}| M \frac{\partial M}{\partial h} = 0 \tag{14.313}$$

$$\frac{\partial R}{\partial \beta} + \frac{\beta}{hR^2}\frac{\partial R}{\partial h} = 0 \tag{14.314}$$

with $R = \beta M/h$. Both equations have the form of *conservation laws* [356], i.e. they are first order nonlinear PDE of the form

$$\frac{\partial u}{\partial \tau} + A(u)\frac{\partial u}{\partial z} = 0 \tag{14.315}$$

with the identifications

Equation	u	τ	z	$A(u)$
(14.313)	M	β	h	$-\|\tilde{J}\|u$
(14.314)	$R = \beta M/h$	β^2	h^2	$1/u^2$

(14.316)

The equation (14.313) is in fact the zero-viscosity limit of Burger's equation, which describes the velocity of a one-dimensional compressible gas at time β and position h. On the other hand, (14.314) has the form of a chromatographic equation for a cylindrical adsorption column. The equations of the form 14.315 have been extensively studied in the literature [475]. For the genuinely (or strictly) nonlinear case [355], that is when $A'(u) \neq 0$ — as is the case in our applications (14.313)/(14.314) — the only possible singularities of the solution are *shocks* which occur only under certain monotonicity behavior of the boundary conditions.

To see this, note that if A where a constant, the solutions of the PDE (14.315) would be waves propagating with velocity A. The most natural guess for the nonlinear case is therefore a wave with nonconstant velocity $A(u)$, that is a solution implicitly defined by

$$u(\tau, z) = g(z - A(u)\tau) \tag{14.317}$$

where $g(z) = u(\tau = 0, z)$ is the initial condition. A simple implicit differentiation shows that $\partial u/\partial\tau = -Ag'/(1 + \tau A'g')$ and $\partial u/\partial z = g'/(1 + \tau A'g')$ and hence (14.317) indeed determines the solution of (14.315) *but only for times* $\tau < \tau_c$ where

$$\tau_c = -\frac{1}{A'g'} \,. \tag{14.318}$$

At such time the solution becomes discontinuous; a shock develops and additional conditions are needed to select the physical solution (jump construction = Maxwell rule). As we are interested in $\tau > 0$, such shock only takes place if $A'g' < 0$. In (14.313)–(14.314) $A' < 0$, hence a shock will develop only if the initial condition g is increasing. This is the case for (14.313) where the $\beta = 0$ (infinite temperature) magnetization is

$$M(\beta = 0, h) = \tanh h \,.$$

The solution (14.317) is therefore

$$M_{MF} = \tanh[h + \beta|\tilde{J}|M_{MF}] \tag{14.319}$$

a well known equation for the mean-field magnetization, and (14.318) implies that the shock "starts" at

$$\beta_c = \frac{1}{|\tilde{J}|}\cosh^2[h_c + \beta_c|\tilde{J}|M_{MF}] \tag{14.320}$$

whose unique solution is

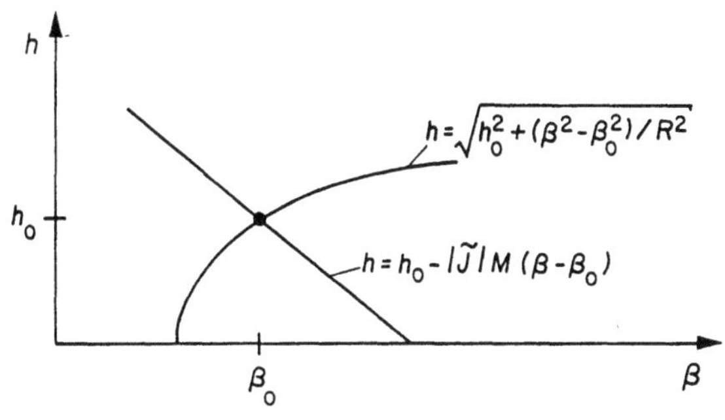

Fig. 14.8. Characteristics of (14.313) and (14.314)

$$h_c = 0 \qquad \beta_c = \frac{1}{|\tilde{J}|}, \tag{14.321}$$

the mean-field critical point.

The systematic way to solve (14.315) is by the method of characteristics. These are curves Γ obtained by integrating, for instance, the system of ODEs

$$\left.\frac{dz}{d\tau}\right|_{\Gamma} = A(u) \tag{14.322}$$

$$\left.\frac{du}{d\tau}\right|_{\Gamma} = 0\,. \tag{14.323}$$

The equations (14.322)–(14.323) are easily integrated; they yield straight lines with slope $A(u)$ along which the solution u is constant. Hence the solution at each point (τ, z) is the value of the boundary condition at the intersection with the characteristic passing through the point (there is only one such characteristic if no shock has yet developed).

In Figure 14.8 we show superimposed characteristics of (14.313) and (14.314) through a point (β_0, h_0). From (14.322)–(14.323) and (14.316), the corresponding equations are

$$h = h_0 - |\tilde{J}| M(\beta_0, h_0)(\beta - \beta_0) \tag{14.324}$$

for the "Burgers" equation (14.313), and

$$h^2 = h_0^2 + \frac{\beta^2 - \beta_0^2}{R(\beta_0, h_0)^2} \tag{14.325}$$

for the "chromatographic" equation (14.314).

14.6.3 Sub- and supra-conservation laws. Extrapolation principles and comparison theorems

We now turn to the PDIs (14.140) and (14.179):

$$\frac{\partial M}{\partial \beta} - |\tilde{J}| M \frac{\partial M}{\partial h} \leq 0 \tag{14.326}$$
$$\frac{\partial R}{\partial \beta} + \frac{\beta}{hR^2} \frac{\partial R}{\partial h} \geq 0 \tag{14.327}$$

satisfied by the exact theory. They have respectively the form of a sub- and a supra-conservation law. that is they have the form

$$\frac{\partial U}{\partial \tau} + A(U) \frac{\partial U}{\partial z} \left\{ \begin{matrix} \leq \\ \geq \end{matrix} \right\} 0 \tag{14.328}$$

with the identifications (14.316). We look for solutions smooth enough on the region $\tau, z > 0$ [we need only once-differentiability, but by Lee-Yang theory [371] we know that M and R are in fact analytic in that region].

If we try to adapt the method of the characteristics to the PDI problem (14.328) we naturally obtain a possible generalization of (14.322)–(14.323) by replacing one of the equal signs by an inequality. Indeed, a curve Γ_I defined by

$$\left.\frac{dz}{d\tau}\right|_{\Gamma_I} = A(U) \tag{14.329}$$

must satisfy, by (14.328)

$$\left.\frac{dU}{d\tau}\right|_{\Gamma_I} \left\{ \begin{matrix} \leq \\ \geq \end{matrix} \right\} 0\,. \tag{14.330}$$

Such a curve Γ_I is not a curve of constant U (and hence not a straight line), but at least we know that along it the solutions U have monotonic behavior. As a consequence, the curve can be used to extrapolate bounds: if $(\tau_0, z_0) \in \Gamma_I$, a $\left\{ \begin{matrix} \text{upper} \\ \text{lower} \end{matrix} \right\}$ bound on $U(\tau_0, z_0)$ is then a $\left\{ \begin{matrix} \text{upper} \\ \text{lower} \end{matrix} \right\}$ bound of U for all points in Γ_I with $\tau \geq \tau_0$. Analogously, the opposite bounds extrapolate to $\tau \leq \tau_0$.

Of course, Γ_I is of little use for practical purposes because to determine it we need to know the actual U. But at least, the discussion suggests a possible line of action: try to find curves

$$\left.\frac{dz}{d\tau}\right|_{\Gamma_s} = s \tag{14.331}$$

where the slope s can be determined without knowing the solution U and along which bounds extrapolate. The total rate of change of a solution of (14.328) along a curve (14.331) is

$$\begin{aligned} \left.\frac{dU}{d\tau}\right|_{\Gamma_s} &= \frac{\partial U}{\partial \tau} + s\,\frac{\partial U}{\partial z} \\ &\left\{ \begin{matrix} \leq \\ \geq \end{matrix} \right\} [s - A(U)]\,\frac{\partial U}{\partial z}\,. \end{aligned} \tag{14.332}$$

From this we immediately conclude:

Proposition 14.12 *If along the curve Γ_s*

$$\operatorname{sgn}[s - A(U)] = \left\{ \begin{matrix} - \\ + \end{matrix} \right\} \operatorname{sgn}\left[\frac{\partial U}{\partial z}\right], \tag{14.333}$$

then the solutions of (14.328) are of monotonic behavior:

$$\left.\frac{du}{d\tau}\right|_{\Gamma_s} \left\{ \begin{matrix} \leq \\ \geq \end{matrix} \right\} 0 .$$

To arrive to more concrete results, we restrict ourselves to cases in which A' and $\partial U/\partial z$ never change signs. [These hypotheses are certainly true for the PDIs (14.326) and (14.327) by second Griffiths and weak GHS inequalities respectively; moreover in both cases $A' < 0$.] Then the requirement (14.333) is equivalent to determine functions $s(\tau)$ such that

$$F(\tau) \equiv s(\tau) - A(U)|_{\Gamma_s} \tag{14.334}$$

has always the same sign. This problem can be reduced to the study of an ODI of the type discussed in Section 14.6.1. Indeed, differentiating (14.334) and using (14.332) we get

$$\begin{aligned} F' &= s' - A'(U) \left.\frac{du}{d\tau}\right|_{\Gamma_s} \\ &\left\{ \begin{array}{ll} \left\{ \begin{matrix} \leq \\ \geq \end{matrix} \right\} s' - A'\dfrac{\partial U}{\partial z} F & \text{if } A' \leq 0 \\ \left\{ \begin{matrix} \geq \\ \leq \end{matrix} \right\} s' - A'\dfrac{\partial U}{\partial z} F & \text{if } A' \geq 0 \end{array} \right. \end{aligned} \tag{14.335}$$

Therefore, if $A' \leq 0$ [as in the PDIs (14.326) and (14.327)] and we demand

$$s' \left\{ \begin{matrix} \leq \\ \geq \end{matrix} \right\} 0 \tag{14.336}$$

then F satisfies the equation

$$F' + \left(A'\frac{\partial U}{\partial z}\right) F \left\{ \begin{matrix} \leq \\ \geq \end{matrix} \right\} 0 . \tag{14.337}$$

If $A' \geq 0$, all the inequality signs must be reversed in (14.336)–(14.337).

If at this point we apply to the ODI (14.337) the Comparison Theorem 14.11 for the case $f_0 = 0 = f^*$, in combination with (14.332), we obtain the following lemma.

Lemma 14.13 *For a differentiable monotonic function $s(\tau)$, let $\Gamma_s = \{(\tau, z(\tau)\}$ be the curve defined by $dz/d\tau|_{\Gamma_s} = s(\tau)$ and the condition $(\tau_0, z(\tau_0)) = (\tau_0, z_0)$. Let U be a solution of*

$$\frac{\partial U}{\partial \tau} + A(U)\frac{\partial U}{\partial z}\left\{\begin{matrix} \leq \\ \geq \end{matrix}\right\}0$$

where A is a monotonic differentiable function. We assume that on the curve Γ_s, $A'\partial U/\partial z$ is of constant sign and locally integrable, and $s'A'\left\{\begin{matrix} \geq \\ \leq \end{matrix}\right\}0$. Then

i) $U(\tau_0, z_0)\left\{\begin{matrix} \leq \\ \geq \end{matrix}\right\}A^{-1}(s(\tau_0))$ implies $U(\tau, z(\tau))\left\{\begin{matrix} \leq \\ \geq \end{matrix}\right\}A^{-1}(s(\tau))$ for all $\tau \geq \tau_0$; moreover if $A'\,\partial U/\partial z \leq 0$ then $U(\tau, z(\tau))$ is $\left\{\begin{matrix} \text{decreasing} \\ \text{increasing} \end{matrix}\right\}$ for $\tau \geq \tau_0$.

ii) $U(\tau_0, z_0)\left\{\begin{matrix} \geq \\ \leq \end{matrix}\right\}A^{-1}(s(\tau_0))$ implies $U(\tau, z(\tau))\left\{\begin{matrix} \geq \\ \leq \end{matrix}\right\}A^{-1}(s(\tau))$ for all $\tau \leq \tau_0$; moreover if $A'\,\partial U/\partial z \geq 0$ then $U(\tau, z(\tau))$ is $\left\{\begin{matrix} \text{decreasing} \\ \text{increasing} \end{matrix}\right\}$ for $\tau \leq \tau_0$.

Proof. We shall prove the upper line when $A' \leq 0$ (and hence $s' \leq 0$); the other cases are analogous. We apply the upper line of Theorem 14.11 [with $f_0 = 0 = f^*$] to the upper equation (14.337). The result is

$$s(\tau_0) - a(U(\tau_0, z_0)) \leq 0 \implies s(\tau) - a(U(\tau, z(\tau))) \leq 0 \;\forall \tau \geq \tau_0 \quad (14.338)$$
$$s(\tau_0) - a(U(\tau_0, z_0)) \geq 0 \implies s(\tau) - a(U(\tau, z(\tau))) \geq 0 \;\forall \tau \leq \tau_0 \quad (14.339)$$

or, as A^{-1} exists and is decreasing

$$A^{-1}(s(\tau_0)) \geq U(\tau_0, z_0) \implies A^{-1}(s(\tau)) \geq U(\tau, z(\tau))\forall \tau \geq \tau_0 \quad (14.340)$$
$$A^{-1}(s(\tau_0)) \leq U(\tau_0, z_0) \implies A^{-1}(s(\tau)) \leq U(\tau, z(\tau))\forall \tau \leq \tau_0 \quad (14.341)$$

as stated in i) and ii) of the lemma. To obtain the monotonicity requirements we must look to the upper inequality in (14.332): if (14.338) holds, then $\partial U/\partial z \geq 0$ implies $dU/d\tau|_{\Gamma_s} \leq 0$; analogous conclusion holds for (14.339) but if $\partial U/\partial z \leq 0$. ∎

The form of this lemma useful to obtain comparison theorems is when Γ_s is a straight line, i.e. $s(\tau) = A(u_0)$.

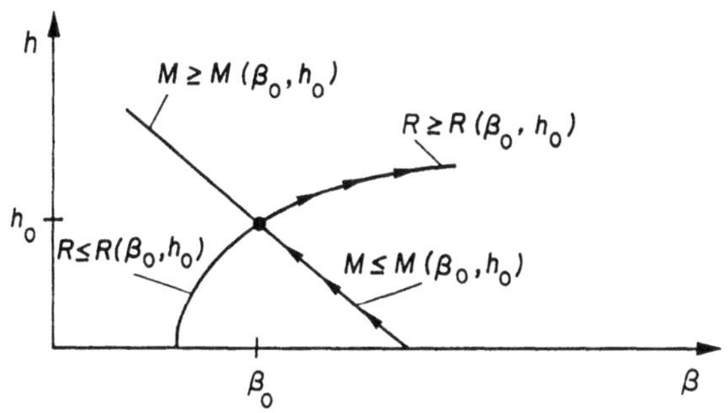

Fig. 14.9. Extrapolation — and monotonicity — for the PDIs (14.326) and (14.327). Arrows indicate directions in which the solutions are known to increase

Lemma 14.14 Extrapolation principles *Let* $\Gamma_{u_0} = \{(\tau, z) \colon z = z_0 + A(u_0(\tau - \tau_0)\}$. *Let* U *be a solution of*

$$\frac{\partial U}{\partial \tau} + A(U)\frac{\partial U}{\partial z} \left\{ \begin{array}{c} \leq \\ \geq \end{array} \right\} 0$$

where A *is a monotonic differentiable function. We assume that on the curve* Γ_{u_0}, $A' \partial U/\partial z$ *is of constant sign and locally integrable. Then*

i) $U(\tau_0, z_0) \left\{ \begin{array}{c} \leq \\ \geq \end{array} \right\} u_0$ *implies* $U|_{\Gamma_{u_0}} \left\{ \begin{array}{c} \leq \\ \geq \end{array} \right\} u_0$ *for* $\tau \geq \tau_0$*; moreover if in addition* $A' \, \partial U/\partial z \leq 0$ *then* $U(\tau, z(\tau))$ *is* $\left\{ \begin{array}{c} \text{decreasing} \\ \text{increasing} \end{array} \right\}$ *for* $\tau \geq \tau_0$.

ii) $U(\tau_0, z_0) \left\{ \begin{array}{c} \geq \\ \leq \end{array} \right\} u_0$ *implies* $U|_{\Gamma_{u_0}} \left\{ \begin{array}{c} \geq \\ \leq \end{array} \right\} u_0$ *for* $\tau \leq \tau_0$*; moreover if in addition* $A' \, \partial U/\partial z \geq 0$ *then* $U(\tau, z(\tau))$ *is* $\left\{ \begin{array}{c} \text{decreasing} \\ \text{increasing} \end{array} \right\}$ *for* $\tau \leq \tau_0$.

Lemma 14.14 can be summarized by saying that along straight lines — that is along characteristics of the PDE — for a sub-conservation law upper bounds extrapolate to larger τ while lower bounds extrapolate to smaller τ. For supra-conservation laws the directions of extrapolation are reversed. Lemmas 14.1 and 14.5 are particular applications of Lemma 14.14[19] for the case $A' < 0$ (plus bounds derived from correlation inequalities). The results of Lemma 14.14 for these cases are summarized in Figure 14.9.

In addition, Lemma 14.13 can be applied to certain non-straight curves $z(\tau)$, which must either be convex [for a sub-conservation law with $A' \geq 0$, or a

[19] The "corner" excluded in Lemma 14.1 corresponds to a point where U is not even continuous.

supra-conservation law with $A' \leq 0$] or concave [for a sub-conservation law with $A' \leq 0$, or a supra-conservation law with $A' \geq 0$]. One example is the curve Γ_I defined by $dz/d\tau = A(U) \equiv A(U(\tau, z(\tau)))$ [cf. (14.329)]: by (14.333) we have that $dU/d\tau\big|_{\Gamma_I}$ is of constant sign, and hence $z''(\tau) = dA(U)/d\tau = A'\,dU/d\tau$ is also of constant sign. However, as mentioned before, this curve is not very useful in practice because the function $U(\tau, z)$ is unknown. A second example is the curve Γ_{II} where U is constant:

$$\left.\frac{dU}{d\tau}\right|_{\Gamma_{II}} = 0\,. \tag{14.342}$$

Note that if $\partial z/\partial\tau|_{\Gamma_{II}} = s$ then by (14.328) we have that for sub- and supra-conservation laws

$$0 \left\{ \begin{matrix} \leq \\ \geq \end{matrix} \right\} [s(\tau) - A(U(\tau))]\,\frac{\partial U}{\partial z}\,.$$

In the case $\partial U/\partial z \geq 0$ this implies

$$\left.\frac{dz}{d\tau}\right|_{\Gamma_{II}} \left\{ \begin{matrix} \leq \\ \geq \end{matrix} \right\} A(U)\,. \tag{14.343}$$

The pair (14.342)–(14.343) provides another possible generalization of the (14.322)–(14.323) in which the equality and inequality signs are in different places respect to (14.329)–(14.330).

Lemma 14.14 immediately yields comparison theorems between solutions of the PDI (14.328) and solutions of the PDE (14.315). In particular, when the initial conditions are given on the τ axis, we obtain the following generalization of Propositions 14.2 and 14.6:

Theorem 14.15 Comparison theorem *Let u^* be the solution of the PDE*

$$\frac{\partial u}{\partial\tau} + A(u)\frac{\partial u}{\partial z} = 0$$

with initial condition $u^*(\tau, z = 0) = k(\tau)$. We assume that A and k are differentiable and strictly monotonic, $A'k' < 0$ and $A(g(\tau)) \neq 0$. Let U be the solution of the PDI

$$\frac{\partial U}{\partial\tau} + A(U)\frac{\partial U}{\partial z} \left\{ \begin{matrix} \leq \\ \geq \end{matrix} \right\} 0$$

with U differentiable and $\partial U/\partial z$ of constant sign . Then:

i) If $A(k(\tau)) < 0$,

$$U(\tau, z = 0) \left\{ \begin{matrix} \geq \\ \leq \end{matrix} \right\} k(\tau) \text{ for } \tau_0 \leq \tau \leq \tau_1 \quad\Longrightarrow\quad U(\tau, z) \left\{ \begin{matrix} \geq \\ \leq \end{matrix} \right\} u^*(\tau, z)$$

at all points in the region

$$\Delta_1 = \left\{ \left[A(k(\tau_0))(\tau - \tau_0)\right]_+ \ \leq \ z \ \leq \ A(k(\tau_1))(\tau - \tau_1)\,; \ \tau \leq \tau_1 \right\} .$$

ii) If $A(k(\tau)) > 0$,

$$U(\tau, z = 0) \left\{ \begin{array}{c} \leq \\ \geq \end{array} \right\} k(\tau) \text{ for } \tau_0 \leq \tau \leq \tau_1 \quad \Longrightarrow \quad U(\tau, z) \left\{ \begin{array}{c} \leq \\ \geq \end{array} \right\} u^*(\tau, z)$$

at all points in the region

$$\Delta_2 = \left\{ \left[A(k(\tau_1))(\tau - \tau_1)\right]_+ \ \leq \ z \ \leq \ A(k(\tau_0))(\tau - \tau_0)\,; \ \tau \geq \tau_0 \right\} .$$

Here we have used the notation $[x]_+ = \max(x, 0)$. The proof of this theorem is an immediate consequence of the fact that the solution of the PDE is constant along the characteristics while the solution of the PDI takes larger or smaller values as stated in Lemma 14.14. The condition $A'k' < 0$ ensures that no shock occurs for $z \geq 0$, while the regions Δ_1 and Δ_2 are defined so that each point inside is connected to the boundary segment $z = 0, \tau_0 \leq \tau \leq \tau_1$ by exactly one characteristic. The requirement $A(g(\tau)) \neq 0$ is equivalent to asking that the τ-axis be nowhere characteristic (otherwise, in general there would be no solution). We remark that u^* is implicitly defined by

$$u^*(\tau, z) \ = \ k\left(\tau - \frac{z}{A(u^*)}\right) .$$

In an analogous way one can obtain the following generalization of Propositions 14.3 and 14.7.

Theorem 14.16 Comparison theorem *Let* u^* *be the solution of the PDE*

$$\frac{\partial u}{\partial \tau} + A(u)\frac{\partial u}{\partial z} = 0$$

with initial condition $u^*(\tau = \tau_c, z) = g(z)$. *We assume that* A *and* g *are differentiable and strictly monotonic. Let* U *be the solution of the PDI*

$$\frac{\partial U}{\partial \tau} + A(U)\frac{\partial U}{\partial z} \left\{ \begin{array}{c} \leq \\ \geq \end{array} \right\} 0$$

with U *differentiable and* $\partial U/\partial z$ *of constant sign . Then:*

i) If $A'g' < 0$,

$$U(\tau = \tau_c, z) \left\{ \begin{array}{c} \geq \\ \leq \end{array} \right\} g(z) \text{ for } z_0 \leq z \leq z_1 \quad \Longrightarrow \quad U(\tau, z) \left\{ \begin{array}{c} \geq \\ \leq \end{array} \right\} u^*(\tau, z)$$

at all points in the region

$$\Delta_3 = \{z_0 + A(g(z_0))(\tau - \tau_c) \ \le \ z \ \le \ z_1 + A(g(z_1))(\tau - \tau_c)\,;\ \tau \le \tau_c\}\ .$$

ii) If $A'g' > 0$,

$$U(\tau = \tau_c, z) \left\{ \begin{matrix} \le \\ \ge \end{matrix} \right\} g(z) \text{ for } z_0 \le z \le z_1 \quad \Longrightarrow \quad U(\tau, z) \left\{ \begin{matrix} \le \\ \ge \end{matrix} \right\} u^*(\tau, z)$$

at all points in the region

$$\Delta_4 = \{z_0 + A(g(z_0))(\tau - \tau_c) \ \le \ z \ \le \ z_1 + A(g(z_1))(\tau - \tau_c)\,;\ \tau \ge \tau_c\}\ .$$

The solution u^* is implicitly defined by

$$u^*(\tau, z) \ = \ g(z - A(u^*\tau))\ .$$

15. Continuum Limits

In this chapter we describe how continuum Euclidean quantum field theories can be obtained as limits of rescaled lattice theories. Our main goal is to give a precise exposition of what is known about the triviality of continuum limits for $d > 4$ and $d = 4$.

15.1 Generalities on continuum limits

15.1.1 What is a continuum limit? What is triviality?

Let $\langle \varphi_{x_1} \varphi_{x_2} \cdots \varphi_{x_n} \rangle_\Gamma$ be the n-point correlation function of an infinite-volume translation-invariant spin model, indexed by some parameters Γ. For example, for the Ising model one might take $\Gamma = (J, h)$, where J is the nearest-neighbor interaction and h is the magnetic field; for the φ^4 model, one might take $\Gamma = (J, \lambda, \tau, h)$, where J and h are as before, λ is the φ^4 coupling, and τ is the φ^2 coupling.

A continuum quantum field theory is obtained from a sequence of lattice theories by rescaling lengths by a factor $\theta \to \infty$ and possibly rescaling field strengths by a factor $\alpha(\theta)$:

$$
\begin{aligned}
S_n^\theta(x_1, \ldots, x_n) &\equiv \alpha(\theta)^n \langle \varphi_{\theta x_1} \cdots \varphi_{\theta x_n} \rangle_{\Gamma(\theta)} && (15.1) \\
S_n^{\text{cont}}(x_1, \ldots, x_n) &\equiv \lim_{\theta \to \infty} S_n^\theta(x_1, \ldots, x_n) && (15.2)
\end{aligned}
$$

Thus, we are observing the lattice models at longer and longer distance scales (of order θ lattice spacings). Equivalently, we are shrinking the lattice so that its lattice spacing $a = \theta^{-1}$, measured in "centimeters", tends to zero. These two points of view are often called the "scaling limit" and "continuum limit", respectively, but there is no real difference between them: they are just two different verbal interpretations of the same limiting process (15.1)–(15.2). It is important to note that the parameters $\Gamma(\theta)$ are allowed to — and in general must — vary with θ.

One way of posing the general problem of quantum field theory is the following: *Consider all possible choices of $\Gamma(\theta)$ and $\alpha(\theta)$; classify all limiting theories $\{S_n^{\text{cont}}\}$ and study their properties.*

For any given choice of $\Gamma(\theta)$ and $\alpha(\theta)$ there are several possible behaviors:

(a) *No limit.* For at least one n, the limit (15.2) fails to exist. This is the typical behavior if $\alpha(\theta)$ is chosen inappropriately.

(b) *Boring limit.* The limiting theory $\{S_n^{\text{cont}}\}$ exists but is completely boring, e.g. identically zero, or identically zero at $x_1 \neq x_2 \neq \ldots \neq x_n$ ("ultra-local theory").

(c) *Gaussian limit.* The limiting theory $\{S_n^{\text{cont}}\}$ is Gaussian, i.e. a generalized free field. This situation is commonly called *triviality*.

(d) *Non-Gaussian limit.* The limiting theory is non-Gaussian. This situation is commonly called *nontriviality*. (We emphasize, however, that non-Gaussianness does *not* necessarily imply that the scattering matrix is different from the identity. For example, the Wick square of a free field is non-Gaussian but nevertheless describes noninteracting particles.)

It is obvious heuristically that, in order to obtain a non-boring continuum limit, the lattice theories $\Gamma(\theta)$ should have correlation lengths of order at least θ, as $\theta \to \infty$ (otherwise, the rescaled theories would have correlation lengths, measured in centimeters, which tend to zero).[1] In particular, the parameters $\Gamma(\theta)$ should approach or sit on the critical surface as $\theta \to \infty$. The two most common ways of achieving this are the following:

- *Massive scaling limit.* The parameters $\Gamma(\theta)$ approach the critical surface in such a way that the rescaled correlation lengths $\xi^\theta \equiv \theta^{-1}\xi(\Gamma(\theta))$ tend to a finite nonzero constant as $\theta \to \infty$. Usually $\alpha(\theta)$ is chosen so that the rescaled susceptibilities

$$\chi^\theta \equiv \int d^d x\, S_2^\theta(0,x) \equiv \theta^{-d} \sum_{x \in \theta^{-1}\mathbb{Z}^d} S_2^\theta(0,x) = \alpha(\theta)^2 \theta^{-d} \chi$$

  also tend to a finite nonzero constant as $\theta \to \infty$. Heuristically this should yield a massive, non-boring continuum theory.[2]

- *Massless scaling limit.* The parameters $\Gamma(\theta)$ sit on the critical surface for all θ, i.e. $\xi(\Gamma(\theta)) \equiv +\infty$. For a suitable choice of $\alpha(\theta)$ this should yield a massless (non-boring) continuum theory.

Of course, other ways of approaching the critical surface are also permissible, provided that the limits (15.2) exist.

---

[1] Indeed, for reflection-positive theories, it is not hard to prove rigorously that if $\theta^{-1}\xi(\Gamma(\theta)) \to 0$, then the limit (15.2) for $n = 2$ either does not exist or else vanishes identically for $x_1 \neq x_2$. If the Griffiths I and Gaussian correlation inequalities hold, then this result also extends to higher n. See Proposition 15.4 for a precise result.

[2] Indeed, one expects heuristically that the continuum susceptibility $\chi^{\text{cont}} \equiv \int d^d x\, S_2^{\text{cont}}(0,x)$ should equal $\lim_{\theta\to\infty} \chi^\theta$, the limit of the rescaled lattice susceptibilities. For a rigorous result, see Proposition 15.8.

Remark on normalization conventions. Consider, for concreteness, a φ^4 lattice model with Hamiltonian[3]

$$H = -J \sum_{\langle xy \rangle} \varphi_x \varphi_y + \sum_x \left(\frac{\lambda}{4} \varphi_x^4 + \frac{\tau}{2} \varphi_x^2 - h\varphi_x \right) , \tag{15.3}$$

where $\langle xy \rangle$ denotes nearest-neighbor pairs of sites (each pair counted once). If we adopt the "continuum limit" viewpoint and imagine this lattice to have a spacing $a = \theta^{-1}$ (in centimeters), then under the identification $\theta^{-d} \sum_x \approx \int d^d x$ we can write

$$H \approx \int d^d x \left[\frac{J_{FT}}{2} (\nabla\varphi)^2 + \frac{\tau_{FT}}{2} \varphi^2 + \frac{\lambda_{FT}}{4} \varphi^4 - h_{FT}\varphi \right] , \tag{15.4}$$

where we have made the definitions

$$J_{FT} = J\theta^{d-2} \tag{15.5}$$
$$\lambda_{FT} = \lambda\theta^d \tag{15.6}$$
$$h_{FT} = h\theta^d \tag{15.7}$$
$$\tau_{FT} = (\tau - 2dJ)\theta^d . \tag{15.8}$$

That is, we can imagine the lattice Hamiltonian (15.3) to be the discretization of the formal continuum Hamiltonian (15.4). Equations (15.5)–(15.8) provides a "translation dictionary" between the statistical-mechanical variables (J, λ, τ, h) and the field-theoretic variables $(J_{FT}, \lambda_{FT}, \tau_{FT}, h_{FT})$.

Note also that the lattice φ^4 model has a trivial rescaling-of-field covariance

$$\langle \varphi_{x_1} \cdots \varphi_{x_n} \rangle_{\alpha^{-2}J, \alpha^{-4}\lambda, \alpha^{-2}\tau, \alpha^{-1}h} = \alpha^n \langle \varphi_{x_1} \cdots \varphi_{x_n} \rangle_{J,\lambda,\tau,h} . \tag{15.9}$$

It follows that the introduction of the explicit field-strength renormalization factor $\alpha(\theta)$ is redundant. We are thus free to impose, without loss of generality, one normalization condition on the redundant set $(\alpha, J, \lambda, \tau, h)$. The following normalization conventions are often used:

(a) $\alpha(\theta) = 1$. The fields φ are then "renormalized fields".

(b) $J_{FT}(\theta) = 1$. The fields φ are then "canonical fields", in the sense that they formally satisfy canonical equal-time commutation relations.

(c) $J(\theta) \to J_c$ (a finite constant) as $\theta \to \infty$. This is the convention usually employed in discussions of critical phenomena in statistical mechanics. (The statistical-mechanical approach to the limit usually keeps λ and τ constant, as well. It is thus less general than the approaches studied in field theory.)

---

[3]Here we are using the field theorists' convention, which takes the *a priori* measure to be Lebesgue measure and places all other factors into the Hamiltonian (which, incidentally, field theorists call the "action"). Earlier in this monograph we put the φ_x^4 and φ_x^2 terms into the single-spin measure dP_x (as is most appropriate for proving correlation inequalities). The difference is purely one of notation.

In our discussion of φ^4 field theory (Section 15.2) *we shall adhere to the convention* $\alpha(\theta) = 1$, as it makes the formulas a bit simpler and easier to understand. In the meantime we shall keep the factor $\alpha(\theta)$ in order to allow for models without the rescaling-of-field covariance (15.9), such as the Ising model with its standard normalization $\sigma = \pm 1$.

Note, finally, that in the φ^4 model the ratios $\widehat{\lambda}_{SM} \equiv \lambda/J^2$ and $\widehat{\lambda}_{FT} \equiv \lambda_{FT}/J_{FT}^2$ are invariant under the field rescaling (15.9); they can thus serve as normalization-independent measures of the bare φ^4 coupling strength. They are related by

$$\widehat{\lambda}_{SM} = \widehat{\lambda}_{FT}\theta^{d-4} . \qquad (15.10)$$

The physics is of course the same whether one uses $\widehat{\lambda}_{SM}$ or $\widehat{\lambda}_{FT}$ as the fundamental variable, but the verbal description is somewhat different — a circumstance which has sometimes caused confusion. As practice in learning to translate between the two languages, let us consider two commonly used ways of taking the continuum limit ($\theta \to \infty$):

(a) $\widehat{\lambda}_{FT}$ *fixed.* This is the standard procedure in superrenormalizable field theory ($d < 4$). From the point of view of field theory, no nontrivial coupling-constant renormalization is being performed; and none need be performed, since the ultraviolet divergences are not so severe in $d < 4$. From the point of view of statistical mechanics, the theory is becoming extremely *weakly* coupled ($\widehat{\lambda}_{SM} \sim \theta^{d-4}$); but the effects of this coupling are amplified by the *infrared* divergences of the φ^4 lattice theory near the critical point in $d < 4$, leading to a non-Gaussian continuum limit.

(b) λ *fixed (hence* $\widehat{\lambda}_{SM} \to$ *a finite constant).* This is the standard procedure in the statistical-mechanical theory of critical phenomena. In this case it is for $d > 4$ that matters are simple: correlation functions are given by a perturbation expansion that is free of infrared divergences. From the point of view of field theory this might seem surprising, since the φ_d^4 field theory for $d > 4$ is perturbatively nonrenormalizable, with horrendous *ultraviolet* divergences; but this is mitigated by the fact that the theory is becoming extremely *weakly* coupled ($\widehat{\lambda}_{FT} \sim \theta^{4-d}$). On the other hand, for $d < 4$ the theory of critical phenomena is very complicated, by virtue of *infrared* divergences; this is reflected in field theory in the fact that the theory is becoming extremely *strongly* coupled ($\widehat{\lambda}_{FT} \sim \theta^{4-d}$), causing perturbation theory to break down even though the internal momentum integrations are ultraviolet convergent.

Clearly it is the factor θ^{d-4} in (15.10) that accounts for the interplay between ultraviolet and infrared as one translates from one viewpoint to the other.

15.1.2 Precise mathematical formulation I: The continuum limit

A continuum Euclidean quantum field theory is, by definition, a set of Schwinger distributions $\{S_n\}$ satisfying the Osterwalder-Schrader axioms. More precisely, one can ask for either of the following:

(a) Schwinger distributions $S_n \in \mathcal{S}'_{\neq}(\mathbb{R}^{nd})$ that satisfy the Osterwalder-Schrader axioms [419, 420, 418]. Here $\mathcal{S}'_{\neq}(\mathbb{R}^{nd})$ is the space of tempered distributions at *noncoinciding* arguments $x_1 \neq x_2 \neq \ldots \neq x_n \in \mathbb{R}^d$. It is the dual of $\mathcal{S}_{\neq}(\mathbb{R}^{nd})$, the space of test functions in $\mathcal{S}(\mathbb{R}^{nd})$ that vanish together with all their derivatives on each hyperplane $x_i - x_j = 0$ $(1 \leq i < j \leq n)$.

(b) Extended Schwinger distributions $S_n \in \mathcal{S}'(\mathbb{R}^{nd})$ that satisfy the extended Osterwalder-Schrader axioms [146, 159]. Here the S_n must be defined also at *coinciding* arguments.

Each of these axiom schemes is "physically natural" for certain purposes:

(a′) A Gårding-Wightman relativistic quantum field theory determines, by analytic continuation, the Euclidean Green's functions (Schwinger functions) only at *noncoinciding* arguments;and conversely, only the Schwinger distributions at noncoinciding arguments play any role in the Osterwalder-Schrader reconstruction of the relativistic quantum field theory.

(b′) Eckmann and Epstein [217, 159] have set out axioms for time-ordered products in relativistic quantum field theory; these axioms are *stronger* than the Gårding-Wightman axioms. An Eckmann-Epstein relativistic quantum field theory determines, by analytic continuation, the Schwinger distributions also at *coinciding* arguments; and conversely, these extended Schwinger distributions (which must satisfy an extended form of Osterwalder-Schrader positivity) are *required* for the reconstruction of the time-ordered vacuum expectation values in Minkowski space.

Thus, what one demands in Euclidean space depends on what one wants to reconstruct in Minkowski space. Of course, from the point of view of probability theory (i.e. classical statistical mechanics), the extended OS axioms are more natural.

We now describe how to obtain continuum Schwinger distributions as a limit of rescaled lattice correlation functions. The first step is to make the lattice theories "live" in the continuum configuration space. The lattice n-point function

$$S_n(x_1, \ldots, x_n) = \langle \varphi_{x_1} \cdots \varphi_{x_n} \rangle \tag{15.11}$$

is an ordinary (pointwise-defined) function, defined for $x_1, \ldots, x_n \in \mathbb{Z}^d$; similarly, the rescaled lattice n-point function

$$S_n^\theta(x_1, \ldots, x_n) = \alpha^n \langle \varphi_{\theta x_1} \cdots \varphi_{\theta x_n} \rangle \tag{15.12}$$

is an ordinary (pointwise-defined) function, defined for $x_1, \ldots, x_n \in \theta^{-1}\mathbb{Z}^d$. On the other hand, a continuum n-point Schwinger distribution is a distribution in the sense of Schwartz, i.e. an element of $\mathcal{D}'(\mathbb{R}^{nd})$ [or $\mathcal{S}'(\mathbb{R}^{nd})$]. There are many ways to associate a distribution in $\mathcal{D}'(\mathbb{R}^{nd})$ with a given function on $\theta^{-1}\mathbb{Z}^d$; the simplest is as a sum of delta functions.[4] That is, we let $S_n^{\theta,\mathrm{dist}}$ be the distribution

$$S_n^{\theta,\mathrm{dist}} = \alpha^n \theta^{-nd} \sum_{x_1,\ldots,x_n \in \theta^{-1}\mathbb{Z}^d} \langle \varphi_{\theta x_1} \cdots \varphi_{\theta x_n} \rangle \, \delta_{(x_1,\ldots,x_n)} \tag{15.13}$$

where δ_y denotes the delta function located at $y \in \mathbb{R}^{nd}$. In other words, if $f_1, \ldots, f_n$ are test functions in $\mathcal{D}(\mathbb{R}^{nd})$, then

$$\begin{aligned} S_n^{\theta,\mathrm{dist}}&(f_1, \ldots, f_n) \\ &= \alpha^n \theta^{-nd} \sum_{x_1,\ldots,x_n \in \mathbb{Z}^d} f_1(\theta^{-1}x_1) \ldots f_n(\theta^{-1}x_n) \, \langle \varphi_{x_1} \cdots \varphi_{x_n} \rangle \, . \end{aligned} \tag{15.14}$$

Clearly $S_n^{\theta,\mathrm{dist}}$ is a well-defined distribution; and if the lattice correlation functions are polynomially bounded in x-space (as they obviously are in any translation-invariant theory), then the $S_n^{\theta,\mathrm{dist}}$ are tempered, i.e. belong to $\mathcal{S}'(\mathbb{R}^{nd})$.

This construction has also a probabilistic interpretation. A lattice field configuration is a sequence $\varphi = \{\varphi_x\}_{x \in \mathbb{Z}^d}$, i.e. an element of $\mathbb{R}^{\mathbb{Z}^d}$. A continuum field configuration is a distribution in the sense of Schwartz, i.e. an element of $\mathcal{D}'(\mathbb{R}^d)$ [or $\mathcal{S}'(\mathbb{R}^d)$]. There are many ways to associate a continuum field configuration with a given lattice field configuration; the simplest is as a sum of delta functions. So let θ and α be strictly positive real numbers; we call θ the *length rescaling factor* (corresponding to the lattice spacing $a = \theta^{-1}$) and α the *field-strength rescaling factor*. We then define a map $P_{\theta,\alpha}$: $\mathbb{R}^{\mathbb{Z}^d} \to \mathcal{D}'(\mathbb{R}^d)$ by

$$P_{\theta,\alpha}\varphi = \alpha\theta^{-d} \sum_{x \in \mathbb{Z}^d} \varphi_x \, \delta_{\theta^{-1}x} \, . \tag{15.15}$$

Note also that $P_{\theta,\alpha}$ maps $s'(\mathbb{Z}^d)$, the space of polynomially bounded sequences, into $\mathcal{S}'(\mathbb{R}^d)$, the space of tempered distributions. Then $P_{\theta,\alpha}$ carries each probability measure μ on $\mathbb{R}^{\mathbb{Z}^d}$ [or $s'(\mathbb{Z}^d)$] into a probability measure $\mu^{\theta,\alpha} = \mu \circ P_{\theta,\alpha}^{-1}$ on $\mathcal{D}'(\mathbb{R}^d)$ [or $\mathcal{S}'(\mathbb{R}^d)$]. The moments of $\mu^{\theta,\alpha}$ are precisely the rescaled lattice Schwinger distributions $S_n^{\theta,\mathrm{dist}}$ defined in (15.13)–(15.14). That is,

$$S_n^{\theta,\mathrm{dist}}(f_1, \ldots, f_n) = \int\limits_{\mathcal{D}'(\mathbb{R}^d)} \varphi(f_1) \ldots \varphi(f_n) \, d\mu^{\theta,\alpha}(\varphi) \, , \tag{15.16}$$

as can be easily verified.

---

[4] Alternatively one could use piecewise-constant interpolation, piecewise-linear interpolation, etc. The only subtle point is the preservation of Osterwalder-Schrader positivity.

Now let $\{\mu_m\}$ be a sequence of lattice probability measures, and let $\{\theta_m\}$ and $\{\alpha_m\}$ be a sequence of length and field-strength rescaling factors. The rescaled lattice Schwinger functions

$$S_n^{(\mu_m,\theta_m,\alpha_m),\text{dist}}(f_1,\ldots,f_n) = \int\limits_{\mathcal{D}'(\mathbb{R}^d)} \varphi(f_1)\ldots\varphi(f_n)\, d\mu_m^{\theta_m,\alpha_m}(\varphi) \qquad (15.17)$$

may or may not converge as $m \to \infty$. If they do converge (for all n), we say that the *continuum limit exists* along the sequence $\{(\mu_m,\theta_m,\alpha_m)\}$ of lattice theories, and we denote the limit S_n^{cont}. We could ask for convergence in either of two senses:

(a) Weak convergence in the space $\mathcal{S}'_{\neq}(\mathbb{R}^{nd})$ of tempered distributions at *noncoinciding* arguments. This means that $S_n^{(\mu_m,\theta_m,\alpha_m),\text{dist}}(f_1,\ldots,f_n)$ converges for all choices of test functions $f_1,\ldots,f_n \in \mathcal{S}(\mathbb{R}^d)$ with *nonoverlapping supports*. The limit is then a distribution $S_n^{\text{cont}} \in \mathcal{S}'_{\neq}(\mathbb{R}^d)$ [457, pp. 74–75].

(b) Weak convergence in the space $\mathcal{S}'(\mathbb{R}^{nd})$ of tempered distributions. This means that $S_n^{(\mu_m,\theta_m,\alpha_m),\text{dist}}(f_1,\ldots,f_n)$ converges for all choices of test functions $f_1,\ldots,f_n \in \mathcal{S}(\mathbb{R}^d)$. The limit is then a distribution $S_n^{\text{cont}} \in \mathcal{S}'(\mathbb{R}^d)$ [457, pp. 74–75].

Technical remark. Weak convergence of sequences (*not* nets!) in $\mathcal{S}'(\mathbb{R}^{nd})$ is equivalent to strong convergence in $\mathcal{S}'(\mathbb{R}^{nd})$ [441, Theorem V.26]; and likewise for $\mathcal{S}'_{\neq}(\mathbb{R}^{nd})$.

The next step is to check the Osterwalder-Schrader (or Eckmann-Epstein) axioms for the continuum theory $\{S_n^{\text{cont}}\}$:

(OS0) *Temperedness and growth condition.* The temperedness is obvious by construction. The growth condition has to be checked case-by-case, but it certainly holds for any model satisfying the Gaussian inequalities [255, 411, 412].

(OS1) *Euclidean invariance.* We show below that the $\{S_n^{\text{cont}}\}$ are translation-invariant whenever the lattice spacings $a_m \equiv \theta_m^{-1}$ tend to zero. This result is not entirely trivial, but it is physically reasonable: the $a_m\mathbb{Z}^d$-translation invariance of the rescaled lattice theories goes over to an $\mathbb{R}^d$-translation invariance in the continuum limit. The rotation invariance of the $\{S_n^{\text{cont}}\}$ is, however, a much more subtle question. The lattice theories are obviously *not* invariant under the rotation group $SO(d)$, but only under its discrete hypercubic subgroup. It is plausible that $SO(d)$ invariance might be restored in the continuum limit, but this is far from assured. Indeed, the restoration of rotation invariance in the continuum limit is a detailed dynamical question: it depends on whether the hypercubic-invariant $SO(d)$-breaking operators are "relevant" or "irrelevant" in the sense of the renormalization group at the appropriate RG fixed point. For "normal" models (such as φ^4), the continuum limit is expected to be rotation-invariant, but there do exist examples of "pathological" models in which the continuum limit seems not to be rotation-invariant [329]. All lattice-based work in constructive quantum field theory must face, therefore, the difficult problem of *proving* rotation invariance of the continuum-limit theory (in those cases where it in fact holds!): this requires rather strong control over the approach to the continuum limit. See [339] for a beautiful idea which works in at least one case. On the other hand, lattice-based work in *destructive* quantum field theory (i.e. triviality proofs) usually need not worry about rotation invariance: one simply proves that the continuum-limit theory is Gaussian *whether or not* it is rotation-invariant.

(OS2) *Osterwalder-Schrader positivity.* If the lattice theories are reflection-positive with respect to the hyperplane $x_1 = 0$, then it is easy to see that the continuum-limit theory will be reflection-positive with respect to this same hyperplane. If, instead, the lattice theories are reflection-positive with respect to the hyperplane $x_1 = -\frac{1}{2}$, then with a little work one can again show reflection-positivity of the continuum-limit theory with respect to the hyperplane $x_1 = 0$, *provided* that the lattice spacings $a_m = \theta_m^{-1}$ tend to zero.

(OS3) *Permutation symmetry.* This is trivial.

We now give the detailed proof of translation invariance:

Proposition 15.1 *Let $\{\mu_m\}$ be a sequence of translation-invariant lattice models, and consider a continuum limit $\{\theta_m, \alpha_m\}$ in which $\lim_{m\to\infty} \theta_m^{-1} = 0$. Suppose that, for some given n, the rescaled n-point function $S_n^{(\mu_m,\theta_m,\alpha_m),\mathrm{dist}}$ converges in $\mathcal{D}'(\mathbb{R}^{nd})$ [resp. $\mathcal{D}'_{\neq}(\mathbb{R}^{nd})$] to a limiting distribution S_n^{cont}. Then S_n^{cont} is translation-invariant.*

Proof. Fix $f_1, \ldots, f_n \in \mathcal{D}(\mathbb{R}^d)$ and $b \in \mathbb{R}^d$; our goal is to show that

$$S_n^{\mathrm{cont}}(f_1, \ldots, f_n) \;=\; S_n^{\mathrm{cont}}(f_1^{(b)}, \ldots, f_n^{(b)}) \,, \tag{15.18}$$

where

$$f^{(b)}(x) \;\equiv\; f(x-b) \,. \tag{15.19}$$

Since $a_m \equiv \theta_m^{-1} \to 0$, there exists a sequence of $b_m \in a_m\mathbb{Z}^d$ with $b_m \to b$. By the $\mathbb{Z}^d$-translation invariance of the lattice theories $\{\mu_m\}$, we have

$$S_n^{(\mu_m,\theta_m,\alpha_m),\mathrm{dist}}(f_1, \ldots, f_n) \;=\; S_n^{(\mu_m,\theta_m,\alpha_m),\mathrm{dist}}(f_1^{(b_m)}, \ldots, f_n^{(b_m)}) \,. \tag{15.20}$$

On the other hand, it is easy to show that $f_i^{(b_m)} \to f_i^{(b)}$ in the topology of $\mathcal{D}(\mathbb{R}^d)$. If we now let $m \to \infty$ in (15.20), we obtain (15.18), by resorting on the right-hand side to a standard lemma from the theory of distributions [446, Theorem 6.18].

The proof for $\mathcal{D}'_{\neq}(\mathbb{R}^{nd})$ is identical: we restrict attention to $f_1, \ldots, f_n$ with nonoverlapping supports, and note that their translates also have this property. ∎

15.1.3 Precise mathematical formulation II: Necessary conditions for non-boring limit

We start by discussing some general results relating "distributional" to "pointwise" convergence; these will play an important role in the remainder of this subsection, as well as in the φ^4 triviality proof (Section 15.2). The first proposition shows that locally uniform pointwise convergence to a continuous function (e.g. zero) implies distributional convergence; the second proposition shows, in a much more restricted setting, that distributional convergence implies pointwise boundedness.

Proposition 15.2 *Let $\{S^{\theta_m}\}$ be a sequence of functions defined on $\theta_m^{-1}\mathbb{Z}^N$, and let $\{S^{\theta_m,\text{dist}}\}$ be the corresponding distributions, i.e.*

$$S^{\theta_m,\text{dist}} = \theta_m^{-N} \sum_{y\in\theta_m^{-1}\mathbb{Z}^N} S^{\theta_m}(y)\delta_y \ . \tag{15.21}$$

Let T be a continuous function on $\mathbb{R}^N$. Now let Ω be an open subset of $\mathbb{R}^N$, and suppose that $\{S^{\theta_m}\}$ converges to T uniformly on compact subsets of Ω, i.e.

$$\lim_{m\to\infty} \sup_{y\in K\cap\theta_m^{-1}\mathbb{Z}^N} |S^{\theta_m}(y) - T(y)| = 0 \tag{15.22}$$

for each compact $K \subset \Omega$. Then $\{S^{\theta_m,\text{dist}}\}$ converges to T in the sense of distributions on Ω, i.e. in the space $\mathcal{D}'(\Omega)$.

Proof. Let f be a test function with compact support $K \subset \Omega$ [i.e. $f \in \mathcal{D}(\Omega)$]. Then

$$\begin{aligned} S^{\theta_m,\text{dist}}(f) - \int T(y)f(y)\,dy = \\ \left[\theta_m^{-N} \sum_{y\in\theta_m^{-1}\mathbb{Z}^N} \Big[S^{\theta_m}(y) - T(y)\Big] f(y)\right] \\ + \left[\theta_m^{-N} \sum_{y\in\theta_m^{-1}\mathbb{Z}^N} T(y)f(y) - \int T(y)f(y)\,dy\right] . \end{aligned} \tag{15.23}$$

The first term tends to zero by the hypothesis (15.22); the second term tends to zero by the usual theorem on the convergence of a Riemann sum to the corresponding integral. ■

Example. Let $N = nd$, and let

$$\Omega = \{(x_1, \dots, x_n)\colon x_i \neq x_j \text{ for all } i \neq j\} \ . \tag{15.24}$$

Then $\mathcal{D}'(\Omega) = \mathcal{D}'_{\neq}(\mathbb{R}^{nd})$, the space of distributions at noncoinciding arguments. Now suppose that $\{S^{\theta_m}\}$ converges to zero uniformly on

$$\Omega_\varepsilon = \{(x_1, \ldots, x_n) \colon |x_i - x_j| \geq \varepsilon \text{ for all } i \neq j\} \tag{15.25}$$

for each $\varepsilon > 0$. Then $\{S^{\theta_m,\text{dist}}\}$ converges to zero in $\mathcal{D}'_{\neq}(\mathbb{R}^{nd})$. This result will be used in the next section.

Remark. Under slightly stronger hypotheses, one can prove convergence in $\mathcal{S}'(\Omega)$, the space of *tempered* distributions on Ω.

Proposition 15.3 *Let $\{\mu_m\}$ be a sequence of nearest-neighbor ferromagnetic models. Consider a continuum limit with length rescaling factors $\{\theta_m\} \to \infty$ and field-strength rescaling factors $\{\alpha_m\}$, and suppose that the rescaled lattice two-point truncated Schwinger distributions $S_{2,T}^{(\mu_m,\theta_m,\alpha_m),\text{dist}}$ converge in the space of distributions at noncoinciding arguments to a limiting distribution $S_{2,T}^{\text{cont}}$. Then the pointwise-defined functions*

$$S_{2,T}^{(\mu_m,\theta_m,\alpha_m)}(x) = \alpha_m^2 \langle \varphi_0 \,;\, \varphi_{\theta_m x} \rangle_{\mu_m} , \tag{15.26}$$

defined for $x \in \theta_m^{-1}\mathbb{Z}^d \subset \mathbb{R}^d$, are bounded uniformly on $|x| \geq \varepsilon$, for any fixed $\varepsilon > 0$ — that is,

$$\sup_m \sup_{\substack{x \in \theta_m^{-1}\mathbb{Z}^d \\ |x| \geq \varepsilon}} \alpha_m^2 \langle \varphi_0 \,;\, \varphi_{\theta_m x} \rangle_{\mu_m} < \infty . \tag{15.27}$$

Proof. The Schrader–Messager–Miracle-Solé inequality (13.66) implies that $S_{2,T}$ is "almost symmetric decreasing", in the sense that if $x, y \in \theta_m^{-1}\mathbb{Z}^d$ with $|y| \leq d^{-1}x$, then

$$\langle \varphi_0 \,;\, \varphi_{\theta_m y} \rangle_{\mu_m} \geq \langle \varphi_0 \,;\, \varphi_{\theta_m x} \rangle_{\mu_m}$$

(see e.g. [481, Appendix A]). In particular, this holds if $x, y \in \theta_m^{-1}\mathbb{Z}^d$ with $|x| \geq \varepsilon$ and $|y| \leq d^{-1}\varepsilon$. Therefore, if there were a subsequence $\{m_i\}$ such that

$$\lim_{i \to \infty} \sup_{\substack{x \in \theta_m^{-1}\mathbb{Z}^d \\ |x| \geq \varepsilon}} \alpha_{m_i}^2 \langle \varphi_0 \,;\, \varphi_{\theta_{m_i} x} \rangle_{\mu_{m_i}} = +\infty ,$$

then we would also have

$$\lim_{i \to \infty} \inf_{\substack{x \in \theta_m^{-1}\mathbb{Z}^d \\ |x| \leq d^{-1}\varepsilon}} \alpha_{m_i}^2 \langle \varphi_0 \,;\, \varphi_{\theta_{m_i} x} \rangle_{\mu_{m_i}} = +\infty ,$$

i.e. $S_{2,T}^{(\mu_{m_i},\theta_{m_i},\alpha_{m_i})} \to +\infty$ *uniformly* on $\{x \colon |x| \leq d^{-1}\varepsilon\}$. But this would contradict the assumed convergence of $S_{2,T}^{(\mu_m,\theta_m,\alpha_m),\text{dist}}$ in the sense of distributions at noncoinciding arguments (just consider a test function $f \geq 0$ supported in the open set $\{x \colon 0 < |x| < d^{-1}\varepsilon\}$). ∎

Remark. This proposition is expected to hold for *any* physically reasonable model. But unfortunately we have been unable to prove it except under the restrictive hypotheses of the Schrader–Messager–Miracle-Solé correlation inequality.

Consider now a sequence of lattice theories $\{\mu_m\}$; we ask which choices $\{(\theta_m, \alpha_m)\}$ of length- and field-strength rescaling factors could lead to sensible continuum limits. The next three propositions deal with this question:

1. It was argued heuristically in Section 15.1.1 that the length rescaling factor θ_m cannot be $\gg$ the correlation length ξ_m of the lattice model (measured in lattice spacings), since otherwise the rescaled theories would have correlation lengths $\xi_m^\theta \equiv \theta_m^{-1}\xi_m$ which tend to zero (measured in centimeters). Proposition 15.4 makes this reasoning precise.

2. The infrared bound places a limitation on the allowable choices of field-strength renormalization: if $\alpha(\theta)$ is too small, then the correlation functions tend to zero. Proposition 15.5 makes this precise.

3. If we insist that the rescaled susceptibilities $\chi_m^\theta \equiv \alpha_m^2\theta_m^{-d}\chi_m$ be *bounded* (as is typical in a *massive scaling limit*), then θ_m cannot be $\ll (J_m\chi_m)^{1/2}$. Proposition 15.6 makes this precise.

Propositions 15.4 and 15.6 together show that we must take

$$(J_m\chi_m)^{1/2} \lesssim \theta_m \lesssim \xi_m \tag{15.28}$$

to obtain a non-boring massive scaling limit. [Note that $(J\chi)^{1/2} \leq \text{const} \times \xi$ by reflection positivity and the infrared bound [cf. (13.88)], and that in dimension $d > 4$ it is expected that $(J\chi)^{1/2} \sim \xi$.] In fact we believe that the only sensible choice for a massive scaling limit (in *any* dimension) is $\theta_m \sim \xi_m$, but we do not know how to prove this.

Proposition 15.4 *Let $\{\mu_m\}$ be a sequence of nearest-neighbor ferromagnetic models, with correlation lengths*

$$\xi_m \equiv \limsup_{|x|\to\infty} \frac{-|x|}{\log\langle\varphi_0\,;\varphi_x\rangle_{\mu_m}} . \tag{15.29}$$

Consider a continuum limit with length rescaling factors $\{\theta_m\} \to \infty$ and field-strength rescaling factors $\{\alpha_m\}$, and suppose that $\lim_{m\to\infty} \theta_m^{-1}\xi_m = 0$. Suppose, further, that the rescaled truncated two-point distributions $S_{2,T}^{(\mu_m,\theta_m,\alpha_m),\text{dist}}$ converge in the space $\mathcal{D}'_{\neq}(\mathbb{R}^{nd})$ of distributions at noncoinciding arguments to a limiting distribution $S_{2,T}^{\text{cont}}$. Then:

(a) $S_{2,T}^{\text{cont}} \equiv 0$ (at noncoinciding arguments).

(b) Suppose, in addition, that the lattice theories are even and satisfy the Griffiths I and Gaussian inequalities. Then, for each $n > 0$, the rescaled n-point function $S_n^{(\mu_m,\theta_m,\alpha_m),\text{dist}}$ converges to zero in the space $\mathcal{D}'_{\neq}(\mathbb{R}^{nd})$ of distributions at noncoinciding arguments.

Proof. (a) Using reflection positivity in lattice hyperplanes (e.g. $x_1 = 0$) and bisector hyperplanes (e.g. $x_1 = -\frac{1}{2}$), it is not hard to show (see Section 13.4) that

$$\langle \varphi_0 \,;\, \varphi_{(x_1,x_2,\ldots,x_d)} \rangle \;\le\; \langle \varphi_0 \,;\, \varphi_{(|x_1|,0,\ldots,0)} \rangle \tag{15.30}$$

and

$$\langle \varphi_0 \,;\, \varphi_{(x_1,0,\ldots,0)} \rangle \;=\; \int_0^{e^{-1/\xi}} d\varrho(\lambda)\, \lambda^{|x_1|} \,, \tag{15.31}$$

where $d\varrho(\lambda)$ is a positive measure. Analogous representations hold in each of the d coordinate directions. It follows from (15.31) and permutations that

$$\langle \varphi_0 \,;\, \varphi_{be_i} \rangle \;\le\; e^{-(|b|-|c|)/\xi} \, \langle \varphi_0 \,;\, \varphi_{ce_i} \rangle \tag{15.32}$$

for $|b| \ge |c|$ and $1 \le i \le d$. Combining this with (15.30) and permutations, we conclude that

$$\langle \varphi_0 \,;\, \varphi_x \rangle \;\le\; e^{-(|x|_\infty - c)/\xi} \max_{1 \le i \le d} \langle \varphi_0 \,;\, \varphi_{ce_i} \rangle \tag{15.33}$$

for $|x|_\infty \equiv \max\limits_{1 \le i \le d} |x_i| \ge c \ge 0$. Now apply this to the functions

$$S_{2,T}^{(\mu_m,\theta_m,\alpha_m)}(x) \;\equiv\; \alpha_m^2 \langle \varphi_0 \,;\, \varphi_{\theta_m x} \rangle_{\mu_m} \,, \tag{15.34}$$

which are defined for $x \in \theta_m^{-1}\mathbb{Z}^d$. Fix $\varepsilon > 0$; then Proposition 15.3 guarantees that $S_{2,T}^{(\mu_m,\theta_m,\alpha_m)}(x)$ is bounded uniformly on $|x| \ge \varepsilon$. Taking $c_m = \lceil \theta_m \varepsilon \rceil \le \theta_m \varepsilon + 1$ and using (15.33), we conclude that

$$S_{2,T}^{(\mu_m,\theta_m,\alpha_m)}(x) \;\le\; \text{const} \times e^{-\theta_m(|x|_\infty - \varepsilon - \theta_m^{-1})/\xi_m} \tag{15.35}$$

uniformly in m, for $|x|_\infty \ge \varepsilon + \theta_m^{-1}$. Since $\lim\limits_{m\to\infty} \theta_m^{-1} = 0$ and $\lim\limits_{m\to\infty} \theta_m \xi_m^{-1} = +\infty$, it follows that

$$\lim_{m\to\infty} S_{2,T}^{(\mu_m,\theta_m,\alpha_m)}(x) \;=\; 0 \tag{15.36}$$

uniformly for $|x|_\infty \ge 2\varepsilon$. But since $\varepsilon > 0$ is arbitrary, Proposition 15.2 implies that $S_{2,T}^{(\mu_m,\theta_m,\alpha_m),\text{dist}} \to 0$ in $\mathcal{D}'_{\neq}$.

(b) The Griffiths I and Gaussian inequalities imply that

$$0 \;\le\; S_n^{(\mu_m,\theta_m,\alpha_m)}(x_1,\ldots,x_n) \;\le\; \sum_{\text{pairings}} \prod S_2^{(\mu_m,\theta_m,\alpha_m)}(x_i,x_j) \,. \tag{15.37}$$

So the claim is an immediate consequence of (15.36) and Proposition 15.2. ∎

The infrared bound (13.86)/(13.87) places a limitation on the allowable choices of field-strength renormalization if we are to have a non-boring continuum limit. For simplicity we restrict attention to nearest-neighbor ferromagnetic models (certain long-range reflection-positive ferromagnetic models could be considered as well). *Formally* the argument goes as follows:

$$\begin{aligned} S_{2,T}^{\mathrm{cont}}(x) &= \lim_{\theta\to\infty} \alpha^2 \langle \varphi_0 ; \varphi_{\theta x}\rangle \\ &\leq \lim_{\theta\to\infty} \alpha^2 \frac{\mathrm{const}}{J|\theta x|^{d-2}} \qquad \text{by the infrared bound} \\ &= 0 \qquad \text{if } \alpha^2 J^{-1}\theta^{2-d} \to 0 \end{aligned} \tag{15.38}$$

For the rigorous argument, all we have to do is smear everything with test functions f and g:

Proposition 15.5 *Let $d > 2$, and let $\{\mu_m\}$ be a sequence of nearest-neighbor ferromagnetic models with nearest-neighbor coupling strengths $\{J_m\}$. Consider now a continuum limit with length rescaling factors $\{\theta_m\} \to \infty$ and field-strength rescaling factors $\{\alpha_m\}$, and suppose that $\lim_{m\to\infty} \alpha_m^2 J_m^{-1} \theta_m^{2-d} = 0$. Then:*

(a) The rescaled truncated two-point functions $S_{2,T}^{(\mu_m,\theta_m,\alpha_m),\mathrm{dist}}$ converge to zero *in $\mathcal{S}'(\mathbb{R}^{2d})$.*

(b) Suppose, in addition, that the lattice theories are even and satisfy the Griffiths I and Gaussian inequalities. Then, for each $n > 0$, the rescaled n-point function $S_n^{(\mu_m,\theta_m,\alpha_m),\mathrm{dist}}$ converges to zero in $\mathcal{S}'(\mathbb{R}^{nd})$.

Proof. The x-space infrared bound (13.87) [which is valid for nearest-neighbor ferromagnets in $d > 2$ with arbitrary single-spin measure] states that

$$0 \leq \langle \varphi_0 ; \varphi_x \rangle \leq \frac{\mathrm{const}}{J(1+|x|)^{d-2}} \,. \tag{15.39}$$

Now let $f, g \in \mathcal{S}(\mathbb{R}^d)$. By definition (15.14), we have

$$S_{2,T}^{\theta,\mathrm{dist}}(f,g) = \alpha^2 \theta^{-2d} \sum_{x,y\in\theta^{-1}\mathbb{Z}^d} f(x)g(y)\, \langle \varphi_{\theta x} ; \varphi_{\theta y}\rangle \,. \tag{15.40}$$

Inserting the bound (15.39), we find

$$\left| S_{2,T}^{\theta,\mathrm{dist}}(f,g)\right| \leq \mathrm{const} \times \alpha^2 J^{-1} \theta^{2-d} \times \left[\theta^{-2d} \sum_{x,y\in\theta^{-1}\mathbb{Z}^d} f(x)\, g(y)\, (\theta^{-1} + |x-y|)^{-(d-2)} \right] . \tag{15.41}$$

As $\theta \to \infty$, the sum in square brackets approaches the integral

$$\int_{\mathbb{R}^d} dx \int_{\mathbb{R}^d} dy\; f(x)g(y)\; |x-y|^{-(d-2)} \,, \tag{15.42}$$

which is finite. Since $\alpha^2 J^{-1}\theta^{2-d} \to 0$ by hypothesis, this proves part (a). Part (b) is proven as in Proposition 15.4. ∎

Remark. This proposition can also be proven using the p-space infrared bound [222, 218]

$$0 \le \widetilde{G}(p) \le \left[2J\sum_{i=1}^{d}(1-\cos p_i)\right]^{-1}, \tag{15.43}$$

which is somewhat more "elementary" (i.e., doesn't require the Schrader–Messager–Miracle-Solé inequality). But the proof is a bit messier.

If we use the convention $\alpha \equiv 1$, then Proposition 15.5 says that $J_{FT} \equiv J\theta^{d-2}$ *must stay bounded* if we are to obtain a non-boring continuum limit.

Proposition 15.6 *Let $d > 2$, and let $\{\mu_m\}$ be a sequence of nearest-neighbor ferromagnetic models with nearest-neighbor coupling strengths $\{J_m\}$ and susceptibilities $\{\chi_m\}$. Consider now a continuum limit with length rescaling factors $\{\theta_m\} \to \infty$ and field-strength rescaling factors $\{\alpha_m\}$. Suppose that $\chi_m^\theta \equiv \alpha_m^2\theta_m^{-d}\chi_m$ is* bounded *as $m \to \infty$, and that $\lim_{m\to\infty} \theta_m(J_m\chi_m)^{-1/2} = 0$. Then $\lim_{m\to\infty} \alpha_m^2 J_m^{-1}\theta_m^{2-d} = 0$, so that*

$$S_{2,T}^{(\mu_m,\theta_m,\alpha_m),\mathrm{dist}} \to 0$$

in $\mathcal{S}'(\mathbb{R}^{2d})$.

Proof.

$$\frac{\alpha^2}{J\theta^{d-2}} = \chi^\theta \frac{\theta^2}{J\chi} \to 0$$

as $m \to \infty$ by hypothesis. The rest follows from Proposition 15.5. ∎

15.1.4 Precise mathematical formulation III: Convergence of summed quantities

The next question is whether the rescaled *summed* lattice quantities converge to their continuum counterparts. For example, it is natural to expect that the rescaled lattice susceptibilities

$$\chi^\theta \equiv \int d^dx\; S_{2,T}^\theta(0,x) \equiv \theta^{-d} \sum_{x\in\theta^{-1}\mathbb{Z}^d} S_{2,T}^\theta(0,x) \equiv \alpha^2\theta^{-d}\chi \tag{15.44}$$

should converge to the continuum susceptibility

$$\chi^{\text{cont}} \equiv \int d^d x \, S_{2,T}^{\text{cont}}(0,x) \, . \tag{15.45}$$

Do they?

The first problem is to make sense of the expression (15.45): since the constant function does not belong to $\mathcal{S}(\mathbb{R}^d)$, it is not *a priori* valid to integrate the distribution $S_{2,T,\text{diff}}^{\text{cont}}(x) \equiv S_{2,T}^{\text{cont}}(0,x)$ over all space. However, let us suppose that:

(a) The lattice theories satisfy $S_{2,T}^{\theta} \geq 0$ (e.g. by the Griffiths II or FKG inequality).

(b) The convergence $S_{2,T}^{\theta} \to S_{2,T}^{\text{cont}}$ takes place in $\mathcal{D}'$, the space of distributions defined at *all* (not just noncoinciding) arguments.

Then it follows that $S_{2,T}^{\text{cont}} \geq 0$; but since every positive distribution is a positive *measure* [457, p. 29], we can integrate *any* positive function on it (although the result may be $+\infty$). It follows that χ^{cont} is well-defined and satisfies $0 \leq \chi^{\text{cont}} \leq +\infty$.

Under these same hypotheses, we can study the convergence of χ^{θ} to χ^{cont}. The main tool is the following measure-theoretic lemma:

Lemma 15.7 *Let $\{\nu_m\}$, ν be positive measures on $\mathbb{R}^k$, and suppose that $\nu_m \to \nu$ in $\mathcal{D}'(\mathbb{R}^k)$ [this means that $\int f \, d\nu_m \to \int f \, d\nu$ for all f which are infinitely differentiable and of compact support]. Then:*

(a) $\int f \, d\nu_m \to \int f \, d\nu$ for all f which are continuous and of compact support.

(b) $0 \leq \int f \, d\nu \leq \liminf_{m\to\infty} \int f \, d\nu_m$ for all f which are nonnegative and continuous.

(c) Let f be continuous, and suppose that there exists a function $F \geq 0$ which goes to $+\infty$ at infinity, such that $\limsup_{m\to\infty} \int |f| F \, d\nu_m < \infty$. Then $\int f \, d\nu_m \to \int f \, d\nu$.

[Here (a) says that, for positive measures, distributional convergence implies vague convergence; (b) is a kind of "Fatou lemma" for measures; and (c) is a sufficient condition for the "non-escape of mass to infinity".]

Proof. (a) Let K be any compact set in $\mathbb{R}^k$, and let $h_K \in \mathcal{D}$ be a nonnegative function which is equal to 1 on K. Then $\nu_m(K) \leq \int h_K \, d\nu_m \to \int h_K \, d\nu < \infty$, so the sequence $\{\nu_m(K)\}$ is bounded. Similarly, $\nu(K) \leq \int h_K \, d\nu < \infty$. So for each compact K, there exists $M_K < \infty$ such that $\nu_m(K) \leq M_K$ for all m and $\nu(K) \leq M_K$.

Now let f be continuous and of compact support, and let K be a compact set whose interior contains the support of f. Then for any $\varepsilon > 0$, there exists $g \in \mathcal{D}$ with $\operatorname{supp} g \subset K$ such that $\|f - g\|_\infty \leq \varepsilon$. We then have

$$\begin{aligned}\left|\int f\,d\nu_m - \int f\,d\nu\right| &\le \left|\int g\,d\nu_m - \int g\,d\nu\right| + \left|\int (f-g)\,d\nu_m\right| + \left|\int (f-g)\,d\nu\right| \\ &\le \left|\int g\,d\nu_m - \int g\,d\nu\right| + 2\varepsilon M_K\,,\end{aligned}$$

so that $\limsup\limits_{m\to\infty} |\int f\,d\nu_m - \int f\,d\nu| \le 2\varepsilon M_K$. Since ε is arbitrary, we have $\int f\,d\nu_m \to \int f\,d\nu$.

(b) Let f be nonnegative and continuous, and let $0 \le g_1 \le g_2 \le \ldots$ be an increasing sequence of continuous functions of compact support such that $g_n \uparrow f$ pointwise. Then $\int f\,d\nu = \lim\limits_{n\to\infty} \int g_n\,d\nu = \sup\limits_n \int g_n\,d\nu$ by the monotone convergence theorem. On the other hand, by (a) we have for each n

$$\int g_n\,d\nu = \lim_{m\to\infty} \int g_n\,d\nu_m \le \liminf_{m\to\infty} \int f\,d\nu_m\,,$$

which proves the claim.

(c) It is not hard to see that there exists a *continuous* function $G \le F$ having all the properties required of F. [Let $g_1(r) = \inf\limits_{|x|\ge r} F(x)$, which is nonnegative, increasing and tends to $+\infty$ as $r \to +\infty$; let $g_2(r) = \left\{\begin{array}{ll} \min\bigl(r, g_1(r)\bigr) & \text{if } r \ge 0 \\ 0 & \text{if } r < 0 \end{array}\right\}$, which in addition is everywhere finite; and let $g_3(r) = \int_{r-1}^{r} g_2(r')\,dr'$, which in addition is continuous. Now let $G(x) = g_3(|x|)$.] So we can assume without loss of generality that F is continuous. It then follows from (b) that $\int |f| F\,d\nu \le \limsup\limits_{m\to\infty} \int |f| F\,d\nu_m \equiv M < \infty$.

Now fix $\varepsilon > 0$, and choose $R < \infty$ such that $F(x) \ge 1/\varepsilon$ whenever $|x| > R$. Let

$$g(x) = \begin{cases} f(x) & \text{for } |x| \le R \\ (R+1-|x|)f(x) & \text{for } R \le |x| \le R+1 \\ 0 & \text{for } |x| \ge R+1 \end{cases}$$

Clearly g is continuous and of compact support, and satisfies $|f(x) - g(x)| \le \varepsilon |f(x)| F(x)$ for all x. We then have

$$\begin{aligned}\limsup_{m\to\infty}\left|\int f\,d\nu_m - \int f\,d\nu\right| &\le \limsup_{m\to\infty}\left[\left|\int g\,d\nu_m - \int g\,d\nu\right| + \left|\int (f-g)\,d\nu_m\right| + \left|\int (f-g)\,d\nu\right|\right] \\ &\le 0 + \varepsilon M + \varepsilon M \\ &= 2\varepsilon M\,,\end{aligned}$$

where we have used (a) to conclude that $\lim\limits_{m\to\infty} |\int g\,d\nu_m - \int g\,d\nu| = 0$. Since ε is arbitrary, we have $\int f\,d\nu_m \to \int f\,d\nu$.

Applying this lemma to $S^{\theta}_{2,T,\text{diff}}$ and $S^{\text{cont}}_{2,T,\text{diff}}$ we obtain:

Proposition 15.8 *Suppose that $S^{\theta}_{2,T} \geq 0$ and that $S^{\theta,\mathrm{dist}}_{2,T} \to S^{\mathrm{cont}}_{2,T}$ in $\mathcal{D}'$. Then:*

(a) $0 \leq \chi^{\mathrm{cont}} \leq \liminf_{\theta\to\infty} \chi^{\theta}$.

(b) Assume that there exists a function $F \geq 0$ which goes to $+\infty$ at infinity, such that $\limsup_{\theta\to\infty} \int d^dx\, F(x) S^{\theta}_{2,T}(0,x) < \infty$. Then $\chi^{\theta} \to \chi^{\mathrm{cont}}$.

In a *massive* scaling limit, we typically choose the length and field-strength rescaling factors so that both the rescaled susceptibilities χ^{θ} and the rescaled second-moment correlation lengths

$$
\begin{aligned}
\xi_2^{\theta} &\equiv \left(\frac{\int d^dx\, |x|^2 S^{\theta}_{2,T}(0,x)}{\int d^dx\, S^{\theta}_{2,T}(0,x)} \right)^{1/2} \\
&= \left(\frac{\theta^{-d} \sum\limits_{x\in\theta^{-1}\mathbb{Z}^d} |x|^2 S^{\theta}_{2,T}(0,x)}{\theta^{-d} \sum\limits_{x\in\theta^{-1}\mathbb{Z}^d} S^{\theta}_{2,T}(0,x)} \right)^{1/2} \\
&= \theta^{-1}\xi_2
\end{aligned}
$$

have limits (or are at least bounded). In this case we can apply Proposition 15.8 with $F(x) = |x|^2$ to conclude that $\chi^{\theta} \to \chi^{\mathrm{cont}}$. Of course, to conclude that $\xi_2^{\theta} \to \xi_2^{\mathrm{cont}}$ (as we would also like to do), we would need to assume the boundedness of some higher-moment correlation length ξ_{ϕ}^{θ} with $\phi > 2$.

Summarizing this discussion, we have proven:

Proposition 15.9 *Suppose that $S^{\theta}_{2,T} \geq 0$ and that $S^{\theta,\mathrm{dist}}_{2,T} \to S^{\mathrm{cont}}_{2,T}$ in $\mathcal{D}'$. Suppose further that for some $\phi > 0$, the rescaled correlation lengths $\xi^{\theta}_{\phi} \equiv \theta^{-1}\xi_{\phi}$ are bounded. Then:*

(a) $\chi^{\theta} \to \chi^{\mathrm{cont}}$.

(b) $\xi^{\theta}_{\phi'} \to \xi^{\mathrm{cont}}_{\phi'}$ *for all* $\phi' < \phi$.

We can also apply Lemma 15.7 to truncated n-point functions $u_n \equiv S_{n,T}$ with $n \geq 3$, with the aim of showing that $\bar{u}_n^{\theta} \to \bar{u}_n^{\mathrm{cont}}$. In particular, for $n = 3, 4$ we can control $\bar{u}_n^{\mathrm{cont}}$ by Lemma 15.7(c) combined with correlation inequalities.[5] Consider first $n = 3$ with $h \geq 0$, and suppose that $u_3^{\theta,\mathrm{dist}} \to u_3^{\mathrm{cont}}$ in the sense of distributions. Now the GHS and Ginibre inequalities (13.53)/(13.52) state that

$$
0 \;\geq\; u_3(x_1,x_2,x_3) \;\geq\; -2M \min[\langle \varphi_{x_1};\varphi_{x_2}\rangle,\, \langle \varphi_{x_1};\varphi_{x_3}\rangle,\, \langle \varphi_{x_2};\varphi_{x_3}\rangle]\,. \quad (15.46)
$$

[5]The basic technique is taken from [481, Section 2.3], with simplifications inspired by [5, Section 15].

Let us fix $x_1 = 0$, multiply both sides by

$$F(x_2, x_3) = |x_2|^\kappa + |x_3|^\kappa \tag{15.47}$$

with $\kappa > 0$, and then sum over x_2, x_3. We get

$$\begin{aligned} 0 \geq \sum_{x_2,x_3} u_3(0,x_2,x_3)F(x_2,x_3) &\geq -4M \sum_{\substack{x_2,x_3 \\ |x_2|\leq|x_3|}} \langle\varphi_0\,;\varphi_{x_3}\rangle F(x_2,x_3) \\ &\geq -\text{const} \times M \sum_x |x|^{d+\kappa} \langle\varphi_0\,;\varphi_x\rangle \\ &= -\text{const} \times M\chi(\xi_{d+\kappa})^{d+\kappa} \end{aligned} \tag{15.48}$$

Now insert the appropriate rescaling factors θ and α. If the rescaled magnetization $M^\theta \equiv \alpha M$, the rescaled susceptibility $\chi^\theta \equiv \alpha^2\theta^{-d}\chi$ and the rescaled ϕ^{th}-moment correlation length $\xi_\phi^\theta \equiv \theta^{-1}\xi_\phi$ are all bounded (for some $\phi > d$), then Lemma 15.7(c) implies that $\bar{u}_3^\theta \to \bar{u}_3^{\text{cont}}$.

Consider next $n = 4$ with $h = 0$, and suppose that $u_4^{\theta,\text{dist}} \to u_4^{\text{cont}}$ in the sense of distributions. The Lebowitz and Ginibre-Newman-Aizenman inequalities (13.36)/(13.35) state that

$$0 \geq u_4(x_1,x_2,x_3,x_4) \geq -2\min[\langle\varphi_{x_1}\varphi_{x_2}\rangle\langle\varphi_{x_3}\varphi_{x_4}\rangle,\ \text{permutations}]\,. \tag{15.49}$$

Let us fix $x_1 = 0$, multiply both sides by

$$F(x_2,x_3,x_4) = |x_2|^\kappa + |x_3|^\kappa + |x_4|^\kappa \tag{15.50}$$

with $\kappa > 0$, and then sum over x_2, x_3, x_4. We get

$$\begin{aligned} 0 &\geq \sum_{x_2,x_3,x_4} u_4(0,x_2,x_3,x_4)F(x_2,x_3,x_4) \\ &\geq -12 \sum_{\substack{x_2,x_3,x_4 \\ |x_3|\leq|x_2| \\ |x_3|\leq|x_4-x_3|}} \langle\varphi_0\varphi_{x_2}\rangle\langle\varphi_{x_3}\varphi_{x_4}\rangle F(x_2,x_3,x_4) \\ &\geq -12 \sum_{\substack{x_2,x_3,x_4 \\ |x_3|\leq|x_2|^{1/2}|x_4-x_3|^{1/2}}} \langle\varphi_0\varphi_{x_2}\rangle\langle\varphi_{x_3}\varphi_{x_4}\rangle F(x_2,x_3,x_4) \\ &= -\text{const} \times \left(\sum_x |x|^{\frac{d}{2}+\kappa}\langle\varphi_0\varphi_x\rangle\right)\left(\sum_x |x|^{d/2}\langle\varphi_0\varphi_x\rangle\right) \\ &= -\text{const} \times \chi^2(\xi_{d/2})^{d/2}(\xi_{(d/2)+\kappa})^{(d/2)+\kappa}\,. \end{aligned} \tag{15.51}$$

Now insert the appropriate rescaling factors θ and α. If the rescaled susceptibility $\chi^\theta \equiv \alpha^2\theta^{-d}\chi$ and the rescaled ϕ^{th}-moment correlation length $\xi_\phi^\theta \equiv \theta^{-1}\xi_\phi$ are both bounded (for some $\phi > d/2$), then Lemma 15.7(c) implies that $\bar{u}_4^\theta \to \bar{u}_4^{\text{cont}}$. In summary:

Proposition 15.10

(a) Consider a sequence of lattice models satisfying the GHS and Ginibre inequalities (e.g. ferromagnetic models in the EMN class). Suppose that $u_3^{\theta,\text{dist}} \to u_3^{\text{cont}}$ in $\mathcal{D}'$. Suppose further that $M^\theta \equiv \alpha M$, $\chi^\theta \equiv \alpha^2\theta^{-d}\chi$ and $\xi_\phi^\theta \equiv \theta^{-1}\xi_\phi$ are all bounded, for some $\phi > d$. Then $\bar{u}_3^\theta \to \bar{u}_3^{\text{cont}}$.

(b) *Consider a sequence of lattice models satisfying the Lebowitz and Ginibre-Newman-Aizenman inequalities (e.g. ferromagnetic models in the BFS class, in the symmetric regime). Suppose that* $u_4^{\theta,\mathrm{dist}} \to u_4^{\mathrm{cont}}$ *in* $\mathcal{D}'$. *Suppose further that* $M^\theta \equiv \alpha M$, $\chi^\theta \equiv \alpha^2\theta^{-d}\chi$ *and* $\xi_\phi^\theta \equiv \theta^{-1}\xi_\phi$ *are all bounded, for some* $\phi > d/2$. *Then* $\bar{u}_4^\theta \to \bar{u}_4^{\mathrm{cont}}$.

An even simpler situation arises in proofs of triviality: suppose that u_n has constant sign [e.g. $u_3 \leq 0$ for $h \geq 0$ by the GHS inequality, $u_4 \leq 0$ for $h = 0$ by the Lebowitz inequality, or $(-1)^k u_{2k} \leq 0$ for $h = 0$ by the Shlosman inequality] and that

$$\bar{u}_n^\theta \equiv \alpha^n \theta^{-(n-1)d} \sum_{x_2,x_3,\ldots,x_n} u_n(0, x_2, x_3, \ldots, x_n) \tag{15.52}$$

tends to zero. Then $u_n^{\theta,\mathrm{dist}} \to 0$ in the sense of distributions (i.e. $u_n^{\mathrm{cont}} \equiv 0$). [In fact, if $\{\nu_m\}$ is a sequence of finite positive measures with total mass tending to zero, then $\int f\, d\nu_m \to 0$ for every bounded measurable function f.] This condition is often convenient for proving the triviality of massive scaling limits (note that it is a sufficient but *not* necessary condition). Indeed, if the length and field-strength rescalings are chosen so that χ^θ = const and ξ_2^θ = const, then $\bar{u}_4^\theta = -\mathrm{const} \times g_2$, where $g_2 \equiv -\bar{u}_4/\chi^2\xi_2^d$ is the dimensionless renormalized 4-point coupling constant. So $g_2 \to 0$ is a sufficient condition for $u_4^{\mathrm{cont}} \equiv 0$ — which, as we shall see shortly, is a sufficient condition for triviality (Propositions 15.11 and 15.12).

15.1.5 Some criteria for triviality

In this section we record some useful sufficient conditions for the triviality of a quantum field theory.

We recall that a quantum field theory is called a *generalized free field* (GFF) if all its truncated Schwinger functions $S_{n,T}$ of order $n \geq 3$ vanish at noncoinciding arguments. One example is the *free field of mass m*, in which $S_2 = (-\Delta + m^2)^{-1}$. Colloquially we use the adjective "trivial" as a synonym for "generalized free field".

The goal of a *triviality theorem* is to show that some class of quantum field theories — often some class of continuum limits of lattice theories — are necessarily GFFs. Usually this is done by "hard analysis", *i.e.*, by detailed estimates based typically on either correlation inequalities or renormalization-group arguments. However, there are also a number of "soft" theorems which give sufficient conditions for a quantum field to be a GFF, and these can sometimes be useful.

One class of theorems asserts that if $S_{n,T} \equiv 0$ for "sufficiently many" $n \geq 3$, then in fact $S_{n,T} \equiv 0$ for all $n \geq 3$. The best *general* theorem of this kind (known to us) is that of Baumann [52]:

Proposition 15.11 [52] *Consider a quantum field theory satisfying all of the Gårding-Wightman (or equivalently, Osterwalder-Schrader) axioms. Suppose that for at least one* even $n \geq 4$, *the truncated n-point Schwinger function* $S_{n,T}$ *vanishes identically at noncoinciding arguments. Then the field is a generalized free field.*

Remarks. 1) This theorem is actually proven in Minkowski space. We have restated it in Euclidean space, using the Osterwalder-Schrader reconstruction theorem. It would be interesting to find a purely Euclidean proof.

2) This theorem assumes *all* of the Wightman axioms, including Lorentz invariance. Therefore, it is applicable to continuum limits of lattice theories only if Euclidean invariance holds.

3) It would be interesting to try to prove the following conjectured extension of Baumann's theorem: Suppose that $S_{n,T} \equiv 0$ at noncoinciding arguments for at least one *odd* $n \geq 3$. Then $S_{n,T} \equiv 0$ at noncoinciding arguments for *all odd* $n \geq 3$.

There also exist theorems of this kind which avoid the assumption of Lorentz invariance, at the price of dealing with much more restricted classes of quantum field theories. One result of this kind is due to Newman [411, Theorem 10]:

Proposition 15.12 *Consider a continuum quantum field theory* $\{S_n^{\rm cont}\}$ *obtained as the limit in* $\mathcal{S}'$ *of rescaled lattice theories* $\{S_n^\theta\}$ *which are* even *(i.e.* $S_n^\theta \equiv 0$ *for n odd) and satisfy the Lee-Yang theorem.*[6] *Suppose that for at least one even* $n \geq 4$, *the truncated n-point Schwinger function* $S_{n,T}^{\rm cont}$ *vanishes identically. Then* $S_{n,T}^{\rm cont} \equiv 0$ *for all* $n \geq 3$, *i.e. the field is a generalized free field.*

A similar result was proven by Aizenman [5, Corollary 12.1] for models of Griffiths-Simon class.

In zero magnetic field, the theorems of Baumann and Newman are convenient but not indispensable, because the same methods (i.e. Aizenman-Fröhlich inequality) which show that $S_{4,T}^\theta \to 0$ can also be adapted to show directly that $R_{2n}^\theta \equiv S_{2n}^\theta - \sum_{\rm pairings} \prod S_2^\theta \to 0$. However, in nonzero magnetic field, the higher-point correlation functions are very difficult to work with; therefore, the theorem of Baumann may play an important role in future work, by reducing everything to the study of $S_{4,T}$. Unfortunately, at present we are unable to say anything about $S_{4,T}$ in nonzero magnetic field; indeed, we have only rather weak results for $S_{3,T}$!

A quite different class of triviality theorems asserts that if the two-point function $S_{2,T}$ has sufficiently "soft" ultraviolet (high-momentum) behavior, then the theory is a generalized free field. The most important of these theorems is due to Jost, Schroer and Pohlmeyer:

[6] This class includes ferromagnetic φ^4 models with $N \leq 3$ components (and their limiting cases, the Ising, plane-rotator and classical Heisenberg models). See [148, 371].

Proposition 15.13 [320, 433] *Consider a quantum field theory satisfying all of the Gårding-Wightman (or equivalently, Osterwalder-Schrader) axioms. Suppose that the truncated two-point function $S_{2,T}$ equals (at noncoinciding arguments) that of a free field of mass m, namely,*

$$S_{2,T} = \text{const} \ \times \ (-\Delta + m^2)^{-1}$$

for some $m \geq 0$. Then the field is a free field of mass m.

Remarks. 1) This theorem was first proven by Jost and Schroer [320] for the case $m > 0$; see also [179, 488]. Some years later, Pohlmeyer [433] proved, by a different method, the case $m = 0$; his proof actually handles all cases $m \geq 0$ in a unified way.

2) The $m = 0$ case of this theorem says, in particular, that a nontrivial *scale-invariant* quantum field theory must have *noncanonical dimensions*, i.e. $S_{2,T} = \text{const} \ \times \ (-\Delta)^{-p}$ with $p > 1$.

3) All known proofs of the Jost-Schroer-Pohlmeyer theorem work only in Minkowski space. It would be interesting to find a purely Euclidean proof.

A stronger result was proven recently by Baumann [53]:

Proposition 15.14 [53] *Consider a quantum field theory in d-dimensional space-time satisfying all of the Gårding-Wightman axioms. Suppose that the truncated two-point Wightman distribution $W_{2,T}$ decays exponentially in momentum space, in the sense that $e^{\alpha\sqrt{p^2}}\widetilde{W}_{2,T}(p) \in \mathcal{S}'(\mathbb{R}^d)$ for some $\alpha > 0$, where $p^2 \equiv p_0^2 - \mathbf{p}^2$. Then, if either*

(a) $d \geq 3$, or

(b) $d = 2$ and there are no zero-mass states in the energy spectrum,

the field is a generalized free field.

Another theorem of vaguely related type is the ICAR theorem of Powers [439, 145, 54] for Fermi fields satisfying canonical anticommutation relations, and its analogue for Bose fields [55, 56]. These theorems also state (though in a very different sense) that a quantum field theory with sufficiently soft ultraviolet behavior is necessarily a generalized free field.

15.1.6 Summary of what is known

Nontriviality Nearly all the known results on nontriviality are for models that are *perturbatively superrenormalizable* [$P(\varphi)_1$, $P(\varphi)_2$ and φ_3^4] or at least *asymptotically free* [Gross-Neveu$_2$, φ_4^4 with $\lambda < 0$, Yang-Mills$_4$]. Only quite recently has progress been made in constructing certain *perturbatively nonrenormalizable* models, notably the three-dimensional Gross-Neveu model at large N. Finally, the continuum limit of the two-dimensional Ising model has been constructed by quite different methods.

1) $P(\varphi)_1$. These models are just anharmonic oscillators, and can be constructed and analyzed by standard methods of nonrelativistic quantum mechanics [463].

2) $P(\varphi)_2$. These models require only Wick ordering to make them well-defined. Their rigorous construction was the first major achievement of the Constructive Quantum Field Theory program (1964–1972), and is described in detail in the books of Glimm and Jaffe [260] and Simon [462].

3) φ^4_3. This model requires a nontrivial mass renormalization (as well as vacuum-energy renormalizations). The first rigorous construction, due to Glimm, Jaffe, Feldman, Osterwalder, Magnen and Sénéor (1972–75), used deep methods of phase-space-cell localization [254, 193, 380]. Different but closely related methods have been proposed by Gallavotti and collaborators [230, 231, 59, 60], Balaban [35], Battle and Federbush [49, 50, 48, 174], and Feldman-Magnen-Rivasseau-Sénéor [380, 190, 192] these recent works stress the connection with renormalization-group ideas. A much simpler (though much weaker) proof of existence and nontriviality for weakly-coupled φ^4_2 and φ^4_3 was given by Brydges, Fröhlich and Sokal [97, 96], using skeleton inequalities derived from the BFS random-walk representation (see Section 6.3).

4) *Gross-Neveu*$_2$. The Gross-Neveu [398, 283] model of an N-component fermion with $(\overline{\psi}\psi)^2$ interaction is asymptotically free in $d = 2$. This model has been rigorously constructed by Feldman *et al.* [191] and by Gawedzki and Kupiainen [243] using renormalization-group ideas. The correlation functions are the Borel sum of the renormalized perturbation expansion [191], so the theory is nontrivial (at least for weak coupling).

5) φ^4_4 $(\lambda < 0)$. The φ^4_4 model with the "wrong" sign of the coupling constant $(\lambda < 0)$ can be defined by analytic continuation from $\lambda > 0$, and it is asymptotically free. Gawedzki and Kupiainen [245] constructed the continuum limit of a hierarchical version of this model, and proved that renormalized perturbation theory is asymptotic; in particular, the correlation functions are nontrivial (at least for $|\lambda|$ small). With more work, the full (non-hierarchical) theory could presumably be constructed by similar methods. However, φ^4_4 with $\lambda < 0$ is not an honest quantum field theory: the correlation functions are almost certainly *not* OS-positive. Its value is as a "toy model" on which to practice dealing with asymptotic freedom. The same goes for Rivasseau's construction [444] of the planar version of this model (which is equivalent to the $N \to \infty$ limit of an $N \times N$ matrix φ^4 model).

6) *Yang-Mills*$_4$. Balaban [34, 36, 37, 39, 41, 40, 42, 43, 44, 45] has recently constructed the continuum limit of four-dimensional pure lattice gauge theory in finite volume; see also Federbush [171, 178, 172, 175, 176, 173]. The correlation functions of this theory are presumably nontrivial, although the proof of this fact is not yet complete.

7) *Perturbatively nonrenormalizable models.* By extending the work on asymptotically free theories, several authors have succeeded in constructing per-

turbatively nonrenormalizable quantum field theories that are controlled by a non-Gaussian fixed point "sufficiently close" to the Gaussian one. The first results in this direction concerned a hierarchical model at small ε or large N [242, 249]; see also [342] for an extension to $\varepsilon = 1$, $N = 1$. Later, Gawedzki and Kupiainen [246] constructed the Gross-Neveu model in dimension $d = 2$ with the modified propagator $\not{p}/|p|^{2-\varepsilon}$, for ε sufficiently small; this mimics the ordinary Gross-Neveu model in dimension "$d = 2 + \varepsilon$". By similar methods, Felder [182] constructed the planar φ_4^4 theory with a propagator $1/p^{2-\varepsilon/2}$ at negative coupling; this mimics the negative-coupling planar $\varphi_{4+\varepsilon}^4$ theory. However, none of these are honest quantum field theories: OS-positivity almost certainly fails. Very recently, however, de Calan *et al.* [119] have constructed the first perturbatively nonrenormalizable quantum field theory satisfying all the OS axioms: the Gross-Neveu model in dimension $d = 3$ with a large number N of components.

8) *Ising*$_2$. The scaling limit of the two-dimensional ferromagnetic Ising model is known to be nontrivial. For a nearest-neighbor interaction, the continuum-limit correlation functions can be computed explicitly [1, 387]. For general finite-range interactions, the nontriviality can be proven by a very simple and beautiful geometric argument using the random-current representation [5, Section 8].

Triviality. The principal triviality theorems concern the continuum limit of the *symmetric* phase of φ^4 and Ising models in dimension $d > 4$ and $d = 4$.

1) $d > 4$. Here the continuum limit is trivial (or boring or nonexistent) no matter what renormalizations $J(\theta)$, $\lambda(\theta)$, $\tau(\theta)$ are chosen.

2) $d = 4$. The continuum limit is believed by most physicists to be trivial (or boring or nonexistent) no matter what renormalizations are chosen, but *this has not yet been proven.* The existing triviality theorems cover only some tantalizing special cases:

(a) If the field-strength renormalization constant $\alpha_{FT}^{-2} J_{FT}(\theta)$ tends to *zero* as $\theta \to \infty$ ("infinite field-strength renormalization").

(b) If the continuum two-point function has *non-canonical short-distance behavior*, i.e. $\lim_{p\to\infty} p^2 \widetilde{S}_2^{\text{cont}}(p) = \infty$.

(c) If the continuum two-point function is *scale-invariant*, i.e., $\widetilde{S}_2^{\text{cont}}(p) = \text{const} \times |p|^{-\alpha}$ for some α.

(d) If the rescaled susceptibilities χ^θ are bounded ("massive scaling limit") and $\chi^{-2}\, \partial\chi/\partial J \to 0$ ("logarithmic corrections to mean-field theory").

(e) If the bare coupling constant $\widehat{\lambda}_{FT}(\theta) = \lambda_{FT}(\theta)/J_{FT}(\theta)^2$ remains *bounded and sufficiently small* as $\theta \to \infty$.

The proofs of (1) and (2a) are based on the Aizenman-Fröhlich inequality; (2b) is an easy consequence of (2a); (2c) follows from (2b) together with the

Jost-Schroer-Pohlmeyer theorem; and (2d) is a consequence of the Aizenman-Graham inequality. A detailed exposition of these proofs is the principal aim of this chapter. The proof of (2e) uses renormalization-group methods [290, 298].

These partial results on φ_4^4 leave us in a rather strange situation:

(i) The "traditional" view of φ_4^4 (circa 1950), based on renormalized perturbation theory, is that φ_4^4 is nontrivial provided that one makes suitable infinite field-strength, coupling-constant and mass renormalizations. This scenario is ruled out by (2a).

(ii) The "modern" (post-1971) view of how φ_4^4 could potentially be nontrivial, based on renormalization-group ideas, is that the φ_4^4 theory may be controlled by some (as-yet-undiscovered) nontrivial RG fixed point. Since the theory *at* such a fixed point would be scale-invariant, this scenario is ruled out by (2c): there does *not* exist a non-trivial RG fixed point that can be reached from φ^4 lattice theories.

(iii) The only possibility left open for nontrivial φ_4^4 theory is the *least* plausible one: finite field-strength renormalization, with no clear connection to RG theory.

This strange situation, in which we have solved the supposedly "hard" case (infinite field-strength renormalization) but failed to solve the "easy" case (finite field-strength renormalization), is presumably an artifact of present methods of proof rather than a real feature of φ_4^4. But until someone proves a complete triviality theorem for φ_4^4 — or its negation — the question remains open. We emphasize, also, that very little is known about the *non-symmetric* regime of φ_d^4, even for $d > 4$.

Recently, attempts have been made to extend the random-walk representation and the triviality theorems to scalar quantum electrodynamics [88, 149] and to four-fermion theories [312]. However, no rigorous triviality theorems have yet been proven. Another important contribution for QED is [375]. The practical consequences of triviality for elementary-particle physics are reviewed in [102].

15.2 Triviality of φ_d^4 ($d \geq 4$) in the symmetric phase

The triviality (Gaussianness) of the continuum limit for φ^4 quantum field theories in dimension $d > 4$ (in the symmetric phase) was proven independently and almost simultaneously by Aizenman [5] and Fröhlich [213]. These papers also proved some partial results in dimension $d = 4$, which were later strengthened by Aizenman and Graham [15] and Aragão de Carvalho, Caracciolo and Fröhlich [28]. The purpose of this section is to give a detailed exposition of all these proofs.

In this section we shall adhere to the normalization convention $\alpha(\theta) \equiv 1$. The reader who prefers another convention can easily put back all the α's.

The key to the triviality proofs is the Aizenman-Fröhlich correlation inequality (13.40)

$$S_{4,T}(x_1,x_2,x_3,x_4) \geq -\sum_z S_2(x_1,z)S_2(x_3,z)\left[\delta_{z,x_2}+\sum_{z_1} J_{zz_1}S_2(z_1,x_2)\right] \times \left[\delta_{z,x_4}+\sum_{z_2} J_{zz_2}S_2(z_2,x_4)\right] - \{2\text{ permutations}\} \tag{15.53}$$

combined with the Lebowitz inequality

$$S_{4,T}(x_1,x_2,x_3,x_4) \leq 0\ . \tag{15.54}$$

To see what is going on, let us first consider a simple special case: a "massive scaling limit" in which $J(\theta)$, $\lambda(\theta)$ and $\tau(\theta)$ are chosen so that the rescaled susceptibilities

$$\chi^\theta \equiv \int d^dx\, S_2^\theta(0,x) \equiv \theta^{-d} \sum_{x\in\theta^{-1}\mathbb{Z}^d} S_2(0,x) = \theta^{-d}\chi \tag{15.55}$$

are bounded as $\theta \to \infty$. We use the "summed" version (13.42)/(13.37) of the Lebowitz and Aizenman-Fröhlich inequalities, namely[7]

$$0 \geq \overline{S}_{4,T} \geq -3\mathcal{J}^2\chi^4\left(1+\frac{1}{\mathcal{J}\chi}\right)^2 \tag{15.56}$$

where

$$\overline{S}_{4,T} \equiv \sum_{x_2,x_3,x_4} S_{4,T}(0,x_2,x_3,x_4) \tag{15.57}$$

and

$$\mathcal{J} \equiv \sum_j J_{ij} = 2dJ \tag{15.58}$$

for a nearest-neighbor interaction of strength J. Inserting the definitions

$$J_{FT} = J\theta^{d-2} \tag{15.59}$$

$$\chi^\theta = \theta^{-d}\chi \tag{15.60}$$

$$\overline{S}^\theta_{4,T} = \theta^{-3d}\overline{S}_{4,T} \tag{15.61}$$

into (15.56), we obtain

$$0 \geq \overline{S}^\theta_{4,T} \geq -\text{ const } \times \left[J_{FT}^2\theta^{4-d}(\chi^\theta)^4 + \theta^{-d}(\chi^\theta)^2\right]\ . \tag{15.62}$$

Now, by assumption, χ^θ is bounded as $\theta \to \infty$. On the other hand, by Proposition 15.5, $J_{FT}(\theta)$ must be *bounded* as $\theta \to \infty$ if we are to obtain a non-boring continuum limit.

---

[7]In the preceding chapters we have used the letter J to denote the *matrix* of couplings $\{J_{ij}\}$, which in the translation-invariant case reduces to a *vector* of couplings $\{J(i-j)\}$; and we have used the symbol $|J|$ to denote the l^1 norm of this vector. In the present chapter, however, we are using the letter J to denote the *nearest-neighbor* coupling strength. Therefore, to avoid confusing formulas like $|J| = 2dJ$, we use the symbol $\mathcal{J}$ for what was formerly called $|J|$.

We conclude that if either
(a) $d > 4$
or
(b) $d = 4$ and $\lim_{\theta\to\infty} J_{FT}(\theta) = 0$,
then

$$\lim_{\theta\to\infty} \overline{S}^{\theta}_{4,T} = 0 \, . \tag{15.63}$$

But this implies, by the discussion surrounding (15.52), that $S^{\theta,\text{dist}}_{4,T}$ converges to *zero* in the space $\mathcal{S}'(\mathbb{R}^{4d})$. Hence

$$S^{\text{cont}}_{4,T} \equiv 0 \, , \tag{15.64}$$

i.e. the continuum-limit truncated four-point function vanishes!

To finish the proof of triviality, we must show that $S^{\text{cont}}_{2n,T} \equiv 0$ for *all* $n \geq 2$. This can be done in either of three ways:

(a) a direct argument using a generalization of the Aizenman-Fröhlich inequality to $2n$-point functions [5];

(b) invocation of Newman's theorem (Proposition 15.12); or

(c) invocation of Baumann's theorem (Proposition 15.11), which applies *if* the continuum-limit theory satisfies all the Osterwalder-Schrader axioms (in particular, rotation invariance). [Of course, if the continuum-limit theory *fails* to satisfy all the OS axioms, then it is of no interest for quantum field theory anyway!]

From now on we concentrate on proving that $S^{\text{cont}}_{4,T} \equiv 0$, as this is really the heart of the problem.

Now let us return to the general case, in which we allow an *arbitrary* sequence of bare parameters $J(\theta)$, $\lambda(\theta)$, $\tau(\theta)$ subject only to the requirement that the theories lie in the ferromagnetic single-phase region and that the rescaled two-point functions $S^{\theta,\text{dist}}_2$ converge in the sense of distributions at noncoinciding arguments to some limit S^{cont}_2.

Let us first recall Proposition 15.5: for any sequence of theories in which $J_{FT} \equiv J\theta^{d-2}$ tends to *infinity* as $\theta \to \infty$, the rescaled n-point functions $S^{\theta,\text{dist}}_n$ converge to zero in $\mathcal{S}'(\mathbb{R}^{nd})$ — in particular, $S^{\text{cont}}_{4,T} \equiv 0$. So we can restrict attention henceforth to sequences of theories in which J_{FT} is *bounded*: for if we can prove $S^{\text{cont}}_{4,T} \equiv 0$ in this case too, then the general result (for *arbitrary* sequences of theories) follows from these two special cases by some abstract nonsense about sub-subsequences.

Consider the set

$$\Omega_\varepsilon = \{(x_1, x_2, x_3, x_4) \in \mathbb{R}^{4d} \colon \; |x_i - x_j| \geq \varepsilon \;\text{ for }\; i \neq j\} \, . \tag{15.65}$$

Our strategy will be to show that $S^{\theta}_{4,T}(x_1, x_2, x_3, x_4)$, considered as a pointwise-defined function for $(x_1, x_2, x_3, x_4) \in \theta^{-1}\mathbb{Z}^{4d}$, converges to *zero* (as $\theta \to \infty$)

uniformly on Ω_ε, for each $\varepsilon > 0$. By Proposition 15.2 and the remark following it, this implies that $S_{4,T}^{\theta,\text{dist}}$ converges to zero in the space $\mathcal{S}'_{\neq}(\mathbb{R}^{4d})$ of tempered distributions at noncoinciding arguments.

For $x_1, x_2, x_3, x_4 \in \mathbb{R}^d$, let $D_{i,\varepsilon}$ be the open ball of radius $\varepsilon/2$ centered at x_i ($i = 1,2,3,4$), i.e.

$$D_{i,\varepsilon} = \{y \in \mathbb{R}^d\colon\ |y - x_i| < \varepsilon/2\}\ . \tag{15.66}$$

Note that if $(x_1, x_2, x_3, x_4) \in \Omega_\varepsilon$, then the balls $D_{i,\varepsilon}$ are disjoint. Now let us study the principal terms in the Aizenman-Fröhlich inequality (15.53): this means that we shall temporarily neglect the terms involving δ_{z,x_2} and δ_{z,x_4}, and we shall temporarily approximate z_1, z_2 by z. (We shall return later to the error terms caused by these approximations.) We obtain

$$\begin{aligned}
&-S_{4,T}^{\theta}(x_1,x_2,x_3,x_4)\\
&\quad= -S_{4,T}(\theta x_1, \theta x_2, \theta x_3, \theta x_4)\\
&\quad\lesssim 3\mathcal{J}^2 \sum_{z\in\mathbb{Z}^d} S_2(\theta x_1, z) S_2(\theta x_2, z) S_2(\theta x_3, z) S_2(\theta x_4, z)\\
&\quad= 12 d^2 J_{FT}^2 \theta^{4-d} \sum_{z\in\theta^{-1}\mathbb{Z}^d} \theta^{-d} S_2(\theta x_1, \theta z) S_2(\theta x_2, \theta z) S_2(\theta x_3, \theta z) S_2(\theta x_4, \theta z)\\
&\quad= 12 d^2 J_{FT}^2 \theta^{4-d} \sum_{z\in\theta^{-1}\mathbb{Z}^d} \theta^{-d} S_2^{\theta}(x_1, z) S_2^{\theta}(x_2, z) S_2^{\theta}(x_3, z) S_2^{\theta}(x_4, z)
\end{aligned} \tag{15.67}$$

We consider $(x_1, x_2, x_3, x_4) \in \Omega_\varepsilon$, and break up the sum over z into five pieces: $z \in D_{i,\varepsilon}$ ($i = 1,2,3,4$) and $z \in \left(\bigcup_{i=1}^4 D_{i,\varepsilon}\right)^c$.

Case $z \in D_{i,\varepsilon}$: The factors $S_2^\theta(x_j, z)$ for $j \neq i$ are uniformly bounded, by Proposition 15.3, since $|x_j - z| > \frac{\varepsilon}{2}$. The factor $S_2^\theta(x_i, z)$ is controlled by the x-space infrared bound (13.87),

$$0 \leq S_2^\theta(x_i, z) \leq \frac{\text{const}}{J_{FT}(\theta^{-1} + |x_i - z|)^{d-2}}\ . \tag{15.68}$$

The contribution to (15.67) is therefore bounded by

$$\text{const} \times J_{FT}\theta^{4-d} \sum_{\substack{y\in\theta^{-1}\mathbb{Z}^d\\ |y|<\frac{\varepsilon}{2}}} \theta^{-d}(\theta^{-1} + |y|)^{-(d-2)}\ . \tag{15.69}$$

As $\theta \to \infty$, the sum approaches the integral

$$\int_{\substack{y\in\mathbb{R}^d\\ |y|<\frac{\varepsilon}{2}}} dy\ |y|^{-(d-2)}\ , \tag{15.70}$$

which is finite. Therefore, if either $d > 4$, or $d = 4$ with $J_{FT} \to 0$, this contribution vanishes in the limit. Moreover, the convergence is *uniform* for $(x_1, x_2, x_3, x_4) \in \Omega_\varepsilon$.

Case $z \in \left(\bigcup_{i=1}^{4} D_{i,\varepsilon}\right)^c$: For the factors $S_2^\theta(x_1, z)$ and $S_2^\theta(x_2, z)$ we use the uniform bound from Proposition 15.3. For the factors $S_2^\theta(x_3, z)$ and $S_2^\theta(x_4, z)$ we use the infrared bound (15.68).[8] The contribution to (15.67) is therefore bounded by

$$\text{const} \times \theta^{4-d} \sum_{\substack{z \in \theta^{-1}\mathbb{Z}^d \\ |z-x_3| \ge \frac{\varepsilon}{2} \\ |z-x_4| \ge \frac{\varepsilon}{2}}} \theta^{-d} |z - x_3|^{-(d-2)} |z - x_4|^{-(d-2)} \,. \tag{15.71}$$

For $d > 4$, the sum is convergent, and it is not hard to see that it is bounded *uniformly* in θ ($\geq$ const), x_3 and x_4. It follows that, for $d > 4$, this contribution to (15.67) vanishes in the limit; and the convergence is *uniform* for $(x_1, x_2, x_3, x_4) \in \Omega_\varepsilon$.

To control this term for $d = 4$, we make the *additional assumption* that

$$S_2^\theta(0, x) \leq C|x|^{-\kappa} \quad \text{for} \quad |x| \geq R \tag{15.72}$$

for some constants $C, R < \infty$ and $\kappa > 0$, *uniformly in* θ. This is a stronger notion of convergence to the continuum limit than the distributional convergence we have been assuming until now; but it is still very weak. Note also that if (15.72) holds for *some* $R < \infty$, then in fact it holds for *all* $R > 0$ (no matter how small) with a suitable $C = C(R)$; this is a consequence of the known uniform bound on S_2^θ (Proposition 15.3). So let us use (15.72) with $R = \varepsilon/2$ to bound $S_2^\theta(x_1, z)$, $S_2^\theta(x_2, z)$ and $S_2^\theta(x_3, x)^\delta$, where δ is a small positive number (≤ 1) to be chosen later; and let us use the infrared bound (15.68) to control $S_2^\theta(x_3, z)^{1-\delta}$ and $S_2^\theta(x_4, z)$. The contribution to (15.67) is therefore bounded by

$$\text{const} \times J_{FT}^\delta \theta^{4-d} \sum_{\substack{z \in \theta^{-1}\mathbb{Z}^d \\ |z-x_i| \ge \frac{\varepsilon}{2}\ \forall i}} |z - x_1|^{-\kappa} |z - x_2|^{-\kappa} |z - x_3|^{-\delta\kappa - (1-\delta)(d-2)} |z - x_4|^{-(d-2)} \tag{15.73}$$

This sum is convergent for $d = 4$ provided that we take $\delta < 2\kappa/(2 - \kappa)$, and in this case it is not hard to see that the sum is bounded *uniformly* in θ ($\geq$ const), x_1, x_2, x_3 and x_4. On the other hand, by taking $\delta > 0$ we gain a factor J_{FT}^δ, which vanishes if $J_{FT} \to 0$.

We now return to the terms from the Aizenman-Fröhlich inequality that were neglected in (15.67). The first error comes from the fact that z_1 and z_2 are not *equal* to z, as in (15.67), but are rather nearest neighbors of z, so that $|z_1 - z| = |z_2 - z| = \theta^{-1}$. This change is obviously irrelevant when $\theta^{-1} \ll \varepsilon$. The other error terms come from the δ's; these terms come in two kinds, depending on whether we take one δ or two. A typical contribution to $-S_{4,T}^\theta$ with one δ is

$$\sum_{z \in \mathbb{Z}^d} S_2(\theta x_1, z) S_2(\theta x_3, z) \delta_{z, \theta x_2} \sum_{z_2} J_{z z_2} S_2(\theta x_4, z_2)$$

[8] To motivate this tactic, recall that the Aizenman-Fröhlich inequality (15.53) has a prefactor J_{FT}^2, while each infrared bound (15.68) carries a factor J_{FT}^{-1}. Since J_{FT} is bounded above but *could go to zero*, we need to end up with J_{FT} raised to a *nonnegative* power. So we can use the infrared bound at most *twice*.

$$\approx \mathcal{J} S_2(\theta x_1, \theta x_2) S_2(\theta x_3, \theta x_2) S_2(\theta x_4, \theta x_2)$$
$$= 2dJ_{FT}\theta^{2-d} S_2^\theta(x_1, x_2) S_2^\theta(x_3, x_2) S_2^\theta(x_4, x_2) \,. \qquad (15.74)$$

The S_2^θ factors are uniformly bounded provided that $(x_1, x_2, x_3, x_4) \in \Omega_\varepsilon$; similarly, J_{FT} is bounded; and so this contribution vanishes as $\theta \to \infty$ for any $d > 2$. A typical contribution to $-S_{4,T}^\theta$ with two δ's is

$$\sum_z S_2(\theta x_1, z) S_2(\theta x_3, z) \delta_{z,\theta x_2} \delta_{z,\theta x_4} \,, \qquad (15.75)$$

which is *zero* if $x_2 \neq x_4$. We conclude that the "error terms" in the Aizenman-Fröhlich inequality are indeed irrelevant.

Let us make one final technical remark. The foregoing argument shows that $S_{4,T}^{\theta,\text{dist}} \to 0$ for test functions $f \in \mathcal{S}(\mathbb{R}^{4d})$ with support in some set Ω_ε, i.e. with support *strictly away* from the hyperplanes $x_i - x_j = 0$. Convergence in $\mathcal{S}'_{\neq}(\mathbb{R}^{4d})$ requires, however, convergence on a slightly larger class of test functions, namely those that *vanish together with all their derivatives* on the hyperplanes $x_i - x_j = 0$. (For example, a sequence of delta functions approaching such a hyperplane, with *exponentially* growing strengths, would converge to zero in the first sense but not in the second.) With a little extra technical work, this stronger convergence can be proven, using the assumed convergence in $\mathcal{S}'_{\neq}(\mathbb{R}^{2d})$ of the rescaled two-point functions together with the Lebowitz and Ginibre-Newman-Aizenman inequalities (13.36)/(13.35).

To summarize the results obtained so far, we have proven the following two theorems[9]:

Theorem 15.15 *Let $d > 4$, and let $\{\mu_m\}$ be a sequence of nearest-neighbor ferromagnetic φ^4 or Ising models in the single-phase region. Consider a continuum limit with arbitrary length rescaling factors $\{\theta_m\} \to \infty$ and field-strength rescaling factors $\{\alpha_m\}$. Assume that the rescaled lattice two-point Schwinger distributions $S_2^{(\mu_m,\theta_m,\alpha_m),\text{dist}}$ converge in the space of distributions at noncoinciding arguments. Then*

$$S_{4,T}^{(\mu_m,\theta_m,\alpha_m),\text{dist}} \to 0$$

in the space of tempered distributions at noncoinciding arguments.

Theorem 15.16 *Let $d = 4$, and let $\{\mu_m\}$ be a sequence of nearest-neighbor ferromagnetic φ^4 or Ising models in the single-phase region, with nearest-neighbor coupling strengths $\{J_m\}$. Consider a continuum limit with length rescaling factors $\{\theta_m\} \to \infty$ and field-strength rescaling factors $\{\alpha_m\}$. Assume that the rescaled lattice two-point Schwinger distributions $S_2^{(\mu_m,\theta_m,\alpha_m),\text{dist}}$ converge in the*

[9]We restore the α's for the convenience of those readers who dislike our convention $\alpha \equiv 1$.

space of distributions at noncoinciding arguments, and in addition that they satisfy the uniform upper bound (15.72). If

$$\lim_{m\to\infty} J_m \theta_m^{d-2} \alpha_m^{-2} = 0 , \tag{15.76}$$

then

$$S_{4,T}^{(\mu_m,\theta_m,\alpha_m),\mathrm{dist}} \to 0$$

in the space of tempered distributions at noncoinciding arguments.

Remark. Results closely related to Theorem 15.15, but of a more probabilistic flavor ("scaling limits of block spins"), have recently been obtained by DeConinck and Newman [124].

We can now derive some easy — but extremely important — corollaries of Theorem 15.16. Recall first that the continuum free field (of any mass $m \geq 0$) has the short-distance behavior

$$S_2^{\mathrm{cont}}(0,x) \sim |x|^{-(d-2)} \quad \text{as} \quad |x| \to 0 . \tag{15.77}$$

Such a field is said to have *canonical short-distance behavior*. If S_2^{cont} is more singular than this, i.e. if

$$\limsup_{|x|\to 0} |x|^{d-2} S_2^{\mathrm{cont}}(0,x) = \infty , \tag{15.78}$$

then the field is said to have *non-canonical short-distance behavior*. (By the Källén-Lehmann representation [442, Section IX.], it is impossible for a quantum field theory satisfying all the OS axioms to have short-distance behavior which is *less* singular than that of a free field, unless $S_2^{\mathrm{cont}} \equiv 0$.) Theorem 15.16 has the consequence that if the continuum φ_4^4 theory has *non-canonical* short-distance behavior, then it is *trivial*. More precisely:

Corollary 15.17 *Let $d = 4$, and let $\{\mu_m\}$ be a sequence of nearest-neighbor ferromagnetic φ^4 or Ising models in the single-phase region. Consider a continuum limit with length rescaling factors $\{\theta_m\} \to \infty$ and field-strength rescaling factors $\{\alpha_m\}$. Assume that the rescaled lattice two-point Schwinger distributions $S_2^{(\mu_m,\theta_m,\alpha_m),\mathrm{dist}}$ converge in the space of distributions at noncoinciding arguments to a limiting distribution S_2^{cont}, and in addition that they satisfy the uniform upper bound (15.72). Then either*

(a) there exists a constant $C < \infty$ such that

$$0 \leq S_2^{\mathrm{cont}}(0,x) \leq C|x|^{-(d-2)} \tag{15.79}$$

(in the sense of distributions at noncoinciding arguments); or else

(b) the rescaled lattice truncated four-point Schwinger distributions $S_{4,T}^{(\mu_m,\theta_m,\alpha_m),\mathrm{dist}}$ converge to zero in the space of tempered distributions at noncoinciding arguments.

Proof. By the infrared bound (15.68), we have

$$0 \leq S_2^{\theta_m}(0,x) \leq \text{const } \times \alpha_m^2 J_m^{-1} \theta_m^{2-d} (\theta^{-1} + |x|)^{-(d-2)} . \tag{15.80}$$

If, for some subsequence $\{m_i\}$, the sequence $\{\alpha_{m_i}^2 J_{m_i}^{-1} \theta_{m_i}^{2-d}\}$ is *bounded*, then we can pass to the limit and deduce (15.79). Otherwise, we must have

$$\lim_{m \to \infty} \alpha_m^2 J_m^{-1} \theta_m^{2-d} = +\infty , \tag{15.81}$$

in which case Theorem 15.16 implies $S_{4,T}^{(\mu_m,\theta_m,\alpha_m),\text{dist}} \to 0$. ∎

Corollary 15.18 *Let $d = 4$, and consider a quantum field theory satisfying all of the Osterwalder-Schrader axioms, whose two-point and four-point Schwinger distributions are obtainable as a limit of nearest-neighbor ferromagnetic φ^4 or Ising models in the single-phase region as described in Corollary 15.17. If the two-point function is exactly scale-invariant, i.e.*

$$S_2^{\text{cont}}(0,x) = \textit{const } \times |x|^{-p} \tag{15.82}$$

for some p, then the theory is a generalized free field.

Proof. By the Källén-Lehmann representation, we must have $p \geq d-2$. If $p = d-2$, then the Jost-Schroer-Pohlmeyer theorem (Proposition 15.13) implies that the theory is a massless free field. If $p > d-2$, then Corollary 15.17 implies that $S_{4,T}^{\text{cont}} \equiv 0$; and Baumann's theorem (Proposition 15.11) implies that the theory is a generalized free field. ∎

We discuss next the refinements which can be obtained in dimension $d = 4$ by using the Aizenman-Graham inequality in place of the Aizenman-Fröhlich inequality. Let us again consider a *massive scaling limit* in which $J(\theta)$, $\lambda(\theta)$ and $\tau(\theta)$ are chosen so that the rescaled susceptibilities $\chi^\theta \equiv \theta^{-d}\chi$ are bounded as $\theta \to \infty$. We use the "summed" version (13.42)/(13.46) of the Lebowitz and Aizenman-Graham inequalities, namely

$$0 \geq \overline{S}_{4,T} \geq -2\mathcal{J}\chi^2 J \frac{\partial \chi}{\partial J} - 2\mathcal{J}\chi^3 - 2\chi^2 . \tag{15.83}$$

This improves the Aizenman-Fröhlich inequality (15.56) because $0 \leq J\,\partial\chi/\partial J \leq \mathcal{J}\chi^2$ is a corollary of the Lebowitz inequality [see (13.38)]. Inserting the definitions (15.59)-(15.61), we obtain

$$0 \geq \overline{S}_{4,T}^\theta \geq -\text{ const } \times \left[J_{FT}^2 \theta^{4-d} (\chi^\theta)^4 \frac{\partial\chi/\partial J}{\chi^2} + J_{FT}\theta^{2-d}(\chi^\theta)^3 + \theta^{-d}(\chi^\theta)^2 \right] . \tag{15.84}$$

Now, by assumption, χ^θ is bounded as $\theta \to \infty$. On the other hand, by Proposition 15.5, $J_{FT}(\theta)$ must be *bounded* as $\theta \to \infty$ if we are to obtain a non-boring continuum limit. We conclude that if $\chi^{-2}\, \partial\chi/\partial J \to 0$ along the chosen path to the critical surface, then

$$\lim_{\theta\to\infty} \overline{S}^\theta_{4,T} = 0 \ , \tag{15.85}$$

i.e. $S^{\rm cont}_{4,T} \equiv 0$. We have therefore proven:

Theorem 15.19 *Let $d = 4$, and let $\{\mu_m\}$ be a sequence of nearest-neighbor ferromagnetic φ^4 or Ising models in the single-phase region, with nearest-neighbor coupling strengths $\{J_m\}$, susceptibilities $\{\chi_m\}$ and susceptibility derivatives $\{(\partial\chi/\partial J)_m\}$. Consider a continuum limit with length rescaling factors $\{\theta_m\} \to \infty$ and field-strength rescaling factors $\{\alpha_m\}$, and suppose that the rescaled susceptibilities $\chi^\theta_m \equiv \alpha_m^2 \theta_m^{-d} \chi_m$ are bounded. Assume further that*

$$\lim_{m\to\infty} \chi_m^{-2}\, (\partial\chi/\partial J)_m \;=\; 0 \, . \tag{15.86}$$

then the rescaled lattice truncated four-point Schwinger distributions $S_{4,T}^{(\mu_m,\theta_m,\alpha_m),\rm dist}$ converge to zero in the space $\mathcal{S}'(\mathbb{R}^{4d})$ of tempered distributions.

The conventional wisdom, as predicted by perturbative renormalization-group calculations [77], is that for the statistical-mechanical approach to the critical surface ($\lambda > 0$ *fixed*, τ fixed and $J \uparrow J_c$), the susceptibility behaves as a mean-field law modified by a logarithm[10]:

$$\chi \;\sim\; (J_c - J)^{-1}\, |\log(J_c - J)|^{1/3} \tag{15.87}$$

and hence

$$\chi^{-2} \frac{\partial\chi}{\partial J} \;\sim\; |\log(J_c - J)|^{-1/3} \;\to\; 0 \, . \tag{15.88}$$

More generally, this behavior is expected to hold for any approach to the critical surface in which $\widehat{\lambda}_{SM} \equiv \lambda/J^2$ is bounded away from zero. *If* this prediction is correct, then Theorem 15.19 shows that the continuum limit is trivial along such paths. On the other hand, it can be proven rigorously that the continuum limit is trivial along any path with $\widehat{\lambda}_{SM} \to 0$: this is a consequence of the Hara-Tasaki theorem discussed below, or more simply, of the first-order skeleton inequality [5, 97].

Finally, we state without proof a theorem due to Hara and Tasaki [298, Proposition 1.2], which shows that the continuum limit is trivial (or boring or nonexistent) for any approach to the critical surface in which $\widehat{\lambda}_{SM} \equiv \lambda/J^2$ is *bounded and sufficiently small.* This theorem is proven by controlling the renormalization-group flow in a small neighborhood of the Gaussian fixed point.

[10]This behavior has been *proven* by Hara and Tasaki [290, 298] provided that the bare coupling constant $\widehat{\lambda}_{SM} \equiv \lambda/J^2$ is *sufficiently small.*

Theorem 15.20 *There exists $\varepsilon > 0$ such that the following statement is true: Let $d = 4$, and let $\{\mu_m\}$ be a sequence of nearest-neighbor ferromagnetic φ^4 or Ising models in the single-phase region, with nearest-neighbor coupling strengths $\{J_m\}$ and bare φ^4 coupling strengths $\{\lambda_m\}$. Consider a continuum limit with length rescaling factors $\{\theta_m\} \to \infty$ and field-strength rescaling factors $\{\alpha_m\}$, and assume that*

$$\lambda_m / J_m^2 \leq \varepsilon \tag{15.89}$$

for all m. Then, in the continuum limit ($m \to \infty$), either

(a) all the correlation functions diverge ("nonexistent limit")*; or*

(b) all the correlation functions tend to zero ("boring limit")*; or else*

(c) the limit is a generalized free field ("triviality").

15.3 Non-symmetric regime for $d > 4$

In this section we discuss what is known about continuum limits for φ^4 and Ising models in dimension $d > 4$ from the *non-symmetric regime* ($h \neq 0$ or $\beta > \beta_c$). These results fall into two classes:

- *Bounds relating the one-point function (magnetization) and two-point function.* We show that for a large class of approaches to the critical surface — those avoiding a "high-temperature horn" — it is *impossible* to have a broken-symmetry continuum theory with finite magnetization and nonzero two-point function (Section 15.3.1).

- *Bounds on the connected three-point function u_3.* We show that for a large class of approaches to the critical surface — those avoiding a "low-temperature horn" — the dimensionless renormalized three-point coupling constant $g^{(3)} \equiv -\bar{u}_3/(\chi^{3/2}\xi^{d/2})$ tends to zero. This implies that certain massive scaling limits have $u_3^{\text{cont}} \equiv 0$ (Section 15.3.2).

Unfortunately, though our general results (Propositions 15.21 and 15.22) are straightforward to prove, we are unable to verify the *hypotheses* of these results except by relying on the *non-universal* bounds of Sections 14.4.2 and 14.5. As a consequence, we are able at present to control the scaling limit only for *Ising* (not φ^4!) models. The key stumbling block to further progress appears to be the lack of a GS-class version of the AFe inequality (12.166).

15.3.1 Cases of non-existence of broken-symmetry phase

The standard scenario for broken discrete symmetry in quantum field theory starts from a sequence of lattice theories approaching the critical point from the broken-symmetry phase (e.g. $\beta \downarrow \beta_c$ at $h = 0^+$, or $h \downarrow 0$ at $\beta = \beta_c$), and constructs in the usual way a massive scaling limit (i.e. using a length rescaling factor $\theta \sim \xi$); the result is thought to be a massive continuum quantum field theory with finite nonzero magnetization $S_1^{\text{cont}} \equiv M^{\text{cont}} \equiv \lim_{\theta\to\infty} M^\theta$ and a finite not-identically-zero truncated two-point function $S_{2,T}^{\text{cont}} = \lim_{\theta\to\infty} S_{2,T}^{\theta,\text{dist}}$.

In this section we show that such a scenario *cannot* occur in dimension $d > 4$, at least for Ising models. More generally, we show that for a large class of approaches to the critical surface — those avoiding a "high-temperature horn" — either the rescaled magnetizations $M^\theta \equiv \alpha M$ tend to *infinity*, or else the rescaled truncated two-point functions $S_{2,T}^{\theta,\text{dist}}$ tend to *zero* (or both).

Our proof is based on the properties of the broken-symmetry phase derived in Section 14.5 from the GHS, ABF and AFe inequalities. In particular, it was shown there that for all approaches to the critical surface avoiding a "high-temperature horn", the quantity $\mathcal{J}^2 M^2 \chi$ is bounded below away from zero. We now show that along any such approach to the critical surface, the boundedness of M^θ implies that $S_{2,T}^{\theta,\text{dist}} \to 0$.

Proposition 15.21 *Consider a scaling limit of nearest-neighbor ferromagnetic models in which $M^\theta \equiv \alpha M$ is* bounded, *and $J^2 M^2 \chi$ is* bounded below *away from zero.*

(a) If $J\chi/\theta^{d/2} \to 0$, then $\chi^\theta \to 0$ (and hence $S_{2,T}^{\theta,\text{dist}} \to 0$ in $\mathcal{S}'$).

(b) If $J\chi/\theta^{d-2} \to 0$, then $\alpha^2 J^{-1}\theta^{2-d} \to 0$ (and hence $S_{2,T}^{\theta,\text{dist}} \to 0$ in $\mathcal{S}'$).

Hypotheses (a) and (b) hold, in particular, if $d > 4$ and $\theta \geq \text{const} \times \xi_\phi$ for some $\phi > 0$.

Proof. (a) By definition,

$$\begin{aligned}\chi^\theta \equiv \alpha^2\theta^{-d}\chi &= \frac{(M^\theta)^2}{M^2}\theta^{-d}\chi \\ &= (M^\theta)^2(J^2M^2\chi)^{-1}\theta^{-d}(J\chi)^2 .\end{aligned}$$

By hypothesis, M^θ and $(J^2M^2\chi)^{-1}$ are bounded, and $\theta^{-d}(J\chi)^2 \to 0$. Hence $\chi^\theta \to 0$. It follows trivially that $S_{2,T}^{\theta,\text{dist}} \to 0$ in $\mathcal{S}'(\mathbb{R}^{2d})$ [see the discussion surrounding (15.52)].

(b) By definition,

$$\alpha^2 J^{-1}\theta^{2-d} \equiv \frac{(M^\theta)^2}{M^2}\frac{1}{J\theta^{d-2}} = (M^\theta)^2(J^2M^2\chi)^{-1}\frac{J\chi}{\theta^{d-2}} .$$

By hypothesis, M^θ and $(J^2M^2\chi)^{-1}$ are bounded and $J\chi/\theta^{d-2} \to 0$. Hence $\alpha^2 J^{-1}\theta^{2-d}) \to 0$. It follows from Proposition 15.5 that $S_{2,T}^{\theta,\text{dist}} \to 0$ in $\mathcal{S}'(\mathbb{R}^{2d})$.

The last statement follows from the universal bound $J\chi \leq \text{const}(\phi) \times \xi_\phi^2$, which is a consequence of reflection positivity [see (13.89)]. ∎

The trouble with Proposition 15.21 is that we are unable to verify its key hypothesis — the boundedness below of $J^2M^2\chi$ — except by using the non-universal inequalities of Sections 14.4.2 and 14.5. The stumbling block appears to be the lack of a universal (GS-class) version of the AFe inequality (12.166), and as a result, the lack of a universal lower bound on $J\chi$ in the critical region. Indeed, such a lower bound on $J\chi$ is the *only* missing ingredient: the Landau-Ginzburg lower bound on M, which follows from the ABF inequality, implies a *universal* (GS-class) lower bound on J^2M^3/h outside a "high-temperature horn" (Section 14.4.1); and the universal boundedness below of J^2M^3/h and $J\chi$ implies, by the weak ABF and weak GHS inequalities, the universal boundedness below of $J^2M^2\chi$ [Proposition 14.9(b)]. On the other hand, we are at present rather far from proving such a universal lower bound on $J\chi$. For example, it is not yet proven that $\chi \to \infty$ (or even that χ is bounded below away from zero!) along all paths to the critical point for φ^4 models with *fixed* non-Ising single-spin measure (although we certainly expect that it is true); and it is even less clear that $J\chi$ must tend to ∞ (or even be bounded below) in every reasonable continuum limit of φ^4 models with possibly *varying* single-spin measure.

It should also be noted that the conclusion of Proposition 15.21 cannot be expected to hold without *some* hypothesis (like the boundedness below of $J^2M^2\chi$) that restricts the path of approach to the critical point so as to exclude a "high-temperature horn". Consider, for example, the Gaussian lattice theory (free field): take $t \downarrow 0$ with $h \sim t^{(d+2)/4}$, and use rescaling factors $\theta \sim \xi \sim t^{-1/2}$ and $\alpha(\theta) \sim th^{-1} \sim t^{-(d-2)/4}$. Then the limiting theory is a continuum massive free field with finite nonzero magnetization (and of course finite nonzero susceptibility). Note how this example evades Proposition 15.21:

- For $d \leq 4$, $J^2M^2\chi \sim t^{(d-4)/2} \geq \text{const} > 0$, but hypotheses (a) and (b) fail [i.e. $J\chi/\theta^{d/2}, J\chi/\theta^{d-2} \geq \text{const} > 0$].

- For $d > 4$, hypotheses (a) and (b) hold, but $J^2M^2\chi \sim t^{(d-4)/2} \to 0$.

15.3.2 Partial results on triviality

The result of the previous subsection shows that for Ising models in dimension $d > 4$ (and paths avoiding a "high-temperature horn"), it is impossible to choose the field-strength rescaling factor $\alpha(\theta)$ so as to obtain in the continuum limit a finite magnetization and at the same time a nonzero truncated two-point function. One must therefore choose one or the other. A theory with $S_{2,T}^{\text{cont}} \equiv 0$ is not terribly interesting, as it corresponds to a field φ with no fluctuations. The other possibility is to choose $\alpha(\theta)$ so as to obtain $S_{2,T}^{\text{cont}} \not\equiv 0$; although the magnetizations M^θ tend to infinity, we can subtract out the magnetization and ask whether the *fluctuation fields* $\psi \equiv \varphi - \langle\varphi\rangle$ have rescaled correlation functions with a good continuum limit.[11] If so, we can ask about the properties of this continuum limit, e.g. is it Gaussian?

[11] This is reminiscent of the standard procedure in the central limit theorem for independent random variables with nonzero mean: we must subtract out a mean of order n before applying the rescaling $n^{-1/2}$ to the fluctuation fields.

The *truncated* correlation functions $S_{n,T} \equiv u_n$ of ψ are the same as those of φ (except of course for $n = 1$). So it suffices to ask whether the continuum truncated correlation functions $S_{n,T}^{\text{cont}} \equiv \lim_{\theta\to\infty} S_{n,T}^{\theta,\text{dist}}$ are identically vanishing for $n \geq 3$. We have only a very weak partial result for the *three*-point function: for a large class of *massive* scaling limits — those avoiding a "low-temperature horn" — the truncated three-point function $u_3^{\theta,\text{dist}}$ tends to zero (in fact, the summed three-point function $\bar{u}_3^{\theta}$ tends to zero as well).

Our proof is based on the properties of the broken-symmetry phase for Ising models derived in Section 14.5 from the GHS, ABF and AFe inequalities. In particular, it was shown there that for all approaches to the critical surface avoiding a "low-temperature horn", the quantities $\mathcal{J}^2M^3/h$ and $M/h\chi$ are bounded above. We now show that along any such approach to the critical surface, if χ^θ is bounded (as in a usual massive scaling limit), then the continuum-limit theory is either boring ($S_{2,T}^{\text{cont}} \equiv 0$) or "trivial" ($u_3^{\text{cont}} \equiv 0$).[12]

Proposition 15.22 *Consider a scaling limit of nearest-neighbor ferromagnetic models of GS class in which χ^θ, J^2M^3/h and $M/h\chi$ and hM are all* bounded above. *Then, either*
(a) $\bar{u}_3^\theta \to 0$ (and hence $u_3^{\theta,\text{dist}} \to 0$ in $\mathcal{S}'$); or
(b) $\alpha^2/J\theta^{d-2} \to 0$ (and hence $S_{2,T}^{\theta,\text{dist}} \to 0$ in $\mathcal{S}'$).

Proof. Recall that

$$\bar{u}_3^\theta = \alpha^3\theta^{-2d}\bar{u}_3 \tag{15.90}$$
$$\chi^\theta = \alpha^2\theta^{-d}\chi \tag{15.91}$$

By the GHS inequality (13.54) and the u_3 tree inequality (13.64) [or (13.63) in the Ising case],

$$0 \leq -\bar{u}_3 \leq \text{const} \times \frac{J^2M^4}{h^3}\,. \tag{15.92}$$

Hence

$$0 \leq -\bar{u}_3^\theta \leq \text{const} \times (\chi^\theta)^{3/2} \left(\frac{J^2M^3}{h}\right)^{1/2} \left(\frac{M}{h\chi}\right)^{5/2} \left(\frac{J\chi}{\theta^2}\right) \theta^{-(d-4)/2}\,. \tag{15.93}$$

By hypothesis, χ^θ, J^2M^3/h and $M/h\chi$ are all bounded. Moreover, since $d > 4$, $\theta^{-(d-4)/2} \to 0$ as $\theta \to \infty$. Therefore, there are two possibilities:

(a) If $J\chi/\theta^2$ is *bounded*, then $\bar{u}_3^\theta \to 0$ as $\theta \to \infty$. This implies that $u_3^{\theta,\text{dist}} \to 0$ in $\mathcal{S}'(\mathbb{R}^{3d})$, as discussed around (15.52).

(b) If $J\chi/\theta^2 \to +\infty$, then Proposition 15.6 implies that $\alpha^2/J\theta^{d-2} \to 0$ and hence that $S_{2,T}^{\theta,\text{dist}} \to 0$ in $\mathcal{S}'(\mathbb{R}^{2d})$. ∎

[12] We emphasize that we have *not* proven that $u_n^{\text{cont}} \equiv 0$ for $n \geq 4$, although we expect that it is true.

Again, the trouble is that we are unable to verify the hypotheses of Proposition 15.22 except for Ising models: our proofs of the boundedness of $J^2 M^3/h$ and $M/h\chi$ outside a "low-temperature horn" rely on the AFe inequality (12.166). We need desperately a universal (GS-class) version of this inequality.

References

[1] D. Abraham. n-point functions for the rectangular Ising ferromagnet. Commun. Math. Phys. **60**, 205–213 (1978).

[2] D. B. Abraham, J. T. Chayes, and L. Chayes. Random surface correlation functions. Commun. Math. Phys. **96**, 439–471 (1984).

[3] M. Aizenman. Translation invariance and instability of phase coexistence in the two dimensional Ising system. Commun. Math. Phys. **73**, 83–94 (1980).

[4] M. Aizenman. Proof of the triviality of φ_d^4 field theory and some mean-field features of Ising models for $d > 4$. Phys. Rev. Lett. **47**, 1–4, 886 (E) (1981).

[5] M. Aizenman. Geometric analysis of φ_4^4 fields and Ising models. Parts I and II. Commun. Math. Phys. **86**, 1–48 (1982).

[6] M. Aizenman. Absence of an intermediate phase for a general one-component ferromagnetic models. Phys. Rev. Lett. **54**, 839–842 (1985).

[7] M. Aizenman. The intersection of Brownian paths as a case study of a renormalization group method for quantum field theory. Commun. Math. Phys. **97**, 91–110 (1985).

[8] M. Aizenman. Rigorous studies of critical behavior. In L. Garrido, editor, *Applications of Field Theory to Statistical Mechanics*, pages 125–139. Lecture Notes in Physics Vol. 216, Springer-Verlag, 1985.

[9] M. Aizenman. Rigorous studies of critical behavior II. In D. Szàsz and D. Petz, editors, *Statistical Physics and Dynamical Systems: Rigorous Results.* Birkhäuser, Boston, 1985.

[10] M. Aizenman, D. Barsky, and R. Fernández. The phase transition in a general class of Ising-type models is sharp. J. Stat. Phys. **47**, 343–374 (1987).

[11] M. Aizenman, J. T. Chayes, L. Chayes, J. Fröhlich, and L. Russo. On a sharp transition from area law to perimeter law in a system of random surfaces. Commun. Math. Phys. **92**, 19–69 (1983).

[12] M. Aizenman and R. Fernández. On the critical behavior of the magnetization in high-dimensional Ising models. J. Stat. Phys. **44**, 393–454 (1986).

[13] M. Aizenman and R. Fernández. Critical exponents for long-range interactions. Lett. Math. Phys. **16**, 39–49 (1988).

[14] M. Aizenman and J. Fröhlich. Topological anomalies in the n dependence of the n-state Potts lattice gauge theory. Nucl. Phys. **B235[FS11]**, 1–18 (1984).

[15] M. Aizenman and R. Graham. On the renormalized coupling constant and the susceptibility in φ_4^4 field theory and the Ising model in four dimensions. Nuclear Physics **B225[FS9]**, 261–288 (1983).

[16] M. Aizenman and C. M. Newman. Tree graph inequalities and critical behavior in percolation models. J. Stat. Phys. **36**, 107–143 (1984).

[17] M. Aizenman, private communication.

[18] O. Alvarez. Theory of strings with boundaries. Nucl. Phys. **B216**, 125–184 (1983).

[19] O. Alvarez and P. Windey. Universality in two-dimensional quantum gravity, 1991.

[20] J. Ambjørn and B. Durhuus. Regularized bosonic strings need extrinsic curvature. Phys. Lett. **B188**, 253–257 (1987).

[21] J. Ambjørn, B. Durhuus, and J. Fröhlich. Diseases of triangulated random surface models and possible cures. Nucl. Phys. **B257[FS14]**, 433–449 (1985).

[22] J. Ambjørn, B. Durhuus, J. Fröhlich, and T. Jónsson. Regularized strings with extrinsic curvature. Nucl. Phys. **B290[FS20]**, 480–506 (1987).

[23] J. Ambjørn, B. Durhuus, J. Fröhlich, and P. Orland. The appearance of critical dimensions in regulated string theories. Nucl. Phys. **B270[FS16]**, 457–482 (1986).

[24] D. J. Amit. *Field Theory, the Renormalization Group, and Critical Phenomena* (2nd edition). World Scientific, Singapore, 1984.

[25] N. Andrei and J. H. Lowenstein. Diagonalization of the chiral-invariant Gross-Neveu Hamiltonian. Phys. Rev. Lett. **43**, 1698–1701 (1979).

[26] N. Andrei and J. H. Lowenstein. Derivation of the chiral Gross-Neveu spectrum for arbitrary $SU(N)$ symmetry. Phys. Lett. **90B**, 106–110 (1980).

[27] N. Andrei and J. H. Lowenstein. A direct calculation of the S-matrix of the chiral-invariant Gross-Neveu model. Phys. Lett. **91B**, 401–405 (1980).

[28] C. Aragão de Carvalho, S. Caracciolo, and J. Fröhlich. Polymers and $g\varphi^4$ theory in four dimensions. Nucl. Phys. **B215[FS7]**, 209–248 (1983).

[29] V. I. Arnol'd. *Catastrophe Theory.* Springer-Verlag, Berlin–Heildeberg–New York–Tokyo, 1984.

[30] G. A. Baker Jr. Critical exponent inequalities and the continuity of the inverse range of correlations. Phys. Rev. Lett. **34**, 268–270 (1975).

[31] G. A. Baker Jr. and L. P. Benofy. On the ferromagnetic Ising model in noninteger spatial dimension. J. Stat. Phys. **29**, 699–716 (1982).

[32] T. Balaban. The ultraviolet stability bounds for some lattice σ-models and lattice Higgs-Kibble models. In K. Osterwalder, editor, *Mathematical Problems in Theoretical Physics.* Lecture Notes in Physics #116, Springer-Verlag, Berlin-Heidelberg-New York, 1980.

[33] T. Balaban. Ultraviolet stability for a model of interacting scalar and vector fields. Aarhus preprint 1981/82, No. 19, November 1981.

[34] T. Balaban. Regularity and decay of lattice Green's functions. Commun. Math. Phys. **89**, 571–597 (1983).

[35] T. Balaban. Ultraviolet stability in field theory. The φ_3^4 model. In J. Fröhlich, editor, *Scaling and Self-Similarity in Physics.* Birkhäuser, Boston-Basel-Stuttgart, 1983.

[36] T. Balaban. Propagators and renormalization transformations for lattice gauge theories. I. Commun. Math. Phys. **95**, 17–40 (1984).

[37] T. Balaban. Propagators and renormalization transformations for lattice gauge theories. II. Commun. Math. Phys. **96**, 223–250 (1984).

[38] T. Balaban. Recent results in constructing gauge fields. Physica **124A**, 79–90 (1984).

[39] T. Balaban. Averaging operations for lattice gauge theories. Commun. Math. Phys. **98**, 17–51 (1985).

[40] T. Balaban. Propagators for lattice gauge theories in a background field. Commun. Math. Phys. **99**, 389–434 (1985).

[41] T. Balaban. Spaces of regular gauge field configurations on a lattice and gauge fixing conditions. Commun. Math. Phys. **99**, 75–102 (1985).

[42] T. Balaban. Ultraviolet stability of three-dimensional lattice pure gauge field theories. Commun. Math. Phys. **102**, 255–275 (1985).

[43] T. Balaban. The variational problem and background fields in renormalization group methods in lattice gauge theories. Commun. Math. Phys. **102**, 277–309 (1985).

[44] T. Balaban. Renormalization group approach to lattice gauge field theories. I. Generation of effective actions in a small field approximation and a coupling constant renormalization in four dimensions. Commun. Math. Phys. **109**, 249–301 (1987).

[45] T. Balaban. Renormalization group approach to lattice gauge field theories. II. Cluster expansions. Commun. Math. Phys. **116**, 1–22 (1988).

[46] T. Balaban, J. Imbrie, and A. Jaffe. Exact renormalization group for gauge theories. In G. 't Hooft *et al.*, editor, *Progress in Gauge Field Theory.* Plenum Press, New York–London, 1984.

[47] T. Balaban and A. Jaffe. Constructive gauge theory. In G. Velo and A. S. Wightman, editors, *Fundamental Problems in Gauge Field Theory (Erice 1985).* Plenum Press, New York, 1986.

[48] G. A. Battle III. Non-Gaussian α-positivity of φ_d^{2n} field theories. J. Funct. Anal. **51**, 312–325 (1983).

[49] G. A. Battle III and P. Federbush. A phase cell cluster expansion for Euclidean field theories. Ann. Phys. **142**, 95–139 (1982).

[50] G. A. Battle III and P. Federbush. A phase cell cluster expansion for a hierarchical φ_3^4 model. Commun. Math. Phys. **88**, 263–293 (1983).

[51] G. A. Battle III and P. Federbush. Ondelettes and phase cell cluster expansions, a vindication. Commun. Math. Phys. **109**, 417–419 (1987).

[52] K. Baumann. When is a field theory a generalized free field? Commun. Math. Phys. **43**, 221–223 (1975).

[53] K. Baumann. On the two-point functions of interacting Wightman fields. J. Math. Phys. **27**, 828–831 (1986).

[54] K. Baumann. Three remarks on Powers' theorem about irreducible fields fulfilling CAR. J. Math. Phys. **27**, 2373–2378 (1986).

[55] K. Baumann. On relativistic irreducible quantum fields fulfilling CCR. J. Math. Phys. **28**, 697–704 (1987).

[56] K. Baumann. On canonical irreducible quantum field theories describing bosons and fermions. J. Math. Phys. **29**, 1225–1230 (1988).

[57] R. B. Baxter. *Exactly Solved Models in Statistical Mechanics.* Academic Press, London–New York–etc., 1982.

[58] A. A. Belavin, A. M. Polyakov, and A. B. Zamolodchikov. Infinite conformal theory in two-dimensional quantum field theory. Nucl. Phys. **B241**, 333–380 (1984).

[59] G. Benfatto, M. Cassandro, G. Gallavotti, F. Nicolò, E. Olivieri, E. Presutti, and E. Scacciatelli. Some probabilistic techniques in field theory. Commun. Math. Phys. **59**, 143–166 (1978).

[60] G. Benfatto, M. Cassandro, G. Gallavotti, F. Nicolò, E. Olivieri, E. Presutti, and E. Scacciatelli. Ultraviolet stability in Euclidean scalar field theories. Commun. Math. Phys. **71**, 95–130 (1980).

[61] B. Berg and A. Billoire. A Monte Carlo simulation of random surfaces. Phys. Lett. **139B**, 297–300 (1984).

[62] T. H. Berlin and M. Kac. The spherical model of a ferromagnet. Phys. Rev. **86**, 821–835 (1952).

[63] D. Bessis, C. Itzykson, and J. B. Zuber. Quantum field theory techniques in graphical enumeration. Adv. Appl. Math. **1**, 109–157 (1980).

[64] A. Bilal. 2D gravity from matrix models. Preprint CERN-TH.5867/90, 1990. Invited lecture at the Johns Hopkins Workshop; Debrecen, Hungary, August 1990.

[65] K. Binder. Monte Carlo investigations of phase transitions and critical phenomena. In C. Domb and M. S. Green, editors, *Phase Transitions and Critical Phenomena (Vol. 5B)*. Academic Press, Londom–New York–San Francisco, 1985.

[66] P. M. Bleher. Integration of functions in a space with complex number of dimensions. Teor. Mat. Fiz. **50**, 370–382 (1982). [Theor. Math. Phys. **50**, 243–251 (1982)].

[67] P. M. Bleher and Ya. G. Sinai. Investigation of the critical point in models of the type of Dyson's hierarchical models. Commun. Math. Phys. **33**, 23–42 (1973).

[68] P. M. Bleher and Ya. G. Sinai. Critical indices for Dyson's asymptotically hierarchical models. Commun. Math. Phys. **45**, 247–278 (1975).

[69] N. Bohr and L. Rosenfeld. Zur Frage der Messbarbeit der elektromagnetischen Feldgrössen. Kgl. Danske Vid. Selsk. Mat. Fys. Medd. **12**, no. 8 (1933).

[70] H. J. Borchers and J. Yngvason. Necessary and sufficient conditions for integral representations of Wightman functionals at Schwinger points. Commun. Math. Phys. **47**, 197–213 (1976).

[71] D. V. Boulatov and V. A. Kazakov. The Ising model on a random planar lattice: the structure of the phase transition and the exact critical exponents. Phys. Lett. **B186**, 379–384 (1987).

[72] D. V. Boulatov, V. A. Kazakov, I. K. Kostov, and A. A. Migdal. Analytical and numerical study of a model of dynamically triangulated random surfaces. Nucl. Phys. **B275[FS17]**, 641–686 (1986).

[73] P. Boutroux. Recherches sur les transcendantes de M. Painlevé. Ann. Sci. École Norm. Sup. **30**, 255–375 (1913).

[74] A. Bovier and G. Felder. Skeleton inequalities and the asymptotic nature of perturbation theory for φ^4-theories in two and three dimensions. Commun. Math. Phys. **93**, 259–275 (1984).

[75] A. Bovier, G. Felder, and J. Fröhlich. On the critical properties of the edwards and the self-avoiding walk model of polymer chains. Nuclear Physics **B230[FS10]**, 119–147 (1984).

[76] A. Bovier, J. Fröhlich, and U. Glaus. Branched polymers and dimensional reduction. In K. Osterwalder and R. Stora, editors, *Critical Phenom-*

ena, Random Systems, Gauge Theories (Les Houches 1984). North-Holland, Amsterdam–New York–Oxford–Tokyo, 1986.

[77] E. Brézin, J. C. Le Guillou, and J. Zinn-Justin. Field theoretical approach to critical phenomena. In C. Domb and M. S. Green, editors, *Phase Transitions and Critical Phenomena (Vol. 6)*. Academic Press, London-New York-San Francisco, 1976.

[78] E. Brézin, C. Itzykson, G. Parisi, and J. B. Zuber. Planar diagrams. Commun. Math. Phys. **59**, 35–51 (1978).

[79] E. Brézin and V. A. Kazakov. Exactly solvable field theories of closed strings. Phys. Lett. **B236**, 144–150 (1990).

[80] E. Brézin, V. A. Kazakov, and Al. B. Zamolodchikov. Scaling violation in a field theory of closed strings in one physical dimension. Nucl. Phys. **338B**, 673–688 (1990).

[81] E. Brézin, E. Marinari, and G. Parisi. A non-perturbative ambiguity free solution of a string model. Phys. Lett. **B242**, 35–38 (1990).

[82] E. Brézin and H. Neuberger. Large-N scaling limits of symmetric-matrix models as systems of fluctuating unoriented surfaces. Phys. Rev. Lett. **65**, 2098–2101 (1990).

[83] E. Brézin and J. Zinn-Justin. Spontaneous breakdown of continuous symmetries near two dimensions. Phys. Rev. **B14**, 3110–3120 (1976).

[84] J. Bricmont, A. El-Mellouki, and J. Fröhlich. Random surfaces in statistical mechanics: roughening, rounding, wetting. J. Stat. Phys. **42**, 743–798 (1986).

[85] J. Bricmont, J.-R. Fontaine, and E. Speer. Perturbation about the mean field critical point. Commun. Math. Phys. **86**, 337–362 (1982).

[86] J. Bricmont and A. Kupiainen. Lower critical dimension for the random-field Ising model. Phys. Rev. Lett. **59**, 1829–1832 (1987).

[87] J. Bricmont and A. Kupiainen. Phase transition in the random-field Ising model. Commun. Math. Phys. **116**, 539–572 (1988).

[88] B. Broda. Magnetic random-walk representation for scalar QED and the triviality problem. Phys. Rev. Lett. **63**, 2709–2712 (1989).

[89] D. Brydges. A rigorous approach to Debye screening in dilute classical Coulomb systems. Commun. Math. Phys. **58**, 313–350 (1978).

[90] D. Brydges. Field theories and Symanzik's polymer representation. In P. Dita, V. Georgescu, and R. Purice, editors, *Gauge Theories: Fundamental Interactions and Rigorous Results*. 1981 Brasov lectures, Birkhäuser, Boston-Basel-Stuttgart, 1982.

[91] D. Brydges and P. Federbush. Debye screening. Commun. Math. Phys. **73**, 197–246 (1980).

[92] D. Brydges, J. Fröhlich, and T. Spencer. The random walk representation of classical spin systems and correlation inequalities. Commun. Math. Phys. **83**, 123–150 (1982).

[93] D. Brydges and T. Spencer. Self-avoiding walks in 5 or more dimensions. Commun. Math. Phys. **97**, 125–148 (1985).

[94] D. Brydges and T. Spencer. Self-avoiding random walk and the renormalization group. In L. Garrido, editor, *Applications of Field Theory to Statistical Mechanics (Sitges, 1984)*, Lecture Notes in Physics #216, pages 189–198. Springer-Verlag, Berlin–Heidelberg–New York, 1986.

[95] D. Brydges and H. Z. Yau. Grad ϕ perturbations of massless Gaussian fields. Commun. Math. Phys. **129**, 351–392 (1990).

[96] D. C. Brydges, J. Fröhlich, and A. D. Sokal. A new proof of the existence and nontriviality of the continuum φ_2^4 and φ_3^4 quantum field theories. Commun. Math. Phys. **91**, 141–186 (1983).

[97] D. C. Brydges, J. Fröhlich, and A. D. Sokal. The random walk representation of classical spin systems and correlation inequalities. II. the skeleton inequalities. Commun. Math. Phys. **91**, 117–139 (1983).

[98] D. C. Brydges, C. Giffen, B. Durhuus, and J. Fröhlich. Surface representations of Wilson loop expectations in lattice gauge theory. Nucl. Phys. **B275[FS17]**, 459–487 (1986).

[99] K. Burdzy and G. F. Lawler. Non-intersection exponents for Brownian paths. I. Existence and an invariance principle. Prob. Th. Rel. Fields **84**, 393–410 (1990).

[100] K. Burdzy and G. F. Lawler. Non-intersection exponents for Brownian paths. II. Estimates and applications to a random fractal. Ann. Probab. **18**, 981–1009 (1990).

[101] K. Burdzy, G. F. Lawler, and T. Polaski. On the critical exponent for random walk intersections. J. Stat. Phys. **56**, 1–12 (1989).

[102] D. J. E. Callaway. Triviality pursuit: Can elementary scalar particles exist? Phys. Reports **167**, 241–320 (1988).

[103] A. Cappelli, C. Itzykson, and J. B. Zuber. The A–D–E classification of minimal $A_1^{(1)}$ conformal-invariant theories. Commun. Math. Phys. **113**, 1–26 (1987).

[104] A. Cappelli, C. Itzykson, and J. B. Zuber. Modular invariant partition functions in two dimensions. Nucl. Phys. **B280[FS18]**, 445–465 (1987).

[105] J. L. Cardy. Conformal invariance. In C. Domb and J. L. Lebowitz, editors, *Phase Transitions and Critical Phenomena (Vol. 11)*. Academic Press, London–New York–San Francisco, 1987.

[106] M. Cassandro and G. Jona-Lasinio. Critical point behavior and probability theory. Adv. Physics **27**, 913–941 (1978).

[107] A. H. Chamsedine. A solution to two dimensional quantum gravity: Non-critical strings. Zürich University Preprint, 1990; to be published in Phys. Lett. B.

[108] Z. Cieselski and S. J. Taylor. First passage times and sojourn times for Brownian motion in space and the exact Hausdorff measure of the sample path. Trans. Amer. Math. Soc. **103**, 434–450 (1962).

[109] P. Colella and O. E. Lanford. Sample field behavior for the free Markov random field. In G. Velo and A. Wightman, editors, *Constructive Quantum Field Theory*. Lecture Notes in Physics #25, Springer-Verlag, Berlin, 1973.

[110] S. Coleman, R. Jackiw, and H. D. Politzer. Spontaneous symmetry breaking in the $O(N)$ model for large N. Phys. Rev. **D10**, 2491–2499 (1974).

[111] P. Collet and J.-P. Eckmann. *A Renormalization Group Analysis of the Hierarchical Model in Statistical Mechanics*. Lecture Notes in Physics #74. Springer-Verlag, Berlin-Heidelberg-New York, 1978.

[112] P. Collet and J.-P. Eckmann. *Iterated Maps of the Interval as Dynamical Systems*. Progress in Physics, Vol. 1. Birkhäuser, Boston-Basel-Stuttgart, 1980.

[113] E Date, M. Jimbo, A. Kuniba, T. Miwa, and M. Okado. Exactly solvable SOS models: Local height probabilities and theta function identities. Nucl. Phys. **B290[FS20]**, 231–273 (1987).

[114] F. David. Infrared divergences of membrane models. Phys. Lett. **102B**, 193–195 (1981).

[115] F. David. A model of random surfaces with non-trivial critical behaviour. Nucl. Phys. **B257[FS14]**, 543–576 (1985).

[116] F. David. Randomly triangulated surfaces in -2 dimensions. Phys. Lett. **159B**, 303–306 (1985).

[117] F. David. Loop equations and non-perturbative effects in two-dimensional quantum gravity. Mod. Phys. Lett. A **5**, 1019–1029 (1990).

[118] F. David. Phases of the large-N matrix models and non-perturbative effects in $2D$ gravity. Nucl. Phys. **B348**, 507–524 (1991).

[119] C. de Calan, P. A. Faria da Veiga, J. Magnen, and R. Sénéor. Constructing the three-dimensional Gross-Neveu model with a large number of flavor components. preprint, February 1991.

[120] P. G. de Gennes. Exponents for the excluded volume problem as derived by the Wilson method. Phys. Lett. **A38**, 339–340 (1970).

[121] P. G. de Gennes. *Scaling Concepts in Polymer Physics.* Cornell University Press, 1979.

[122] P. G. de Gennes and C. Taupin. Microemulsions and the flexibility of oil/water interfaces. J. Phys. Chem. **86**, 2294–2304 (1982).

[123] B. de Wit and G. 't Hooft. Non-convergence of the $1/N$ expansion for $SU(N)$ gauge fields on a lattice. Phys. Lett. **69B**, 61–64 (1977).

[124] J. DeConinck and C. M. Newman. The magnetization-energy scaling limit in high dimension. J. Stat. Phys. **59**, 1451–1467 (1990).

[125] J. des Cloizeaux. Private communication, cited in: E. Brézin, "The renormalization group in the theory of critical phenomena", in: *Order and Fluctuation in Equilibrium and Nonequilibrium Statistical Mechanics* (17th Solvay Conference, 1978), edited by G. Nichols, G. Dewel and J. W. Turner, Wiley-Interscience, New York, 1981.

[126] J. des Cloizeaux. The Lagrangian theory of polymer solutions at intermediate concentrations. J. Physique (Paris) **36**, 281–291 (1975).

[127] S. Deser and B. Zumino. A complete action for the spinning string. Phys. Lett. **65B**, 369–373 (1976).

[128] E. d'Hoker and D. H. Phong. The geometry of string perturbation theory. Rev. Mod. Phys **60**, 917–1065 (1987).

[129] R. Dijkgraaf, E. Verlinde, and H. Verlinde. Loop equations and Virasoro constraints in non-perturbative 2-d quantum gravity. Nucl. Phys. **348B**, 435–456 (1991).

[130] R. Dijkgraf and E. Witten. Mean field theory, topological field theory and multi-matrix models. Nucl. Phys. **342B**, 486–522 (1990).

[131] J. Distler and H. Kawai. Conformal field theory and 2-D quantum gravity. Nucl. Phys. **B321**, 509–527 (1989).

[132] R. L. Dobrushin. Existence of a phase transition in the two-dimensional and three-dimensional Ising models. Soviet Phys. Doklady **10**, 111–113 (1965).

[133] R. L. Dobrushin. Gibbs state describing coexistence of phases for a three-dimensional Ising model. Theor. Prob. Appl. **17**, 582–600 (1972).

[134] R. L. Dobrushin. Investigation of Gibbsian states for three-dimensional lattice systems. Theor. Prob. Appl. **18**, 253–271 (1973).

[135] C. Domb, A. J. Barrett, and M. Lax. Self-avoiding walks and real polymer chains. J. Phys. **A6**, L82–L87 (1973).

[136] C. Domb and M. S. Green (Eds.). *Phase Transitions and Critical Phenomena, Vol. 6.* Academic Press, London–New York, 1976.

[137] C. Domb and D. L. Hunter. On the critical behavior of ferromagnets. Proc. Phys. Soc. (London) **86**, 1147–1151 (1965).

[138] C. Domb and G. S. Joyce. Cluster expansion for a polymer chain. J. Phys. **C5**, 956–976 (1972).

[139] Vl. S. Dotsenko and V. A. Fateev. Conformal algebra and multipoint correlation functions in 2D statistical mechanics. Nucl. Phys. **B240[FS12]**, 312–348 (1984).

[140] Vl. S. Dotsenko and V. A. Fateev. Four-point correlation functions and the operator algebra in 2D conformal invariant theories with central charge $c \leq 1$. Nucl. Phys. **B251[FS13]**, 691–734 (1985).

[141] Vl. S. Dotsenko and V. A. Fateev. Operator algebras of two-dimensional conformal theories with central charge $c \leq 1$. Phys. Lett. **B154**, 291–295 (1985).

[142] M. Douglas. Strings in less than one dimension and the generalized KdV hierarchies. Phys. Lett. **B238**, 176–180 (1990).

[143] M. R. Douglas, N. Seiberg, and S. H. Shenker. Flow and instability in quantum gravity. Phys. Lett. **B244**, 381–386 (1990).

[144] M. R. Douglas and S. H. Shenker. Strings in less than one dimension. Nucl. Phys. **B335**, 635–654 (1990).

[145] W. Driessler. A strengthened no-go theorem for Fermi fields obeying CAR. Nuovo Cim. **21A**, 583–591 (1974).

[146] W. Driessler and J. Fröhlich. The reconstruction of local observable algebras from the Euclidean Green's functions of relativistic quantum field theories. Ann. Inst. Henri Poincaré **A27**, 221–236 (1977).

[147] J. M. Drouffe, G. Parisi, and N. Sourlas. Strong coupling phase in lattice gauge theories at large dimension. Nucl. Phys. **B161**, 397–416 (1980).

[148] F. Dunlop and C. M. Newman. Multicomponent field theories and classical rotators. Commun. Math. Phys. **44**, 223–235 (1975).

[149] B. Duplantier. Partition function of magnetic random walks. Phys. Rev. Lett. **66**, 240 (1991).

[150] B. Duplantier and K.-H. Kwon. Conformal invariance and intersections of random walks. Phys. Rev. Lett. **61**, 2514–2517 (1988).

[151] B. Durhuus and J. Fröhlich. A connection between ν-dimensional Yang-Mills theory and $(\nu - 1)$-dimensional non-linear σ-models. Commun. Math. Phys. **75**, 103–151 (1980).

[152] B. Durhuus, J. Fröhlich, and G. Felder. Unpublished.

[153] B. Durhuus, J. Fröhlich, and T. Jónsson. Self-avoiding and planar random surfaces on the lattice. Nucl. Phys. **B225[FS9]**, 185–203 (1983).

[154] B. Durhuus, J. Fröhlich, and T. Jónsson. Critical behaviour in a model of planar random surfaces. Nucl. Phys. **B240[FS12]**, 453–480 (1984).

[155] B. Durhuus, J. Fröhlich, and T. Jónsson. Critical properties of a model of planar random surfaces. Phys. Lett. **137B**, 93–97 (1984).

[156] B. Durhuus, P. Olesen, and J. L. Petersen. Polyakov's quantized string with boundary terms. Nucl. Phys. **B198**, 157–188 (1982).

[157] P. M. Duxbury and S. L. A. de Quieroz. A unifying model of generalised random walks. J. Phys. **A18**, 661–670 (1985).

[158] A. Dvoretsky, P. Erdös, and S. Kakutani. Double points of Brownian motion in n-space. Acta Sci. Math. (Szeged) **12**, 75–81 (1950).

[159] J.-P. Eckmann and H. Epstein. Time-ordered products and Schwinger functions. Commun. Math. Phys. **64**, 95–130 (1979).

[160] S. F. Edwards. The statistical mechanics of polymers with excluded volume. Proc. Phys. Soc. (London) **85**, 613–624 (1965).

[161] T. Eguchi and H. Kawai. The number of random surfaces on the lattice and the large-N gauge theory. Phys. Lett. **110B**, 143–147 (1980).

[162] R. S. Ellis, J. L. Monroe, and C. M. Newman. The GHS and other correlation inequalities for a class of even ferromagnets. Commun. Math. Phys. **46**, 167–182 (1976).

[163] R. S. Ellis and C. M. Newman. Necessary and sufficient conditions for the GHS inequality with applications to analysis and probability. Trans. Am. Math. Soc. **237**, 83–99 (1978).

[164] P. Erdös and S. J. Taylor. Some intersection properties of random walk paths. Acta Math. Acad. Sci. Hungar. **11**, 231–248 (1960).

[165] G. Eyink. Quantum field theory models on fractal spacetime. I. Introduction and overview. Commun. Math. Phys. **125**, 613–636 (1989).

[166] G. Eyink. Quantum field theory models on fractal spacetime. II. Hierarchical propagators. Commun. Math. Phys. **126**, 85–101 (1989).

[167] L. Faddeev. Integrable models in (1+1)-dimensional quantum field theory. In J. B. Zuber and R. Stora, editors, *Recent Advances in Field Theory and Statistical Mechanics (Les Houches 1982)*. North-Holland, Amsterdam–New York–Oxford–Tokyo, 1986.

[168] L. Faddeev and L. A. Tachtajan. *Hamiltonian Models in the Theory of Solitons*. Springer-Verlag, Berlin-Heidelberg-New York, 1987.

[169] L. D. Faddeev. Two-dimensional integrable models in quantum field theory. Physica Scripta **24**, 832–835 (1981).

[170] M. S. Falkovitz, M. Seul, H. L. Frisch, and H. M. McConnell. Theory of periodic structures in lipid bilayer membranes. Proc. Natl. Acad. Sci. USA **79**, 3918–3921 (1982).

[171] P. Federbush. A phase cell approach to Yang-Mills theory. I. Modes, lattice-continuum duality. Commun. Math. Phys. **107**, 319–329 (1986).

[172] P. Federbush. A phase cell approach to Yang-Mills theory. III. Local stability, modified renormalization group transformation. Commun. Math. Phys. **110**, 293–309 (1987).

[173] P. Federbush. A phase cell approach to Yang-Mills theory. VI. Non-abelian lattice-continuum duality. Ann. Inst. H. Poincaré **47**, 17–23 (1987).

[174] P. Federbush. Quantum field theory in ninety minutes. Bull. Amer. Math. Soc. **17**, 93–103 (1987).

[175] P. Federbush. A phase cell approach to Yang-Mills theory. IV. The choice of variables. Commun. Math. Phys. **114**, 317–343 (1988).

[176] P. Federbush. A phase cell approach to Yang-Mills theory. V. Analysis of a chunk. Commun. Math. Phys. **127**, 433–457 (1990).

[177] P Federbush and T. Kennedy. Surface effects in Debye screening, a model example. Commun. Math. Phys. **102**, 361–423 (1985).
[178] P. Federbush and C. Williamson. A phase cell approach to Yang-Mills theory. II. Analysis of a mode. J. Math. Phys. **28**, 1416–1419 (1987).
[179] P. G. Federbush and K. A. Johnson. The uniqueness of the two-point function. Phys. Rev. **120** (1960).
[180] M. Feigenbaum. Quantitative universality for a class of nonlinear transformations. J. Stat. Phys. **19**, 25–52 (1978).
[181] M. Feigenbaum. The universal metric properties of nonlinear transformations. J. Stat. Phys. **21**, 669–706 (1979).
[182] G. Felder. Construction of a non-trivial planar field theory with ultraviolet-stable fixed point. Commun. Math. Phys. **102**, 139–155 (1985).
[183] G. Felder. *Renormalization Group, Tree Expansion, and Non-Renormalizable Quantum Field Theories.* PhD thesis, ETH–Zürich, 1986.
[184] G. Felder. Renormalization group in the local potential approximation. Commun. Math. Phys. **111**, 101–121 (1987).
[185] G. Felder and J. Fröhlich. Intersection properties of simple random walks: A renormalization group approach. Commun. Math. Phys. **97**, 111–124 (1985).
[186] G. Felder and G. Gallavotti. Perturbation theory and non-renormalizable scalar fields. Commun. Math. Phys. **102**, 549–571 (1985).
[187] J. Feldman, T. Hurd, L. Rosen, and J. Wright. *QED: A Proof of Renormalizability.* Lecture Notes in Physics #312. Springer-Verlag, Berlin–Heidelberg–New York, 1988.
[188] J. Feldman, J. Magnen, V. Rivasseau, and R. Sénéor. Bounds on completely convergent Euclidean Feynman graphs. Commun. Math. Phys. **98**, 273–288 (1985).
[189] J. Feldman, J. Magnen, V. Rivasseau, and R. Sénéor. Bounds on renormalized Feynman graphs. Commun. Math. Phys. **100**, 23–55 (1985).
[190] J. Feldman, J. Magnen, V. Rivasseau, and R. Sénéor. Infra-red ϕ_4^4. In K. Osterwalder and R. Stora, editors, *Critical Phenomena, Random Systems, Gauge Theories (Les Houches 1984).* North-Holland, Amsterdam–New York–Oxford–Tokyo, 1986.
[191] J. Feldman, J. Magnen, V. Rivasseau, and R. Sénéor. A renormalizable field theory: The massive Gross-Neveu model in two dimensions. Commun. Math. Phys. **103**, 67–103 (1986).
[192] J. Feldman, J. Magnen, V. Rivasseau, and R. Sénéor. Construction and Borel summability of infrared φ_4^4 by a phase expansion. Commun. Math. Phys. **109**, 437–480 (1987).
[193] J. Feldman and K. Osterwalder. The Wightman axioms and the mass gap for weakly coupled $(\varphi^4)_3$ quantum field theories. Ann. Phys. **97**, 80–135 (1976).
[194] R. Fernández, unpublished.
[195] S. Ferrara, R. Gatto, and A. F. Grillo. Conformal algebra in two space-time dimensions and the Thirring model. Nuovo Cimento **12A**, 959–968 (1972).
[196] D. Fisher, J. Fröhlich, and T. Spencer. Ising model in a random magnetic field. J. Stat. Phys. **34**, 863–870 (1984).
[197] M. E. Fisher. Critical temperatures of anisotropic Ising lattices. II. general upper bounds. Phys. Rev. **162**, 480–485 (1967).

[198] M. E. Fisher. Theory of equilibrium critical phenomena. Rep. Prog. Phys. **30**, 615–730 (1967).

[199] M. E. Fisher. Rigorous inequalities for critical-point correlation exponents. Phys. Rev. **180**, 594–600 (1969).

[200] M. E. Fisher. General scaling theory for critical points. In B. Lundqvist and S. Lundqvist, editors, *Collective Properties of Physical Systems (Proc. 24th Nobel Symposium)*. Academic Press, New York–London, 1973.

[201] M. E. Fisher. Crossover effects and operator expansions. In J. D. Gunton and M. S. Green, editors, *Renormalization Group in Critical Phenomena and Quantum Field Theory: Proceedings of a Conference*, pages 65–72. Temple University, Philadelphia, 1974.

[202] M. E. Fisher. The renormalization group in the theory of critical behavior. Rev. Mod. Phys. **46**, 597–616 (1974).

[203] M. E. Fisher. Walks, walls, wetting and melting. J. Stat. Phys. **34**, 667–729 (1984).

[204] M. E. Fisher. Critical endpoints, interfaces and walls. Physica A **172**, 77–86 (1991).

[205] M. E. Fisher and D. S. Fisher. Wall wandering and the dimensionality dependence of the commensurate-incommensurate transition. Phys. Rev. **B25**, 3192–3198 (1982).

[206] M. E. Fisher and D. S. Gaunt. Ising model and self-avoiding walks on hypercubical lattices on "high-density" expansions. Phys. Rev. **133**, A224–A239 (1964).

[207] M. E. Fisher, S. K. Ma, and B. G. Nickel. Critical exponents for long-range interactions. Phys. Rev. Lett. **29**, 917–920 (1972).

[208] M. E. Fisher and M. F. Sykes. Excluded-volume problems and the Ising model of ferromagnetism. Phys. Rev. **114**, 45–48 (1959).

[209] D. Friedan, Ż. Qiu, and S. Shenker. Conformal invariance, unitarity, and critical exponents in two dimensions. Phys. Rev. Lett. **52**, 1575–1578 (1984).

[210] J. Fröhlich. Phase transitions, Goldstone bosons and topological superselection rules. Acta Physica Austriaca **Suppl. 15**, 133–269 (1976).

[211] J. Fröhlich. The pure phases, the irreducible quantum fields, and dynamical symmetry breaking in Symanzik-Nelson positive quantum field theories. Ann. Phys. **97**, 1–54 (1976).

[212] J. Fröhlich. Some comments on the crossover between strong and weak coupling in $SU(2)$ pure Yang-Mills theory. Phys. Rep. **67**, 137–149 (1980).

[213] J. Fröhlich. On the triviality of $\lambda\varphi_4^4$ theories and the approach to the critical point in $d_{(}\gtrsim_{)}4$ dimensions. Nuclear Physics **B200**[**FS4**], 281–296 (1982).

[214] J. Fröhlich. Quantum field theory in terms of random walks and random surfaces. In G. 't Hooft *et al.* (eds.), editor, *Progress in Gauge Field Theory*. Plenum Press, New York–London, 1984.

[215] J. Fröhlich. Survey of random surface theory. In J. Ambjørn, B. J. Durhuus, and J. L. Petersen, editors, *Recent Developments in Quantum Field Theory*. North-Holland, Amsterdam, 1985.

[216] J. Fröhlich. The statistical mechanics of surfaces. In L. Garrido, editor, *Applications of Field Theory to Statistical Mechanics (Sitges, 1984)*, Lecture Notes in Physics #216, pages 31–57. Springer-Verlag, Berlin–Heidelberg–New York, 1986.

[217] J. Fröhlich, J.-P. Eckman, and H. Epstein. Asymptotic perturbation expansion for the S-matrix and the definition of time-ordered functions in relativistic quantum field models. Ann. Inst. H. Poincaré **25**, 1–34 (1976).

[218] J. Fröhlich, R. Israel, E. H. Lieb, and B. Simon. Phase transitions and reflection positivity. I. General theory and long range lattice models. Commun. Math. Phys. **62**, 1–34 (1978).

[219] J. Fröhlich and E. H. Lieb. Phase transitions in anisotropic lattice spin systems. Commun. Math. Phys. **60**, 233–267 (1978).

[220] J. Fröhlich and C. Pfister. On the absence of spontaneous symmetry breaking and of crystalline ordering in two-dimensional systems. Commun. Math. Phys. **81**, 277–298 (1981).

[221] J. Fröhlich, C. E. Pfister, and T. Spencer. On the statistical mechanics of surfaces. In *Stochastic Processes in Quantum Theory and Statistical Physics*, Lecture Notes in Physics #173, pages 169–199. Springer-Verlag, Berlin–Heidelberg–New York, 1982.

[222] J. Fröhlich, B. Simon, and T. Spencer. Infrared bounds, phase transition and continuous symmetry breaking. Commun. Math. Phys. **50**, 79–85 (1976).

[223] J. Fröhlich and A. D. Sokal. The random-walk representation of classical spin systems and correlation inequalities. III. Nonzero magnetic field. Unpublished.

[224] J. Fröhlich and T. Spencer. The Kosterlitz-Thouless phase transition in the two-dimensional plane rotator and Coulomb gas. Phys. Rev. Lett. **46**, 1006–1009 (1981).

[225] J. Fröhlich and T. Spencer. The Kosterlitz-Thouless transition in two-dimensional Abelian spin systems and the Coulomb gas. Commun. Math. Phys. **81**, 527–602 (1981).

[226] J. Fröhlich and T. Spencer. Massless phases and symmetry restoration in Abelian gauge theories and spin systems. Commun. Math. Phys. **83**, 411–454 (1982).

[227] J. Fröhlich and T. Spencer. The phase transition in the one-dimensional Ising model with $1/r^2$ interaction energy. Commun. Math. Phys. **84**, 87–101 (1982).

[228] J. Fröhlich and T. Spencer. Some recent rigorous results in the theory of phase transitions and critical phenomena. Séminaire Bourbaki #586, Astérisque **92–93** (1982).

[229] J. Fröhlich and T. Spencer. The Berezinskii-Kosterlitz-Thouless transition (energy-entropy arguments and renormalization in defect gases). In J. Fröhlich, editor, *Scaling and Self-Similarity in Physics*. Birkhäuser, Boston–Basel–Stuttgart, 1983.

[230] G. Gallavotti. Some aspects of the renormalization problems in statistical mechanics and field theory. Mem. Accad. Lincei **15**, 23–59 (1978).

[231] G. Gallavotti. On the ultraviolet stability in statistical mechanics and field theory. Ann. Mat. Pura Appl. **120**, 1–23 (1979).

[232] G. Gallavotti. Renormalization theory and ultra-violet stability for scalar fields via renormalization group methods. Rev. Mod. Phys. **57**, 471–561 (1985).

[233] G. Gallavotti and G. Jona-Lasinio. Limit theorems for multidimensional markov processes. Commun. Math. Phys. **41**, 301–307 (1975).

[234] G. Gallavotti, A. Martin-Löf, and S. Miracle-Solé. Some problems connected with the description of coexisting phases at low temperatures in the Ising

model. In A. Lenard, editor, *Statistical Mechanics and Mathematical Problems*, pages 162–204. Springer-Verlag, 1971.

[235] G. Gallavotti and F. Nicolò. Renormalization theory in four-dimensional scalar fields (I). Commun. Math. Phys. **100**, 545–590 (1985).

[236] G. Gallavotti and F. Nicolò. Renormalization theory in four-dimensional scalar fields (II). Commun. Math. Phys. **101**, 247–282 (1985).

[237] D. S. Gaunt and A. J. Guttmann. Asymptotic analysis of coefficients. In C. Domb and M. S. Green, editors, *Phase Transitions and Critical Phenomena (Vol. 3)*. Academic Press, London–New York–San Francisco, 1972.

[238] K. Gawedzki, R. Kotecký, and A. Kupiainen. Coarse graining approach to first-order phase transitions. J. Stat. Phys. **47**, 701–724 (1987).

[239] K. Gawedzki and A. Kupiainen. A rigorous block spin approach to massless lattice theories. Commun. Math. Phys. **77**, 31–64 (1980).

[240] K. Gawedzki and A. Kupiainen. Renormalization group study of a critical lattice model. I. Convergence to the line of fixed points. Commun. Math. Phys. **82**, 407–433 (1981).

[241] K. Gawedzki and A. Kupiainen. Block spin renormalization group for dipole gases and $(\nabla\phi)^4$. Ann. Phys. **147**, 198–243 (1983).

[242] K. Gawedzki and A. Kupiainen. Non-Gaussian fixed points of the block spin transformation. Hierarchical model approximation. Commun. Math. Phys. **89**, 191–220 (1983).

[243] K. Gawedzki and A. Kupiainen. Gross-Neveu model through convergent perturbation expansions. Commun. Math. Phys. **102**, 1–30 (1985).

[244] K. Gawedzki and A. Kupiainen. Massless lattice φ^4_4 theory: Rigorous control of a renormalizable asymptotically free model. Commun. Math. Phys. **99**, 197–252 (1985).

[245] K. Gawedzki and A. Kupiainen. Non-trivial continuum limit of a φ^4_4 model with negative coupling constant. Nucl. Phys. **B257[FS14]**, 474–504 (1985).

[246] K. Gawedzki and A. Kupiainen. Renormalization of a non-renormalizable quantum field theory. Nucl. Phys. **B262**, 33–48 (1985).

[247] K. Gawedzki and A. Kupiainen. Asymptotic freedom beyond perturbation theory. In K. Osterwalder and R. Stora, editors, *Critical Phenomena, Random Systems, Gauge Theories (Les Houches 1984)*. North-Holland, Amsterdam–New York–Oxford–Tokyo, 1986.

[248] K. Gawedzki and A. Kupiainen. Continuum limit of the hierarchical $O(N)$ non-linear σ-model. Commun. Math. Phys. **106**, 523–550 (1986).

[249] K. Gawedzki, A. Kupiainen, and B. Tirozzi. Non-Gaussian scaling limits. Hierarchical model approximation. J. Stat. Phys. **35**, 267–284 (1984).

[250] Y. Gefen, B. Mandelbrot, and A. Aharony. Critical phenomena on fractal lattices. Phys. Rev. Lett. **45**, 855–858 (1980).

[251] I. M. Gel'fand and L. A. Dikii. Asymptotic behavior of the resolvent of Sturm-Liouville equations and the algebra of the Korteweg-de Vries equations. Russian Math. Surveys **30**, 77–113 (1975).

[252] J. Ginibre. General formulation of Griffiths inequalities. Commun. Math. Phys. **16**, 310–328 (1970).

[253] J. Glimm. Boson fields with non-linear self-interactions in two dimensions. Commun. Math. Phys. **8**, 12–25 (1968).

[254] J. Glimm and A. Jaffe. Positivity of the φ_3^4 Hamiltonian. Fortschritte der Physik **21**, 327–376 (1973).
[255] J. Glimm and A. Jaffe. Remark on the existence of φ_4^4. Phys. Rev. Lett. **33**, 440–442 (1974).
[256] J. Glimm and A. Jaffe. φ_2^4 quantum field model in the single-phase region: Differentiability of the mass and bounds on critical exponents. Phys. Rev. **D10**, 536–539 (1974).
[257] J. Glimm and A. Jaffe. Absolute bounds on vertices and couplings. Ann. Inst. Henri Poincaré **A22**, 97–109 (1975).
[258] J. Glimm and A. Jaffe. Particles and scaling for lattice fields and Ising models. Commun. Math. Phys. **51** (1976).
[259] J. Glimm and A. Jaffe. Critical exponents and elementary particles. Commun. Math. Phys. **52**, 203–209 (1977).
[260] J. Glimm and A. Jaffe. *Quantum Physics: A Functional Integral Point of View.* Springer-Verlag, New York, 1981 (2nd edition 1987).
[261] J. Glimm, A. Jaffe, and T. Spencer. Phase transitions for φ_2^4 quantum fields. Commun. Math. Phys. **45**, 203–216 (1975).
[262] J. Glimm, A. Jaffe, and T. Spencer. A convergent expansion about mean field theory. I. Ann. Phys. (NY) **101**, 610–630 (1976).
[263] J. Glimm, A. Jaffe, and T. Spencer. A convergent expansion about mean field theory. II. Ann. Phys. (NY) **101**, 631–669 (1976).
[264] J. Goldstone. Field theories with 'superconductor' solutions. Nuovo Cimento **19**, 154–164 (1961).
[265] T. Goto. Relativistic quantum mechanics of one-dimensional mechanical continuum and subsidiary condition of dual resonance models. Progr. Theor. Phys. **46**, 1560–1569 (1971).
[266] M. Green, J. Schwarz, and E. Witten. *Superstring Theory.* Cambridge University Press, Cambridge etc, 1987.
[267] J. Greensite and M. B. Halpern. Stabilizing bottomless action theories. Nucl. Phys. **B242**, 167–188 (1984).
[268] R. B. Griffiths. Peierls' proof of spontaneous magnetization of a two-dimensional Ising ferromagnet. Phys. Rev. **A136**, 437–439 (1964).
[269] R. B. Griffiths. Ferromagnets and simple fluids near the critical point: Some thermodynamic inequalities. J. Chem. Phys. **43**, 1958–1968 (1965).
[270] R. B. Griffiths. Correlations in Ising ferromagnets II. External magnetic fields. J. Math. Phys. **8**, 484–489 (1967).
[271] R. B. Griffiths. Correlations in Ising ferromagnets. III. Commun. Math. Phys. **6**, 121–127 (1967).
[272] R. B. Griffiths. Rigorous results for Ising ferromagnets of arbitrary spin. J. Math. Phys. **10**, 1559–1565 (1969).
[273] R. B. Griffiths. Thermodynamics near the two-fluid critical mixing point in He^3–He^4. Phys. Rev. Lett. **24**, 715–717 (1970).
[274] R. B. Griffiths. Rigorous results and theorems. In C. Domb and M. S. Green, editors, *Phase Transitions and Critical Phenomena (Vol. 1).* Academic Press, London–New York–San Francisco, 1972.
[275] R. B. Griffiths. Proposal for notation at tricritical points. Phys. Rev. **B7**, 545–551 (1973).

[276] R. B. Griffiths. Critical phenomena and tricritical points in multicomponent systems. Physica **73**, 174–183 (1974).

[277] R. B. Griffiths. Thermodynamic models for tricritical points in ternary and quaternary fluid mixtures. J. Chem. Phys. **60**, 195–206 (1974).

[278] R. B. Griffiths and P. D. Gujrati. Convexity violations for non-integer parameters in certain lattice models. J. Stat. Phys. **30**, 563–589 (1983).

[279] R. B. Griffiths, C. A. Hurst, and S. Sherman. Concavity of magnetization of an Ising ferromagnet in a positive external field. J. Math. Phys. **11**, 790–795 (1970).

[280] R. B. Griffiths and J. C. Wheeler. Critical points in multicomponent systems. Phys. Rev. **A2**, 1047–1064 (1970).

[281] D. Gross and V. Periwal. String perturbation theory diverges. Phys. Rev. Lett. **60**, 2105–2108 (1988).

[282] D. J. Gross and A. A. Migdal. A nonperturbative treatment of two-dimensional quantum gravity. Nucl. Phys. **B340**, 333–365 (1990).

[283] D. J. Gross and A. Neveu. Dynamical symmetry breaking in asymptotically free field theories. Phys. Rev. **D10**, 3235–3253 (1974).

[284] P. D. Gujrati. Magnetic analog of self-avoiding random walks (polymer chains) on a lattice. Phys. Rev. **A24**, 2096–2108 (1981).

[285] J. D. Gunton and M. J. Buckingham. Behavior of the correlation function near the critical point. Phys. Rev. Lett. **20**, 143–146 (1968).

[286] J. D. Gunton and M. J. Buckingham. Correlations at the critical point of the Ising model. Phys. Rev. **178**, 848–853 (1969).

[287] A. J. Guttmann. Asymptotic analysis of power-series expansions. In C. Domb and J. L. Lebowitz, editors, *Phase Transitions and Critical Phenomena (Vol. 13)*, pages 1–234. Academic Press, London–San Diego, 1989.

[288] C. K. Hall. Scaling in the ideal Bose gas. J. Stat. Phys. **13**, 157–172 (1975).

[289] A. Hankey and H. E. Stanley. Systematic application of generalized homogeneous functions to static scaling, dynamic scaling, and universality. Phys. Rev. **B6**, 3515–3542 (1972).

[290] T. Hara. A rigorous control of logarithmic corrections in four-dimensional φ^4 spin systems. I. Trajectory of effective Hamiltonians. J. Stat Phys. **47**, 57–98 (1987).

[291] T. Hara. Mean-field critical behaviour for correlation length for percolation in high dimensions. Prob. Th. Rel. Fields **86**, 337–385 (1990).

[292] T. Hara, T. Hattori, and H. Tasaki. Skeleton inequalities and mean field properties for lattice spin systems. J. Math. Phys. **26**, 2922–2929 (1985).

[293] T. Hara and G. Slade. Self-avoiding walk in five or more dimensions. I. The critical behaviour. In preparation.

[294] T. Hara and G. Slade. Self-avoiding walk in five or more dimensions. II. Convergence of the lace expansion. In preparation.

[295] T. Hara and G. Slade. Mean-field critical behaviour for percolation in high dimensions. Commun. Math. Phys. **128**, 333–391 (1990).

[296] T. Hara and G. Slade. On the upper critical dimension of lattice trees and lattice animals. J. Stat. Phys. **59**, 1469–1510 (1990).

[297] T. Hara and G. Slade. Critical behaviour of self-avoiding walk in five or more dimensions. preprint, January 1991.

[298] T. Hara and H. Tasaki. A rigorous control of logarithmic corrections in four-dimensional φ^4 spin systems. II. Critical behavior of susceptibility and correlation length. J. Stat Phys. **47**, 99–121 (1987).

[299] T. Hara and H. Tasaki. Rigorous renormalization group analysis of φ^4 field theory. Prog. Theor. Phys. Suppl. **92**, 26–45 (1987).

[300] G. Hartsleben. Das 1-Matrix Modell und dessen Anwendung auf nichtstörungstheoretische 2-dimensionale Quantumgravitation. Diplomarbeit, ETH-Zürich, 1990.

[301] T. Hattori. A generalization of the proof of the triviality of scalar field theories. J. Math. Phys. **24**, 2200–2203 (1983).

[302] G. C. Hegerfeldt. Correlation inequalities for Ising ferromagnets with symmetries. Commun. Math. Phys. **57**, 259–266 (1977).

[303] Y. Higuchi. On the absence of non-translationally invariant Gibbs states for the two-dimensional Ising system. In J. Fritz, J. L. Lebowitz, and D. Szasz, editors, *Random Fields: Rigorous Results in Statistical Mechanics and Quantum Field Theory (Esztergom 1979)*. North Holland, Amsterdam, 1981.

[304] G. 't Hooft. A planar diagram theory for strong interactions. Nucl. Phys. **B72**, 461–473 (1974).

[305] G. 't Hooft. Planar diagram field theories. In G. 't Hooft *et al.*, editor, *Progress in Gauge Field Theory*. Plenum Press, New York–London, 1984.

[306] J. Imbrie. Debye screening for jellium and other Coulomb systems. Commun. Math. Phys. **87**, 515–565 (1982).

[307] J. Z. Imbrie. Phase diagrams and cluster expansions for low temperature $P(\varphi)_2$ models. I. The phase diagram. Commun. Math. Phys. **82**, 261–304 (1981).

[308] J. Z. Imbrie. Phase diagrams and cluster expansions for low temperature $P(\varphi)_2$ models. II. The Schwinger functions. Commun. Math. Phys. **82**, 305–343 (1981).

[309] R. B. Israel. *Convexity in the Theory of Lattice Gases*. Princeton University Press, 1979.

[310] C. Itzykson and J. B. Zuber. *Quantum Field Theory*. McGraw-Hill, New York, 1980.

[311] C. Itzykson and J. B. Zuber. Matrix integration and combinatorics of modular groups. Commun. Math. Phys. **134**, 197–207 (1990).

[312] T. Jaroszewicz and P. S. Kurzepa. Intersections of fermionic paths and four-fermion interactions. UCLA preprint UCLA/91/TEP/2, January 1991.

[313] M. Jimbo, A. Kuniba, T. Miwa, and M. Okado. The $A_n^{(1)}$ face models. Commun. Math. Phys. **119**, 543–565 (1988).

[314] M. Jimbo, T. Miwa, and M. Okado. Solvable lattice models with broken $\mathbb{Z}_n$ symmetry and Hecke's indefinite modular forms. Nucl. Phys. **B275[FS17]**, 517–545 (1986).

[315] M. Jimbo, T. Miwa, and M. Okado. Solvable lattice models related to the vector representation of classical simple Lie algebras. Commun. Math. Phys. **116**, 507–525 (1988).

[316] M. Jimbo, T. Miwa, and M. Sato. Holonomic quantum fields — The unanticipated link between deformation theory of differential equations and quantum fields. In *Mathematical Problems in Theoretical Physics*. Lecture Notes in Physics #116, Springer-Verlag, Berlin-Heidelberg-New York, 1980.

[317] G. Jona-Lasinio. The renormalization group: A probabilistic view. Nuovo Cimento **B26**, 99–119 (1975).

[318] G. Jona-Lasinio. Probabilistic approach to critical behavior. In M. Lévy and P. Mitter, editors, *New Developments in Quantum Field Theory and Statistical Mechanics*. Plenum, New York-London, 1977.

[319] B. D. Josephson. Inequality for the specific heat. II. Application to critical phenomena. Proc. Phys. Soc. (London) **92**, 276–284 (1967).

[320] R. Jost. Properties of Wightman functions. In E. R. Caianiello, editor, *Lectures on Field Theory and the Many-Body Problem*. Academic Press, New York, 1961.

[321] R. Jost. *The General Theory of Quantized Fields*. American Mathematical Society, Providence, Rhode Island, 1965.

[322] J. Jurkiewicz. Regularization of one-matrix models. Phys. Lett. B **245**, 178–184 (1990).

[323] L. P. Kadanoff. Scaling laws for Ising models. Physics **2**, 263–272 (1966).

[324] L. P. Kadanoff. Scaling, universality and operator algebras. In C. Domb and M. S. Green, editors, *Phase Transitions and Critical Phenomena (Vol. 5a)*. Academic Press, London–New York–San Francisco, 1976.

[325] L. P. Kadanoff, A. Houghton, and M. Yalabik. Variational approximation for renormalization group transformations. J. Stat. Phys. **14**, 171–203 (1976).

[326] L. Kadanoff *et al.* Static phenomena near critical points: theory and experiments. Rev. Mod. Phys. **39**, 395–431 (1967).

[327] S. Kakutani. On Brownian motion in n-space. Proc. Japan Acad. **20**, 648–652 (1944).

[328] M. Karliner and A. A. Migdal. Nonperturbative 2D quantum gravity via supersymmetric string. Mod. Phys. Lett. **A5**, 2565–2572 (1990).

[329] L. H. Karsten and J. Smit. The vacuum polarization with SLAC lattice fermions. Phys. Lett. **B85**, 100–102 (1979).

[330] B. Kaufman and L. Onsager. Crystal statistics III. Short-range order in a binary Ising lattice. Phys. Rev. **76**, 1244–1252 (1949).

[331] H. Kawai and Y. Okamoto. Entropy of planar random surfaces on the lattice. Phys. Lett. **130B**, 415–419 (1983).

[332] V. A. Kazakov. $U(\infty)$ lattice gauge theory as a free lattice string theory. Phys. Lett. **128B**, 316–320 (1983).

[333] V. A. Kazakov. Bilocal regularization of models of random surfaces. Phys. Lett. **B150**, 282–284 (1985).

[334] V. A. Kazakov. Ising model on a dynamical planar random lattice: exact solution. Phys. Lett. **A119**, 140–144 (1986).

[335] V. A. Kazakov. Exactly solvable Potts models, bond- and tree-like percolation on dynamical (random) planar lattice. Nucl. Phys. B (Proc. Suppl.) **4**, 93–97 (1988).

[336] V. A. Kazakov. The appearance of matter fields from quantum fluctuations of 2D-gravity. Mod. Phys. Lett. **A4**, 2125–2139 (1989).

[337] V. A. Kazakov, I. K. Kostov, and A. A. Migdal. Critical properties of randomly triangulated planar random surfaces. Phys. Lett. **157B**, 295–300 (1985).

[338] V. A. Kazakov and A. A. Migdal. Recent progress in the theory of noncritical strings. Nucl. Phys. **B311**, 171–190 (1988/89).

[339] C. King. The $U(1)$ Higgs model. II. The infinite volume limit. Commun. Math. Phys. **103**, 323–349 (1986).

[340] V. G. Knizhnik, A. M. Polyakov, and A. B. Zamolodchikov. Fractal structure of $2d$ quantum gravity. Mod. Phys. Lett. **A3**, 819–826 (1988).

[341] C. M. Knobler and R. L. Scott. Multicritical points in fluid mixtures: experimental studies. In C. Domb and J. L. Lebowitz, editors, *Phase transitions and critical phenomena (Vol. 9)*. Academic Press, London–Orlando–etc., 1984.

[342] H. Koch and P. Wittwer. A non-Gaussian renormalization group fixed point for hierarchical scalar lattice field theories. Commun. Math. Phys. **106**, 495–532 (1986).

[343] V. E. Korepin. The mass spectrum and the S-matrix of the massive Thirring model in the repulsive case. Commun. Math. Phys. **76**, 165–176 (1980).

[344] I. K. Kostov. Multicolor QCD in terms of random surfaces. Preprint 1983.

[345] I. K. Kostov. Random surfaces, solvable lattice models and discrete quantum gravity in two dimensions. Nucl. Phys. B (Proc. Suppl.) **10A**, 295–322 (1989).

[346] I. K. Kostov and A. Krzywicki. On the entropy of random surfaces with arbitrary genus. Phys. Lett. **187B**, 149–152 (1987).

[347] I. K. Kostov and M. L. Mehta. Random surfaces of arbitrary genus: exact results for $D = 0$ and -2 dimensions. Phys. Lett. **B189**, 118–124 (1987).

[348] A. Kupiainen. On the $1/N$ expansion. Commun. Math. Phys. **73**, 273–294 (1980).

[349] O. E. Lanford III. Smooth transformations of intervals. Séminaire Bourbaki #563, in Lecture Notes in Mathematics #901, Springer-Verlag, Berlin–Heidelberg–New York, 1981, pp. 36–54.

[350] G. F. Lawler. A self-avoiding random walk. Duke Math. J. **47**, 655–694 (1980).

[351] G. F. Lawler. The probability of intersection of random walks in four dimensions. Commun. Math. Phys. **86**, 539–554 (1982).

[352] G. F. Lawler. Intersections of simple random walks. Contemp. Math. **41**, 281–289 (1985).

[353] G. F. Lawler. The probability of intersection of random walks in four dimensions. II. Commun. Math. Phys. **97**, 583–594 (1987).

[354] I. D. Lawrence and S. Sarbach. Theory of tricritical points. In C. Domb and J. L. Lebowitz, editors, *Phase transitions and critical phenomena (Vol. 9)*. Academic Press, London, Orlando, etc, 1984.

[355] P. D. Lax. Hyperbolic systems of conservation laws. II. Commun. Pure Appl. Math. **10**, 537–566 (1957).

[356] P. D. Lax. The formation and decay of shock waves. Amer. Math. Monthly **79**, 227–241 (1972).

[357] H. J. Leamy, G. H. Gilmer, and K. A. Jackson. Statistical thermodynamics of clean surfaces. In J. M. Blakely, editor, *Surface Physics of Crystalline Materials*. Academic Press, New York, 1976.

[358] J. Lebowitz. Coexistence of phases in Ising ferromagnets. J. Stat. Phys. **16**, 463–476 (1977).

[359] J. Lebowitz. Number of phases in one component ferromagnets. In G. F. Dell'Antonio, F. Guerra, and G. Jona-Lasinio, editors, *Mathematical Problems in Theoretical Physics*. Lectures Notes in Physics #80, Springer-Verlag, Berlin–Heidelberg–New York, 1978.

[360] J. Lebowitz and A. Martin-Löf. On the uniqueness of the equilibrium state for Ising spin systems. Commun. Math. Phys. **25**, 276–282 (1972).

[361] J. Lebowitz and O. Penrose. Decay of correlations. Phys. Rev. Lett. **31**, 749–752 (1973).

[362] J. L. Lebowitz. GHS and other inequalities. Commun. Math. Phys. **35**, 87–92 (1974).

[363] T. D. Lee and C. N. Yang. Statistical theory of equations of state and phase transitions. II. Lattice gas and Ising model. Phys. Rev. **87**, 410–419 (1952).

[364] A. Lesniewski. Effective action for the Yukawa_2 quantum field theory. Commun. Math. Phys. **108**, 437–467 (1987).

[365] S. S. Leung and R. B. Griffiths. Thermodynamic properties near the vapor-liquid critical line in mixtures of He^3 and He^4. Phys. Rev. **A8**, 2670–2683 (1970).

[366] P. Lévy. Le mouvement brownien plan. Amer. J. Math. **62**, 487–550 (1940).

[367] P. Lévy. La mesure de Hausdorff de la courbe du mouvement brownien. Giorn. Ist. Ital. Attuari **16**, 1–37 (1953).

[368] B. Li and A. D. Sokal. High-precision Monte Carlo test of the conformal-invariance predictions for two-dimensional mutually avoiding walks. J. Stat. Phys. **61**, 723–748 (1990).

[369] E. H. Lieb. Exactly soluble models. Physica **73**, 226–236 (1974).

[370] E. H. Lieb and A. D. Sokal. Improved rigorous upper bound for the renormalized 4-point coupling. Unpublished. See also: E. H. Lieb, Sharp constants in the Hardy-Littlewood-Sobolev and related inequalities, *Ann. Math.* **118**, 349–374 (1983).

[371] E. H. Lieb and A. D. Sokal. A general Lee-Yang theorem for one-component and multicomponent ferromagnets. Commun. Math. Phys. **80**, 153–179 (1981).

[372] H. E. Lieb. A refinement of Simon's correlation inequality. Commun. Math. Phys. **77**, 127–135 (1980).

[373] M. J. Lowe and D. J. Wallace. Instantons and the Ising model below T_c. J. Phys. **A13**, L381–L385 (1980).

[374] M. J. Lowe and D. J. Wallace. Membrane models and generalized Z_2 gauge theories. Phys. Lett. **93B**, 433–436 (1980).

[375] M. Lüscher. Charge screening and an upper bound on the renormalized charge in lattice QED. Nucl. Phys. **B341**, 341–357 (1990).

[376] M. Lüscher and G. Mack. Unpublished paper, 1976. Reviewed by I. T. Todorov, in *Differential Geometric Methods in Theoretical Physics*, H. D. Doebner and T. D. Palev (eds.), World Scientific, Singapore, 1986.

[377] M. Lüscher, K. Symanzik, and P. Weisz. Anomalies of the free loop wave equation in the WKB approximation. Nucl. Phys. **B173**, 365–396 (1980).

[378] S. K. Ma. The $1/n$ expansion. In C. Domb and M. S. Green, editors, *Phase Transitions and Critical Phenomena (Vol. 6)*. Academic Press, London–New York–San Francisco, 1976.

[379] S. K. Ma. *Modern Theory of Critical Phenomena*. Benjamin, London–Amsterdam–Don Mills, Ontario, 1976.

[380] J. Magnen and R. Sénéor. The infinite volume limit of the φ^4_3 model. Ann. Inst. Henri Poincaré **24**, 95–159 (1976).

[381] J. Magnen and R. Sénéor. Phase space cell expansion and Borel summability for the Euclidean φ^4_3 theory. Commun. Math. Phys. **56**, 237–276 (1977).

[382] J. Magnen and R. Sénéor. The infrared behaviour of $(\nabla\phi)^4_3$. Ann. Phys. **152**, 130–202 (1984).

[383] F. Mansouri and Y. Nambu. Gauge conditions in dual resonance models. Phys. Lett. **39B**, 375–378 (1972).

[384] E. Marinari and G. Parisi. A non-perturbative definition of 2D quantum gravity. Phys. Lett. **B247**, 537–542 (1990).

[385] E. Marinari and G. Parisi. The supersymmetric one-dimensional string. Phys. Lett. **B240**, 375–380 (1990).

[386] O. McBryan and T. Spencer. On the decay of correlations in $SO(n)$-symmetric ferromagnets. Commun. Math. Phys. **53**, 299–302 (1977).

[387] B. M. McCoy, C. A. Tracy, and T. T. Wu. Two-dimensional Ising model as an exactly soluble relativistic quantum field theory: Explicit formulas for n-point functions. Phys. Rev. Lett. **38**, 793–796 (1977).

[388] B. M. McCoy and T. T. Wu. *The Two-Dimensional Ising Model.* Harvard University Press, Cambridge, 1973.

[389] D. S. McKenzie and C. Domb. The second osmotic virial coefficient of athermal polymer solutions. Proc. Phys. Soc. (London) **92**, 632–649 (1967).

[390] M. L. Mehta. A method of integration over matrix variables. Commu. Math. Phys. **79**, 327–340 (1981).

[391] N. D. Mermin. Absence of ordering in certain classical systems. J. Math. Phys. **8**, 1061–1064 (1977).

[392] N. D. Mermin. The topological theory of defects in ordered media. Rev. Mod. Phys. **51**, 591–648 (1979).

[393] N. D. Mermin and H. Wagner. Absence of ferromagnetism or antiferromagnetism in one- or two-dimensional isotropic Heisenberg models. Phys. Rev. Lett. **17**, 1133–1136 (1966).

[394] A. Messager and S. Miracle-Solé. Correlation functions and boundary conditions in the Ising ferromagnet. J. Stat. Phys. **17**, 245–262 (1977).

[395] L. Michel. Symmetry defects and broken symmetry. Configurations. Hidden symmetries. Rev. Mod. Phys. **52**, 617–652 (1980).

[396] A. A. Migdal. Conformal invariance and bootstrap. Phys. Lett. **37B**, 386–388 (1971).

[397] P. K. Mitter and T. R. Ramadas. The two-dimensional $O(N)$ nonlinear σ-model: Renormalisation and effective actions. Commun. Math. Phys. **122**, 575–596 (1989).

[398] P. K. Mitter and P. H. Weisz. Asymptotic scale invariance in a massive Thirring model with $U(n)$ symmetry. Phys. Rev. **D8**, 4410–4429 (1973).

[399] G. Moore. Geometry of the string equations. Commun. Math. Phys. **133**, 261–304 (1990).

[400] H. Müller-Krumbhaar. Kinetics of crystal growth: microscopic and phenomenological theories. In E. Kaldis, editor, *Current Topics in Materials Science, Vol. 1.* North-Holland, Amsterdam, 1978.

[401] J. R. Munkres. *Elements of Algebraic Topology.* Addison-Wesley, Redwood City, California, 1984.

[402] N. Nakanishi. *Graph Theory and Feynman Integrals.* Gordon and Breach, 1971.

[403] Y. Nambu. Lectures at the Copenhagen Symposium (1970).

[404] D. R. Nelson, T. Piran, and S. Weinberg (Eds). *Statistical Mechanics of Membranes and Surfaces.* World Scientific, Singapore, 1989.

[405] E. Nelson. The construction of quantum fields from Markov fields. J. Funct. Anal. **12**, 97–112 (1973).

[406] E. Nelson. The free Markov field. J. Funct. Anal. **12**, 211–217 (1973).

[407] E. Nelson. Probability theory and Euclidean field theory. In G. Velo and A. Wightman, editors, *Constructive Quantum Field Theory*, pages 94–124. Lecture Notes in Physics #25, Springer-Verlag, 1973.

[408] E. Nelson. Quantum fields and Markoff fields. In D. C. Spencer, editor, *Partial Differential Equations.* American Mathematical Society, Providence, R.I., 1973.

[409] C. M. Newman. Shock waves and mean field bounds: Concavity and analyticity of the magnetization at low temperature. Unpublished preprint (1981). Published as Appendix to C. M. Newman, Percolation theory: A selective survey of rigorous results, in *Advances in Multiphase Flow and Related Problems*, G. Papanicolaou, editor (SIAM, Philadelphia, 1986), pp. 163–167.

[410] C. M. Newman. Gaussian correlation inequalities for ferromagnets. Z. Wahrscheinlichkeistheor. Verw. Geb. **33**, 75–93 (1975).

[411] C. M. Newman. Inequalities for Ising models and field theories which obey the Lee-Yang theorem. Commun. Math. Phys. **41**, 1–9 (1975).

[412] C. M. Newman. Moment inequalities for ferromagnetic Gibbs distributions. J. Math. Phys. **16**, 1956–1959 (1975).

[413] C. M. Newman. Fourier transforms with only real zeros. Proc. Am. Math. Soc. **61**, 245–251 (1976).

[414] C. M. Newman. Rigorous results for general Ising ferromagnets. J. Stat. Phys. **15**, 399–406 (1976).

[415] C. M. Newman. Critical point inequalities and scaling limits. Commun. Math. Phys. **66**, 181–196 (1979).

[416] B. G. Nguyen. Gap exponents for percolation processes with triangle condition. J. Stat. Phys. **49**, 235–243 (1987).

[417] L. Onsager. Crystal statistics. I. A two-dimensional model with an order-disorder transition. Phys. Rev. **65**, 117–149 (1944).

[418] K. Osterwalder. Constructive quantum field theory: Scalar fields. In P. Dita, V. Georgescu, and R. Purice, editors, *Gauge Theories: Fundamental Interactions and Rigorous Results.* Birkhäuser, Boston–Basel–Stuttgart, 1982.

[419] K. Osterwalder and R. Schrader. Axioms for Euclidean Green's functions. Commun. Math. Phys. **31**, 83–112 (1973).

[420] K. Osterwalder and R. Schrader. Axioms for Euclidean Green's functions. II. Commun. Math. Phys. **42**, 281–305 (1975).

[421] J. Palmer and C. Tracy. Two-dimensional Ising correlations: Convergence of the scaling limit. Adv. Appl. Math. **2**, 329–388 (1981).

[422] J. Palmer and C. Tracy. Two-dimensional Ising correlations: The SMJ analysis. Adv. Appl. Math. **4**, 46–102 (1983).

[423] G. Parisi. Hausdorff dimensions and gauge theories. Phys. Lett. **81B**, 357–360 (1979).

[424] G. Parisi. *Statistical Field Theory.* Addison-Wesley, Reading, Massachusetts, 1988.

[425] Y. M. Park. Direct estimates on intersection probabilities of random walks. J. Stat. Phys. **57**, 319–331 (1989).

[426] V. Pasquier. Etiology of IRF models. Commun. Math. Phys. **118**, 355–364 (1988).

[427] P. A. Pearce. Mean-field bounds on the magnetization for ferromagnetic spin models. J. Stat. Phys. **25**, 309–320 (1981).

[428] R. Peierls. Ising's model of ferromagnetism. Proc. Cambridge Philos. Soc. **32**, 477–481 (1936).

[429] R. C. Penner. The moduli space of a punctured surface and perturbation series. Bull. Amer. Math. Soc. **15**, 73–77 (1986).

[430] R. C. Penner. Perturbative series and the moduli space of Riemann surfaces. J. Diff. Geom. **27**, 35–53 (1988).

[431] O. Penrose. Phase transitions on fractal lattices with long-range interactions. J. Stat. Phys. **45**, 69–88 (1986).

[432] C. E. Pfister. Interface and surface tension in Ising model. In J. Fröhlich, editor, *Scaling and Self-Similarity in Physics*. Birkhäuser, Boston–Basel–Stuttgart, 1983.

[433] K. Pohlmeyer. The Jost-Schroer theorem for zero-mass fields. Commun. Math. Phys. **12**, 204–211 (1969).

[434] J. Polchinski. Renormalization and effective Lagrangians. Nucl. Phys. **B231**, 269–295 (1984).

[435] J. Polchinski. Scale and conformal invariance in quantum field theory. Nucl. Phys. **B303**, 226–236 (1988).

[436] A. M. Polyakov. Conformal symmetry of critical fluctuations. Zh. Eksp. Teor. Fiz. Pis. Red. **12**, 538–541 (1970). [JETP Lett. **12**, 381–383 (1970)].

[437] A. M. Polyakov. Quantum geometry of bosonic strings. Phys. Lett. **103B**, 207–210 (1981).

[438] V. N. Popov. *Functional Integrals in Quantum Field Theory and Statistical Physics*. D. Reidel, Dordrecht–Boston–Lancaster, 1983.

[439] R. T. Powers. Absence of interaction as a consequence of good ultraviolet behavior in the case of a local Fermi field. Commun. Math. Phys. **4**, 145–156 (1987).

[440] D. Ray. Sojourn times and the exact Hausdorff measure of the sample path for planar Brownian motion. Trans. Amer. Math. Soc. **106**, 436–444 (1963).

[441] M. Reed and B. Simon. *Methods of Modern Mathematical Physics. Vol I. Functional Analysis*. Academic Press, New York–London–Toronto–Sydney–San Francisco, 1972, 2nd edition 1980.

[442] M. Reed and B. Simon. *Methods of Modern Mathematical Physics. Vol II. Fourier Series, Self-Adjointness*. Academic Press, New York–London–Toronto–Sydney–San Francisco, 1975.

[443] N. Yu. Reshetikin and P. B. Wiegmann. Towards the classification of completely integrable quantum field theories (the Bethe-ansatz associated with Dynkin diagrams and their automorphisms). Phys. Lett. **B189**, 125–131 (1987).

[444] V. Rivasseau. Construction and Borel summability of planar 4-dimensional Euclidean field theory. Commun. Math. Phys. **95**, 445–486 (1984).

[445] J. Rosen. Mass renormalization for the $\lambda\varphi^4$ Euclidean lattice field theory. Adv. Appl. Math. **1**, 37–49 (1980).

[446] W. Rudin. *Functional Analysis.* McGraw-Hill, New York, 1973.
[447] D. Ruelle. *Statistical Mechanics.* W. A. Benjamin Inc., 1977.
[448] M. Sato, T. Miwa, and M. Jimbo. Publ. RIMS, Kyoto Univ., **14**, 223 (1978); **15**, 201, 577, 871 (1979); **16** 531 (1980).
[449] R. S. Schor. The particle structure of ν-dimensional Ising models at low temperatures. Commun. Math. Phys. **59**, 213–233 (1978).
[450] R. Schrader. New rigorous inequality for critical exponents in the Ising model. Phys. Rev. **B14**, 172–173 (1976).
[451] R. Schrader. A possible constructive approach to φ_4^4. I. Commun. Math. Phys. **49**, 131–153 (1976).
[452] R. Schrader. A possible constructive approach to φ_4^4. III. Commun. Math. Phys. **50**, 97–102 (1976).
[453] R. Schrader. New correlation inequalities for the Ising model and $P(\varphi)$ theories. Phys. Rev. **B15**, 2798–2803 (1977).
[454] R. Schrader. A possible constructive approach to φ_4^4. II. Ann. Inst. Henri Poincaré **A26**, 295–301 (1977).
[455] B. Schroer. A trip to scalingland. In E. Ferreira, editor, *V Brazilian Symposium on Theoretical Physics (Vol. 1)*, pages 285–322. Livros Técnicos e Científicos Editora S.A., Rio de Janeiro, 1975.
[456] T. D. Schultz, D. C. Mattis, and E. H. Lieb. Two-dimensional Ising model as a soluble problem of many fermions. Rev. Mod. Phys. **36**, 856–871 (1964).
[457] L. Schwartz. *Théorie des Distributions.* Actualités Scientifiques et Industrielles #1245. Hermann, Paris, 1957.
[458] G. Segal and G. Wilson. Loop groups and equations of KdV type, 1985.
[459] E. Seiler. *Gauge Theories as a Problem of Constructive Quantum Field Theory and Statistical Mechanics.* Lecture Notes in Physics #159. Springer-Verlag, Berlin–Heidelberg–New York, 1982.
[460] S. B. Shlosman. Signs of the Ising model Ursell functions. Commun. Math. Phys. **102**, 679–686 (1986).
[461] B. Simon. Positivity of the Hamiltonian semigroup and the construction of Euclidean region fields. Helv. Phys. Acta **46**, 686–696 (1973).
[462] B. Simon. *The $P(\varphi)_2$ Euclidean (Quantum) Field Theory.* Princeton University Press, 1974.
[463] B. Simon. *Functional Integration and Quantum Physics.* Academic Press, New York–San Francisco–London, 1979.
[464] B. Simon. Correlation inequalities and the decay of correlations in ferromagnets. Commun. Math. Phys. **77**, 111–126 (1980).
[465] B. Simon. *The Statistical Mechanics of Lattice Gases (Vol. I).* Princeton Univ. Press, to appear.
[466] B. Simon and R. B. Griffiths. The φ_2^4 field theory as a classical Ising model. Commun. Math. Phys. **33**, 145–164 (1973).
[467] Ya. G. Sinai. Self-similar probability distributions. Theor. Prob. Appl. **21**, 64–80 (1976).
[468] Ya. G. Sinai. Mathematical foundations of the renormalization group method in statistical physics. In G. Dell'Antonio, S. Doplicher, and G. Jona-Lasinio, editors, *Mathematical Problems in Theoretical Physics.* Lecture Notes in Physics #80, Springer-Verlag, Berlin–Heidelberg–New York, 1978.

[469] Ya. G. Sinai. *Theory of Phase Transitions: Rigorous Results.* Pergamon Press, Oxford–New York–etc, 1982.
[470] E. K. Sklyanin, L. A. Takhtajan, and L. D. Faddeev. Quantum inverse problem method. I. Theor. Math. Phys. **40**, 688–706 (1979).
[471] G. Slade. The diffusion of self-avoiding random walk in high dimensions. Commun. Math. Phys. **110**, 661–683 (1987).
[472] G. Slade. Convergence of self-avoiding random walk to Brownian motion in high dimensions. J. Phys. A: Math. Gen. **21**, L417–L420 (1988).
[473] G. Slade. The scaling limit of self-avoiding random walk in high dimensions. Ann. Probab. **17**, 91–107 (1989).
[474] J. Slawny. On the mean field theory bound on the magnetization. J. Stat. Phys. **32**, 375–388 (1983).
[475] J. Smoller. *Shock Waves and Reaction-Diffusion Equations.* Springer-Verlag, New York–Heidelberg–Berlin, 1983.
[476] A. D. Sokal. Unpublished (1983). See G. Felder and J. Fröhlich, Intersection properties of simple random walks: A renormalization group approach, *Commun. Math. Phys.* **97**, 111–124 (1985).
[477] A. D. Sokal. Unpublished (1980–82).
[478] A. D. Sokal. A rigorous inequality for the specific heat of an Ising or φ^4 ferromagnet. Phys. Lett. **71A**, 451–453 (1979).
[479] A. D. Sokal. More inequalities for critical exponents. J. Stat. Phys. **25**, 25–56 (1981).
[480] A. D. Sokal. Rigorous proof of the high-temperature Josephson inequality for critical exponents. J. Stat. Phys. **25**, 51–56 (1981).
[481] A. D. Sokal. An alternate constructive approach to the φ_3^4 quantum field theory, and a possible destructive approach to φ_4^4. Ann. Inst. Henri Poincaré **37**, 317–398 (1982).
[482] A. D. Sokal. Mean-field bounds and correlation inequalities. J. Stat. Phys. **28**, 431–439 (1982).
[483] A. D. Sokal, unpublished (1983). Cited in A. Berretti and A. D. Sokal, New Monte Carlo method for the self-avoiding walk, *J. Stat. Phys.* **40**, 483–531 (1985).
[484] A. D. Sokal, *Math. Rev.*, Vol. 86, p. 4311 (review 86i:81075).
[485] E. Speer. Dimensional and analytic renormalization. In G. Velo and A. S. Wightman, editors, *Renormalization Theory.* Plenum, 1977.
[486] H. E. Stanley. Spherical model as the limit of infinite spin dimensionality. Phys. Rev. **176**, 718–722 (1968).
[487] H. E. Stanley. *Introduction to Phase Transitions and Critical Phenomena.* Oxford Univ. Press, New York–Oxford, 1971.
[488] R. F. Streater and A. S. Wightman. *PCT, Spin and Statistics, and All That.* Benjamin/Cummings, Reading, Massachusetts, 1964, 2nd edition 1978.
[489] F. Strocchi. Locality and covariance in QED and gravitation: General proof of Gupta-Bleuler type formulations. In W. Brittin, editor, *Mathematical Methods in Theoretical Physics.* Lectures in Theoretical Physics, Vol. 14B, Gordon and Breach, New York etc, 1973.
[490] S. J. Summers. On the phase diagram of a $p(\varphi)_2$ quantum field model. Ann. Inst. Henri Poincaré **34**, 173–229 (1981).

[491] G. S. Sylvester. Inequalities for continuous-spin Ising ferromagnets. J. Stat. Phys. **15**, 327–341 (1976).

[492] G. S. Sylvester. The Ginibre inequality. Commun. Math. Phys. **73**, 105–114 (1980).

[493] K. Symanzik. Euclidean quantum field theory. I. Equations for a scalar model. J. Math. Phys. **7**, 510–525 (1966).

[494] K. Symanzik. Euclidean quantum field theory. In R. Jost, editor, *Local Quantum Theory.* 1968 Varenna Lectures, Academic Press, New York–London, 1969.

[495] D. Szàsz. Correlation inequalities for non-purely-ferromagnetic systems. J. Stat. Phys. **19**, 453–459 (1978).

[496] S. J. Taylor. The Hausdorff α-dimensional measure of Brownian paths in n-space. Proc. Cambridge Philos. Soc. **49**, 31–39 (1953).

[497] S. J. Taylor. The exact Hausdorff measure of the sample path for Brownian motion. Proc. Cambridge Philos. Soc. **60**, 253–258 (1964).

[498] H. N. V. Temperley. Combinatorial problems suggested by the statistical mechanics of domains and of rubber-like molecules. Phys. Rev. **103**, 1–16 (1956).

[499] H. B. Thacker. Exact integrability in quantum field theory and statistical systems. Rev. Mod. Phys. **53**, 253–285 (1981).

[500] G. Toulouse and P. Pfeuty. *Introduction au Groupe de Renormalisation et à ses Applications.* Presses Universitaires de Grenoble, 1975.

[501] H. van Beijeren. Exactly solvable model for the roughening transition of a crystal surface. Phys. Rev. Lett. **38**, 993–996 (1977).

[502] D. J. Wallace. The ε-expansion for expoennts and the equation of state in isotropic systems. In C. Domb and M. S. Green, editors, *Phase Transitions and Critical Phenomena (Vol. 6)*, pages 293–356. Academic Press, London–New York–San Francisco, 1976.

[503] J. D. Weeks, G. H. Gilmer, and H. J. Leamy. Structural transition of the Ising-model interface. Phys. Rev. Lett. **31**, 549–551 (1979).

[504] F. Wegner. Duality in generalized Ising models and phase transitions without local order parameters. J. Math. Phys. **12**, 2259–2272 (1971).

[505] F. J. Wegner. The critical state, general aspects. In C. Domb and M. S. Green, editors, *Phase Transitions and Critical Phenomena (Vol. 6).* Academic Press, London–New York–San Francisco, 1976.

[506] D. Weingarten. Non-planar diagrams in the large N limit of $U(N)$ and $SU(N)$ lattice gauge theories. Phys. Lett. **90B**, 285–288 (1980).

[507] S. G. Whittington. Statistical mechanics of polymer solutions and polymer adsorption. Advances in Chemical Physics **51**, 1–48 (1982).

[508] B. Widom. Equation of state in the neighborhood of the critical point. J. Chem. Phys. **43**, 3898–3905 (1965).

[509] B. Widom. Surface tension and molecular correlations near the critical point. J. Chem. Phys. **43**, 3892–3897 (1965).

[510] B. Widom. The critical point and scaling theory. Physica **73**, 107–118 (1974).

[511] A. S. Wightman. Convexity and the notion of equilibrium state in thermodynamics and statistical mechanics. Introduction to R. B. Israel, *Convexity in the Theory of Lattice Gases*, Princeton University Press, 1979.

[512] A. S. Wightman. Quantum field theory in terms of vacuum expectation values. Phys. Rev. **101**, 860–866 (1956).

[513] A. S. Wightman. La théorie quantique locale et la théorie quantique des champs. Ann. Inst. Henri Poincaré **A1**, 403–420 (1964).

[514] A. S. Wightman and L. Gårding. Fields as operator-valued distributions in relativistic quantum theory. Ark. för Fysik **28**, 129–184 (1964).

[515] K. Wilson. Confinement of quarks. Phys. Rev. **D10**, 2445–2459 (1974).

[516] K. G. Wilson. Renormalization group and critical phenomena. I. Renormalization group and the Kadanoff scaling picture. Phys. Rev. **B4**, 3174–3183 (1971).

[517] K. G. Wilson. Renormalization group and critical phenomena. II. Phase space cell analysis of critical behavior. Phys. Rev. **B4**, 3184–3205 (1971).

[518] K. G. Wilson. Quantum field-theory models in less than 4 dimensions. Phys. Rev. **D7**, 2911–2926 (1973).

[519] K. G. Wilson and M. E. Fisher. Critical exponents in 3.99 dimensions. Phys. Rev. Lett. **28**, 240–243 (1972).

[520] K. G. Wilson and J. Kogut. The renormalization group and the ε expansion. Phys. Rep. **12C**, 75–200 (1974).

[521] E. Witten. On the structure of the topological phase of two-dimensional gravity. Nucl. Phys. **340B**, 281–332 (1990).

[522] Z. Wizimirski. On the existence of the field of operators in the axiomatic quantum field theory. Bull. Acad. Polon. Sci. Ser. Math. Astr. Phys. **14**, 91–93 (1966).

[523] C. N. Yang. The spontaneous magnetization of a two-dimensional Ising model. Phys. Rev. **85**, 808–816 (1952).

[524] C. N. Yang. The spontaneous magnetization of a two-dimensional Ising model. Phys. Rev. **85**, 808–816 (1952).

[525] A. B. Zamolodchikov. On the entropy of random surfaces. Phys. Lett. **117B**, 87–90 (1982).

[526] A. B. Zamolodchikov. Conformal symmetry and multicritical points in two-dimensional quantum field theory. Sov. J. Nucl. Phys. **44**, 529–533 (1986).

[527] R. K. P. Zia and D. J. Wallace. Euclidean group as a dynamical symmetry of surface fluctuations: the planar interface critical behavior. Phys. Rev. Lett. **43**, 808–812 (1980).

Index

MIX
Papier aus verantwortungsvollen Quellen
Paper from responsible sources
FSC® C105338

If you have any concerns about our products,
you can contact us on
ProductSafety@springernature.com

In case Publisher is established outside the EU,
the EU authorized representative is:
Springer Nature Customer Service Center GmbH
Europaplatz 3, 69115 Heidelberg, Germany

Printed by Libri Plureos GmbH
in Hamburg, Germany